Human Evolution Beyond Biology and Culture

Evolutionary Social, Environmental and Policy Sciences

Both natural and cultural selection played an important role in shaping human evolution. Since cultural change can itself be regarded as evolutionary, a process of gene–culture coevolution operates. The study of human evolution – in the past, present and future – is therefore not restricted to biology. An inclusive comprehension of human evolution relies on integrating insights about cultural, economic and technological evolution with relevant elements of evolutionary biology. In addition, proximate causes and effects of cultures need to be added to the picture – issues which are at the forefront of social sciences such as anthropology, economics, geography and innovation studies.

This book highlights discussions on the many topics to which such generalised evolutionary thought has been applied: the arts, the brain, climate change, cooking, criminality, environmental problems, futurism, gender issues, group processes, humour, industrial dynamics, institutions, languages, medicine, music, psychology, public policy, religion, sex, sociality and sports.

Jeroen C. J. M. van den Bergh is ICREA Professor at the Institute of Environmental Science & Technology of Universitat Autònoma de Barcelona (2007–present), and full Professor of Environmental & Resource Economics at VU University Amsterdam (1997–present). He is Editor-in-Chief of the journal *Environmental Innovation and Societal Transitions* and served on the Netherlands' Energy Council. He is cited more than 17 000 times in Google Scholar, and has received the Royal/Shell Prize 2002, IEC's Sant Jordi Environmental Prize 2011 and an ERC Advanced Grant.

Human Evolution Beyond Biology and Culture

Evolutionary Social, Environmental and Policy Sciences

JEROEN C. J. M. VAN DEN BERGH

Universitat Autònoma de Barcelona,

VU University Amsterdam and

ICREA, Barcelona

CAMBRIDGE
UNIVERSITY PRESS

Shaftesbury Road, Cambridge CB2 8EA, United Kingdom

One Liberty Plaza, 20th Floor, New York, NY 10006, USA

477 Williamstown Road, Port Melbourne, VIC 3207, Australia

314–321, 3rd Floor, Plot 3, Splendor Forum, Jasola District Centre, New Delhi – 110025, India

103 Penang Road, #05–06/07, Visioncrest Commercial, Singapore 238467

Cambridge University Press is part of Cambridge University Press & Assessment, a department of the University of Cambridge.

We share the University's mission to contribute to society through the pursuit of education, learning and research at the highest international levels of excellence.

www.cambridge.org
Information on this title: www.cambridge.org/9781108470971

DOI: 10.1017/9781108564922

First published 2018

A catalogue record for this publication is available from the British Library

Library of Congress Cataloging-in-Publication data
Names: Bergh, Jeroen C. J. M. van den, 1965– author.
Title: Human evolution beyond biology and culture : evolutionary social, environmental and policy
    sciences / Jeroen C.J.M. van den Bergh, Universitat Autònoma de Barcelona.
Description: Cambridge, United Kingdom ; New York, NY : Cambridge University Press, 2018. |
    Includes bibliographical references and index.
Identifiers: LCCN 2018017450 | ISBN 9781108470971 (hardback : alk. paper) |
    ISBN 9781108456883 (paperback : alk. paper)
Subjects: LCSH: Human evolution–Social aspects. | Social change. | Technological innovations. |
    Evolution (Biology)
Classification: LCC GN281 .B48 2018 | DDC 599.93/8–dc23
LC record available at https://lccn.loc.gov/2018017450

ISBN   978-1-108-47097-1   Hardback
ISBN   978-1-108-45688-3   Paperback

Contents

Preface

In the summer of 2015, my family and I visited Charles Darwin's *Down House* in the UK. While zigzagging along narrow roads through dense forests, our beloved car navigation system provided guidance. Since it had occasionally misled us in the past, we decided to double-check whether we were going in the right direction. As our station wagon crawled past the van of a young gardener, we called through the open window: 'Down House', followed by 'Darwin'. Although he gave the impression of being a local, neither name seemed to ring any bells. Indeed, occasionally one can bump into Earthlings who are unaware of Darwin and the great intellectual leap he made. A Darwinian evolutionary explanation of life represents one of the most counterintuitive results of science. That's why, more than 150 years after Darwin's magnum opus *On the Origin of Species* (...) appeared, many still refuse to accept it. Others have gone far beyond acceptance, offering evolutionary accounts of numerous phenomena outside the realm of biology. This had already begun in Darwin's time and, more recently, has advanced considerably, with the aim of better understanding all sorts of complex non-living processes and systems.

Many take for granted that the evolutionary history of humans has, like that of animals, completely been determined by genetic evolution. But there is increasing evidence that, once culture emerged in human groups, a combination of natural and cultural selection started to shape the course of human evolution. Culture as information obtained from other individuals through social transmission processes has given rise to cumulative social learning beyond generations. It has affected the evolution of human physiology, brain–mind and behaviour, for at least the last 100 000 years. This worked through cultural traits, such as cooperation with non-kin, sharing of food, exchange of products, labour division, technological innovation, religion and even cooking. In other words, the old and widespread idea that genetic evolution caused culture in a unidirectional way is erroneous. Given that cultural change takes the form of an evolutionary process, as defended in this book, a genuine process of gene–culture coevolution operates, meaning that genetic and cultural evolution exerted mutual influences. The study of human evolution – in the past, present and future – is therefore not the sole domain of biology. A comprehensive picture of human evolution relies on successfully integrating insights about cultural evolution with relevant elements of human evolutionary biology. To this, we need to add information about proximate causes and effects of cultures as studied in social sciences, such as anthropology, economics, geography and innovation studies. These two tasks characterise what this book sets out to do.

Its approach is consistent with the modern belief that the main difference between humans and the most intelligent other primates is not overall multipurpose intelligence, but the unique human ability for social learning and precise imitation. Culture is the ultimate expression of this ability, and has made humans in cultural groups immensely more intelligent than they would be if they had, hypothetically, lived a solitary life – in which case, they would have lacked frequent feedback and advice from experienced individuals since birth. Humans would then not possess cultural knowledge, tools, language, books, science and education. In fact, if one could erase social learning and hence culture from human history, the intellectual ability of humans would probably not exceed that of chimpanzees by much. Given that this is a far cry from reality, the study of human evolution requires a broader scientific approach as sketched.

Another theme guiding this book is that evolution is everywhere, what has been occasionally called 'universal Darwinism', 'generalised evolution' or an 'extended synthesis'. While a broad range of topics to which evolutionary thinking has been applied will receive attention, the treatment includes a strong emphasis on social science issues and public policy challenges. Evolutionary concepts are inextricably woven into the fields of sociology, anthropology, organisation studies, political science, economics, technological innovation studies and environmental sciences. Some of these can even be said to have a genuine 'evolutionary branch', as is definitely true for anthropology and economics. Admittedly, the term 'evolution' is not consistently used in all studies. Authors sometimes speak instead of multi-agent or agent-based modelling, population theories, heterogeneous agents or even complexity theory. In many cases, though, these reflect a foundation in evolutionary principles. My motivation for writing this book is that a comprehensive account of evolutionary thought in the social, environmental and policy sciences is utterly missing. By filling this gap, I hope to contribute to transdisciplinary exchange and learning, among the social sciences, as well as with the natural sciences. In addition, I intend to reach the well-educated reader who is interested in how genetic and non-genetic evolution affect culture, technology, the economy, the environment and climate, and even politics and policies.

The notion of 'generalised evolution', which characterises the book's approach, emphasises that a similar evolutionary framework is employed in distinct study areas. One might call this the V-S-I-R approach, referring to the combination of four basic components and processes, namely $\underline{v}$ariation, $\underline{s}$election, $\underline{i}$nnovation and $\underline{r}$eplication. Without submitting to 'biology envy', proponents of evolutionary approaches in a wider domain can unquestionably learn a great deal from biology, given its 150-year history of evolutionary reasoning. It is undeniable that evolutionary biology is far ahead of the crowd in exploring evolutionary concepts, models and experiments. Therefore, this book will devote attention to transferring concepts and insights from biology to the social sciences – and to some extent also vice versa. Social scientists can learn from debates in biology, such as regarding the levels of selection, the role of groups versus individuals, or the importance of modular evolution. This does not mean that one has to rely solely on biological metaphors in expanding the reach of evolutionary thinking. It should be reassuring to know that mathematics has already come up with generalised metaphors, namely in the form of evolutionary models and algorithms. These are frequently used nowadays, not only in biology but also in economics, innovation studies, operations research, computer science, robotics and artificial intelligence.

It would be an understatement to claim that evolutionary thinking in the social sciences is not accepted by everyone. Averting and defensive responses take the form of evolutionary social studies not being required, not being relevant, or being politically incorrect. No offence, but such rejection often takes the form of knocking down a straw man. Indeed, one frequently encounters a lack of understanding of basic evolutionary principles, let alone of advanced notions, among social scientists rejecting evolutionary social science thinking. Regrettably, many evolutionary biologists who dislike notions of social or cultural evolution don't do much better. Their motivations frequently witness a misapprehension of non-genetic evolution and how it differs from, as well as resembles, genetic evolution. A growing number of researchers, though, seem to accept the usefulness of evolutionary thinking to the social sciences. This book will, in various places, most systematically in Section 1.4, address the concerns of the sceptics and critics of evolutionary social science approaches. It will clarify that genetic and non-genetic evolution share many similarities – justifying the term evolution – while they also differ in important ways. Non-genetic evolution is neither an extrapolation nor a simple analogy of biological evolution. Specific differences depend on whether we are talking about cultural, economic, technological or other types of non-genetic evolution. Furthermore, the two can be interactive, ranging from reinforcing to opposing each other's tendencies and outcomes, as is central to theories of dual inheritance or gene–culture coevolution. As we will see, such extended evolutionary thinking can provide surprising insights about many aspects of the modern world, including music, sports, economic development, cooking, language, medicine, criminal law, the role of sex and gender in society, religions and even humour.

I expect this book to offer stimulating ideas to different readers. Biologists may be intrigued by the details of evolutionary studies in the social sciences. Social scientists can learn from advanced theories in evolutionary biology, as well as about evolutionary thinking in social science disciplines other than their own. Unlike many other treatises, I attempt to give a balanced and fair account of ideas and theories, allowing space for pro- and contra-arguments and, when useful, adding a personal judgement. Those motivated by topical issues will be happy to find practical evolutionary outlooks on environmental problems, climate change and, more generally, the design of public policies. The book is, moreover, aimed at reaching laypersons and experts alike. Readers do not require much background as concise accounts of generalised evolutionary thinking and evolutionary biology are offered, including non-technical treatments of advanced topics. Moreover, while the book is scholarly in depth and scope, I have tried to write in accessible language. To further improve readability, a 'box format' is used throughout with the objective of separating illustrative and advanced themes from the main text.

Since my late teens, I have been intrigued by anything associated with evolutionary reasoning. It struck me as surprising and meaningful, and I felt everyone should know about it. Its focus on ultimate causes was effective in removing the mystery around many fundamental questions about origins. I miss such an approach in the social sciences, which was one motivation to write this book. Some three decades ago, I started reading seriously into evolutionary social sciences and wrote about them for the first time almost 20 years ago. Given that I conducted research on a variety of themes in environmental economics and innovation studies, and worked regularly on projects with biologists, it was almost inevitable that evolutionary thinking and modelling would enter my academic research. I have learnt a great deal from my former PhD students while working on evolutionary and related themes in environmental science, behavioural economics and innovation economics, notably Joelle Noailly, Karolina Safarzynska, Julian Garcia, Volker Nannen, Paolo Zeppini, Elisabeth Gsottbauer, Juliana Subtil Lacerda and Ardjan Gazheli. I further had the luck to collaborate in evolution-oriented studies with resourceful colleagues: Guszti Eiben, Albert Faber, Koen Frenken, John Gowdy, Annemarth Idenburg, Giorgos Kallis, Frans Oosterhuis, Christian Rammel, Sigrid Stagl and Cees Withagen. In addition, this book has profited from my teaching of evolutionary economics in Amsterdam, Barcelona and Vienna, as well as at the Max Planck Institute of Economics in Jena, Germany. The latter hosted a unique evolutionary economics unit directed by Ulrich Witt, from whom I have always received great intellectual and moral support. I am very grateful to other colleagues for spending the time and intellectual energy to comment on particular chapters or parts thereof: Jan Boersema, Jeffrey Funk, Frank Geels, John Gowdy, Fjalar de Haan, Rutger Hoekstra, Javier Martínez-Picado, Sergio Rossi, Karolina Safarzynska, Victor Sarto Monteys and David Stern. I am especially indebted to Nico van Straalen for very detailed comments on Chapters 3 and 4, to Karl Frost for critical feedback on Chapter 7, and to Eric Galbraith for closely reading Chapters 11 and 16. In addition, Matthew Kelly and Tessa Dunlop provided excellent language suggestions, while Miklós Antal helped me raise the resolution of two figures. I am grateful to the team of Cambridge University Press for such proficient support during the various stages of the production process: Dominic Lewis, Aleksandra Serocka, Jenny van der Meijden and Judith Shaw. Others inadvertently influenced my thinking about evolution. My son, Django, at times pointed me to internet videos or television documentaries featuring unusual organisms – one of his fascinations. A number of these creatures have made it into this book. From him and my daughter, Gaia, I learned that evolution can be fascinating to youngsters. My wife, Rosa, served as an unmerited sounding board and insistently encouraged me to finish the manuscript. Without her, I would probably still be halfway.

Some material previously appeared in another form. Chapter 6 is a revised and extended version of an article in *Journal of Economic Behavior and Organization* (2009), 72(1), pp. 1–20. Section 8.5 contains revised and extended text parts from my chapter on 'Evolutionary modelling' in J. Proops and P. Safonov (eds), 2004, *Modelling in Ecological Economics*, Edward Elgar, Cheltenham. Some material in Sections 13.2 to 13.4 is adapted and revised from an article in *Journal of Evolutionary Economics* (2007), 17(5), pp. 521–549, and text parts in Sections 14.3 and 14.5 from an article in *Technological Forecasting and Social Change* (2013), 80(1), pp. 11–23. Chapter 15 is a revised and extended version of an article that appeared in *Journal of Bioeconomics* (2013), 15(3), pp. 281–303. Permission was granted by the associated publishers, Edward Elgar Publishing, Elsevier and Springer.

In the spirit of evolutionary thought, I want to end with two disclaimers. My personal flaws – of a physiological (bad eyesight), intellectual (selective reading) and social (sporadic hermit-like behaviour) nature – serve as a proximate explanation for any mistake the reader may find in this book. The ultimate explanation is, of course, that natural selection is imperfect, having failed to wipe out such flaws in my ancestors. So, in all earnestness, at the end of the day evolution is to be blamed for all remaining errors.

Part I

Prevue

1 Making the Improbable Probable

1.1 Generalised Evolution

The amount of attention that academia and popular media have given to evolutionary thought in recent years is unprecedented. More text is being published now on evolution than on theology, signifying a turning point in intellectual history. The mere mention of evolution tends to evoke strong reactions in many of us, ranging from enthusiasm to resistance. This should not come as a surprise. Accepting the evolutionary tale of life and humans has far-reaching implications for how we perceive the biosphere and our place within it. This book exposes the idea that a generalisation and extended application of evolutionary thinking can lead to equally astonishing conclusions. It has the capacity to contribute to a renewed and profound understanding of human cultures, economies, technologies, institutions and even public policies.

Partly triggered by its success in the field of biology, evolutionary thinking has become prevalent in other areas. This is manifest in terms such as non-genetic, social, cultural, symbolic, memetic, exogenetic and exosomatic evolution. These are all instances of what may be called 'generalised evolution'. It encapsulates the flexible nature of evolutionary features and concepts, which can be recognised in applications to many different phenomena. A related term, 'generalised Darwinism', is also appropriate, given that Darwin was writing before the development of genetics, so was not aware of the genetic basis of organic evolution. His ideas about evolution therefore possess a certain level of universality, which enables their transfer to non-genetic realities.

Evolutionary thinking explains things that were hitherto mysterious, or that were explicated in an unsatisfactory way, such as through the infamous design, or watchmaker, argument (Box 1.1). The reason is that evolution is a stunningly effective mechanism for producing dynamic complexity, to begin with in living systems. But we are now able to apply evolutionary principles to solving many other puzzles. This book collects theoretical and empirical insights supporting the idea that evolutionary thinking is our best chance to come to grips with all kinds of real-world complex systems and their dynamics. The fact that evolutionary thought was first fully elaborated theoretically, empirically and experimentally in the domain of biology does not preclude its broader application. One way to regard evolution is as an algorithm or set of general mechanisms that apply to biological and various non-biological phenomena in equal measure (Dennett, 1995). The relevance of such an algorithmic interpretation is supported by the successful formalisation of evolution in mathematics (Nowak, 2006)

Box 1.1 The watchmaker argument double wrong

In 1802, William Paley proposed the famous 'watchmaker argument' in his book *Natural Theology: or, Evidences of the Existence and Attributes of the Deity, Collected from the Appearances of Nature*. Its thesis was that the complexity of living organisms suggests design and therefore a designer, in the same way that a complex clock is designed by a watchmaker. Design arguments had already been proposed by earlier theologians, notably Thomas Aquinas in the thirteenth century. They were considered to be the most important pieces of 'evidence' for the existence of an 'Almighty God'. While, in pre-Darwinian times, the design analogy might have been excused, current evidence for an evolutionary history of life (summarised in Section 3.5) makes the design-of-life-forms a difficult theory to defend. In a creative twist, Richard Dawkins (1986, p. 5) dubbed organic evolution the 'blind watchmaker', to emphasise its unconscious and unplanned character.

What is wrong with the design analogy? For one thing, it reflects a lack of imagination. Living organisms are far more complex than any watch. This is clear when one considers the number of active protein molecules in a single cell of a living organism, which run into the millions (Milo, 2013). By comparison, the most complicated pocket and wristwatches have at most a few thousand components, not all of them even active. But the design analogy is erroneous for another reason. No one single person or firm designed the watches we use or the automobiles we drive from scratch – or any other technology for that matter. Instead, they are the outcomes of an enduring process of cumulative inventions that counted on the participation of numerous individuals and firms. Over time, many ideas arose and either competed or merged. From the diversity of designs at each moment, only a subset carried on to the next phase, not guided by a grand designer but by evolutionary principles of diversity creation, selection and imitation. The proximate explanation for any watch is that it was made, though not designed, by a particular watchmaker. As an ultimate explanation, watch design came about through a long historical process of technological evolution. Section 2.4 considers design versus evolution in more detail, while Chapter 10 elaborates the evolutionary view of technological innovation.

---

and evolutionary computation (Mitchell and Taylor, 1999; Eiben and Smith, 2003). According to Blackmore (1999, p. 11), 'Algorithms are "substrate-neutral", meaning that they can run on a variety of different materials.'

But what is evolution, precisely? This question has no adequate brief answer. Moreover, the answer will vary between the contexts to which evolutionary thinking is applied, as later chapters will attest. Following a generalised outline of evolution in Chapter 2, the book delves into evolutionary thought in biology in Chapters 3 and 4. Following on, Chapters 5 to 7 look at evolution as it relates to society and culture, while subsequent chapters will cover the economic, organisational, technological, political and public policy realms. For now, I include some key points about generalised evolution to whet the reader's appetite. 'Generalised evolution' denotes a continuous interaction among three general mechanisms affecting variety in some biological, cultural or technological population: selection, innovation and replication, jointly referred to in this book as the V-S-I-R algorithm. Specific labels differ among disciplines. In biology, selection is also known as differential survival, while innovation is referred to as variety creation, and replication is often referred to as heritability, inheritance or reproduction. In economics and innovation studies, variety creation is commonly called innovation. The interaction of the V-S-I-R mechanisms causes seemingly unlikely constellations and structures to become surprisingly likely. One could say that evolution makes the improbable probable, without any preconceived plan or purpose.

Ostensibly a simple process, the implications of evolution are remarkable. In its entirety, though, the process of evolution is certainly not simple, and definitely more complex than many of its critics can imagine. Interpretations of evolution, such as coincidence, randomness or blind chance, are often misconstrued. The reason is that only one of the three V-S-I-R processes (note that V is a component, not a process), namely innovation (or variety creation), is random. The other two, selection and replication, represent rather deterministic processes, which has been summed up as: 'Mutations propose but selection decides' (Cavalli-Sforza and Cavalli-Sforza, 1995, p. 95). More accurately, even variation creation is not genuinely random. As we will see in Chapter 4, it is better to characterise it as 'undirected' given that it involves many understandable, and even predictable, processes. In the process of evolution as a whole, variety creation then serves as the creative part, environmental selection provides the directive force, and replication or inheritance adds the finishing touch by guaranteeing endurance and stability. The combination of the three processes (S, I and R, working upon V) gives rise to an endless accumulation of adaptations, in biological, cultural or other units, to the respective environment. So, while evolution is unguided by any intermediate or end goals, it would be misleading to describe it as wholly random. If it were, evolution would be unable to produce the unlikely yet functional constellations that it does. Section 2.1 looks more systematically at the specific interaction of the V, S and R factors, including cases in which only one or two of them are present. This helps to better understand their complementarity in generating complexity and diversity.

Evolutionary thinking has two other main features that distinguish it from other approaches found in the natural and social sciences. One is a population approach, which allows for an explicit and comprehensive analysis of variation among agents and changes therein. It provides an alternative to theories that employ average, representative or aggregate agents (Kirman, 1992). A population is a group of individuals or elements generally alike but differing in some subtle ways. In biological contexts, many populations include individuals belonging to the same species and having offspring through sexual intercourse. In social-cultural settings, they consist of similar types of individuals or entities, such as consumers or firms, which interact in terms of imitation, information exchange or competition for customers. The advantage of a population approach is that it offers a suitable foundation for studying distributional or inequity issues, a central concern of many social sciences. In addition, social interactions and networks, common in recent social studies, can be easily integrated into a population approach.

A second important feature of evolutionary thinking is that it enables researchers to identify ultimate causes, next to proximate ones. These refer to why things exist and how they work, respectively. Proximate causes involve physiological, behavioural-psychological, economic, institutional or technical factors. Ultimate causes relate to evolutionary processes, whether they are of a biological, cultural, economic or technological nature. These processes help to explain how the proximate factors came about. Ultimate causes can be seen as a subcategory of a wider class of long-term, historical causes. The latter include geographical and climatological factors, which tend to change little over prolonged periods. Proximate and ultimate causes are complementary and need to be studied in tandem to achieve a comprehensive understanding of complex systems. Evolution is perfectly capable of describing the mechanisms through which ultimate causes give rise to proximate causes. This explains why biology has a solid grasp of ultimate causes, unlike the social sciences, which tend to focus on proximate ones (Scott-Phillips et al., 2011). Linking the two types of causes through evolutionary thinking could enrich the social sciences. In turn, this would enable policy studies, as part of the wider social sciences, to design more effective solutions for social and economic problems, namely through a profound understanding of ultimate causes and consequences.

This volume zooms in on the social sciences, setting it apart from other treatises of universal evolution. It covers sociology, anthropology, political science, economics, organisational science, innovation studies, history, policy studies, environmental science and climate studies. In other words, it applies an evolutionary approach to social, environmental and policy sciences. While it would be too ambitious to claim that the full spectrum of possible evolutionary applications is covered, it is fair to say that a broad set of topics will receive attention – including those at the boundary between the natural and social sciences. Examples are evolutionary medicine, evolutionary criminal law, evolutionary mineralogy, the role of sex and gender in society, and the evolution of language, science, religion, music and humour.

Four specific goals motivate this book: (1) to inform the general, well-educated reader about evolutionary thinking in a variety of areas; (2) to convince social scientists that evolutionary approaches deserve our serious attention and that they already exist in productive forms within various social science fields; (3) to inform biologists that evolutionary concepts and methods serve a useful role in fields outside of biology, from which they can possibly learn or to which they can contribute original thoughts; and (4) to show that biological concepts and theories can be fruitfully transferred to the evolutionary social, environmental and policy sciences, enriching their scope and explanatory power.

1.2   Human Evolution Beyond Biology

The notion of evolution is one of the most impactful ideas that science has ever generated. It represents what philosopher Thomas Kuhn called a paradigm shift or what another philosopher described as 'Darwin's dangerous idea' and 'a universal acid' that destroys old ideas (Dennett, 1995). This is because evolutionary explanations encapsulate cause–effect mechanisms that can be applied to all kinds of structural change phenomena. Darwinism has already moved beyond well-trodden paths to advance new ideas on such diverse issues as language, medicine, knowledge formation, ethics, religion, computation and Darwin machines, such as the immune system and the brain (Buskes, 2006). Evolutionary approaches can also be found in the majority of social sciences (White, 1998). Indeed, it is undeniable that human cultures, economies, organisations, institutions and technologies are related through historical descent. This validates efforts to use evolutionary thinking to better understand them.

The transition from biological to cultural evolution occurred through various stages of social evolution, i.e. evolution of social behaviours, in primates and humans. Chapter 5 clarifies the origin, functions and dynamics of social behaviour and groups, while providing some detail on social insects and primates. During the 1970s, sociology and anthropology were conquered by the fear that invoking evolutionary arguments implies some form of biological determinism. This was largely an emotional response to an emerging biological area of research baptised as sociobiology (E.O. Wilson, 1975). The ensuing debate was complicated by politically sensitive societal topics, such as racism, sexism, homosexuality, aggression and crime (Ruse, 1979; Nielsen, 1994). Over time, the research continued, wisely adopting new names, such as behavioural ecology and evolutionary psychology. Nowadays, evolutionary thinking is widely accepted in the fields of anthropology and psychology. It partly denotes a revival of the old ideas, and partly an integration of previously separate fields. For example, evolutionary anthropology integrates insights from comparative analysis of primates and humans, behavioural ecology and ethology, human evolutionary biology and genetics, palaeontology and archaeology, evolutionary psychology and linguistics, and broader cognitive science. This practice of collecting many distinct, mutually consistent insights to reach unequivocal ultimate explanations is common in evolutionary studies. Indeed, it already characterised Darwin's way of working (see Section 3.2).

As will be discussed extensively in Chapter 7, once culture emerged in human groups, due to the unusual ability of humans for social learning and imitation, a combination of natural and cultural selection started to direct human evolution. Here culture should be broadly interpreted, comprising language, cooperation with non-kin, redistribution and sharing, exchange and trade, invention and perfection of tools, specialisation and division of labour, and institutions such as rules, norms, law and religion. For example, in Section 4.8, we will consider the role of economic processes in shaping human evolution, while in Section 11.2 the role of cooking will receive attention. The presence of cultural selection or evolution that affects human behaviour, brain–mind and even physiology, undercuts the common idea that genetic evolution caused or shaped culture. Instead, a two-way process of mutual influences of genetic and cultural evolution, or gene–culture coevolution, is a more apt characterisation of human evolution since the first appearance of cultural elements in human groups. Moreover, it also holds the key to adequately explaining current and future human evolution (Chapter 16). Hence, the study of human evolution is not the sole domain of biology, but requires a synthesis of human evolutionary biology with insights about culture and cultural dynamics from the various social sciences.

Biologist E.O. Wilson (1998) has suggested that the social sciences can achieve greater progress by striving for consilience. By this, he means that results from the social sciences should be consistent with relevant and widely accepted insights from other disciplines, notably the natural sciences. This view is supported by many scientists today, notably behavioural ecologists, ethologists, primatologists and psychologists (de Waal, 1996). As stressed by Wilson (1998), the boundary between, on the one hand, the natural sciences and, on the other, the social sciences is very incompletely studied, and to make progress, both need to be involved and even cooperate. According to Richard Nelson (2007), 'universal Darwinism' is dominated by contemporary biological evolutionary theory, which contributes to evolution in the social sciences being seen as analogous to biological evolution. He considers Dawkins (1976), notably his proposed meme thinking (addressed in Section 7.6), and Daniel Dennett (1995) as the principal representatives of this so-called 'biological imperialism'. His preferred alternative is a specific social science approach to evolution, such as associated with the ideas of the philosopher David Hull (1988) and the anthropologists Boyd and Richerson (1985). Nelson claims that a generalised formulation of evolution in the social sciences has to be unbiased towards biology. I personally believe that fear of biological imperialism is unwarranted. In various places in this book, I will argue why biology-inspired notions can sometimes offer fresh angles and deserve to be given a fair chance in the social sciences.

As later chapters testify, evolutionary social sciences encompass both genetic and non-genetic evolution. Genetic evolution applies to the living world, including humans, also those living in cultural settings. Non-genetic evolution applies mainly to humans and their cultural, social, economic and technological realities. The study of human genetics includes the disciplines of evolutionary psychology (Barkow et al., 1992) and genetics (Carey, 2002). The first combines evidence from genetics, physiology, human palaeontology and primatology to understand the evolutionary history of human behaviour. The second synthesises human evolutionary theory and behavioural experiments to understand how the mind, brain and human culture have co-adapted. It also looks at how humans are currently maladapted to their environment. Evolutionary analysis in anthropology, sociology, economics and technology studies focuses on non-genetic evolution, even though sometimes genetics is included, such as when studying the evolution of pest resistance due to agricultural practices. Some studies move one step further by assessing the dual influence of genetic and non-genetic factors and associated gene–culture coevolution (Sections 7.3 and 7.4).

This book aims to offer a fair and balanced account of the main evolutionary ideas and theories in the social sciences, including those that cross the natural–social science divide. This complements other publications on social, cultural or economic evolution, which often emphasise or promote one particular theory. At best, these mention alternatives in a tangential and superficial way, or provide disproportionate criticism to sweep aside competing approaches. While certainly not eschewing critiques, the treatment here will consist of constructive accounts and comparisons of all relevant theories. This starts in Chapter 5 with approaches to clarify social behaviour from an evolutionary angle. The various evolutionary perspectives on the role of groups in evolution, a much-debated topic, are considered in Chapter 6. This is complemented in Chapter 7 with a broad range of evolutionary theories proposed in sociology and anthropology, covering themes such as dual inheritance, gene–culture coevolution, learning, religion, music and humour. In Section 7.10, we will systematically and critically compare the different evolutionary approaches used to study social-cultural systems, that is, sociobiology, evolutionary psychology, dual inheritance (including gene–culture coevolution) and memetic thinking. Next, Chapters 8 to 10 extend the cultural evolutionary palette by addressing evolutionary economics, organisational evolution and innovation studies. These integrate insights from behavioural economics, consumer analysis, firm organisational studies, financial economics and recently also macroeconomics. Evolutionary approaches are further encountered in geography and political science; these receive attention in Chapters 8 and 15. Chapters 11 and 12 present an evolutionary angle on long-term history, focusing on the evolution of *Homo sapiens*, the emergence of agriculture, and the rise of modern science and industry.

1.3   Practical Values of Evolutionary Reasoning

Evolution as a mode of thinking has begun to enter disciplines that aim to improve the human world. We will consider this in more detail in Chapters 13 and 14, where environmental and climate policies are examined from the angles of evolutionary biology, economics and technology, and in Chapter 15, which scrutinises more generally the design of public policies and the dynamics of political processes. The notion of evolutionary or Darwinian policy draws attention to the relevance of distinguishing between proximate and ultimate causes of human problems. In addition, it brings to light questions as to what are relevant evolutionary criteria for public policy design, and raises the much-debated issue of evolutionary progress, including in socioeconomic and technological contexts. Incidentally, the policy-related term 'policing' is used in evolutionary biology to refer to detection and punishment mechanisms in groups of genetically weak or unrelated organisms.

The practical value of evolutionary thinking is clear in the realm of the biomedical sciences. This is particularly evident in evolutionary medicine and clinical evolutionary psychology and psychiatry (Nesse and Williams, 1994). The traditional approaches in these fields focus on the direct or proximate reasons for, and physiological processes underlying, human sickness and disease. Evolutionary approaches add insights about ultimate factors behind human diseases as well as about the nature of defence mechanisms in the human body. Explanations include imperfections arising from trade-offs among adaptation to different selection conditions, maladaptation to cultural environments, and relationships between genes and molecular processes underlying diseases. In particular, our understanding of cancer and HIV/AIDS has improved considerably thanks to evolutionary analysis. The recognition that these diseases are effectively a battle against evolution has given rise to more effective cures. Box 1.2 explains this in some detail.

Box 1.2 Cancer and HIV/AIDS infection as evolutionary processes

Cancer can be regarded as an undesirable outcome of an evolutionary process internal to a multicellular organism. It occurs when a population of mutant cells competes with healthy cells for scarce space and resources. Mutant cells are selected as they are better adapted to the micro-ecological environment. Martin Nowak (2006, Chapter 12, p. 209) states that cancer is typical for multicellular organisms, since it means the 'evolution of defection' and a 'breakdown of cellular cooperation'. He notes that while computers are sensitive to viral infections, they are immune to cancer. The micro-selection environment of cancer cells varies between cells, even within a small region. In particular, proximity to blood vessels influences how cells are affected by chemotherapy. For example, more distant cells can escape the strong effects of cancer treatment. The connection between evolution and cancer is further strengthened by the fact that most, if not all, cancers have genetic causes. Activation of 'oncogenes', often transferred by viruses, triggers cell proliferation and deactivates tumour-suppressing genes. Attempts to cure cancer can avoid the process of selection for resistance by taking heed of underlying evolutionary processes. Treatments following this approach have become more effective over the last decade in controlling certain types of cancer. Techniques include multi-drug therapies that alter the level of competition between mutant and benign cells so as to improve the strength of the latter (Merlo et al., 2006). From an evolutionary perspective, so-called 'targeted therapies' that help to obstruct the proliferation of tumour cells, appear to be an effective and less damaging alternative to standard chemotherapy. The latter aims to kill cancerous cells, but with significant collateral damage to the human body. Considering tumours from the angle of selection points to novel remedies such as blocking the growth of new blood vessels to tumours. Another idea is to treat a tumour only when it is growing and leaving it alone when it is not, as this might be selecting for cells that have a slow life-history strategy (Aktipis et al., 2013). We can learn more about combating tumours by studying the mechanisms that the body uses to defend itself against them. In addition, one can derive insights from comparing animal species that are more and less prone to cancer.

Acquired immunodeficiency syndrome or AIDS is the expression of the HIV retrovirus that can infect and kill so-called CD4 cells of the immune system. Vaccines have so far proven ineffective, as the virus quickly responds by mutating to escape their selective pressures (Nowak, 2006, Chapter 10). The evolutionary origins of these viruses are simian immunodeficiency viruses (SIV) that infect many primate species, without leading to a disease. When they jump to other species, however, they can induce AIDS. Probably the most viral HIV virus, HIV-1, crossed the species barrier from the chimpanzee and gorilla to humans multiple times, giving rise to four groups of the virus (M, N, O and P). The first transmission occurred around 1910, and may have been related to 'bushmeat activities' combined with the emergence of larger, colonial African cities that brought with them higher levels of promiscuity and prostitution, increasing the risk of HIV transmission. It has further been suggested that railways have played a role in wider diffusion of AIDS (Cohen, 2014). The

---

Box 1.2 (*cont.*)

evolution of HIV in the body creates a diversity of antigens over time. An antigen is the part of the virus that causes the immune system to generate antibodies. HIV is unique in that it has an unusually fast replication cycle and high mutation rate. This results in a high diversity of HIV antigens over time, causing the immune system to become overloaded and, also because of losing CD4 cells, incapable of controlling the virus. Although there is no definite cure for HIV/AIDS, effective antiretroviral therapies exist. One combines various medicines from six drug classes to create a complex selection environment that controls the virus for a long time, thus both extending patients' lives and improving their quality of life. In the short run, HIV quickly evolves in individual humans during infections, that is, in the absence of antiviral treatment. Its long-run fate is yet uncertain. In fact, a remarkable 8% of our human genome is accounted for by inactive remnants of ancient retroviral infections. For a highly pathogenic retrovirus, such as HIV-1, evolution towards a less virulent state would likely take a long time – probably at least several millennia. This suggests a high death toll before HIV turns into a silent passenger in our genome or the HIV-host becomes resistant. HIV may, though, not follow the same evolutionary pattern as other retroviral infections, given that it moves through a global population and is subject to anti-retroviral control. Effectiveness of such control over a long time period, especially reducing transmissibility of the virus, is essential to making HIV less destructive.

The policy sciences can likewise benefit from evolutionary angles, not only regarding genetic evolution, such as that associated with the occurrence of pesticide resistance in modern agriculture, but also with respect to non-genetic evolution, such as in the areas of technological innovation (Chapter 10), environmental regulation (Chapter 13) or inequality and poverty (Chapter 15). Since climate change ties many of these issues together, evolutionary thinking can help in guiding a sociotechnical transition to a sustainable and low-carbon economy (Chapters 13 and 14). Incidentally, natural history abounds with transitions, which have received much attention in evolutionary biology (Section 3.6). One can thus expect rewarding cross-disciplinary spillovers in this respect.

Evolutionary medicine and evolutionary environmental policy share several themes. Foremost, that humans are in some ways maladapted to the current environment, also known as 'evolutionary mismatch'. In this respect, an improper diet and lack of exercise are often mentioned as the consequence of Pleistocene humans living in a modern technology-driven world. According to evolutionary approaches to environmental science, humans are maladapted even as a species, since environmentally damaging technologies, overconsumption and a large population push them to live beyond the material and energy limits of local and global environments. As a result, humans are altering basic environmental conditions, in turn worsening the degree of their maladaptation. Both evolutionary medicine and policy analysis recognise that an understanding of proximate causes of problems is insufficient for identifying effective solutions to relevant health or environmental problems. One has to understand their ultimate causes.

To illustrate this, a fundamental insight of evolutionary medicine is that sickness can be beneficial for overall long-term health. This is because getting sick triggers defence mechanisms to eliminate disease-causing microorganisms, which builds up resistance to future attacks. Coughing, diarrhoea, vomiting and fever are all part of the body's strategies to fight disease. Repressing such symptoms with medication may lengthen the duration of a disease and prevent resistance formation that is useful during later outbreaks. Moreover, medicines will make sick people feel better than they actually are, weakening incentives to take a rest, which in turn impedes a quick and complete recovery. In addition, evolutionary medicine suggests that mental states, such as melancholy, sadness and even depression, should not always be treated with medication as a first resort. These conditions often serve an adaptive function, by signalling to overworked persons that they should take things easy and lower their ambitions. Another explanation is that such mental states help people to make good decisions by enabling the brain to focus attention on a problem or conflict at hand.

Further insights of evolutionary medicine are that the incorrect use of antibiotics can act as a delay tactic that eventually leads to antimicrobial resistance. Recommendations for the use of antibiotics include that a high short-term dosage is better than mild doses over a long time period. This is because the short bursts will not give illness-causing bacteria time to build up resistance. If a particular antibiotic proves ineffective, one should shift to a distinctly different one. A combination of antibiotics or other medicines can be used to treat a bacterial infection to create multiple selection pressures from which bacteria find it difficult to escape.

Ageing is another focus of evolutionary medicine. The idea is that some genes function perfectly when they are younger but confer disadvantages at an old age. An example is maintaining high calcium levels in the body at a young age to help bones heal quickly after accidents; but high calcium levels contribute to a silting of arteries (atherosclerosis) in later years. Another fundamental issue is illustrated by sickle-cell anaemia found predominantly in black people. This creates a trade-off between two effects, namely better protection against malaria and a higher risk of dying at a young age due to an inability of red blood cells to bind to oxygen. As a result, in malaria risk areas in Africa, heterozygotes have a higher chance of survival than dominant and recessive homozygotes, causing the allele coding for sickle-cell anaemia to persist. This is an example of an antagonistic form of balancing selection.

Modern human maladaptation takes several forms. Humans use technologies, such as cars, televisions and computers, which inhibit regular movement and exercise during the day. They eat too many fatty, salty and sweet foods. These habits were rare when humans were living in natural environments, but have become prevalent in modern cultural contexts with a greater accessibility to machinery and processed foods. Both phenomena contribute to various lifestyle diseases, or diseases of longevity and civilisation, such as obesity, diabetes, atherosclerosis, cancer, stroke and dementia. In fact, such diseases are now among the main causes of death, not just in developed countries, but in the world as a whole.

A more innocuous issue relates to the fashionable removal of pubic hair, facilitated by technologies such as electric shavers, depilatory creams and waxing. This can have negative effects that are largely unknown to the general public. Pubic hair serves various functions, such as protection against pathogens or providing a cushion against friction to avoid skin irritation and damage (Gibson, 2012). In addition, it signals sexual maturity and hence fertility, which relates to an evolutionary ultimate factor. The removal of pubic hair may also have undesirable cultural effects. For instance, Tiggeman and Hodgson (2008) warn that it can contribute to sexual objectification of young girls. More harmful cultural practices associated with the genital region of the body include circumcision (removal of the foreskin from the penis without an urgent medical reason) and female genital mutilation (partial or total removal of the external female genitalia). Both practices are prevalent in various countries and religions. Yet many overlook the fact that the foreskin and external female genitalia contribute to numerous physiological and psychological functions. In fact, some suggest that having a foreskin may have as many as 20 distinct advantages supported by scientific evidence.[1]

Interaction with dangerous technologies entails more than transportation. As part of the modern lifestyle, excessive sunbathing is common, even though it elevates the risk of skin cancer as people are not evolutionarily adapted to so much radiation. Many people further suffer from allergic reactions of the skin and respiratory system (such as asthma). This occurs when their immune system is underutilised and thus weakened under modern conditions of extreme sanitation through daily showering and frequent cleaning of living and working spaces. Specific problems arise for women owing to a deviation from the natural pattern of reproduction and breastfeeding due to, among others, the use of contraceptives. As a result, the frequency of menstrual cycles is altered, in turn affecting the likelihood of breast, ovarian and uterine cancer. Evolutionary explanations for human interaction with such technologies can help to formulate effective policies to reduce derived risks.

1.4   Resistance to Evolutionary Social Science

Generalisation of evolutionary thinking beyond biology is not uncontroversial. In particular, there has been considerable resistance to the application of evolutionary thought in the social sciences, by biologists as well as social scientists. Here we consider their main concerns. Where appropriate, the reader is referred to subsequent chapters for more details.

Some biologists take the view that there is only one type of evolution – gene-based. Accordingly, they perceive 'cultural evolution' to be a misplaced term and notion. This dictates that any system or process possessing evolutionary features and mechanisms but lacking a direct genetic basis should not be called evolution. It negates similarity of processes and patterns in different realms, characterised by populations with diverse units, selection, innovation (variety creation) and replication. While reproduction is a critical element of biological evolution, imitation is essential to cultural evolution – but both are merely special cases of replication. With the right adjective, one can distinguish

---

[1] See http://www.noharmm.org/advantage.htm for arguments and references.

between different types or domains of evolution – whether biological, genetic, social, cultural, economic, technological or non-genetic. The adjective clarifies that they are distinct instances of a general phenomenon called 'evolution'. In this respect, evolution is an appropriate and informative term. It signals a population approach, a consistent distinction between proximate and ultimate factors, and application of the V-S-I-R algorithm. Unless someone comes up with a better term, let us use 'evolution' as a common denominator. This will also aid the transferral of insights among researchers in different disciplines using evolutionary approaches.

Many social scientists seem to fear the incorporation of evolutionary thinking into the social sciences. But why would one reject a rich and well-established approach that adds an original perspective on, and can enlighten our understanding of, complex systems and phenomena, such as human culture, the economy and technology? Associated fears are often the result of limited knowledge of evolutionary concepts and methods. Biologist D.S. Wilson (2007) expresses little patience for 'evolutionary analphabetism' in the social sciences, comparing it to creationism (Section 2.4) resisting evolutionary biological explanations of organic life: 'Some intellectuals rival young-earth creationists in their rejection of evolution when it comes to human affairs' (p. 3). Earlier, Durham (1991, p. 31) used the term 'cultural creationism' for this. Wilson stresses that the equivalent of an integrative concept like evolution, connecting the many specialisations within biology, is lacking in the social sciences. But he is hopeful, and points out that in an influential journal such as *Behavioral and Brain Sciences*, approximately one-third of the articles between 2000 and 2014 used the term 'evolution'. The survey of social, environmental and policy sciences in this book support his optimism.

A staunch critic of evolutionary thinking in the social sciences, Bryant (2004, p. 460), thinks that 'evolutionary theory in the social sciences continues to operate in a largely metaphorical idiom'. This may be true for particular sub-disciplines or areas, but it does not apply to the full range of evolutionary studies encountered in the social sciences, from anthropology to economics and innovation studies. The latter, instead, are guided by a genuine evolutionary V-S-I-R algorithm of generalised evolution. Besides, the use of metaphors or analogies is not necessarily bad (see Section 2.3 towards the end). Indeed, many theories in the social sciences use concepts with a metaphorical element derived from previous uses in other areas. Think of central notions such as function, rationality, agent, equilibrium, stability, system and cycle – the list is virtually endless. An essential part of modern evolutionary analysis is based in mathematical models and computer programming. In other words, the substrate of evolutionary processes can cover not only biophysical, behavioural and social-cultural but also mathematical dimensions. One might claim that evolutionary biology and evolutionary social sciences are simply specific cases of mathematical evolution, or that they use mathematical analogies. Hence, the application of evolutionary ideas in the social sciences – while sometimes metaphorical – is generally best characterised as an algorithmic approach.

A common criticism of evolutionary social sciences is that there is no equivalent to the genotype/phenotype dichotomy that is found in biological organisms. Three counter arguments are as follows. First, replication and variation in social science domains do not depend on a genotype notion. Instead, they arise from imitation, intelligence/

memory, artefacts like books and computers, interpersonal transfer and teaching, and individual and organisational creativity and innovation (further clarified in Chapters 7 to 10). Second, one can in fact argue that a cultural equivalent to the genotype or a replicator at the cultural level exists. A convincing example of this is the meme, associated with the unique human capacity for imitating others. This has given rise to a meme-based theory of cultural evolution, which has already generated surprising and testable hypotheses (see Section 7.6). Third, a sharp genotype/phenotype distinction is not a necessary condition for a generalised evolutionary V-S-I-R-type process to function, as discussed in Section 2.1. Fourth, a blueprint for a cultural entity, such as an architectural design of a house, is not unlike a genotype. Both are codified plans that provide the essential basis for physical structures. Just as phenotypes of living organisms are not entirely determined by the genotype but codetermined by environmental factors during development, social-cultural-technological phenotypes tend to be affected by external environmental factors. In the case of a house, many environmental factors play a role during the building process: e.g. the particular builders and their habits, availability of materials and components, and governmental regulations.

Some scientists have suggested that social science phenomena tend to resemble an epidemiological process more than an evolutionary one. An example of the first is the spread of a pathogen, such as a bacteria or virus. A motivation for this view is that spread due to imitation is characteristic of cultural phenomena. This does, however, not mean that epidemiology is a better or more logical approach to social science than evolutionary thinking. In fact, an evolutionary approach is widely accepted as adding to our understanding of the dynamics of contagious diseases such as HIV/AIDS (Box 1.2). Compared to an evolutionary approach, an epidemiological model is a cruder, less flexible and more aggregate option as it disregards a significant proportion of micro-level heterogeneity. This is not to deny that an epidemiological model may be a good approximation for certain social processes. To claim, though, that it is a sufficiently general approach goes too far. Many cultural, economic and technological changes are not analogous to the spread of a pathogen – whether in a single host body or among several host bodies. The subtleties of social diffusion processes require building in microbehaviour characterised by bounded rationality, social interaction, diversity and relative performance (equivalent to 'evolutionary fitness'). This underpins the relevance of evolutionary thinking, not epidemiology. Social science phenomena tend moreover to involve multiple populations, like firms and consumers, changing in concert, through mutual selection. This means coevolution is a more adequate conceptualisation. Anyhow, if epidemiological types of models were so appropriate, they would have experienced a widespread diffusion in the social sciences, which is not the case. Admittedly, technological diffusion studies have employed epidemiological models, but even then, evolutionary models are more widespread as they are descriptively richer (see Sections 8.5 and 10.3). With regard to religion, Dawkins (1993) has suggested the 'virus of the mind' (see Section 7.6). Not denying its clarifying character, it is arguably better operationalised in an evolutionary than a strictly epidemiological model as unlike the second, the first allows for competition between viruses (e.g. ideologies or religions) and recognises that one can carry only one religious virus at the time.

Another objection comes from Pinker (1997), who suggests that instead of adopting an evolutionary basis, the social sciences require a psychological foundation. He states that 'nothing in culture makes sense except in the light of psychology' (p. 201) – harking back to a famous statement by one of the most influential evolutionary biologists of the twentieth century (Dobzhansky, 1973): 'Nothing in biology makes sense except in the light of evolution'. Few scientists nowadays would argue against the relevance of psychology, or more generally the behavioural sciences, for the social sciences. Illustrative of this is the integration of psychology and economics through the emergence of the field of behavioural economics. Nevertheless, the two options sketched by Pinker are not mutually exclusive. For example, evolutionary economics and economic psychology are actually two sides of the same coin. If one intends to describe a single individual, insights from psychology may suffice. But to describe social interactions between multiple individuals, like consumers imitating each other or participating in a status-seeking game, a population of socially interacting agents makes more sense. Likewise, dual-inheritance theories of cultural evolution integrate many aspects of social psychology (Section 7.3). These examples signal the usefulness of marrying social evolutionary thinking and the behavioural-cognitive sciences.

While modern understanding sees biological evolution as (Neo-)Darwinian, social-cultural evolution is often suggested to be 'Lamarckian' (after French biologist Jean-Baptiste Lamarck – see Section 3.1) and therefore not fitting the modern definition of evolution. This view is common among biologists and social scientists. This book makes a clear distinction between Lamarckism on the one hand, and the ideas of Darwin, Weismann and Baldwin on the other. Section 7.5 will argue that Lamarckism is not an apt label of social, cultural, economic or technological evolution, for various reasons. Perhaps most decisively, Lamarckian inheritance is restricted to a biological context with a sharp genotype/phenotype dichotomy, as it effectively says, in modern genetic terms, that phenotype changes translate to genotype changes. Of course, neo-Darwinism teaches us instead that evolutionary causality runs in the opposite direction, from genotype to phenotype. This is associated with the Weismann barrier, which states that germ cells are separated from somatic cells and thus are generally not affected by somatic (body) changes. The term 'Lamarckism' is often, disputably, used to signal that social-cultural evolution depends on human characteristics such as being conscious, intelligent, forward looking, and using planning and design. These appear to contradict that evolution is without plan or purpose. But the fact that humans or their organisations possess some capacity to anticipate and plan does not signify that cultural change is directed and planned. One should not overestimate the effectiveness of individual and social planning. According to cognitive science, many human decisions result from unconscious processes in the brain. We further know that organisations often fail to forecast and plan well. Many authors believe evolution requires individuals in an evolving population to be dumb or passive. But this is a grave misunderstanding. Evolution can select alternatives, regardless of whether they derive from conscious or unconscious processes. In addition, aggregation of planned and unplanned decisions by populations summing up to billions of people and millions of firms resembles more an evolutionary than a centrally planned process. Witness also that culture, technology and science histories are packed with instances of trial and error, failure and wastage, reflecting the unmistakable fingerprint of evolution.

An associated criticism of cultural evolution says that cultural variation is less random and undirected than genetic variation, suggesting that cultural evolution cannot work. But selection works upon any variation, independently of how it came about. Anyhow, new genetic variation is limited in scope because it is never purely random, being limited by initial genetic structures (see Section 4.2). Even a common replication mechanism in social systems, like imitation, is frequently subject to errors ('mutations'). Hence, it not only contributes to homogeneity but also to new diversity. These points mean there is no fundamental difference between genetic and cultural variation. But even if it would be true that cultural variation is less random and more directed, this does not hinder the functioning of any selection process. This view is supported by many respected writers on evolution outside biology. For instance, Daniel Dennett has noted that 'non-random variation does not prevent the algorithmic process of natural selection from grinding out its well-tested innovations', and Peter Richerson has stated that 'Natural selection works on any pattern of heritable variation'.[2]

Some worry that combining social science and evolutionary thinking inevitably results in a kind of Social Darwinism. This is a derogatory term for an ideology which holds that the strongest or fittest humans should survive and replicate to further socioeconomic progress. This stance has been criticised for its justification of harsh, laissez-faire policies and perpetuating an inequality of opportunities and outcomes (Hodgson, 2004). The association with Social Darwinism has impeded the development of an evolutionary social science – Hofstadter (1944) was particularly influential in this regard. Incidentally, the name 'Social Darwinism' does a disservice to the well-founded ideas of Darwin. A more appropriate term would be 'Social Lamarckism' or 'Social Spencerism', as the respective ideas were largely propounded by Spencer, who in effect applied something more closely resembling Lamarckism than Darwinism to social realities (see Box 7.1).

The criticism that evolutionary social science gives rise to Social Darwinism reduces evolutionary social science approaches to a normative interpretation, instead of distinguishing clearly between normative and explanatory evolutionary approaches. Social Darwinism is a strictly normative approach which presumes that evolution in a social context is equivalent to progress, something which lacks a scientific foundation (Section 15.2). Explaining something with evolutionary arguments, on the other hand, generally does not imply a justification of it. In other words, 'is' does not imply 'ought to be'. Of course, both evolutionary and non-evolutionary explanations of social phenomena can, and are often intended to, influence future responses in terms of policies and institutional design. But this is something fundamentally different from justifying a status quo that is the outcome of a historical biological or cultural evolutionary process. Unfortunately, rather than carefully distinguishing between 'is' and 'ought to be', many critics of evolutionary thinking in the social sciences mix them up – also known as committing the 'naturalistic fallacy'.

---

[2] Citations are from online responses by Dennett and Richerson to an essay by Steven Pinker: http://edge.org/conversation/steven_pinker-the-false-allure-of-group-selection.

To better understand this point, notice that evolution tends to generate inequalities by amplifying favourable circumstances for some and bad luck for others. In addition, evolutionary path dependence (Section 10.5) hampers continuous progress, and may result in a socially undesirable lock-in of technologies or institutions from which it is difficult to escape. Rather than accepting such outcomes of cultural or economic evolution, one can formulate corrective policies informed by a thorough understanding of the evolutionary mechanisms and ultimate factors involved. In this respect, Penn (2003, p. 294) notes 'What many fail to understand is that evolutionary perspectives on humanity can help to challenge the erroneous assumptions of the political right as well as the left.' As later chapters will illustrate, positive evolutionary social and economic analyses can be politically unbiased, while their normative counterparts allow examination of how certain political goals – whether left or right wing – are served most effectively by which policies. In this sense, evolutionary approaches to social science do not differ from non-evolutionary social science approaches: either can add normative goals to its positive framework to develop public policy. This book will argue that the assessment and design of public policy in areas as diverse as social welfare, employment, health, environment and climate can benefit from the adoption of an evolutionary angle, without resorting to the ideology of Social Darwinism. If an evolutionary process is at play in the creation of undesirable social outcomes, then a greater understanding of it will help in formulating effective remedial or preventive measures. Chapters 13 to 15 give weight to this argument. A clear distinction will be made between positive (descriptive and historical) and normative (evaluative and planning) aspects of the evolutionary perspective on public policy, institutions and politics.

My hope is that the above arguments convince the reader of the many reasons to invest time and effort in an evolutionary approach to the social sciences. This, however, should not be misconstrued as a blanket proposal to explain all culture, economics and technology in evolutionary terms, or to replace all existing social science theories by evolutionary counterparts. Suggesting that evolutionary thinking is an all-or-nothing approach is to attack a straw man. As this book shows in subsequent chapters, evolutionary and other approaches can sometimes compete and contradict each other, and at other times be combined in ingenious ways to form a credible description of, or explanation for, particular social phenomena. According to Geoffrey Hodgson (2002), universal Darwinism, or general evolutionary mechanisms, can and should be combined with discipline-specific insights and theories to develop comprehensive explanations of phenomena studied within each relevant discipline.

This book argues that evolutionary thinking can enrich the social sciences, sharpen its precision and enable it to tackle new problems. As Part IV of this book illustrates, we are beyond the test phase. The evolutionary social sciences are already thriving.

2 The World According to Evolution

2.1 The V-S-I-R Algorithm

The power of evolution lies in the interaction between its core components: variety (V), selection (S), innovation (I) and replication (R). Whereas innovation introduces a random, creative element, selection adds a deterministic and directional one, completed by replication, which ensures endurance and accumulation. Here we will consider each of these components, which jointly affect variety over time.

Selection is not a law, like gravity, but a collection of rather heterogeneous processes. These generally reduce variety and tend to be directive and equilibrating. Selection causes better-adapted units to replicate (reproduce or be imitated) more than less well-adapted ones, increasing their share in the population. In addition, selection can simply remove maladapted units from the system. Examples of selection from biological and cultural settings include competition for food, sexual partners, funding and clients, and predator pressure, scarcity of space and public policy pressure. While sustained and stable selection tends to reduce diversity, some forms of selection can increase variation. For instance, disruptive (also known as diversifying or centrifugal) selection benefits the extreme cases rather than those at the average (Figure 2.1). It can arise due to continued cyclical environmental change, or because of genetic drift in small, isolated (founder) species. It thus sometimes marks the start of a process of speciation.

In biology, the term natural selection is common. Coined by Darwin, the term was inspired by, but also intended to be distinguished from, the notion of artificial selection. Selection is often used as a synonym for evolution, which, strictly speaking, is not correct as evolution is broader, also including variety creation and replication. The term 'selection' can be confusing, as it incorrectly suggests a conscious or even purposeful action by an intelligent, human-like being. In biology, the term differential survival is often used as well and is arguably clearer.

Innovation is complementary to selection in that it creates new variety. It thus functions as a disequilibrating and partly random force. Innovation often works by uniting what is already there to form new entities. A special case of this is modular or compositional evolution, which combines already complex and pre-adapted components, thus fostering rapid and non-gradual evolutionary change. Endogenous retroviruses, for example, make up a part of the genetic material of many organisms (Section 4.3), while crossovers and transboundary work have proven to be a source of major innovations in technology, science or even music. One can observe equally in

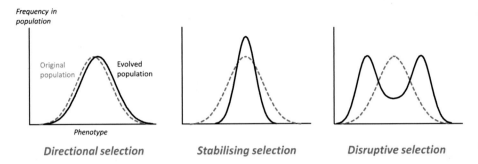

Frequency in population

Original population

Evolved population

Phenotype

Directional selection   Stabilising selection   Disruptive selection

Figure 2.1 Change in phenotype distribution under three selection modes.
(*A black and white version of this figure will appear in some formats. For the colour version, please see the plate section.*)

biological, cultural and technological evolution an enhancement of the effectiveness and scope of the innovation component. This is the outcome of an evolution of innovation systems themselves, which has allowed evolution to span an expanding space of creations. In nature, this is reflected in the emergence of genetic recombination, sexuality and learning modules in organisms. In socioeconomic systems, it is expressed through the continuous improvement of methods of public education, research and development (R&D) in firms and basic research at universities.

Generalised evolution involves information transmission through replication. This assures that populations at different points in time are related. Whereas evolution in nature operates mainly (but not only: see Section 4.2) through genetic replication, cultural-economic-technological systems lack such a dominant device. Instead, they realise 'intertemporal continuity' through multiple transmission channels: blueprints, individual memory, teaching and education, various types of learning, imitation, formal and informal norms and rules, and a number of ways to store information (on paper or electronically). Related to this, social systems are based on horizontal (peers), oblique (tutors) and vertical (parents) transmission processes, whereas biological evolution is dominated by vertical transmission from genetic parents to offspring.

Table 2.1 lists the consequences of different combinations of selection, innovation and replication for variety, covering seven cases in total. If one or two of the three components are absent, processes that are considerably simpler than evolution appear. For instance, variety creation tends to dominate if selection is not present, which tends to result in a largely random process (especially Cases 4 and 6). Without replication, cumulative adaptation is missing (Cases 2, 5 and 6). This scenario shares more similarities with evolution but represents merely a stripped-down version as stability and accumulation are lacking. Pure selection forces, as in Case 5, can generate a process of self-organisation, as observed in many physical and chemical systems (see also Sections 9.5 and 9.6). As an illustration, pebbles can be reshaped under the selective forces of a flowing river or seawater. This produces a tendency towards homogeneity as all stones are rendered smooth. Such homogeneity can be avoided only if a new source of variation appears. Then we obtain Case 2. To illustrate, we can extend the previous

Table 2.1 Characterising distinct combinations of evolutionary V-S-I-R components

Case	Evolutionary components			Outcome	Examples
	Selection	Innovation	Replication		
1	x	x	x	Evolution resulting in environmentally well-adapted units through an accumulation of adaptive complexity over time	Biological and social-economic-technological evolution
2	x	x		Changing variation without stability and accumulation, and thus limited adaptation	Pseudo-evolution of minerals (see Box 2.1)
3	x		x	Decrease of initial variety owing to repeated selection until an equilibrium state is reached (similar to Case 5, although replication may enhance the speed of convergence to equilibrium)	Autocratic systems (e.g. certain religions, dictatorships) eradicating diversity of opinions and triggering conformism
4		x	x	Randomly changing variation; possibly developmental constraints guide the direction of phenotypic evolution (Nei, 2007)	Genetic drift (Section 4.2), when selection forces are weak or absent
5	x			Decrease of initial variety due to selection until an equilibrium state is reached	Characteristic of many self-organising physical processes – see examples in the text and Box 2.1
6		x		Variation changing continuously without any direction	Prize-winning numbers in lotteries over time
7			x	Initial variety is replicated so that variety is constant over time	Probably no real-world examples: there is always a constraint/force that creates selection pressure/new variation

Note: Case 1 characterises genuine evolution while Cases 2 and 4 are also found in evolutionary systems, albeit less frequently; Cases 3 and 5–7 typify non-evolutionary systems.

example by including rocks near the river being broken by ice in winter, which acts as source of variation because it adds unpolished stones to the river.

It is fair to say that a full explication of evolution goes far beyond the V-S-I-R algorithm, for various reasons. First, V, S, I and R can adopt many different forms, not just beyond biology, but even within organic evolution itself (van der Steen, 2000). Variety has multiple dimensions: the number of distinct elements in the population, their frequency distribution (balance), and their disparity or degree of difference (Stirling, 2007). In

Box 2.1 Pseudo-evolution of minerals and life on Earth

While evolution is generally associated with living organisms, including humans and their cultures, evolutionary thinking has also been applied to non-living systems and processes in astronomy, physics and chemistry. Diversity of elements in this sense refers to such physical features as energy, speed, shape and position. Selection factors involve the basic physical forces, such as the 'strong force' responsible for binding nuclei, the electromagnetic force that operates between electrically charged particles, and gravity. However, as the notions of population and replication are not evident in such non-living contexts, the term evolution can only be used in a loose sense. In this case, it would be better to speak of pseudo-evolution.

In this vein, some authors have proposed the idea of mineral evolution, meaning that the mineralogy of planets and moons is subject to a process of evolution. Hazen et al. (2008) argue that this involves turning uniform distributions of atoms and molecules into more heterogeneous ones and multiple selection pressures that affect mineralisation, notably physical pressure, temperature and activities of water, carbon dioxide and oxygen.

Taking such a pseudo-evolution approach helps to distinguish a number of stages (Figure B2.1.1), each with its own particular physical, chemical and biological selection pressures, as well as basic molecular elements. Stage 1 occurred in the stellar nebula, where the earliest generations of meteorites, called chondrites (covering 60 different minerals), were formed. Chondrites were clumped together through gravitational pull into larger and larger planetesimals. This altered the aqueous and thermal conditions leading to Stage 2 which included some 250 minerals. The formation of early planets led to Stages 3, 4 and 5, which encompassed over 1000 minerals. It was during these later stages that new selection pressures on mineral formation appeared, such as volcanism, plate tectonics and many other processes affecting pressure and fluid–rock interactions. In Stage 6, biological processes appeared and began to influence surface mineralogy, resulting in a more than twofold

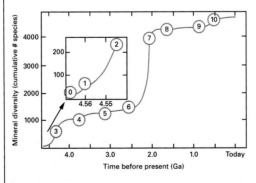

Figure B2.1.1 Multistage evolution of minerals.
Note: Biological processes appear in Stage 6.
Source: http://hazen.carnegiescience.edu/research/mineral-evolution. With permission of R.M. Hazen.
(*A black and white version of this figure will appear in some formats. For the colour version, please see the plate section.*)

Box 2.1 (*cont.*)

increase in the number of minerals on Earth. A major change occurred in Stage 7 through the 'great oxidation event', which gave rise to multicellular life and skeletal biomineralisation. Finally, Stage 10 was marked by the biological innovations of shells, teeth and bones, and the expansion of terrestrial life. Among others, this contributed to the extensive formation of clay minerals in soils brought about by vegetation roots breaking down rocks.

The emergence of new minerals triggered by biological evolution altered the conditions for further organic evolution, which in turn affected the evolution of minerals. One might therefore say that minerals have coevolved with life. In fact, as the figure illustrates, the majority of 'mineral species' would not exist without the presence of organic life (on this, see also Westbroek, 1991). This explains why Earth has the largest variety of minerals of the four so-called rocky or terrestrial planets in the solar system, which also include Mercury, Venus and Mars.

One may regard evolutionary thinking about minerals as stretching the meaning of evolution too much. At best, we are talking here about pseudo-evolution of Case 2 in Table 2.1, rather than genuine evolution as in Case 1. According to Hazen et al. (2008), taking an evolutionary approach can answer questions about the diversity, relative abundance, rarity and geographical features of minerals, as well as about the historical conditions and (near surface) environments under which minerals appeared. Insights obtained can aid in investigating whether other planets have hosted life processes.

---

addition, selection can be categorised in many ways. One is between stabilising, directional and (aforementioned) disruptive selection (Figure 2.1). Directional selection signifies that a more extreme phenotype is favoured over other phenotypes. This is common in artificial selection (plant and animal breeding). Stabilising or centripetal selection means that selection reduces variety by eliminating individuals that depart significantly from optimum phenotypes. The associated genetic distribution will then concentrate around the mean. This is probably the most common mode of natural selection. It explains why nature is stable and evolutionary change is predominantly gradual.

Under the influence of evolution, the environment changes itself, not only because its abiotic part is subject to geological and weathering processes, but also as evolution alters its biotic part. The reason is that evolution of one species generally triggers, and responds to, evolutionary changes in closely interacting species. This is captured by the notion of coevolution, denoting the mutual selection between living organisms that interact in distinctive ways, such as through predator–prey, flower–insect and herbivore–plant relationships. It is here that one can really begin to appreciate now how complex evolution really is. It involves a complicated system of interactive individuals, populations, species and mechanisms in which virtually all components can and will change (Section 4.6). Incidentally, biotic changes due to evolution can also change certain abiotic conditions. This is illustrated for minerals in Box 2.1.

Diversity is both an input to, and output of, evolution. Innovation outcomes often depend on diversity in that they are formed through the recombination of existing elements and pre-adapted modules. Selection can produce better outcomes if it works upon more variation. The dependence of both innovation and selection on diversity implies that a system of feedback mechanisms is at work in evolutionary systems. This adds to the complexity resulting from the many distinct units that make up a population. As a result, the intricacy of evolution is easily underestimated. But it can be seen to match the sophistication of evolutionary consequences – in terms of living organisms, ecosystems, culture and technology. In other words, evolution is a complex process able to generate complex outcomes. To capture and understand all this complexity and the associated evolutionary dynamics, scientists have developed a variety of mathematical models (Nowak, 2006).

A further complication is the role of history in the process of evolution and the constraints this poses on evolutionary paths and outcomes. This is referred to in different disciplines by a variety of terms: path dependence, hyper-selection, lock-in, historical or development constraints, and bauplan limits. It involves irreversibility, which has been argued to be stronger in biological than cultural and technological evolution (Mokyr, 1990, p. 285). Because of its detailed, disaggregate description of reality, an evolutionary approach can explain path dependence, namely by connecting it to the extent of and change in diversity within a population of agents or artefacts. A minimum amount of diversity helps to avoid or escape a situation of lock-in, whereas dominance of one characteristic or a sharply skewed distribution of characteristics in a population is often an indication of a path-dependent process heading towards a lock-in.

The population approach typical of evolutionary analysis can be seen as an alternative to the representative-agent approach which is common in the social sciences, notably economics. Although the representative agent is an integral part of standard microeconomics, adopting a population approach could be seen as a more genuine micro-approach. This is because its focus on populations and their internal diversity enables one to study interactions among agents, such as information diffusion and social learning – issues which remain outside the traditional representative-agent framework. In addition, a population focus provides a fertile basis for analysing distribution in all its subtle details, including aspects of economic inequality, poverty and social or political polarisation.

For now, this explanation of evolution is still relatively straightforward. A more complete set of mechanisms will be introduced to the reader in due course: for the biological context in Chapters 3 and 4, and for the socioeconomic-technological context in subsequent chapters.

2.2 Darwin's Legacy in the Social Sciences

Is it bold to say that evolutionary theory might have come to maturity in economics before it did in biology? In this case, non-genetic evolution could have risen to fame before genetic evolution, and biologists would now be accused of using economic metaphors rather than the other way around. Darwin built upon earlier evolutionary ideas in biology, but he was also influenced in several ways by aspects of early social sciences. One might consider the maturation of evolutionary theory in biology, therefore, as a historical coincidence. Indeed, premature forms of economics and sociology missed their chance to develop into evolutionary sciences.

The early economist Thomas Malthus (1766–1834) and early sociologist Herbert Spencer (1820–1903) had expressed ideas that contained elements of evolutionary thought before it became broadly accepted in biology. Both influenced the thinking of Charles Darwin (1809–1882). Spencer wrote about evolution in 1852. From him, Darwin adopted the phrase 'survival of the fittest' as a substitute for 'natural selection' in later versions (from 1869 onwards) of his magnum opus *On the Origin of Species by Means of Natural Selection, or the Preservation of Favoured Races in the Struggle for Life* (first published in 1859), better known in an abbreviated formulation as *On the Origin of Species*. Spencer favoured the term 'evolution', while Darwin preferred '(natural) selection'.

Malthus had a more profound influence on Darwin's thinking, namely through his famous work *An Essay on the Principle of Population* which appeared in 1798. In it, he expressed concern about the effect of human population growth on the subsistence of poor people. He argued that while such growth follows a geometrical pattern – what we would now call 'exponential growth' – food supplies can at best increase arithmetically (i.e. linearly). The resulting incongruence would lead to famine, disease and conflicts in his view. That is, unless people – especially the poor –resorted to self-restraint, in terms of postponing marriages and engaging in less sexual intercourse. Darwin acknowledged in later versions of his *Origin* that reading Malthus' work sparked his idea that resource scarcity acts as a selective pressure on the survival and reproduction of organisms and, indirectly, on the population distribution of their characteristics. Moreover, while Darwin initially wanted to concentrate on competition between populations that belong to different species, Malthus' *Essay* inspired him to shift his focus to the struggle between individuals belonging to a single species.

Darwin also makes direct references to economics. For example, in his *Origin* he says, 'a more general principle, namely, that natural selection is continually trying to economise in every part of the organisation. [...] for it will profit the individual not to have its nutriment wasted in building up a useless structure.' (Darwin, 1859, p. 114). In Chapter 4, he further argues that natural selection belongs to the utilitarian doctrine in that it causes organisms to accumulate adaptations that have utility for them, in the sense of improving their chances of survival and replication in a particular environment. With this economic interpretation in mind, Darwin's thought has been coined 'a work of economics, of nature-as-economy' (Jeff Wallace, Introduction to reprint,1998).

It is not surprising then that Darwin, accepting the close conceptual and analytical connection between studying the economy and nature, already hinted at the potential of evolutionary theory to render insights beyond biology. In particular, he proposed the application of evolutionary thought to understand the modification and appearance of language and the development of moral ideas. For instance, in a letter to Asa Gray at Harvard University (17 February 1862), he spoke of a 'community of descent of allied languages'. He also indicated having read Max Müller's (1861) *Lectures on the Science of Language*, which proposed an analogy between diversification of species and languages. In addition, Darwin recognised that morals could be studied from an evolutionary angle. In fact, he initiated this line of research in his book *The Descent of Man, and Selection in Relation to Sex* (1871), building upon the notion of 'social instincts'.

It cannot be excluded that the parallel founder of evolutionary biology, Alfred Russel Wallace (1823–1913), was influenced by economics as well. His ideas were informed by class struggle, which he experienced first-hand as a land surveyor and pondered upon while reading the socialist writings of Robert Owen. He showed a deep interest in political economy and was familiar with the works of Adam Smith, David Ricardo, John Stuart Mill and Stanley Jevons. Like Darwin, he read Malthus' *Essay* before arriving at his conjecture of natural selection. But Wallace was not simply informed about established economics of the day, he openly criticised free trade, supported minimum wage, and espoused his views on finance, land property and the nationalisation of essential industries. He was, moreover, critical of exporting scarce natural resources (such as coal) and even noted the risk of environmental degradation overseas due to trade practices (Coleman, 2001). Wallace's exposure to economic and political debate of the time is likely to have influenced his views on evolution. The fact that the two independent 'discoverers' of natural selection both lived in England during a time of significant social and economic change, while the discipline of economics was growing more prominent and knowledge about evolution-related topics was accumulating (Section 3.1, notably Table 3.1), is perhaps no coincidence.

The interaction between biology and economics goes even further back in time. Already in the first half of the eighteenth century, long before Darwin's time, the early economist Bernard Mandeville and philosopher David Hume used notions that were evolutionary in spirit: the first was to describe warship technology as the slow but steady accumulation of incremental changes over time, and the second, to explain how law and political institutions were formed (Hodgson, 1993a; Nelson, 1995). During Darwin's era, various thinkers in social science were charmed by evolutionary and population concepts. One might claim that evolution was in the air. Given that socioeconomic-technological evolution is often easier to observe in a short time span due to the relatively high speed with which it operates, one can witness it during one's lifetime. In light of this, it has been noted that Darwin and other early evolutionary thinkers were affected, like the early English economists, by rapid changes that characterised the Industrial Revolution in England during the nineteenth century. Darwin also knew various industrialists personally who were part of the intellectual circle of his family. For example, Josiah Wedgwood (1730–1795), the grandfather he shared with his wife Emma Wedgwood (they were cousins), is considered to be a founder of industrialising ceramic pottery manufacturing. Moreover, his father, Robert Darwin, invested in stocks of canal and road infrastructure that were essential for industrialisation and mass production. In different ways, therefore, the Industrial Revolution may have had an influence on the development of modern evolutionary thought. Let us also not forget that Darwin had the freedom to work on his evolutionary theory for decades, without interruptions or restrictions imposed on him by a grantee or a university. The reason is that he enjoyed financial independence, benefitting indirectly from the industrial success of his grandfather Wedgwood. Ruse (1999, pp. 247–248) makes a further connection between evolution and industrialisation by suggesting that no serious theories of evolutionary biology sprang from Russia due to its emphasis on bureaucracy and planning.

Adam Smith's *An Inquiry into the Nature and Causes of the Wealth of Nations* from 1776 is generally considered to mark the birth of economic thought. Economic historians still debate whether Smith was aware of industrialisation and of the major economic changes it involved. Most think not, pointing to the emphasis he placed on free markets and his neglect of the topic of technological change. He proposed a non-evolutionary theory which affected the direction that the discipline of economics took. Ofek (2001, p. 27) argues that despite differences, the natural selection and market economy frameworks developed by Darwin and Smith share fundamental features. They both recognise that something which seems designed is the outcome of a process of spontaneous self-organisation, driven by competition at the level of individuals and order at the level of populations.

One of the founders of neoclassical economic analysis, Alfred Marshall (1890, p. xiv), is famed for having stated: 'The Mecca of the economist lies in economic biology rather than in economic dynamics.' According to Hodgson (1993b), a discipline of evolutionary economics did not materialise at the time because negative sentiments associated with Social Darwinism along Spencerian lines stood in the way. Another reason is the difficulty of evolutionary, compared with mechanistic, analysis, as Marshall himself had already recognised – witness the sentence immediately following the previous citation: 'But biological conceptions are more complex than those of mechanics.'

In 1898, Thorstein Veblen proposed that economics needed to follow an evolutionary course. But his idea had to wait until the work of Nelson and Winter in the 1960s and 1970s before it gained traction. This blended with earlier ideas of Joseph Schumpeter from the 1930s and 1940s, providing an impetus for the so-called neo-Schumpeterian school of evolutionary economics. Shortly afterwards, in the 1980s, evolutionary game theory was imported from biology to influence an approach within analytical economics that holds the middle between evolutionary and equilibrium analysis. Non-evolutionary game theory was developed by John von Neumann. He and Oskar Morgenstern applied it to economics in the 1940s, providing the basis for its elaboration in an evolutionary context by George R. Price and John Maynard Smith in the 1970s (for more discussion, see Section 8.5). One can see that methodological spill-over between biology and economics has been bi-directional, suggesting that both would have developed differently in isolation.

The idea that evolutionary thinking has a bright future in economics is argued convincingly by Potts (2000). He regards economic systems as complex 'hyper-structures', i.e. nested sets of connections among components. Economic development he then sees as a process of evolutionary change in these connections, driven by selection and innovation. In this vein, he calls for a new type of microeconomics based on the technique of discrete, combinatorial mathematics, operationalised through graph theory and multi-agent, population models. The far-reaching implication is that the many brands of heterodox economics could be joined together into a broad evolutionary approach to economics (elaborated in Section 8.4).

2.3   Evolutionary Philosophy

Evolution as Knowledge Formation

For a long time, the philosophy of science had difficulties in responding to evolutionary thinking. Traditional epistemology was dominated by physics, the study of which differs from that of evolutionary sciences in terms of complexity, historical dynamics and feasibility of experimentation, to name just a few. Currently, there is some body of philosophical knowledge that deals with evolution, involving contributions from both philosophers (e.g. Dennett, 1995) and evolutionary scientists (e.g. Mayr, 1988). Here we examine what it has to say about knowledge formation in relation to (generalised) evolution. Understandably, much of this is motivated by experiences with evolutionary biology.

A fundamental idea in the epistemology of evolutionary thought is that adaptation through selection is a process of knowledge formation (Buskes, 2006). Natural selection can be seen as a search on an adaptive landscape. It scans the physical and chemical characteristics of an environment and responds to these by stimulating improved adaptations of organisms living in it. One can see evolution as a problem-solver that uses the strategy of trial and error by creating many parallel, diverse experiments. This insight is being applied successfully in the area of evolutionary computation to solve ill-defined optimisation or design problems (Eiben and Smith, 2003). As evolution solves a challenge posed by an environmental mismatch of an organism, knowledge accumulates in its revised design. Hence, birds and flying insects reflect implicit knowledge about aerodynamics, and river and sea animals about hydrodynamics. Predators reveal secrets about hunting strategies, and prey about hide and escape tactics. Likewise, social organisms echo details about the nature of public goods, and so on. The design details of organisms thus contain indirect information about the characteristics of the environment in which they live.

An interesting viewpoint on acquiring knowledge is offered by ethology. This is the scientific study of animal behaviour, which has strong roots in evolutionary biology. It took off in the 1930s, with observations and experiments by Konrad Lorenz, Niko (Iaas) Tinbergen and Karl von Frisch, who were jointly awarded a Nobel Prize for physiology/medicine. This work provided many insights about instincts, individual and social learning, and nature versus nurture. Ethology considers behaviour as an adaptation resulting from repeated natural selection. Tinbergen (1963) stipulated that ethology should deliver four complementary explanations of behaviour, namely in terms of:

- *adaptive function*: examining how behaviour affects survival and reproduction;
- *proximate causation*: associating behaviour with immediate neural or hormonal stimuli and learning;
- *development (ontogeny)*: relating behaviour to age and early experiences; and
- *evolutionary history or ultimate causation (phylogeny)*: comparing behaviour across related species or species living in similar environments, and assessing their historical, phylogenetic origin.

Motivated by these 'four why questions' of Tinbergen, Richard Dawkins (1976), who was his PhD student at Oxford, mentions four 'who questions', namely who profits from evolution at the levels of the replicator, the vehicle, the species and an eventual designer (who might derive use or non-use value from evolving life forms)? His answers are: the gene, the organism, the gene pool and no one, respectively.

An important finding of ethology is that animals, including humans, possess innate behavioural characteristics and knowledge in the form of instincts and learning capacity.

Chris Buskes (2006) establishes a link with epistemology — by noting that both empiricists — starting with Aristotle — and rationalists — starting with Descartes — have erred: the first by assuming that there is not innate knowledge, and the second by postulating overly specific innate ideas and concepts.

Both ethology and evolutionary psychology (Section 5.4) stress that many adapted behaviours are only valid in the specific environment where they emerged through natural selection and thus can become maladapted in others. Preferences for fatty, sweet and salty foodstuffs are an example. Other adaptations relate to a broad range of environments, such as the capacity to recognise general dangers. Notably, the perception of real causality that involves danger, inevitably through an exchange of energy, has evolutionary advantages. Examples include being bitten, hit by a fist or weapon, a fire burning a forest or lightning followed by thunder.

Humans undertaking scientific research are often confronted with situations that are detached from their natural environment and normal observation capacity. This is the case whether they study the infinitesimal microcosm (particle physics, chemistry and cell biology), the extra-terrestrial macrocosm (astronomy), or long-run history (geology and evolutionary biology). Here, our behavioural habits and tendencies may fall short. W.V. Quine (1969) noted that early science was limited as it was based on the meagre sensory input of humans. He coined the term 'epistemology naturalised' to stress that only by being critical on our intuition and initial thoughts, through repeated testing and cooperation with other humans, including standing on the shoulders of earlier researchers, can we overcome our behavioural limits and contribute to scientific progress. In addition, advanced measurement technologies are essential, such as equipment for microscopic and telescopic observation, or for detection of chemical substances and radiation.

Science as an Evolutionary Process

A next step in evolutionary philosophy is to consider science itself as an evolutionary process. A staunch defender of this view is the philosopher David Hull (1988). He regards evolution as a powerful knowledge theory – or epistemology – that provides a credible alternative to the dominant view on the process of science formation, namely 'neo-positivism'. This is a blend of logical empiricism and logical positivism. It suggests that the dynamics of science can be — by and large — understood as the outcome of testing theories against empirical facts. Originally such testing was focused on 'verificationism', but in following the ideas of Karl Popper, scientific focus shifted to falsification. Neo-positivism is still dominant, despite the efforts of philosophers such as Imre Lakatos, Paul Feyerabend, Thomas Kuhn and, to a greater degree, Jorgen Habermas and Michel Foucault, to argue that science, notably social science, is predominantly driven by social constructions and subjective political motives. An evolutionary explanation of science as discussed below integrates such elements.

According to Richard Nelson (2007), an evolutionary perspective on how science makes progress originates from Karl Popper's work (1959). As fitness in his theory of science serves the fact that theories survive, that is, they are employed and 'not falsified'. Donald Campbell (1960, 1969, 1974) elaborated on the idea that science follows an evolutionary pattern, and called this 'evolutionary epistemology'. He characterised it as consisting of endless interactions of local innovations (mutations) and imperfect selection through a range of cultural selection factors: powerful individuals, theoretical schools, influential refereed journals, university politics, publishers and academic funding agencies. David Hull (1988) has further expanded the idea of science as a Darwinian struggle for 'academic survival' and intellectual reproduction through citations and influential students. Scientists compete with each other for scarce journal space, funding, jobs and assistant researchers. The ideas of scientists undergo change, through coincidental mutations and recombination by gathering ideas from the literature. In fact, great innovations often result from the recombination of ideas that reside at the boundaries between different disciplines – witness many Nobel Prizes that have been awarded to crossover research. Individual scientists form an integral part of a larger scientific society. They group in schools and associations, leading to assortment, driven by conformism, collaboration and limited information – by not reading the ideas from all schools. This suggests that not only individual but also group selection (Chapter 6) is a mechanism relevant to understand the dynamics of science.

Evolution occurs through competition as well as through collaboration, including a limited degree of altruism. The latter commonly takes the form of reciprocity, since favours are (expected to be) returned, as in the case of serving as a reviewer for journals or funding organisations, or providing comments on papers by colleagues. Some of this reciprocity is indirect, associated with one's reputation in academia. It takes a long time to build up such a reputation, which is aided by spreading one's ideas and influencing the thinking of others. In other words, individual researchers are motivated to increase, if not maximise, their 'intellectual fitness'.

More elements can be added to complete an evolutionary theory of science. Notably, it should account for path dependence and lock-in, which hinder change and progress. This was captured by Kuhn's notion of 'scientific paradigm'. The role of social context and politics in science, as stressed – though conceivably oversold – by social constructivists, also fits this picture. On the whole, scientific knowledge production can be best categorised as evolutionary trial and error, as it appears to be a process far removed from one of careful and optimal planning.

On Evidence, Falsifiability, and Hypotheses Versus Theories

According to Ernst Mayr (1988), philosophers of science have found it difficult to give evolutionary biology a place in epistemology. This is because the uncertainty and complexity involving evolutionary theorising confound simple hypotheses and perfectly repeatable phenomena to which researchers in physics and chemistry are accustomed. Instead, historical phenomena are to be dealt with, giving rise to conditional statements that are not easily falsified as the conditions cannot be replicated. This applies equally to the realities studied by biological and social sciences.

In evolutionary biology, as in other scientific disciplines, a clear distinction must be made between the notions of 'theory' and 'hypothesis'. Theory does not signify a simple guess or hunch, but rather an accumulation of insights supported by logical reasoning, including mathematical analysis, and empirical and experimental findings. One should realise then that, to date, no single observation or experiment has with any certainty been assessed as incompatible with an evolutionary explanation of life. After more than 150 years of countless experiments and empirical studies (including nine categories of methods; Section 3.5), evolutionary biology comes out as unrefuted.

This then leads to the inevitable question: is evolutionary theory in biology actually falsifiable? The answer is yes — in many ways. Some concrete examples are as follows:

- if genetic and phenotypic (physiologic and behavioural) diversity were lacking in any population or species;
- if there were an absence of, or an unstable, relationship between genes and phenotypes;
- if a species permanently were badly adapted to their environment without becoming extinct;
- if mechanisms through which new variation was created were missing;
- if the physical basis (DNA) to store and accumulate information did not exist;
- if the Earth should prove to be too young to permit an evolutionary unfolding of life;
- if unexplainable huge jumps or missing links in natural history were discovered, such as humans having appeared without any fossil primate ancestor having been identified.

There is obviously no conclusive evidence of a process that occurs on the scale of the biosphere and has a history of billions of years. Even if all the people on Earth were engaged in biological research, we would still be unable to undertake all possible experimental tests and empirical investigations. Conclusive evidence is restricted to logic and mathematics. One should also accept that science is a social process, characterised by peer review and mutual criticism. In view of this, the fact that virtually all academic biologists accept an evolutionary history of life on Earth lends significant weight to the argument.

Evolutionary biology is further appealing as it is fully compatible with the rich set of insights from the fields of physics and chemistry. It has found that living structures are built on the same scientific laws as non-living structures with, as an additional principle, the organisation laws of systems biology. In other words, the transition from non-life to life, or from chemistry to biology, is characterised by a high degree of continuity, driven by the logic of physical and chemical processes, complemented by evolutionary mechanisms. This extends to the origin of life as a process of chemical evolution (Section 4.5).

Analogies and Metaphors

The transferral of ideas from evolutionary biology to the social sciences may involve the use of analogies and metaphors. While some judge this as negative, according to Hodgson (1993a), the use of analogies and metaphors helps to approach reality from various angles, recognising that concepts have distinct and subjective interpretations as well as a social and academic history. Metaphorical thinking can be inspiring and

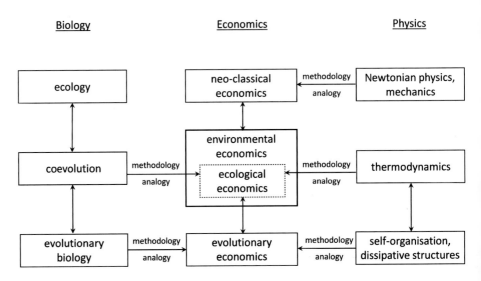

Figure 2.2 Economics between biology and physics.
Source: Mulder and van den Bergh (2001, Figure 1, p. 118).

stimulate creative processes in the human mind. Many models in economics can be traced back to modelling traditions in physics and engineering. The use of biological metaphors in the social sciences, notably economics, is not new either (Khalil, 1998). Of course, one has to be wary of the potential for metaphors to mislead if they are incomplete or inaccurate.

The mutual influence of biology, physics and economics has a long history. This is illustrated in Figure 2.2. They do not just share analogue concepts, they also exchange methods. Notice that environmental and ecological economics are assigned a central position because they take account not only of economic and technological evolution but also of biological evolution, namely in relation to ecological-environmental problems that are caused by, or have repercussions for, economic activities (Chapters 13 and 14).

2.4   Design Versus Evolution

The main alternative view to an evolutionary explanation of life on Earth is creationism. It claims that a god – some conscious, intelligent and even benevolent designer – created or designed the universe and Earth, as well as all existing and extinct life forms. Box 1.1 in Chapter 1 discussed the basic design or watchmaker argument associated with 'Natural Theology'. Specific types of creationism are encountered in all major religions, including Christianity, Islam, Hinduism and Judaism. The most fervent resistance to evolution, though, has come from Christians, notably Protestants. This has, particularly in the United States, gone so far as to witness Christian parents filing lawsuits against public schools teaching evolutionary biology. In line with this, many states have, or had at some point in the past, legislation that hampered the diffusion of evolutionary knowledge and required public education curricula to include a creationist perspective.

This design perspective is unscientific, though, as it relies on statements that build on undefined or unmeasurable concepts, such as god, heaven, souls and angels. Some creationists have suggested that Satan invented evolution to foster the moral degradation of humans (Schadewald, 1983, p. 295). More generally, creationism rejects explanations of reality on the basis of only naturalism and instead assumes some form of supernaturalism, unbounded by the laws of physics, chemistry and biology. Whereas evolutionary thinking means adopting an explicit temporal perspective involving cumulative causation and unfolding, built around proximate and ultimate causes and effects, creationist design entails an atemporal account devoid of cause–effect logic. In the absence of a satisfactory explanation for the complexity of life on Earth, it is understandable that humans arrived at a creationist-like viewpoint. However, once a convincing evolutionary theory appeared, this excuse evaporated. About this, Richard Dawkins (1986, pp. 5–6) once said: 'I could not imagine being an atheist at any time before 1859, when Darwin's *Origin of Species* was published [...] Darwin made it possible to be an intellectually fulfilled atheist.' In fact, this statement by Darwin himself goes a long way in the direction of atheism: 'why is it more irreligious to explain the origin of man as a distinct species by descent from some lower form, through the laws of variation and natural selection, than to explain the birth of the individual through the laws of ordinary reproduction.' (1871, *The Descent of Man, and Selection in Relation to Sex*, Chapter XXI).

Creationism as anti-evolutionism has a long history, documented well in Numbers (2006). One distinction is between 'Young Earth creationism', which posits that a god created the Earth within the last 10 000 years, and 'Old Earth creationism', according to which the universe was created by a god while its age and that of the Earth may be in line with insights from astronomy and geology. Both question evolutionary explanations of organic life and humans. Other branches of creationism (for a full account see Kitcher, 2007) suggest that a god created all species, which subsequently evolved under his guidance. Some creationists accept a restricted version of evolution, such as selection that brings about minor changes, motivated by strong evidence. But they reject the feasibility of speciation, arguably as it is difficult to imagine and has an extremely slow pace, and hence is invisible. The only form of creationism that combines belief in a god with evolution without any conflict is 'evolutionary creationism', also known as 'theistic evolution'. According to it, god created living organisms indirectly by creating the process of evolution. This position can be seen as part of 'deism', the belief that a god made the universe but since then has not intervened in its functioning, so that it entirely runs on the basis of natural laws created by it.

Most types of creationism are anti-science because they attack a straw man that represents a caricature of evolutionary theory. This takes the form of incomplete presentation or misinterpretation of facts, or using quotes from evolutionary biologists out of context, to suggest incorrectly that they have doubts about an evolutionary history of life. Publications in this vein claiming essential weaknesses in evolutionary biology, such as Behe (2006) and Fodor (2007), reflect a lack of basic understanding of modern biology. An example is Fodor's misinterpretation and overgeneralisation of the spandrel or exaptation argument (Section 4.4). Not surprisingly, these publications have

Figure 2.3 Place of homo creationist in the march of evolutionary progress.
Source: Bill Day. With permission of Cagle Cartoons Inc.

received only criticism from respected scientists and philosophers and, accordingly, must be regarded as pseudo-science.

Focusing on gaps or missing links in the fossil record has been a popular tactic of creationists to counter evolutionary theory. This is turned on its head in the 'march of progress' joke in Figure 2.3. The reality is that no 'big holes' exist in evolutionary biology. Of course, there are unresolved issues in evolutionary research, but this is inherent to the large study area. Creationists have proposed that fossils were implanted in Earth by a god to test the religious faith of humans. He did a good job then, in this case, as the deepest strata contain the more primitive organisms, consistent with an evolutionary history. Another creationist argument is that fossils are remnants of the classic flood in Noah's lifetime, as described in the Old Testament. But this 'flood or deluge theory' cannot explain systematic patterns in the fossil record. Some creationists even go as far as claiming that multiple catastrophes in history were followed by divine creations. To debunk these myths, the eminent palaeontologist David Raup (1983) provides a detailed account and criticism of the geological and palaeontological arguments of creationism.

Irreducible Complexity

Perhaps the most attention-grabbing strategy by creationists to cast doubt on a complete evolutionary explanation of life has been to identify cases characterised by so-called irreducible or 'unevolvable' complexity. This supposedly implies that some components of organisms are too complex to have evolved from simpler ones. Common examples cited by creationists are the eye, a bird's wing, the flagella movement of certain bacteria, the process of blood clotting, animal hearts and the immune system. Nevertheless, all of these examples can be explained comprehensively through the concept of gradual evolution. It must be noted, however, that 'gradual' should not be interpreted too literally: it includes mutations that involve exchanges of larger components between genes – an example of 'modular evolution' – that can alter organisms or their functioning drastically (see Section 4.3).

One argument that creationists employ in this context is that only fully complete and complex systems can function properly, while intermediate stages of an organism's supposed evolution do not. If this were true, then such transitional stages could not have appeared as a result of evolution through adaptive selection. This, however, is refuted by considerable research indicating that the intermediate stages of the above-mentioned examples *do* possess adaptive functionality. This is illustrated in Box 2.2 for the eye, an oft-mentioned example by creationists of so-called unevolvable complexity. Regarding wings, another supposed case of 'unevolvability', Pinker (1997, p. 170) proposes potential intermediate stages, namely a proto-wing that allows animals to glide between trees or escape danger more rapidly, or that initially served a function distinct from flying, such as cooling. So neither the eye nor the wing is intractable to evolutionary processes. For more discussion and examples, see the discussions on adaptation and exaptation in Section 4.4.

Anti-explanation and Stupid Design

It is easy to criticise creationism, as it involves so many untestable assumptions (Kitcher, 1982). The hard-core creationist should realise that using the concept of god or multiple gods in an effort to explain real-world complexity begs the question: 'who, or what, created god(s)?' or even 'did gods evolve?' These questions do not, of course, have a rational answer, since the concept of god suggests a degree of complexity incomparable to anything we know. This is unsatisfactory as an explanation of a much more modest complexity, namely organic life. One can therefore regard invoking a god not just as a failure to explain but as an anti-explanation, since it accomplishes the opposite of what we generally expect from an explanation.

An insightful contrast between rational, scientific explanations and non-rational, creationist explanations of life comes from Dennett (1995): namely as 'cranes' versus 'skyhooks'. Both approaches explain why certain ideas 'hang high up in the sky', a metaphor for complexity. However, a skyhook is an imaginary means of suspension in the sky. That is, it appears out of nowhere, raising the question about origins. Cranes, on the other hand, are built beginning with small ones, which can be used to gradually construct bigger ones, and so on. This process explains anything that is hanging up in the sky, no matter how high. Creationism employs a skyhook by proclaiming that a god created life on Earth. Evolution, instead, can be regarded as a mechanism to construct higher cranes, or more organic complexity – without the need to invoke supranational powers.

Box 2.2 Does the eye reflect unevolvable complexity?

The eye is often mentioned by creationists as a complex organ that could not have evolved gradually. We now know, however, that it was subject to recurrent evolution in that it evolved independently some 50 times, at least. This is not surprising, as the selective advantage of vision is enormous. Evolution of the eye always began with the appearance of light-sensitive cells that enjoyed a selection advantage in the respective organisms. Darwin devoted attention to this theme – see the summary of his Chapter 6 in Table 3.1 later in this volume.

Most animals possess two eyes to ascertain depth and estimate distance — useful for finding food and avoiding danger. Spiders have evolved multiple eye pairs, with four sets of pairs being the most common. Jumping spiders have three fixed eyes in addition to a principal, movable eye. The fixed ones are of lower quality and merely serve to detect movement, while all eyes together allow for an almost 360-degree view. This goes some way to explaining why such spiders show agile, almost cat-like predator behaviour.

Some of the most impressive optical organs are found in the mantis shrimp, which is not technically a shrimp but a stomatopod type of marine crustacean. Its eyes are compound and can move and focus independently from one another. Moreover, they are capable of observing ultraviolet and polarised light, as well as an unusual range of colours in the visual field. Speculations on the purpose of such singular eyes are twofold: the need to coordinate vision with equally special claws or spears that swing swiftly at prey or an enemy; and sexual selection, given that males perform courtship dances that involve variation in the colour and brightness of their body, possibly using polarised light as a secret communication system.

Nilsson and Pelger (1994) developed a simulation model of the evolution of a three-layer slab of virtual skin with pigmented cells at the bottom, light-sensitive cells in the middle and translucent cells on top. Through different stages, the curvature of the simulated tissue gradually becomes stronger. This involves changes in the refractive index of the transparent tissue and the formation of a separate lens in the transparent tissue. The latter shrinks, letting a flat iris form and the focal length of the lens shorten until it equals the distance to the retina. The process is completed with the formation of a full-blown, complex eye capable of sharp focus. This all happens in fewer than 400 000 generations, given relatively pessimistic assumptions and conservative parameter estimates regarding heritability, mutations and selection forces. Figure B2.2.1 summarises the simulation. It is evident that eye evolution is not something that happened overnight, but involved numerous steps.

So what to make of the popular creationist argument that an intermediate stage of the eye could not have evolved through natural selection, because it would not have functioned? Each stage in the figure contributes to utility of the individual that possesses the respective tissue through signalling for: danger is present – flee; it is getting colder – find a warmer place; a potential mate looks attractive – try to copulate with it; or the day ends – time to rest or look for food, depending on

Box 2.2 (*cont.*)

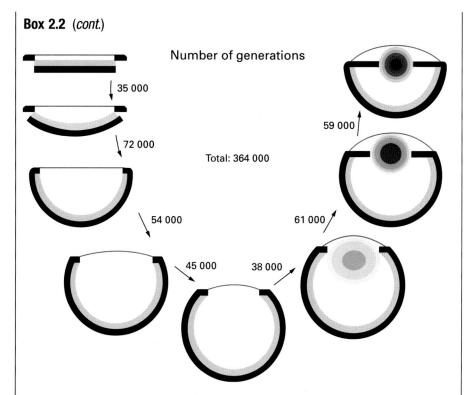

Figure B2.2.1 Computer simulations of eye evolution.
Source: Land and Nilsson (2002, Figure 1.6, p. 9). By permission of Oxford University Press.

whether you belong to a nocturnal or diurnal species. All of these signals and associated actions translate into an improved chance of survival and reproduction. From each generation to the next there is more representation of the genes coding for the intermediate forms of the eye. In other words, the intermediate stages are consistent with a process of adaptive evolution. All taken together, the eye is not illustrative of 'irreducible complexity'.

From a theoretical angle, an evolutionary explanation of complexity comes down to exploiting a dynamic probability model, such as the aforementioned one for simulating eye evolution. Now most people find static probability theory already hard to handle. Indeed, our intuition often errs in responding to abstract 'probability puzzles' (less so, though, when challenged by probabilistic real-world situations). Imagine how easily it is to misunderstand the outcomes, or underestimate the power, of repeatedly interacting evolutionary chance mechanisms.

In the 1980s, a number of creationists decided that it would be a better strategy to replace the term 'creationism' with that of 'intelligent design'. It suggests a non-religious, scientific approach that offers an alternative to an evolutionary explanation of life. But it is nothing of the kind – it is just creationism in disguise, a pseudo-scientific trick to carry on with creationist explanations of life. In the Netherlands, the notion of intelligent design was subject to considerable public debate and media attention in 2003, after a Cabinet Minister of Education, Culture and Science, a Christian, suggested that it provided a credible alternative that deserved serious attention in education. This resulted in a wave of criticism. The biologist Edi Gittenberger commented on the re-emergence of intelligent design concepts in an interview with the Dutch newspaper *Trouw*[1], saying that it is a typically human tendency to fill lacunae in one's world view with a supra-intelligent power. He referred to the god of 'holes' or 'putty', expressing his frustration with the process whereby biology fills gaps that could previously support the creationist view, and instead of admitting that evolutionary biology solved a problem that seemed unsolvable, the creationists move on to another unsolved hole. He noted that as time passed many lacunae have been filled – such as the step from dead matter to the first life forms, the evolution of complicated molecules, or the transition from ape to human. He raised the pertinent question: is intelligent design only needed for the difficult issues that require more research and time to be resolved by biological science? Similar statements have been made by others. Michael Shermer, director of the Sceptics Society and Magazine, coined creationism as the 'god of the gaps'.

Evolution is the only design mechanism for which there is credible evidence. Evolution, moreover, has generated considerably more complex systems than humans, the only (other) designer of complex systems whose existence is a proven fact. Consider a human brain: it consists of a number of active elements (i.e. molecules) with a magnitude in the order of 10^{22}, while a modern computer developed by humans consists of less than 10^{12} active elements, namely transistors. Evolution is a successful designer because it makes use of an immense number of parallel experiments and then selects the best-adapted outcomes. On the basis of these, it generates a huge number of additional parallel experiments, and so on, indefinitely – accumulating useful changes along the way. Perhaps more so than human design, evolution is limited by historical constraints, including already evolved components that can be recombined. However, evolution is less hindered by any preconceptions or theories about what might work or what might be the ideal solution. Instead, evolution uses the force of large numbers and will try anything that is possible. This goes along with a huge number of failures, in terms of wasted organisms and even entire species being selected out. Similarly, the history of human inventions is characterised by many failures – indeed, human technology is subject to a process of generalised evolution (Chapter 10).

Many evolved organisms or technologies are sufficiently effective perhaps but still imperfect or inefficient. This is sometimes referred to as 'stupid design', to distinguish it from supposedly perfect design by a hypothetical god-like intelligence. One might say that they reflect survival of the 'sufficiently fit' rather than the 'fittest'. This is due to evolution not anticipating changes needed for later functions and circumstances. Pinker (1997, p. 166) gives the following example. The seminal ducts in men do not go directly from testicles to penis, but take a detour, entering the body and passing over the ureter

---

[1] See http://www.trouw.nl/tr/nl/4324/Nieuws/archief/article/detail/1767010/2003/05/10/Op-zoek-naar-de-nieuwe-Darwin.dhtml

before passing down to the penis. This reflects evolutionary inefficiency. While our reptilian, cold-blooded ancestors had testicles inside their bodies, mammals did not. To avoid the higher body temperature, which hampers sperm production, the testicles descended into a scrotum. So, in this case, owing to a particular history – or a lack of foresight – natural selection was unable to arrive at the shortest-route solution for mammals. One might conclude that human inventors are superior in this respect, but many designs are restricted by their particular technological histories as well. For example, take the early Windows PC operating system that possessed limited functionality because it was built on top of a previously dominant DOS operating system.

Intelligent design explanations overlook the fact that invoking a god or some other super-intelligent designer is a bad explanation for the waste and imperfection found in natural history – such as extinct species and half-adapted, or maladapted, organisms. Unfinished business provides further evidence for evolution, as it is constantly underway in forming new states and performing what in optimisation theory is known as 'local optimisation'. In fact, it would undercut evolutionary theory to find no imperfections in nature. The only way to reconcile a supernatural god with these imperfections is to assume that it deliberately tried to deceive humans that all life was generated through a process of organic evolution.

A Harmonising Solution?

Some scientists have defended the idea that science and religion are 'non-overlapping magisteria' (cf. Gould, 2002). This makes one wonder why creationists then have fought so vigorously against evolutionary biology. Adopting a historical perspective, Harrison (2015) contends that they have inhabited both partly overlapping and partly non-overlapping territories. The non-overlap and lack of conflict is consistent with the view that a god does not necessarily deny an evolutionary history of life on planet Earth because it may have merely created the conditions for life to evolve. This represents the earlier mentioned deist viewpoint, considered a golden mean between theism and atheism by some (Ruse, 2000).

Yet, religion makes statements, as in the Christian Old and New Testaments, about natural and supernatural events that supposedly took place in history. Some of these enter the realm of biology, notably the creation of living organisms, including humans (Adam and Eve), or Noah's flood, reducing diversity in all species to a minimum (one male and female), or the ten odd plagues God inflicted upon Egypt in response to Israeli slavery. In addition, evolution offers a clear explanation for morality, which may conflict with religious views on it. In fact, many religious people see human morals and ethical norms as the unique domain of religion, whereas evolution depicts them as being widespread because they have had survival value for human individuals and groups in the past (Chapters 5 and 6).

Some would add the dating of the age of the Earth according to the Bible to this list, as it arguably implies insufficient time for evolution to explain natural history. Life on Earth began over 3.5, and possibly even more than 4, billion years ago. This is consistent with current knowledge from physics, astronomy and geology, which suggests that the Earth is a little over 4.5 billion years old. However, the Bible does not mention any concrete age of the Earth. It is true that in the seventeenth century Archbishop James Ussher in Ireland interpreted the time line of the Bible to arrive at 4004 BC as the year of the creation of the Earth. This had a significant influence on how others subsequently tried to derive the Earth's age from the Bible. Young-Earth creationists often mention roughly 6000 years, based on the belief that there are 2000 years between Adam and Abraham, and that the latter lived in 2000 BC. To be precise, five more days should be added, as Adam was created on the sixth day. A serious complication is that 'days' as mentioned in Genesis should not necessarily be interpreted in a literal sense, but may well signify long periods of times.[2] In science in general, the Bible is accepted as open to multiple interpretations and not an accurate account of historical events.

A major blow to the idea of 'non-overlapping magisteria' comes from insights about human evolution. The special place assigned to humans and their 'souls' in major world religions conflicts directly with the evolutionary insight that humans are merely intelligent animals that descended from ape-like predecessors. Further evidence against non-overlap is the low percentage of religious scientists, which is in fact far below the fraction of religious people in the general population. The large majority of evolutionary biologists are self-proclaimed atheists who see religion as a social phenomenon, while religion is virtually absent among Nobel Prize laureates in the sciences (Swaab, 2014, Section 16.1). In sum, many facts undercut the view that science and religion have non-overlapping scopes.

In defending the non-overlap view, some claim that the Catholic Church has accepted evolutionary biology.[3] This is, however, a widespread misunderstanding. In 1996, the Vatican declared that while the human body could be attributed to evolution, the human soul was placed in each human being by a god. This is, of course, contrary to what we know about human evolution. For one thing, it raises the question: at which stage during human evolution, from apes to modern humans, did God decide to insert a soul? (Buskes, 2006, p. 292). This quasi-acceptance of evolution by the Catholic Church is understandable. How can one expect religious people, whose belief system rests on the idea that they are special beings on this planet who can count on an afterlife, to fully accept that human 'souls' (or technically 'minds') are the result of evolution from an ape-like ancestor species? Evolution suggests that the human species is one among

---

[2] For an informative comparison of what the Bible and science suggest about the age of the Earth and life, see: http://godandscience.org/youngearth/age_of_the_earth.html

[3] In this context one can find frequent reference to the pseudo-scientific book *The Phenomenon of Man* (1955) by Teilhard de Chardin, a French palaeontologist and Jesuit. Using a speculative philosophical approach, he suggested evolution to be a teleological process creating "organic unity in reality" and leading to the "noosphere" with the "Omega Point" as a higher consciousness. This allowed the author to reconcile his evolutionary and religious beliefs. Nobel laureate Peter Medawar said of the book: "Yet the greater part of it, I shall show, is nonsense, tricked out with a variety of metaphysical conceits, and its author can be excused of dishonesty only on the grounds that before deceiving others he has taken great pains to deceive himself" http://bactra.org/Medawar/phenomenon-of-man.html).

many that coincidentally evolved on a coincidental planet.[4] Life is now explained in terms of replicating molecules. That is, without resorting to a 'vital force' that is independent of physical and chemical processes. Along these lines, science accepts that the mind and consciousness are fully explained by neural or chemical interactions among brain cells. The idea of a metaphysical soul can thus be discarded. We now have a consistent picture of the historical, interrelated changes of brain and behavioural patterns of animals and (pre-)humans. 'Vital forces', 'soul', and 'separation of soul and body' are outdated notions, painfully inconsistent with evidence and insights about brain evolution. Fields like evolutionary psychology and cognitive science stress the evolution of the mind, which is the brain at work, and its co-adaptation with the rest of human body (Section 11.3).

The notion of an eternal soul runs into logical problems anyway, regardless of whether one accepts the insights of evolution and cognitive science. For example, does an unconscious foetus dying in the womb, or even a baby dying soon after birth, have a soul that exists forever in some hereafter? Or, what about a person in an advanced stage of Alzheimer's disease or someone whose brain is severely damaged in a traffic accident? Which soul are we talking about then: before or after the disease or accident? What about the souls of people with Down syndrome, autism or more serious mental impairments: will their souls eternally reflect their mental condition? If all of these factors do not matter to the eternal soul, then we have to assume that the time of death, and thus life experiences, parental care or education, have no impact on the soul. But this contradicts the idea that all these factors influence our personalities. Hence, one has to assume the soul does not reflect our personality but is devoid of any individuality. One should also realise that many adult animals, not just primates, have more consciousness than human babies. So why would the former lack a soul? All these considerations attest to the fact that the notion of a soul is unfounded and illogical.

Dennett (2006) draws attention to the adult and childhood psychology of creationism. He argues that humans have three possible ways of confronting complex reality: physical stance (science), design stance (a god-like explanation) and intentional stance (everything has purpose). The first, of which evolutionary thinking is part, is the hardest to grasp. It requires work and time, deep thinking and lots of reading (of science). The other two are easy short-cuts that save time and effort. The intentional stance is typical of young children, who tend to see purpose in everything, even when it is not the case (Bloom and Weisberg, 2007). A possible explanation for this 'intention bias' is that the human brain evolved so as to adapt to larger social groups. This involved many social interactions with other humans who acted with intention. In such a setting, there is an evolutionary advantage in being able to properly identify another person's intentions – either good or bad.

---

[4] As argued in more detail in Section 11.2, the evolution of modern humans (*Homo sapiens sapiens*) from ape-like predecessors is a fact. The evidence for it is multidimensional, including: temporal logic as visualised in the partial evolutionary tree of apes (Figure 11.1); spatial regularity reflected by the overlap among living spaces in Africa between species; human genes overlapping with those of extant species of primates (98.8% with chimpanzees and about 93% with monkeys); and humans and apes sharing many physiological and behavioural traits (Table 11.1).

Responding to Science Denial

The psychologist John Cook (2017) notes that while common wisdom says that communicating more science should be the solution to science denial, a growing body of evidence indicates that such an approach can actually entrench people's existing beliefs, because the information threatens their worldview. Cook argues that a more effective response to this 'worldview backfire effect' is to focus teaching on 'debunking misconceptions about the science' instead of just teaching the science. Such an approach would involve understanding the characteristics of science denial, such as giving equal or more weight to non-experts than to experts, questioning the honesty of scientists (ad hominem attack) and depending on a variety of logical fallacies: emphasising minority opinions and suppressing evidence (cherry picking), misrepresentation (straw-man attack), hasty generalisation and jumping to conclusions, diverting attention away from main issues (red herring), creating a false dilemma or shifting the burden of proof, etc.

Creationism receives significant institutional support, not only from various religions, but also in the form of university-based theology. Not everyone is enthusiastic about this. Daniel Dennett once referred to a statement by the philosopher Ronald de Sousa that philosophical theology is 'intellectual tennis without a net', to convey that defenders of religion will use logical reasoning as long as it supports their beliefs while discarding it as soon as it goes against these. It is clear then that theology cannot be considered a part of science. Add to this that most of its researchers are biased in favour of religious views, associated with the religion of their particular theological faculty. This clearly is a far cry from independent, unprejudiced and liberal research. The study of religion with its rich palette of psychological, economic, sociological, political and even evolutionary (see Section 7.7) dimensions is too important to be left to theology. It deserves a more serious treatment. Richard Dawkins (1993) has forcefully argued that theology is self-referencing and empty, lacking any scientific approach and insight. In spite of this, theology still exerts considerable impact on society, namely by contributing to the replication of dogmatic falsehoods and fostering the denial of scientific insights that undercut these. Cynically, theology does this under the umbrella of science. Indeed, it is tolerated, if not respected, in many academic institutions around the world. But to any unbiased observer, theology can only be viewed as an anachronism. It is as similar to genuinely academic religion studies as astrology is to astronomy.

Evolutionary biologist Richard Lewontin noted in his preface to Godfrey (1983, p. xxvi) that evolution and creationism are incompatible as world views because: 'We cannot live simultaneously in a world of natural causation and of miracles, for if one miracle can occur, there is no limit.'

Part II

Evolutionary Biology

3 Pre-Darwinism, Darwinism and Neo-Darwinism

3.1 Evolutionary Thinking Until Darwin

Evolutionary biology is, in many ways, the crown on the scientific revolution that started around the middle of the second millennium with the development of astronomy by Nicolaus Copernicus and Galileo Galilei. This proceeded with mechanical physics developed by Isaac Newton and, just before Charles Darwin appeared on the scene, the emergence of the new discipline of geology. The central thread of this revolution was that non-living nature is governed by laws, regular processes which can be understood and accurately described in great detail. In line with this, evolutionary explanations of life involved law-like processes of changes in, and even the appearance of new, species. Altogether, the natural sciences represent a materialistic paradigm shift away from the Aristotelian teleological worldview that everything had a purpose. They suggest physical and biological processes do not have any goal, purpose or meaning, but function without an almighty supervisor (Buskes, 2006).

Understanding of the natural world pre-Darwin led to species' classification schemes – nowadays known as 'systematics' – based on morphological and physiological similarities between species. Some of these were later found to make much sense in the context of an evolutionary understanding of natural history. Classification had already begun before Plato and Aristotle, who, like others before them, used the notion of species as an essential discrete unit or dividing line in the natural world. Aristotle developed the notion of a *Scale of Nature* to depict a hierarchical and even progressive view of nature, from inanimate matter through lower life forms to mammals and finally humans. This idea maintained a strong position in human thinking about nature, reflected by such notions as Ladder of Nature and Great Chain of Being (Strickberger, 1996, p. 7). This hierarchical view fitted in times in which human society was strongly hierarchically organised in religion, society and politics.

In contrast with evolutionary classifications and views of the world, non-evolutionary ones are in effect static. Carl Linnaeus (1707–1778), the founder of modern taxonomy, was mainly guided in his choices by morphological characteristics of species. His hierarchy consisted of species grouped downward into classes, orders and genera. The definition of a species has always been debated. Before genetics, a precise definition was impossible. Species were instead regarded as fixed entities. Linnaeus defined the term on the basis of common descent, which has elements of evolution in it. Comte de Buffon (1707–1788) defined a species as a group of interbreeding organisms in the modern sense, recognising reproductive isolation in the form of barriers to crossbreeding among groups. At the same time, this contributed to a false impression that species were fixed and static.

Although it is sometimes suggested that several ancient writers had already raised the possibility of an evolutionary pattern in nature, this mostly concerned general references to qualitative change, without identifying any particular evolutionary mechanism. Aristotle supposedly referred to 'adaptation to an end' but in terms of animal development, not natural selection. One might say that evolution, in a specific, algorithmic sense – as clarified in Section 2.1 – is such a difficult idea that it was unlikely to be coincidentally stumbled upon. Of course, once discovered, one may find such a process of evolution fairly simple. However, it is not trivial, as is illustrated by the fact that one can find many different interpretations and definitions of it. Evolution really should be regarded as being among the more complex ideas that humans have generated.

In the seventeenth and at the beginning of the eighteenth centuries, early scientists such as Gottfried Leibniz (1646–1716) from Germany, Charles Bonnet (1720–1793) from Geneva and James Burnett (1714–1799) from Scotland suggested the idea of evolution, by pointing to fundamental connections between different generations of living organisms. Others, such as Jean-Baptiste-Claude Delisle de Sales (1770), even went as far as to suggest – risking their freedom or life – that humans were connected to apes and monkeys.

Jean-Baptiste de Lamarck (1744–1829) is generally regarded as the first serious evolutionist. He was curator of the invertebrate collection at the Natural History Museum in Paris. His *Philosophie Zoologique* (1809) proposed that species gradually change because individual organisms adapt their appearance and behaviour to gradual changes in their environment. He believed that species distinctions were human artefacts and that intermediate forms existed, even if yet undiscovered. In line with this, he argued that species had evolved or branched from each other, starting with simple organisms. Like many earlier thinkers, he thought that species do not become extinct. Even though Lamarck lacked an explanation of evolution, his ideas had a great influence on later thinkers. Best known are his suggestions about the transmission of characteristics of living organisms: use and disuse and inheritance of acquired characteristics. These were at the basis of his theory that modifications in organisms due to use and disuse are passed on to offspring, such as his famous example that giraffes stretched their neck to reach leaves higher up in trees.[1]

---

[1] While Lamarck's explanation of acquired characteristics is firmly rejected nowadays, it is still not settled whether a giraffe's long neck is the result of food or sexual selection (Mitchell et al., 2013). The first explanation, known as the 'competing browsers hypothesis', was suggested by Charles Darwin and is to some extent supported by the fact that females also have long necks, and that a long neck goes along with serious costs in terms of breathing, stability and nutrient uptake. The sexual selection hypothesis – females preferring males with longer necks, and males entering in neck combats resulting in more reproductive success for the winner – was proposed by Simmons and Scheepers (1996). Females' long necks are then explained by the absence of selection against it and genetic information for long necks being shared. Generally, similarly extreme features in other animals are caused by sexual selection, as it fosters a biological arms-race type of dynamics. Supporting the sexual selection hypothesis is the fact that male giraffes reach nearly a metre higher than females. On the other hand, this gives females and their calves a free feeding range, which might be seen as supporting the food selection hypothesis.

Although these ideas were ridiculed later on, Lamarck's insights were far ahead of their time and stimulated an understanding of the complexity of nature without resorting to supernatural causes. Moreover, although Darwin later came up with a more convincing set of mechanisms to explain evolution, given that genetics had not yet been developed, he was unable to offer a convincing alternative for heredity of characteristics of organisms. At some point, Darwin referred to an old theory (pangenesis) according to which all body parts contribute information ('pangenes') to sex cells, so that physical changes in body parts would be transmitted to the offspring. This is in fact not so far removed from Lamarck's 'inheritance of acquired characteristics'. Two other, related theories during the early nineteenth century were 'reformationism' and 'emboitement or encasement' theory. The first expressed the idea that each embryonic organism is conceived as a perfect, miniature replicate of the adult structure. This is a special instance of the view that nature lacks novelty and merely shows changes in scale. The second theory expressed the idea that the initial member of a species (generally thought to have been created by a god) already included all the germs of future generations.

In addition, other views influenced thinking about evolution. The geologists James Hutton (1726–1797) and Charles Lyell (1797–1875) suggested a uniformitarian perspective on the dynamics of the Earth, arguing that geological changes are gradual and have taken place over a tremendously long period of time. Somewhat earlier, the founder of palaeontology, the French anatomist Georges Cuvier (1769–1832), discovered that the history of life is recorded in the Earth in a systematic way, and that deeper or older layers generally contain flora and fauna that is further removed in a morphological sense from current life.

Charles Robert Darwin was well aware of the writings of these early thinkers on evolution, or what at the time was often called 'transmutation' or 'transformism'. Since evolution and natural selection are not easy ideas, it is no wonder that a person like Darwin arose only after being able to benefit from the ideas of many others, including Lamarck, Lyell, the economist Thomas Malthus, his grandfather and his teachers in Edinburgh and Cambridge. In an appendix titled 'An historical sketch of the recent progress of opinion on the origin of species' to an 1861 edition of his *On the Origin of Species*, as well as in his autobiography of 1876, Darwin carefully documented the influences on him of particular individuals.

According to Ruse (1999, p. 41), some of the ideas generally attributed to Lamarck were in fact already put down on paper some 15 years earlier by Charles Darwin's grandfather, Erasmus Darwin (1731–1802), in his *Zoonomia; or, The Laws of Organic Life*, written around 1794. He can thus be considered an important early thinker about evolution. Unlike him, Charles could benefit from the young science of geology and the associated development of palaeontology. Some suggest that the influence of Erasmus on Charles was negligible, perhaps because the grandfather had died before his grandson was born. But *Zoonomia* must have brought the young Charles into contact with the mere idea of evolution. – it simply is too coincidental that among the initial writers on evolution were a grandfather and his grandson. Both Darwins employed industrial metaphors, stimulated by the Industrial Revolution in England, as well as by contacts with leading industrialists. Interestingly, geology not only created a firm basis for the development of evolutionary theory, but also played a crucial role in the Industrial Revolution itself, through application of its insights to mining of fuels and minerals.

To complete the family picture, note that Charles Darwin's son, George, was a student of William Thomson (the later Lord Kelvin), who calculated that the Earth might be as young as 25 million years. Charles was aware that this was at odds with the idea that natural selection had been operative for a much longer period. The physicist was wrong, and the correct age of the Earth, as later assessed by geologists, proved to be 4 to 5 billion years, implying sufficient time for evolution to produce observable nature (Ruse 1999, p. 65).

This brief historical overview illustrates that, like all scientists, Darwin stood on the shoulders of others. Table 3.1 provides an overview of main thinkers who had a direct or indirect influence on Darwin. Section 2.2 has already discussed the influence of the classical economist Malthus.

The year 1858 was critical to the development of modern science, as Charles Darwin (1809–1882) and Alfred Russel Wallace (1823–1913) simultaneously presented their ideas on natural selection to the Linnaean Society of London. Responsible for this gathering were Darwin's two intellectual supporters and close friends, the leading British botanist Joseph Hooker and Charles Lyell, the main geologist at the time. The two wanted to assure that the significance and originality of Darwin's ideas were broadly recognised. Without Wallace having come up with similar though less elaborated ideas on evolution, it is possible that Darwin would never, or at least only much later, have published his ideas on evolution. Darwin's contribution to the 1858 papers presented in London included excerpts from his 'Manuscript on Species' (later titled *On the Origin of Species* (...)) and from a letter to Professor Asa Gray from Harvard University, another of his intellectual supporters.[2] In introducing the first excerpt, a telling note was added stating: 'This MS. work was never intended for publication, and therefore was not written with care. – C.D. 1858.'

Darwin told Lyell that Wallace used core terminology that was consistent with the titles of various chapters in his *Origin* manuscript. He added that if Wallace had had access to his manuscript, he could not have written a better summary. Of course, Darwin had worked for a much longer time on his theory and so its exposition contained endlessly more details and examples to support evolution. Wallace never showed any jealousy of Darwin achieving such public fame as the inventor of the modern theory of evolution.[3]

Central to Darwin's (and Wallace's) theory of natural selection is the insight that many more individuals of each species are born than can possibly survive and reproduce. In many quickly reproducing species fewer than one out of 10 000, and in some bird species, fewer than one out of 100 individuals born, will survive to the age of

---

[2] Two other important supporters of Darwin were Thomas Henry Huxley, known as 'Darwin's bulldog', and early sociologist Herbert Spencer (see Box 7.1).

[3] Costa (2014) gives a detailed account of the overlap and differences between Darwin's and Wallace's ideas on biological evolution. One crucial distinction concerned the human brain and mind. Unlike Darwin, Wallace was convinced that it could not have evolved as he considered it far too sophisticated, i.e. not in line with the so-called 'principle of utility', according to which natural selection cannot produce a structure that is overdesigned or with greater perfection than needed for an organism in its environment. Darwin's intuition was in line with what we know now about brain–mind evolution (Section 11.3).

Table 3.1 Early evolutionists and other intellectual influences on Darwin

1650	1700	1750	1800	1850	1900

Leibniz (1646–1716)

Taxonomist Linnaeus (1707–1778)

Geologist Hutton (1726–1797) — *Theory of Gradualism 1795*

Erasmus Darwin (1731–1802) — *Zoonomia; or, The Laws of Organic Life 1794*

Biologist Lamarck (1744–1829) — *Theory of Evolution 1809*

Economist Malthus (1766–1834) — *Essay on the Principle of Population 1798*

Palaeontologist Cuvier (1769-1832)

Geologist Lyell (1797–1875) — *Principles of Geology 1830*

Early sociologist Herbert Spencer (1820–1903) — *Progress: Its Law and Cause 1857*

Darwin (1809–1882) — *Voyage of the Beagle 1831–1836* — *On the Origin of Species (…) 1838–1859*

49

reproduction. This is indicative of the enormous selection pressure that works upon variation of individuals within all species. Such variation is based partly in environmental and partly in genetic differences. The latter were unknown at the time of Darwin. Indeed, the experimental work by Gregor Mendel (1822–1884), which provided the foundation of genetics, was not recognised until later. Incidentally, it is ironic that while the acceptance of evolutionary biology undercuts a fundamental basis for religious belief, namely the mysterious origin of organic and human life, Darwin originally went to Cambridge to study theology so as to become a clergyman, while Mendel was a monk who ultimately became a priest. Of course, this just illustrates that in their days intellectualism and religious power were firmly connected.

3.2   The Significance of Darwin

Would evolutionary biology have followed a similar historical path without Darwin? This question is considered in Bowler's (2013) counterfactual story *Darwin Deleted: Imagining A World Without Darwin*. It argues that Darwin's theory of natural selection was so complete that it created an intellectual revolution that invited strong resistance from religion. If instead society had had more time to slowly adapt to the idea of evolution, arguably anti-evolutionary spirits would have been frailer and the evolutionary synthesis might have happened earlier, fostered by biometricians such as Karl Pearson. Bowler thinks it is unlikely that Wallace would have adopted Darwin's role, suggesting that instead Lamarck's thought would have lingered on, complemented by the ideas of the German biologist and naturalist Ernst Haeckel (1834–1919). Of course, this is all mere speculation. As the next section argues, the theoretical and methodological contributions by Darwin were so extraordinary that it is hard to imagine that others in the nineteenth century could have accomplished them in his absence.

Darwin's role in the emergence of modern evolutionary thinking has been multifold. He was both an accomplished theorist and empiricist, a multidisciplinary thinker with eclectic interests, and a generalist as well as a specialist. Darwin was an expert on such distinct species as flowers (the English cowslip and orchids), insectivorous plants, worms and barnacles (an aquatic crustacean). He experimented with the latter over the course of eight years, which resulted in two books that were published before he finished his foundational work on natural selection. In addition, Darwin had a superb network of scientific contacts at the time and, moreover, was innovative in developing a range of methods to provide support for evolution. Above all, Darwin showed an unusual quality of covering and connecting different areas of knowledge.

His theoretical ideas were born during an expedition around the world with the H.M.S. *Beagle*, a British naval ship used for geographical surveying. Darwin was only 22 when this journey started in December 1831, but had already gained a reputed knowledge of animals and plants. Because of this, he was enthusiastically recommended by Cambridge botanist and geologist John Stevens Henslow to the similarly young, 27-year-old *Beagle* captain Robert Fitzroy – who later would found the UK's Meteorological Office, as well as became a governor of New Zealand. Although it was planned to last just two years, the voyage continued for five long years, to map the coastlines of South America for the British Navy. A report of Darwin's travel experiences appeared as *Journal of Researches (into the Geology and Natural History of the Various Countries Visited by H.M.S. Beagle)*. The popular version, better known as *Voyage of the Beagle*, was published in 1839 and sold well. Before the voyage, Darwin had considered a career in the church – after his return he was strongly interested in science while doubting God and religion.

During the time of his voyage on the *Beagle*, Darwin collected plants and animals, notably birds, insects and invertebrates. He sent all specimens with explanatory letters to Henslow. Through the latter's diffusion of the findings, Darwin acquired considerable fame among leading natural scientists in England, even before his return. Darwin spent most of the time during the five-year voyage on land, and thus came to note the geographical particularities and distribution of living and fossil species, as later studied by the sub-field of biogeography. He discovered that species in temperate parts of South America were taxonomically closer to those in tropical zones of the same continent than to similar species in Europe. In particular, Darwin's experiences on the equatorial Galapagos Islands, located about 900 km west of the South American mainland, were critical to the advent of his thinking on evolution. Darwin devoted much time to the study of the Galapagos' finches, several closely related species of small birds that differ, especially with regard to their beaks, and occupied niches that elsewhere were occupied by other species. Initially, he thought the birds he had collected were from distinct species, namely blackbirds, grosbeaks and finches, but ornithologist John Gould identified them as 12 different species of finches (currently 15 are distinguished). This stimulated Darwin to develop the idea of adaptation to local environments, which explained why isolated islands or other isolated environments would host unique species. Darwin's finches, for instance, had beaks and dwelling places – e.g. on the ground, on cactuses or in trees – adapted to available diets on the respective islands – such as seeds, insects and their larvae, cactus fruit and buds, nectar and pollen of flowering plants. The study of these finches made Darwin believe that through long-term isolation subpopulations of a single species would be able to accumulate adaptations that could ultimately give rise to separate species. The Galapagos case was particularly informative as finches could, due to evolution, occupy a variety of ecological niches because the islands did not host other species that normally fill these.

By the early 1840s, Darwin had elaborated his notes into a theory of natural selection, which presented the main mechanism of evolution. The first formulation likely originates from 1838, only a few years after the trip with the *Beagle* (Ruse, 1999, p. 57). That it took so long for him to publish his work, ultimately stimulated by the joint publication of a number of his letters and a paper by Alfred Russel Wallace on natural selection, was mainly due to the worry that his theory would meet with much resistance. Hence, he chose to wait in order to gather more empirical support. It should be noted that, in Darwin's time, biology was mainly the domain of clergyman. This explains why most naturalists stuck to a tradition of Natural Theology (see Box 1.1).

Soon after the presentation of Darwin's and Wallace's ideas on natural selection in London, Darwin finished his book *On the Origin of Species* (published in November 1859). Table 3.2 briefly summarises what must be considered one of the most important

Table 3.2 A summary of Charles Darwin's *On the Origin of Species* (1859)

Chapter	Goal and main conclusion (following closely Darwin's wording)
1. Variation under domestication	Considers artificial selection of domesticated breeds of animals and plants. The cumulative action of selection is the dominant cause of observed variability in them. Various domesticated plant species, such as apples, dahlias, hyacinths, potatoes and wheat, and animal species, such as cattle, dogs, ducks, pigs, poultry, sheep, rabbits and horses, are mentioned. Based on his belief that one can learn much from studying one example in detail, Darwin reports insights derived from keeping domestic pigeons. He claims he kept every breed which he could purchase or obtain, and consulted many treatises on pigeons, communicated with several eminent fanciers, and was allowed to join two London Pigeon Clubs. The chapter acknowledges that little is known about the precise origin and history of domesticated pigeons. Distinguishing between varieties and species of domesticated breeds is noted to be difficult. Selection is methodical or unconscious. Artificial selection to augment any peculiarity will almost certainly modify other parts of the structure, owing to the mysterious laws of the correlation of growth. Domestic varieties, when run wild, gradually but certainly revert in character to their aboriginal stocks, implying that deductions from domestic races to species in a state of nature are difficult if not impossible.
2. Variation under nature	In nature one can observe diversity in useful characteristics, where the distinction between species and varieties is not sharp and clear. An inordinate amount of variation can be found in some species, notably 'protean' or 'polymorphic', about which naturalists can hardly agree which forms to rank as a species or as varieties. Species of large genera present a strong analogy with varieties, which can be understood if species have once existed as varieties, and were thus originated instead of created. Varieties are species in the process of formation, or incipient species. Darwin on speciation:
	Hence I look at individual differences, though of small interest to the systematist, as of high importance for us, as being the first step towards such slight varieties as are barely thought worth recording in works on natural history. And I look at varieties which are in any degree more distinct and permanent, as steps leading to more strongly marked and more permanent varieties: and at these latter, as leading to subspecies, and to species.
3. Struggle for existence	Clarifies how the struggle for existence affects natural selection. All organic beings are subject to severe competition and scarcity of life factors. Food and space are scarce, abiotic conditions are harsh, and animals kill each other to eat. Every individual strives for geometric growth, but populations do not realise geometric powers of increase due to the existence of natural checks. Lighten any check and the population will increase to large numbers. The amount of food gives the extreme limit, but the serving as prey to

Table 3.2 (*cont.*)

Chapter	Goal and main conclusion (following closely Darwin's wording)
	other animals determines the average numbers of a species. The struggle of life is most severe among individuals and varieties of the same species, and sometimes between species of the same genus. Not only animals but also plants struggle for existence and reproduction, as they generate many more seeds than can come to maturity.
4. Natural selection	The principle of selection also applies to nature. Natural selection is the preservation of favourable variations and the rejection of injurious variations. Variations neither useful nor injurious are unaffected by natural selection. Selection of variations is based on survival, reproduction and inheritance of characteristics. Sexual selection is important, and is based on a struggle between the males for possession of the females, translating in success in terms of number of offspring. As a result, sexual selection leads to formation in males of characteristics like 'weapons', force and attractive physical features. Repeated natural selection results in divergence of character, i.e. the augmentation of smaller into larger differences. The more diversified are the descendants of a species, the easier it will be for them to occupy a wide range of habitats that in turn allow them to increase their numbers. Or the more living beings diverge in structure, habits and constitution, the more can live in the same area. Selection is the only way to explain that all organic beings are clustered around genera, families, species, etc., namely by explaining that all life forms throughout all time and space are related to each other, in a tree of life. Natural selection belongs to the utilitarian doctrine because all characters have arisen for the good of its possessor; selection will never produce in a being characters that are injurious to it.
5. Laws of variation	Variability is mainly due to the conditions of life to which ancestors have been exposed during several generations. Habit, use and disuse have played a considerable part in the modification of organisms, but are largely combined with, and sometimes overmastered by, natural selection of innate differences. Modifications accumulated solely for the good of the young or larva will affect the structure of the adult. A part developed in any species in an extraordinary degree or manner relative to allied species tends to be highly variable, owing to it having undergone an extraordinary amount of modification in the past. Beings low in the scale of nature are more variable than those that are higher, and related, less specialised parts are more variable than more specialised ones. Multiple parts are variable in number and structure, possibly because of little specialisation to any particular function. Rudimentary parts are apt to be highly variable, since natural selection has no power to check deviations in their structure. Generic characters inherited from a remote period and shared by several species are less variable than specific characters that contribute to uniqueness of a species. Secondary sexual

Table 3.2 (*cont.*)

Chapter	Goal and main conclusion (following closely Darwin's wording)
	characters, which appear when animals become sexually mature, are very variable due to the power of sexual selection.
6. Difficulties on theory	Darwin realised that the reader might raise a number of objections against his theory. He examines and tries to solve the following difficulties: the absence or rarity of transitional forms; the origin and transitions of organic beings with particular habits and structure (swim bladder in aquatic animals converted into air-breathing lung in terrestrial animals); the formation of organs of extreme perfection and complication (the eye is discussed) and organs of little apparent importance (the tail as a fly-flapper is discussed); acquiring and modification of instincts through natural selection; and that the crossing of species leads to sterile offspring, while crossing of varieties results in fertile offspring. The old canon in natural history, *natura non facit saltum*, follows from the theory of natural selection. Moreover, the law Unity of Type follows directly from unity of descent.
7. Instinct	Instincts are of the highest importance to each animal, which helps organisms to multiply, which in turn allows natural selection to be effective in modifying them. Complex instincts can be produced through slow and gradual accumulation of numerous, slight, yet profitable, variations. Instincts are variable, shown by birds nesting or birds' fear. Natural instincts are sometimes lost under domestication. Among others, the cases of the cuckoo laying eggs in other birds' nests and slave-making ants are considered in detail.
8. Hybridism	Is in response to the belief of many naturalists at the time that species, when intercrossed, have been specially endowed with the quality of sterility, in order to prevent the confusion of all organic forms. The sterility of hybrids, the offspring of the unity of two species, has been neglected. Sterility of hybrids is not an advantage to them, and therefore not the direct result of selection, but is incidental to other acquired characteristics. The sterility of two species when crossed is not to be confused with the sterility of the hybrids produced from them. The latter is, unlike the first, caused by the reproductive organs being functionally impotent. Nevertheless, the degrees of sterility correspond, which is consistent with the fact that they both depend on the amount of difference between the species crossed.
9. On the imperfection of the geological record	Responds to the noted problem that not all transitional links between the many species that now exist and those known to have existed are found in the successive formation in the crust of the Earth. The answer is quite simple, and phrased as an analogy, as follows: the natural geological record is a history of the world which is imperfectly kept, and written in a changing dialect; of this history we possess the last volume alone, relating only to two or three countries; of this volume, only here and there a short chapter has been preserved; and of each page, only here and there a few lines.

Table 3.2 (*cont.*)

Chapter	Goal and main conclusion (following closely Darwin's wording)
10. On the geological succession of organic beings	Continues the previous chapter, focusing on the question whether the geological facts and rules better accord with the then-common view of the immutability of species, or with that of their slow and gradual modification, through descent and natural selection. The geological record is extremely imperfect. Only a small portion of the globe has been geologically explored with care. Only certain classes of organic beings have been largely preserved in a fossil state. The number of specimens in museums is absolutely nothing compared with the incalculable number of generations which must have passed away. There has probably been more extinction during the periods of subsidence, and more variation during the periods of elevation, and during the latter the record will have been least perfectly kept. The duration of each formation is, perhaps, short compared with the average duration of specific forms. Varieties have, at first, often been local. All these causes taken conjointly must have tended to make the geological record extremely imperfect. Many facts of palaeontology seem to confirm the theory that: species slowly and successively appear in geological time; groups of species follow the same general rules in their (dis) appearance as do single species; species lost do not appear again; in the long run all species undergo modification to some extent; the extinction of old forms is the almost inevitable consequence of the production of new forms; the succession of the same types of structures within a single area is explained by inheritance.
11. Geographical distribution 12. Continued	These two chapters discuss the geographical distribution of species. Similarities and dissimilarities cannot be accounted for by climate and other physical conditions. For example, despite parallel conditions in the New and Old World, the species found in each differ greatly. All leading facts of geographical distribution are explicable on the theory of migration, together with subsequent modification and the multiplication of new forms. We can thus understand the importance of barriers, whether land or water, which separate zoological and botanical provinces. Barriers of any kind act as obstacles to free migration, and are closely related to the differences between the productions of various regions. This holds first and foremost for the separation of continents. No single mammal is common to Europe and Australia or South America, but certain plants are found in more than one continent. This can be explained by animals being unable to migrate, while plant seeds are now and then transported over long ranges by wind and birds. It also holds within continents for mountain ranges, great deserts, and, sometimes, even for large rivers. Moreover, marine faunas separated by land are distinct, as illustrated by the east and west shores of South and Central America. Change of climate must have had a powerful influence on migration. The theory of barriers and migration also explains why oceanic islands have few inhabitants, but, of these, a great number is endemic or

Table 3.2 (*cont.*)

Chapter	Goal and main conclusion (following closely Darwin's wording)
	peculiar. Darwin cites A.R. Wallace in this chapter: 'every species has come into existence coincident both in space and time with a pre-existing closely allied species'.
13. Mutual affinities of organic beings; morphology; embryology; rudimentary organs	Finds support for evolution in the fact that all organic beings are found to resemble each other in descending degrees, so that they can be classified in a hierarchical system. In fact, past and current classifications turn out to be consistent with elements of descent. All facts in morphology become intelligible from the perspective of evolution: from homologous parts and organs to different species of a class. The similarity of body plan and organisation is remarkable. Similarly, the homologous nature of different parts or organs in a single individual is remarkable, and also follows logically from descent by modification. Examples are the anterior and posterior limbs in each member of the vertebrates; and the parts of flowers seen as metamorphosed leaves. Rudimentary organs are common throughout nature, and always bearing the stamp of inutility. Embryology offers further support, in particular by showing that the embryos of distinct animals within the same class are often strikingly similar and that the larvae of many insects resemble each other much more closely than do the mature insects. In addition, the idea is mentioned that ancient animals resemble to a certain extent the embryos of more recent animals of the same class.
14. Recapitulation and conclusion	Opens with the famous statement: 'this whole volume is one long argument'. Darwin acknowledges that many grave objections may be advanced against the theory of descent with modification through natural selection. He adds, 'I have endeavoured to give to them their full force.' He argues that the core of his theory consists of three propositions: that gradations in the perfection of any organ or instinct either do now exist or could have existed, each good of its kind; that all organs and instincts are, in ever so slight a degree, variable; and that there is a struggle for existence leading to the preservation of each profitable deviation of structure or instinct. The belief that species were immutable does not hold up against the knowledge that the Earth has existed for at least hundreds of millions of years. All animals and plants have probably descended from one primordial form, into which life was first breathed. We can dimly foresee a considerable revolution in natural history. Light will be thrown on the origin of man and his history.

Note: This is based on the Wordsworth edition of 1998, which reflects the first edition from 1859.

books in human intellectual history. In hindsight, it is amazing how profound was Darwin's comprehension of evolution. The phrase 'origin of species' in the title of Darwin's book is actually a bit confusing, as his work is mainly about selection and variation. Only his Chapter 10 deals, to some extent, with the origin of species, albeit at an abstract and palaeontological level of discussion. Darwin consistently refers to his theory as 'theory of descent with modification through natural selection'.

In addition to his observations during this voyage, Darwin drew lessons from a long history of artificial selection in selective breeding of domesticated animals and plants. Chapter 1 in *On the Origin of Species* addresses this topic. It discusses the selection strategies employed by various breeders, distinguishing between systematic, goal-oriented and unconscious selection, the causes of variability, the character of domestic varieties and the difficulty of distinguishing between domesticated varieties and species. In addition, this chapter pays particular attention to domestic pigeons, talking about their great variety in size, beaks, crops, skeletons and plumage (wings, tail-feathers and head). Darwin notes, 'I am fully convinced that the common opinion of naturalists is correct, namely, that all have descended from the rock-pigeon (Columba livia).' Two particular notions addressed are 'effects of habit' and 'correlation of growth'. The first is explained, 'Not a single domestic animal can be named which has not in some country drooping ears; and the view suggested by some authors, that the drooping is due to the disuse of the muscles of the ear, from the animals not being much alarmed by danger, seems probable.' The modern explanation is that it – along with brains and teeth that are smaller in domesticated animals than in their wild ancestors – is the result of artificial selection for calmer, less 'fighty and flighty', specimens. On the second notion, Darwin says 'Variability is governed by many unknown laws, more especially by that of correlation of growth.' He summarises that 'if man goes on selecting, and thus augmenting, any peculiarity, he will almost certainly unconsciously modify other parts of the structure, owing to the mysterious laws of the correlation of growth'.

Darwin's work was well-known in his time and received great support in the scientific community. The first edition of his book immediately sold out. Five revised and expanded editions followed, some containing detailed responses to critiques. After his death, the popularity of Darwin's theory waned somewhat, mainly as it lacked a convincing hereditary mechanism. Only after the 1910s, through the work of the influential evolutionary biologists Fisher, Haldane and Wright, was Darwin's thought reassessed. The resulting modern synthesis during the 1920s – also known as the synthetic theory of evolution, or neo-Darwinism – was based on fusing Mendelian genetics and Darwinian selection theory ('Darwinism'). After the 1960s, this was extended with insights from molecular biology. These developments allowed a clear separation between individual development and evolution, which were until then confused, notably in the persistent Lamarckian view that developmental changes could become hereditary and thus steer the direction of evolution. The widespread use of the term (Neo-)Darwinism to refer to modern evolutionary theory reflects that Darwin's ideas were largely correct and that he accomplished so much in both theoretical and empirical aspects. Coyne (2009) states that this name-giving is unusual, noting that similar terms like 'Newtonism' or 'Einsteinism' are not in use, even though no one denies Newton's and Einstein's fundamental and pervasive influences on modern physics.

Darwin was extremely productive during his life, being author of, among others, *The Voyage of the Beagle*; *The Structure and Distribution of Coral Reefs;* works on *Cirripedia (Barnacles); On the Origin of Species; The Descent of Man and Selection in Relation to Sex; The Expression of the Emotions in Man and Animals; The Power of Movement in Plants;* and *The Formation of Vegetable Mould through the Action of Worms*. In addition, he wrote some 15 000 letters to colleagues, showing an unusual interest in testing his ideas against, and learning from, other experts at the time. He himself attributed his high productivity to his life-long illness, which kept him at home and away from distractions. His ailment involved regular stomach problems, such as cramps and vomiting, severe tiredness, skin problems and anxiety, all of which is still a mystery. Two main explanations, also suggested by Darwin himself, are nervousness about the implications of his evolutionary theory, notably a possible conflict with religion, and an insect bite – during his voyage with the *Beagle* – transmitting a virus, possibly causing Chagas disease (also known as American trypanosomiasis). In support of the latter is the fact that the first symptoms appeared soon after his return to England. Other factors that have been proposed to possibly play a role are lactose intolerance, chronic arsenic poisoning, and even a complicated relationship with a dominant father who criticised him as a young man for being idle. Likely, a combination of factors was at work, in turn having psychosomatic repercussions in the form of an overly sensitive stomach.

3.3   The Rise of Genetics

The renowned experiments by Gregor Mendel signify the dawn of genetics. Mendel was a monk from Czechoslovakia who, in 1865, undertook empirical research on varieties of peas, observing features such as stem length, and colour, shape and position of flowers, seeds and pods. He showed that biological inheritance was based on discrete information units, later called genes, about which he formulated two principles. The principle of segregation states that through a cell division process known as meiosis, diploid organisms – possessing a pair of genes (called alleles) for each characteristic – produce sex cells (gametes) that have a single allele. The alleles, which are said to segregate in the meiosis, can contain identical information (in homozygote organisms) or different information (in heterozygotes), in which case one allele is often dominant and the other recessive, relative to each other. However, codetermination has also been found: here both alleles are expressed leading to a phenotype that mixes the features of homozygotes. The principle of independent assortment states that segregations of alleles for different traits are independent. Later, it was shown that both principles do not always apply. An important reason is that particular alleles are linked together on the same chromosome. In addition, some characteristics are not transmitted through the chromosomes in the cell nucleus – so-called 'nuclear inheritance' – but through the cytoplasm of the egg – known as extra-nuclear or maternal inheritance; this holds for organelles such as mitochondria and chloroplasts, which have their own genetic material, due to the fact that they have a distinct evolutionary origin. Later, it was found that chromosomes can also exchange information, called crossover. These features imply that Mendelism was somewhat more purist than the later genetics. Nevertheless, the power of Mendel's insights is that they were able to explain so much of the logic of modern-day genetics, despite an incomplete understanding of the underlying physical-molecular processes.

One may consider it unfortunate that Darwin never knew about Mendel's experiments, because he might have developed a more complete theory. On the other hand, it would have distracted him and might have complicated things to such an extent that he perhaps would never have finished his magnum opus. Mendel's laws of inheritance were independently rediscovered by Hugo de Vries and Carl Correns, who in turn influenced others, notably William Bateson, to coin the term 'genetics' (Ruse 1999, p. 81).

An major catalyst in the development of genetics was August Weismann's 'germ plasm theory' (1883), according to which the body consists of germ and somatic cells. The first can transmit hereditary information. This is based on the process of meiosis, what Weismann called 'reduction division of chromosomes', which introduces variation. In line with this, he recognised that natural selection is active at the level of the resulting heterogeneous germ cells. In contrast, the somatic cells are not involved in the hereditary transfer of information. Hence, Weismann argued, acquired characters cannot be inherited as the germ cells are separated and thus protected from somatic cells, so that they are not affected by 'body events' or somatic changes. This is called the 'Weismann barrier'. Note that the absolute, general validity of this barrier is questioned, or at least nuanced, by recent research on epigenetic inheritance. This suggests that certain epigenetic tags, activating certain genes under particular environmental conditions, can be transmitted to the next generation (Section 4.2).

After these advances, the young field of genetics was then – rather quickly – integrated into the main body of evolutionary biology during the 1930s. Thomas Hunt Morgan played a crucial role in this process during the 1920s by discovering genes and their location on chromosomes in fruit flies (*Drosophila*) and showing they form the physical basis of heredity. Next, in the 1930s, Ronald A. Fisher, J.B.S. Haldane and Sewall G. Wright took evolutionary science to a more precise level of analysis by developing mathematical models of population genetics. The integration, or evolutionary synthesis, took some time, as early Mendelians opposed Darwinian natural selection, while Darwinian 'selectionists', also known as 'biometricians', were critical of Mendelism. Moreover, the synthesis itself required conceptual innovations as well as elegant mathematical modelling of the long-run impact of interactions among heterogeneous individuals – in particular mating – on the gene pool. Central notions include dynamic equilibrium, shifting balance, adaptive landscapes and genetic drift. A famous controversy was between Fisher and Wright on the importance of genetic drift, also known as the 'Sewall Wright effect'. Fisher did not assign it a main role, as opposed to Wright, who connected it to adaptive landscape and shifting balance theories. Nowadays, it is recognised as a key force in evolution (Section 4.2).

The mathematical formalisation of evolutionary theory helped to convince a wider group of scientists of its explanatory and predictive value. For example, questions about sex ratios could be dealt with in a more accurate and definitive way. The next generation – with the godfather Theodosius Dobzhansky (1900–1975) and prominent colleagues like Ernst Mayr, George G. Simpson and G. Ledyard Stebbins – ensured that this work was translated and elaborated in teaching and research programmes, through activities such as founding journals on evolutionary biology. Possibly the most influential publication was Dobzhansky's textbook *Genetics and the Origin of Species* (1937).

In addition, through writing popular science books, they promoted evolutionary thinking to a wider audience. In England, Julian Sorell Huxley, grandson of 'Darwin's bulldog' Thomas Henry Huxley, contributed to the momentum of evolutionary biology. His *Evolution: The Modern Synthesis* (1942) is considered one of the best accounts of knowledge about evolution at the time.

Dobzhansky continued the work of Wright, notably the shifting balance theory and the notion of adaptive landscapes with peaks and valleys. His most important contribution was the idea of superior heterozygote fitness, which explained that, despite selection pressure and adaptation, variation within populations can continue because heterozygotes are fitter than both dominant and recessive homozygotes. This was generalised in his shifting balance theory, which stated that forces stimulating genetic similarity and those stimulating diversity are in balance. A well-known example in humans is sickle-cell anaemia, which predominantly affects black people. Dominant homozygotes have a higher probability of dying due to red blood cells not binding oxygen well. Heterozygotes, however, are better protected against malaria than recessive homozygotes. The two effects jointly cause heterozygotes to perform relatively well, which allows the dominant allele to continue in the population. In contrast, the classical position was that there is only little genetic variation within a population owing to continuous selection pressure, causing rare mutations to end in either complete success or extinction.

Probably both positions hold some part of the truth since either process can be found in reality.[4] Research on diversity based on applying chemical techniques showed that there is much variation in species, regardless of the type of species (Lewontin, 1974). Note, however, that there is a third explanation for maintenance of genetic diversity. This is based on the idea that much diversity is invisible as it does not directly affect phenotypes. If this is the case, then genetic drift can occur, increasing genetic diversity or compensating for its loss due to selection. Lewontin suggested a fourth idea, which has subsequently attracted more attention, namely genomes are wholes, which cannot be separated into genes that are subject to independent selection.

Genetics is the study of traits and their inheritance as well as of changes in gene frequencies in populations. A gene is a sequence of DNA (or RNA) that codes for the construction of one or several proteins. The term 'gene'[5] derives from the Greek word 'gonos' (γόνος) meaning offspring, and was proposed in 1905 by Danish botanist and geneticist Wilhelm Johannsen, along with the terms 'genotype' and 'phenotype'. Mendelian genetics pertaining to the level of individuals was generalised to the population level, leading to so-called population genetics. A core assumption of Mendelian genetics is the so-called 'law of segregation', which claims that copies of genes pass on to offspring without alteration, except for small (point) mutations. This became the starting point of a theoretical approach in population genetics in which the collection of sex cells in a population is considered as one large gene pool from which random combinations form new individuals. This then allows for the application of

---

[4] Ruse (1999, p. 164) notes that classic theory often leads to support for eugenics – aimed at improving the genetic (and phenotypic) quality of humans – whereas balance theory generally fosters support for diversity.

[5] Distinct from 'pangene', which is part of the old 'pangenesis' theory adhered to by Darwin (see Section 3.1).

mathematical probability theory. One notable result is the Hardy–Weinberg principle (or law), which states that in a large isolated population with randomly mating individuals the equilibrium of genotype frequencies derives from gene (allele) frequencies (Maynard Smith 1989). Note that this is merely an outcome of recombination in the absence of selection and mutation. In smaller populations, chance effects may dominate, which is the basis for genetic drift (Section 4.2).

The year 1953 is significant as Francis Crick and James Watson deduced the molecular and spatial structure of genetic information from X-ray crystallography photos made by Rosalind Franklin, who built upon research started by Maurice Wilkins. This was called deoxyribonucleic acid, better known as DNA. Their work created a stepping stone for clarifying how genetic instructions were represented in chemical assemblies and copied in cells. This ultimately gave rise to a full understanding of the genetic code, and more recently to research on functional genomics, aimed at identifying the functions of all genes of a certain organism. The most significant meaning of the discovery of the DNA structure was that it turned out to be similar in all organisms, pointing at a joint evolutionary origin (see Section 4.5).

DNA has a double-helix structure, resembling a ladder in the shape of a corkscrew, with two strings of DNA. Each string has nucleotides, as basic building blocks, with nitrogenous bases – containing cytosine, guanine, adenine and thymine, or C, G, A, T, respectively – which are directed at complementary nucleotides on the other string (that is, forming the pairs C–G and A–T), representing the rungs of the ladder. The unique order of these bases in the DNA represents information about the formation of amino acids, which are the basic components of proteins. Separation of the strings occurs, among others, in the process of meiosis to form gametes or germ cells. In addition, a string can be read to create amino acids or proteins composed of these – through transcription of a single stranded ribonucleic acid (RNA) – or for replication through mitosis for body repair or growth. The metabolism of the cell and thus the organism, and its reproduction, both start with the separating and handling of the strings of DNA. Figure 3.1 shows images of the chemical and three-dimensional structures of DNA and the associated RNA.

To summarise, the abstract notion of genotype denotes the total genetic information in cells, unique to each species, and similar in general structure between members of a species. The genetic information is organised in sets of units, known as genes, with alleles as their variants encountered in diploid organisms. The genes are physically present in immense molecules known as chromosomes, which contain the genetic code for an individual's ontogenetic development, metabolism, behavioural features, etc. The chromosomes consist of a conglomerate of molecules at the centre of which is DNA, surrounded by proteins such as histones and methyl markers that control the form and the winding of the DNA to create the typical shape of a chromosome. Genes are associated with sequences of base pairs in the DNA molecules. Another oft-used term is genome, which integrates the terms 'gene' and 'chromosome', to denote all coding and non-coding genetic (heritable) elements of the complete DNA of an organism. In other words, whereas genotype is an information concept, genome refers to its physical or molecular counterpart.

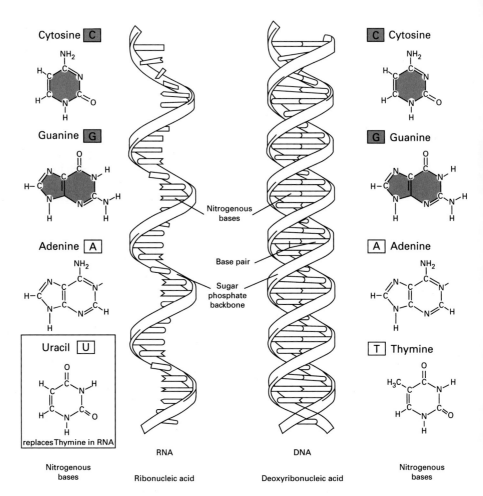

Figure 3.1 Chemical and spatio-physical structure of DNA and RNA.
Source: National Human Genome Research Institute.

The relationship between genes and phenotypic expressions is complex. One gene may cause a single effect, although this is unlikely. However, it is possible – and this is not the same – that a mistake in one gene has major consequences. This is a logical place to start searching for gene–behaviour–culture links. Most common is that multiple genes, known as polygenes – usually between a few and in the order of tens – jointly determine a phenotypic characteristic. It is also possible that one gene contributes to various characteristics – known as pleiotropy – simply because the coded protein participates in multiple processes. Some phenotypic expressions are fixed and show little to no variation. This is the result of severe or enduring selection that has completely reduced genotype variation and associated phenotype variation. Examples abound – witness general features of humans, such as two legs, ten fingers, position of body parts, the reflexes, or the visual and acoustic senses. The reason is that any variation in these would seriously harm fitness.

Evolution since the 1960s has seen rapid development, becoming the synthesis framework for specialised biology sub-disciplines, connected to genetics and molecular biology at one extreme and to ecology at the other. In addition, development and evolutionary biology have been integrated – known as 'evo-devo' – which involves a focus on the expression of genes during development, notably in embryonic phases (Müller, 2007). The most recent branch is the ambitious study of whole genomes or genetic sequences to understand the structure and behaviour of organisms. This has already seen many applications, notably in medicine and agriculture. It has even turned out to be of use for environmental science, as it helps to identify the low genetic quality of species that are at risk of extinction, namely through the appearance of asymmetric features (Lens et al., 2002). The next chapter will provide more information on recent developments.

Another key element of evolutionary theory since the 1960s has been the study of behaviour in a social context using straightforward neo-Darwinism and genetic theory. Authors such as William (Bill) Hamilton, Robert Trivers, John Maynard Smith and Edward O. Wilson, defined a research programme for genetic evolution of social behaviour and societies, covering both animals and humans. Initially, it met with strong opposition in biology, social science and public media. Later on, many biologists, psychologists and anthropologists recognised the value of linking the analysis of social behaviour and systems to genetic evolution. In addition, it is increasingly accepted, by biologists, sociologists and economists alike, that non-genetic evolution is relevant at the level of culture, economy and technology as well as that understanding its interaction with genetic evolution can improve our understanding of human history. These issues are elaborated in Parts III and IV of the book.

3.4 Central Concepts and Mechanisms of Neo-Darwinism

The V-S-I-R Algorithm in Evolutionary Biology

The generalised V-S-I-R algorithm, as discussed in Section 2.1, applies to biological evolution, as this involves variety creation, (natural) selection and replication. It reflects two seemingly opposite features of genes: constancy and variability. Constancy results from the transmission of genetic information in stable, chemical forms (i.e. chromosomes) from generation to generation. Variability is caused by the imperfection of the replication of genetic information due to recombination and point mutations, which can be seen as processes of diversity creation or innovation.

A (point) mutation can be defined as a small change on a chromosome. It can result from various factors, for example, substitution of small elements such as one nucleotide base for another, duplication of chemical sequences within a chromosome, loss of chromosomal material, reversal of chromosomal gene order, or transfer of material to a chromosome in another allele pair. Indirect causes include ageing of individuals (egg cells in females), chemical or toxic pollution, and radiation.

Recombination denotes the mixing of chromosomes by merging the gametes of two parents to create a new and unique set of chromosomes ultimately leading to a unique individual. In addition, recombination is used to refer to crossover, i.e. exchange of genes between chromosomes during meiosis, the cell division that produces gametes. Recombination is the main source of variation in sexual individuals. Through causing new diversity, it has accelerated evolution.

Selection thus works from generation to generation upon slightly altered genetic diversity. Fisher's (1930) fundamental theorem is worth mentioning here: 'The greater the genetic variability upon which selection for fitness may act, the greater the expected improvement in fitness.' As a result of selection, a proportion of the population survives and reproduces, meaning that only the genes of the surviving individuals are transferred to the next generation. Selection thus changes genetic diversity in a population or its gene pool. Since selection generally reduces variety, on its own it would ultimately lead to a state where there was no variation left. While in scarcity situations selection dominates, in affluent situations diversity creation dictates.

Natural selection is based on competition between individuals in a population of a species – 'the struggle for existence' in Darwin's words. It involves the interaction between individuals and their environment, as well as the interaction among individuals of the same population. The first includes searching for food or avoiding predators, and the second finding a sexual partner. Hence, selection is not a particular process, but an umbrella term that covers a variety of processes causing differential survival and reproduction. It selects for greater reproductive success, known as fitness, owing to a phenotype of an organism being better adapted to its environment – in the sense of finding food, avoiding predators, encountering a sexual mate, etc. Sometimes selection types are distinguished based on life-cycle stages. An important division here is between survival selection and reproductive or fertility selection. A key component of the latter is sexual selection, which is considered in some detail, including its cultural implications, in Section 4.7.

Selection can be understood from the angle of population ecology. Competition among individuals of the same species or population for scarce, limiting resources is central to this. It causes the population distribution of characteristics to change. Selection pressure is due to both abiotic and biotic factors. The first include space, water, food, nutrients, light, temperature and ecological stability. Biotic factors include intraspecific (competition, mating) and interspecific interactions (competition, predation, herbivore–plant interactions, parasitism, cooperation). Selection pressure is felt immediately from the beginning of a new life. Moreover, different life phases of an organism are characterised by specific selection pressures: competition between gametes, competition between offspring for parental attention and care, or competition between adults for sexual mates.

Usually, selection is used to denote external factors that constrain the variety of a certain phenotypic, physical or behavioural, expression. But sometimes the notion of 'internal selection' is used, denoting that an organism performs worse if its internal or intrinsic stability is imperfect. An extreme example is when a serious error at the chromosomal or genetic level has occurred that causes the propensity to survive to be low, irrespective of the prevailing environmental conditions.

Notions like historical 'bauplan limits' or 'development constraints' mean that selection is bounded by internal constraints or the current design ('bauplan'). So adaptations are restricted and cannot always go in the most straightforward or easiest direction. A species might even end in an adaptive cul-de-sac, in which case stasis or even extinction are possible.

Natural selection works upon phenotypes, which are the physical expression of the interaction between genotype and the environment. Organisms in a population show phenotypic variations (or variability or diversity) with regard to physiology, morphology and behaviour that relate to genotypic variations. Phenotypic variation results from variation in genotypes and in environmental conditions.

A necessary condition for natural selection is heritability or replication or, in Darwin's terminology, 'descent'. Inheritance is the transmission of the chromosomes. In sexual species, this happens through the production of gametes (sex cells) through the process called meiosis, and subsequent fusion of the genetic material in gametes from two parents. Instead, in asexual species, reproduction and inheritance occur through regular cell division, known as mitosis.

Evolution can thus be seen as the outcome of an interaction between diversity creation, natural selection and environmental changes. Environmental change and new variations due to mutations and recombination imply that evolution will not end up in some equilibrium but is a never-ending process of change.

Fitness, Adaptation and By-product Explanation

Several other concepts play a central role in neo-Darwinian evolution and deserve further inspection. First and foremost is the notion of fitness of an individual, sometimes referred to as adaptive or selection value. This is a measure of reproductive success relative to other individuals in the same population. Defined at the genotype level, it denotes the relative rate at which genotypes are replicated and appear in later generations. Two specific meanings of fitness can be derived from internal and external selection (Heylighen, 1996): fit as strong or robust is related to internal selection and can be regarded as absolute; fit as adapted is related to external selection and can be regarded as relative. Overall, fitness should be seen as a relative concept since it depends on the genotype variation present in the relevant population. A characteristic that performs well in one population may not do so in another. Moreover, the fitness of an individual or characteristic can alter owing to a mere change in the distribution of similar characteristics in the population, hence not merely due to changes in the wider ecosystem or abiotic environment.

The result of natural selection is the adaptation of organisms to their environment. Adaptation can be seen as a change in genotype that *ex post* implies an improved phenotype adjustment to prevailing biotic and abiotic environmental conditions. An old representation of adaptation, conceived by Wright (1932), is as ascending a hill when moving in an 'adaptive or fitness landscape' with adaptive peaks and valleys. Adaptation is a multi-interpretable concept. It can be viewed from an engineering, design or functional angle, where selection is the designer or causal process. Alternatively, one can consider it from the angle of fitness or reproductive success. Specific terms have been used to distinguish between such distinct interpretations. 'Adaptedness' is a concept that stresses the *ex post* situation more than the process leading to it. However, adaptedness might also be just coincidental or the consequence of processes other than natural selection, which is reflected in the idea of 'exaptation' (Vbra and Gould, 1982). In addition, the concept 'pre-adaptation' has been used to denote characteristics that resulted from adaptation to certain environmental conditions and subsequently provided the starting point or initial stage for a second adaptation under different environmental conditions. But such pre-adaptations are just historical coincidences that only later on generated or contributed to new functions, since natural selection is blind and cannot foresee the future. The issue of exaptation is discussed further in Section 4.4.

Gould and Lewontin (1979) have argued that what are considered functional adaptations are often mere by-products or coincidences. They illustrate their case by referring to decorative, non-functional spandrels at the top of pillars in churches, mentioning St. Mark's Basilica in Venice as a concrete example. Others have criticised this, though (see Section 4.4). The truth is that in many cases there is an adaptive story to tell, but that it is difficult to provide a solid proof of it, due to incomplete information about the historical conditions prevailing during the period when the evolutionary process took place. The best one can do then is to try to separate good from bad adaptive stories – the latter are known as adaptationist fallacies or 'just so stories'.[6] Such a separation should be done based on a maximum use of information, preferably involving multiple theoretical and empirical angles, as discussed in Section 3.5. Unravelling good from bad evolutionary stories requires two steps: the adaptive function of an observed feature needs to be clearly identified; and the biological physiological mechanism underlying this function needs to be reverse engineered. The latter involves clarifying why adaptive selection rather than by-product explains the observed feature.

Adaptation is limited by the environment in which organisms live. Animals could never have developed wheels because the ancient world in which they evolved was characterised by uneven terrain with heterogeneous texture and many barriers, lacking roads or rails on which to easily roll wheels. Under such selection conditions, legs were a more logical solution to allow early amphibians to move (Pinker, 1997, p. 10). From these, reptiles and mammals later evolved. They were limited by the historical or bauplan constraints of amphibians, explaining why legs were here to stay.

Speciation

Another core evolutionary notion is 'species', which refers to a group of organisms that can breed with each other to produce fertile offspring but cannot achieve the same outcome with organisms outside the group. For example, horses, donkeys and zebras are separate species, since their joint offspring (a mule, hinny, zorse, zonkey, etc.) are generally infertile.

This brings us to speciation or the emergence of a new species, 'that mystery of mysteries, the replacement of extinct species by others', as astronomer John Herschel called it in 1836, after having read Charles Lyell's *Principles of Geology*. Repeated selection and variety creation over generations of individuals can lead to speciation,

---

[6] This was inspired by the title of the children's book by Rudyard Kipling, *Just So Stories for Little Children* (1902). It told, fictitiously, how certain animals got their unique features.

especially if environmental conditions change, due to abiotic changes or migration of an animal or plant population. This is a branching process in which populations split off from existing ones and gradually become different.

Various reproductive isolation theories have been developed to explain speciation. Darwin hinted at the relevant mechanisms (see Table 3.2, notably the summaries of his Chapters 11 and 12 on geographical distribution). Reproductive isolation means the creation of barriers between groups of individuals within a population or species, based on the presence of behavioural, morphological or physiological barriers. There are two main categories of reproductive isolation processes, linked to allopatric and sympatric speciation.

Permanent spatial isolation means that subpopulations are spatially separated by water or land barriers, leading to allopatric ('allos'=other, 'patra'= native country) speciation. This is part of the theory of (Island) Biogeography (item v in Section 3.5). Two particular forms are distinguished: peripatric speciation, occurring in meta-populations at the edge of (large) populations occupying a large territory; and parapatric speciation, occurring in large populations where subpopulations are subject to distinct selection environments.

All other isolation mechanisms relate to the same geographic location ('sym'= same), and contribute to sympatric speciation:

- Temporal isolation, which can result from mating occurring in different seasons.
- Ecological isolation, which can occur when subpopulations occupy different niches in an ecosystem (e.g. different food habits).
- Mechanistic isolation, such as coevolution of flowering plants and pollinating insects.
- Behavioural or sexual isolation, which means that even if technically populations belong to the same species (i.e. individuals from different populations can produce fertile offspring), their sexual behaviour prevents sexual intercourse owing to distinct courtship preferences, such as creating or observing particular sounds, colours or smells.

Whereas early on in the evolutionary synthesis, biologists judged allopatric speciation to be more common than sympatric speciation, this is no longer believed. They are probably of equal importance.

Speciation can be linked to adaptation and the adaptive landscape of Wright (1932). The evolutionary search for improvements means that populations divide into local subpopulations, known as demes, which try out different strategies by searching in distinct directions in the adaptive landscape and becoming isolated. As a result, some of them will decrease while others will increase in fitness. It is possible that an initial fitness decrease may, in the long run, turn out to lead to the highest fitness of all subpopulations.

Speciation involves splitting and co-existence of overlapping species. Furthermore, speciation extends over an enormous amount of time and hence cannot be observed. This all contributes to speciation and species being fuzzy concepts, meaning that they are not well bounded. It is impossible to pinpoint the exact time when one species became two species. In this respect, one may note that Darwin realised that the boundary between species and varieties can be vague (see the summary of his Chapter 2 in Table 3.2).

Historical accidents matter to speciation. An example is a small deer-like mammal wading in coastal waters some 50 million years ago, and transforming over the course of millions of years to an animal spending its entire life in the water, namely the first whale, which later evolved into various whale species. That this particular evolutionary history was highly unlikely to unfold explains why the overwhelming majority of sea animals are descended from earlier sea animals. Incidentally, the hippopotamus, a mammal that spends a great deal of its time in water, is the closest living relative of the whale.

A Systems View on Biological Evolution

The previous collection of evolutionary components or mechanisms, resulting in various feedback loops, is illustrated in Figure 3.2. It suggests a rather gradual and microlevel approach, characterised by an endless 'adaptation cycle' involving the steps in the V-S-I-R algorithm. In the next chapter, we will discuss additional components to complete the picture of biological evolution, which will considerably complicate the system of feedbacks.

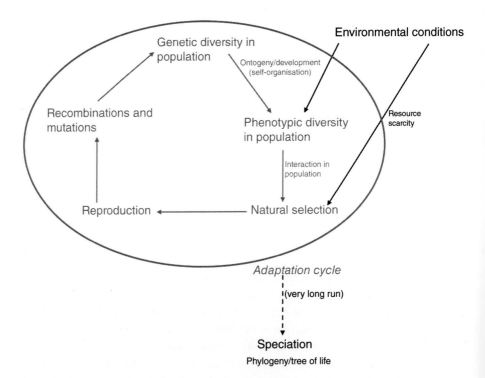

Figure 3.2 Neo-Darwinian evolution as a dynamic system.
(*A black and white version of this figure will appear in some formats. For the colour version, please see the plate section.*)

3.5 Nine Types of Evidence

Biological evolution is now widely accepted in science, as well as in society in general. The reason is that there is a broad range of supportive facts for an evolutionary history of life. The main ones are identified hereafter. Many of these appeared in Darwin's *On the Origin of Species* (1859), as becomes clear when reading Table 3.2. In fact, it has been suggested that Darwin's work is equally significant to evolutionary biology for its theoretical innovation and quality as for its methodological richness and amount of novelties. According to Richmond (2013), Darwin revolutionised the methodology of the life sciences and was the first to rigorously employ a hypothesis-driven research programme, now called the 'hypothetico-deductive method'. His motivation was to maximise use of different types of facts obtained with independent methods to ensure his theory was unlikely to be false.

(i) Morphological Similarities: Analogous, Homologous and Rudimentary Features

Many similarities among certain groups of species have been identified: with regard to anatomy (body plan), physiology, behaviour and embryology. This has served as the basis for species taxonomy. In particular, comparative anatomy and embryology have allowed tracing many gradual, step-wise changes in virtually all physiological character-istics of plants and animals. A key distinction is between analogous and homologous characteristics. The latter denote organs or other features that are related through common descent but possibly different in function and form – such as wings of birds and forelimbs of mammals. In contrast, analogous organs perform the same function but have different origins. Examples are wings of birds, bats and insects, and eyes of mammals and octopuses. Such analogous features are the result of convergent evolution, which is due to species having lived for a long time in similar environments and thus having been subject to similar selection pressures. Analogies can go beyond features and organs and even pertain to entire organisms. For instance, sharks are descended from a fish species whereas whales are descended from a mammal species, but they are much alike. Both homologous and analogous characteristics provide support for a process of biological evolution. A homologous character points at a common history or ancestor. Analogous characteristics reflect that similar problems often invite for similar solutions, as these work the best or are the easiest to reach through an evolutionary process.

In addition, we see organs or body parts that evolved for a certain function, which later was partly or completely lost after the original selection pressure was absent for a long time. What is often left then are so-called rudimentary or vestigial attributes, a special case of homologous characteristics. Examples abound in all species, as this is an integral part of evolution: rudimentary hind legs in whales, tiny skin-covered eyes in blind mole rats, and rudimentary wings in island-dwelling and other flightless birds. In humans, well-known examples are the tailbone, body hair, wisdom teeth, muscles of the external ear, the appendix (connected to the caecum, which in herbivores serves to break down plant cellulose as part of digestion); and goose bumps which in human ancestors were meant to raise the body's hair in stress situations, to look more impressive.

(ii)   ## Parallels Between Ontogeny and Phylogeny

Ernst Haeckel (1834–1919) argued that ontogeny – i.e. the physical, embryonic development of the individual – is a concise and compressed replay of phylogeny – i.e. the ancestral sequence of species, often captured by the image of a 'tree of life' (Figure 3.3). The development of an organism can be regarded as climbing up its family tree, variously referred to as the 'ontogeny–phylogeny analogy', 'recapitulation theory' or 'biogenetic law' (Ridley 1997, p. 228). This explains why species closer to each other resemble one another for a longer time during embryonic development. Haeckel was the first, in 1892, to illustrate this point well with drawings of fish, salamander, tortoise, chick, pig, calf, rabbit and human embryos. These are now considered to exaggerate the similarities, as illustrated by a comparison undertaken using photos by Richardson et al. (1998). Nevertheless, these authors conclude that Haeckel was basically correct in the sense that all vertebrates develop a similar body plan which reflects a shared evolutionary history. This is consistent with the fact that overlapping genetic information steers their development. Add to this that embryos in an early stage are rather similar across many species, while embryos of closely related species resemble one another until a late phase of embryonic development. In other words, differences appear relatively late, as they did in evolutionary history.

Molecular science provides support for the idea that in similar stages of ontology and phylogeny similar proteins are active: simpler molecules appear in early ontogeny and phylogeny and provide the necessary basis for more complex molecules in later stages. In line with this, ontogenesis is characterised by increasingly complex organs and further division of labour within living organisms. The combination of developmental and molecular biology thus provides additional support for an evolutionary history of molecular processes underlying organ and tissue development, chemical interactions among cells (signalling) in tissue development, and the processes that stimulate cells to divide, die or become of one or another (specialised) type. It makes sense that ontogeny, an unfolding in chemical space, follows a path from simple to complex, like phylogeny or natural history. It would indeed be surprising if these patterns were extremely different, since ontogeny is the result of an accumulation of adaptations in evolutionary history, so is likely to reflect that history. A complete lack of similarity between ontogeny and phylogeny would suggest independence of evolution and ontogeny, which would challenge evolutionary theory.

We can see ontogenetic development occurring over time – as any parent can testify. The complexity of the human body is clear through the sheer numbers involved: the human fertilised egg (zygote) undergoes about 50 cell divisions to create about 1 quadrillion cells in a new-born infant, forming about 260 cell types in the body in the process. This involves an extremely complex network that could only have developed and be stable over time due to an equally complex supporting network of genes, RNA, proteins and other molecules. We now know that individual development is a complex chemical process. It is not vague or mystic, it happens following a clear physical, chemical and spatial (three-dimensional) logic. The problem with phylogenetic evolution is that, unlike ontogeny, it is hidden in the sense that humans cannot observe it during their lifetime. If one can accept that the complexity of an organism is the result of its

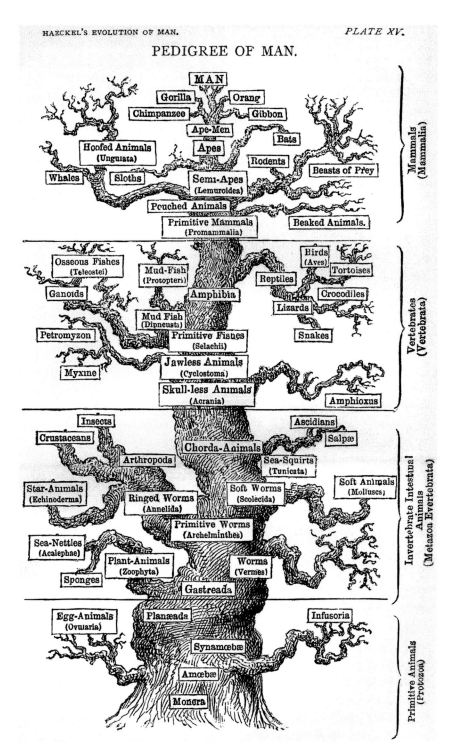

Figure 3.3 Haeckel's tree of life.
Source: Haeckel (1879).

ontogenetic development from a single cell containing the genetic material, one should be able to accept that a temporally connected set of competing chemical processes, known as spontaneous evolution, was the cause of complex living organisms and their ontogeny.

(iii)   Palaeontology and the Fossil Record

Since Darwin's days, an important area of support for evolution has been palaeontology, the study of life in natural history as reflected in the fossil record. It is in turn supported by geology, which sketches an environmental-geological historical context and offers several methods to assess the age of fossils. Frequent discovery of fossils initially posed many problems for religious beliefs. It suggested that the Earth was about 4.5 billion years old, much older than literal interpretations of biblical statements suggested. Palaeontology thus contributed to the understanding of the long-term horizon involved in evolution. Second, it was shown that many species have become extinct and new ones have appeared during historical times, posing a challenge to the view that species were fixed. In fact, it turned out that species extinction is as common as speciation. The fossil record conveys that over 99 per cent of all species that once lived on Earth did become extinct. Notice, however, that extinction includes both dying of all individuals of a species and evolution of a species into a new species – or more correctly, splitting into two or more new species.

Another implication of the fossil record is that many major, though certainly not all, gaps between species are filled, giving support to the idea of the continuity of evolution – not to be confused with gradualism. Referring to such gaps as missing links in the fossil record has been a popular means of attack of evolutionary theory, notably by creationists (Section 2.4). It is, of course, impossible to recover every single species that ever existed in the fossil record. Still, most major connections have been solved, for example, the links between (ancient) fish and amphibians and between reptiles and mammals. A classic transition, between dinosaurs and birds, as recognised already by Darwin, was the *Archaeopteryx* – in German language 'Urvogel' (= prehistoric or original bird) – found in 1861 in Bavaria. Whereas gaps in the fossil record do not disprove evolution, fossils that show clear inconsistency with an evolutionary explanation of life's history would. However, such fossils have not been discovered. Instead, the huge amount of fossils found, representing a large number of species, is completely in line with everything we understand about evolution as well as the other categories of evidence reported in the current section.

Finally, palaeontology supports the idea that evolution is an ongoing process, slowly but continuously changing species. If one goes back in time long enough, currently existing species are not found, while in their place predecessor species with different features are encountered. For instance, fossils of modern humans (*Homo sapiens sapiens*) from up to 100 000 years ago have been found, but such fossils do not appear when going back more than 1 million years. Instead, one only finds fossils of predecessors of humans.

(iv)   Molecular Genetic Relationships Among Species

Molecular analysis of chromosomes goes beyond mere morphological (phenotype) similarities. It finds genetic similarities and historical connections across species. The most fundamental finding is that all living species – bacteria, animals, plants, etc. – have a common genetic code, suggesting that all current life forms have a common origin. Collection and analysis of enormous amounts of data shows consistency with genetic and evolutionary theory, as well as with palaeontological data from the fossil record. In general, the closer (more distant) species are the more similar (distinct) their DNA is. For instance, the genetic similarities among apes and humans suggest common roots; and various human diseases that are sex-related and linked to the X-chromosome have identical positions on X-chromosomes of many other mammalian species as well, suggesting a common ancestor. Examples of this type abound, and accumulate as more and more scientists generate molecular data for a wide range of species. The massive study of the human genome is the most recent development in this direction. (See further discussion on chemical evolution in the next section.)

(v)   Biogeography and Spatial Regularities

When considering the geographical distribution of species, one can observe spatial regularities and similarities, variation and even unique species in certain locations. Biogeography, the study of the spatial distribution of living organisms, offers an explanation for this. It covers the distribution of species and ecosystems in space, as well as the long-run dynamics of these (i.e. over geological time). Central to its approach is the recognition of evolutionary effects of isolation (with islands as an extreme case), latitude, elevation (mountains), and the associated gradients in climates, on habitats, ecosystems and species.

A basic idea in biogeography is that groups of organisms of a species can become isolated, and then evolve differently from the original species, owing to founder effects, which ultimately can give rise to a new species. Isolation can occur through a number of factors, such as mountain ranges, rivers and seas (think of islands). In developing his thoughts along these lines, Darwin was inspired by the unique flora and fauna he discovered on remote islands, notably the Galapagos Islands. Spatial isolation can already be observed on a small geographical scale: rivers isolate areas, so that some species may be different on either side of a river (e.g. squirrels in the Grand Canyon; or monkeys in the Amazon area). On the other hand, rivers do not create a barrier for birds, which is in line with rivers not having been found to separate bird species. Isolation in combination with favourable climatic conditions also explains why one of the largest islands on the planet, Madagascar, is known for hosting such a great amount of unique biodiversity, and why Luzon, the largest island in the Philippines, has the highest number of endemic mammal species in the world (52 endemic out of a total of 56 mammal species).

Because of a mixture of isolation and random genetic drift, founder effects – also known as founding accidents or population bottlenecks – can appear. This denotes that species which have originated from a small number of individuals and become isolated, usually spatially, are sensitive to random genetic drift and non-random mating, like inbreeding or selection of mates with specific phenotypes. Consequently, the resulting subpopulations can respond quickly to environmental changes, causing them to possess considerable evolutionary – genetic rather than phenotypic – adaptive capacity in the face of strong selection pressure. For founder effects to be significant, a combination is needed of spatial isolation and environmental change (or an environment different from the place of origin). Founder effects have probably been essential to speciation over the course of evolutionary history of life on Earth.

The sometimes peculiar geography of evolution is illustrated well by the adaptive radiation of marsupial mammals, such as kangaroos, wallabies, koalas and possums, in Australia. Marsupials carry their young in a pouch, and are the main category of mammals next to placental mammals. Because larger placental mammals had become extinct in Australia more than 50 million years ago (the reason is still not known), marsupials could prosper and diversify, filling many different niches and showing various cases of convergent evolution with placental mammals elsewhere. The only mammals that survived the expansion of marsupials were the echidnas (spiny anteaters) and duck-billed platypus. The reason is likely that they lived in ecological niches that marsupials, with their particular reproductive strategy, were unable to occupy. These species are the only current egg-laying mammals, while the platypus is one of the few mammals that produce venom.

There is also a clear spatial pattern of ecosystems that consist of species remarkably adapted to one another as well as to their abiotic environment. Another remarkable spatial regularity is the distinct evolution of monkeys in the Old and New Worlds (Africa and the Americas). In addition, extinction patterns have been found to differ among continents, partly due to climatic events and human presence, associated with hunting and deforestation pressures. For example, human origins in Africa caused animal species here to be better adapted to – due to having coevolved with – humans than species on other continents. There humans arrived later in the evolutionary history so that species were not adapted to them and could be more easily hunted to extinction.

Biogeography has a long history in biology, going back at least to the works by Carl Linnaeus, Alexander von Humboldt, Charles Darwin and Alfred Russel Wallace. The latter was responsible for identifying the faunal divide, now known as the Wallace Line. This is the separation in the Indonesian archipelago between a western region, with animals mainly being of Asian ancestry, and an eastern region, with animals having more Australasian origins. Biogeography, with its focus on geological time, is able to connect evolution to the history of continental drift (plate tectonics), clarifying why some species are related even though they are currently divided by oceans. It is further able to clarify spatial regularities including the demarcation of biogeographical regions within each continent. Robert MacArthur and Edward O. Wilson's *The Theory of Island Biogeography* (1967) is a classic book that stimulated systematic research leading to the sub-discipline of biogeography.

(vi) **Artificial Selection**

Humans have, for a long time, practised artificial selection or selective breeding of traits in domesticated animals, farm crops and flowers. Darwin recognised that its study could offer fundamental insights to evolutionary theory – witness the summary of Chapter 1 of his *Origin* in Table 3.2. For instance, the enormous variation in the races of dogs – i.e. populations of a species that are quite different in appearance from each other – indicates the propensity for artificially selecting phenotypic variation. Dogs are descended from one species, namely wolves, which were domesticated by humans at least 14 000 years ago (Section 11.5). Many other domesticated animals have been altered (see Table 11.2), but the physical differences are less obvious since selection was guided by utilitarian rather than aesthetic objectives and was started later (about 9000 years ago or afterwards). In addition, many plant crops have been bred since the rise of agriculture (Section 11.6). Some of these even derive from a single wild species: e.g. artificial selection of wild mustard has, due to different selection criteria, produced cabbage (through suppression of intermodal length), broccoli (suppression of flower development) and cauliflower (creating sterility of flowers). More recently, laboratory experiments with fruit flies have shown that it is possible to produce a new species through artificial selection.

(vii) **Observable Evolution**

Natural selection can be observed in some cases. The best examples are microorganisms causing a variety of diseases and insects responding to actions such as pesticide use in agriculture by humans. These cause selection and adaptation processes that are rather quick and therefore can be observed. Another example is air pollution during early industrialisation in England, which put a black layer of soot on buildings and trees. This caused selection pressure against the white peppered moth: it lost its camouflage and became more susceptible to predation by birds. In turn, a black variant became dominant. Section 13.3 offers details.

We can also observe speciation progressing, meaning that groups of one species are on the verge of becoming separate species. Often this means they may still mate and produce fertile offspring, but they do not do this because of visual differences (colours) or behavioural strategies. An example is cichlids in Lake Victoria in Africa, shared by Tanzania, Kenya and Uganda. These are small percoid fish that have formed hundreds of new species at a comparably high rate over the past 10 000 to 15 000 years. The probable cause of this is climate change variation, which caused the water level of the lake to drop, leading to transformation of the lake into multiple smaller, isolated lakes. The original fish population (cichlids) thus became divided among these sub-lakes giving rise to spatial separation for several thousands of years. Hence, populations of the same fish species could slowly evolve in different directions, causing adaptive radiation, i.e. a multitude of branching processes. When the lakes were merged again after climate conditions altered and water levels rose, populations of similar fish often did not 'integrate', due to evolved physical or behavioural differences.[7]

(viii)   Experiments in Laboratories

Since the early 1970s, laboratories have experimented with genetic mutations, recombinant DNA techniques and selection. This has, over the course of only two decades, given rise to a standard set of techniques known as 'genetic engineering'. Species such as quickly reproducing fruit flies and bacteria (notably the gut bacteria *Escherichia coli*) have been subject to many selection pressures over many generations, leading to considerable changes in their phenotype. One may now consider genetic engineering or modification of organisms, despite initial resistance, widely applied in agriculture and medicine, as an addition to the tools of artificial selection, even though strictly seen it is closer to the mutation/variation than selection mechanism of evolution. The rapid development and use of such techniques and processes has generated much debate about whether to accept, strictly regulate or stop the development of modern biotechnology (for an overview of arguments, see van den Bergh and Holley, 2002). Adversaries of such techniques tend to use the pejorative term 'genetic manipulation' and might claim that such techniques come down to playing god, whereas the traditional breeding was merely 'playing Mendel'. Genetically modified organisms illustrate that the functional value of evolutionary and molecular biology and genetics is beyond any doubt – merely providing additional support for the theory. Experiments have gone beyond individuals and single species. They range from the testing of group, community and ecosystem selection processes (see Section 6.4) to the study of ecological interactions emerging among bacterial populations during an almost 30-year long experiment, involving some 60 000 generations (Plotkin, 2017).

(ix)   Mathematical Models and Numerical Computer Experiments

Modelling of evolution, theoretically/analytically or by using digital computers for numerical experiments, allows for the calculation of the effects of combining various evolutionary mechanisms. In this sense, unique experiments can be undertaken over many artificial generations, the equivalent of which is impossible under real-world or laboratory conditions (Nowak, 2006). The results show various interesting phenomena that characterise evolution in reality, such as diffusion, spatial regularity, trade-offs between distinct features, path dependence, environmental adaptation, coevolution and

---

[7] This pattern of evolution was roughly disturbed by intentional introduction in the late 1950s of the non-endemic Nile perch in the lake from the Ugandan side. This invasive species diffused rapidly which benefitted local fishing industries, which boomed in the 1980s. In the process, the Nile perch exterminated a large number of evolutionarily unprepared cichlid species. This story is well told by Tijs Goldschmidt (1996). Recently, it was found that current evolutionary change is probably affected by the lake's water quality, which has been lowered by agricultural run-off and debris from deforestation. In cloudy water parts, hybridisation of semi-species is occurring because the sexual selection that kept them separate is no longer effective. Worsened light conditions weaken female choosiness, previously triggered by males' colours.

emergence of complexity. Lipson and Pollack (2000) report the results of a study that, starting from basic building blocks, evolved simple electromechanical systems through computer simulations, and coupled its outcomes to automatic, robotic fabrication to test if the winning design actually worked. See Box 2.2 for an illustration of modelling the evolution of an eye.

Taken together, the previous nine categories of evidence create overwhelming support for an evolutionary history of life. Sceptics of biological evolution should note that, despite so many studies having been undertaken in these nine areas of investigation, no factual evidence against an evolutionary history of life on Earth has been found.

Two other major issues that have to be resolved to arrive at a complete theoretical-empirical picture of evolution are the major transitions in natural history, addressed in the next section, and the origin of life, dealt with in Section 4.5.

3.6　　Major Transitions in Natural History Along Three Dimensions

The classification of organisms can be done in many ways. In an evolutionary classification, special consideration is given to evolutionary (genealogical) relationships and evolutionary distance, rather than to morphological relationships. Thus groups can consist of morphologically quite different species, which is uncommon in non-evolutionary classification schemes, such as originated by Linnaeus (1707–1778). Evolution recognises that species branch and that all species change. Names should reflect this. Mammals and birds, for instance, descended from reptile and dinosaur species, respectively. This means that they cannot be regarded as groups (classes) at the same level as reptiles, as this would suggest that the two derive from a single pre-reptilian species. This purist approach to classification based on branching is also known as 'cladistics'. Taxonomic categories that have been maintained since the first systematic classification by Linnaeus include a downward hierarchy: kingdom, phylum, class, order, family, genus and species. In many cases, this hierarchy is consistent with the branching genealogy of the tree of life.

A concise overview of natural history that forms the foundation for systematics is based on knowledge obtained through the fossil record, genetics and molecular research. There are three ways to summarise natural history, namely in terms of main groups of species appearing over time, in terms of major evolutionary innovations generated, and in terms of the abstract information communication nature of such innovations. We consider each of these below.

Main Groups of Species Appearing Over Time

A logical starting point in time is 4.6 billion years ago, the approximate age of the Earth. Most of the time since then, from 3.5 to 0.6 billion years ago, was dominated by unicellular life. A sub-period from 3.5 to 1.5 billion years ago was the age of simple prokaryotes with a single chromosome, which included Bacteria (previously called

Eubacteria) and Archaea (previously called Archaebacteria, even though they are not bacteria). Evolution was rather slow owing to the small size of the genome, a limited capacity for recombination (absence of sexual organisms), and a low concentration of oxygen in the atmosphere that limited energy-rich processes. Aggregation of cells was possible but complex body plans did not evolve. Archaea still make up about 20 per cent of microbial plankton in the world's oceans.

A major transition was set in motion 2.3 billion years ago when oxygen started to accumulate in the atmosphere due to cyanobacteria – also known as blue-green algae, even though they are not algae – which produced oxygen through photosynthesis. Cyanobacteria are still critical for global carbon and nitrogen cycles. They can be found in habitats with extreme conditions, such as underwater volcanoes, salty lakes, hot springs, or extremely hot or cold places. The 'oxygen revolution' or 'great oxidation event' triggered by cyanobacteria reduced the number of anaerobic organisms and ultimately created the basic conditions for more advanced eukaryotic cells, which emerged about 2 million years ago, to radiate about 1.5 billion years ago. Eukaryotes differ from prokaryotes in that they do not have a rigid outer cell wall but have an amoeba-like structure allowing for motion, a complex system of internal membranes, and a nucleus containing the genetic material contributing to more accurate cell division (mitosis) and genetic replication. It is now hypothesised that a cell nucleus was formed through repeated replication and evolution of viruses within bacteria, known as 'viral eukaryogenesis'.

The probable starting point for eukaryotes is the loss of the outer cell wall in prokaryotes, which allowed the cell to incorporate particles and other, smaller organisms (Maynard Smith and Szathmáry, 1995, p. 122). This in turn was followed by symbiosis which gave rise to inclusion of certain organelles. Endosymbiosis (or symbiogenesis), between Archaea and Bacteria species led to cells like Protista, amoebae, and photosynthetic algae with specialised internal bodies (organelles or plastids), such as mitochondria – oxygen-based energy factories that store energy in adenosine triphosphate (ATP), the 'energy currency' of life – and chloroplasts – responsible for photosynthesis. As a result, eukaryote cells are larger and can better accumulate energy in sugars than prokaryotes. All fungi, plants and animals consist of eukaryotic cells. One may wonder why it took evolution so long, about 2 billion years, to move from the prokaryote to the eukaryote stage. One explanation is that many conditions had to be fulfilled to make the new eukaryotes competitive with the old prokaryotes, notably to compensate for any fitness loss due to the lack of a rigid cell wall.

A subsequent major transition occurred 0.6 to 0.7 billion years ago, when larger, soft-bodied multicellular animals appeared. Multicellular plants arrived somewhat later, probably following the existence of colonial and terrestrial (wetland) algal organisms. The main advantage of multicellular organisms is that they allow for specialised cells and a division of labour, enabling the formation of more complex structures and functions. An important event in the radiation of multicellular life forms is the 'Cambrian explosion' a little more than 500 million years ago. Here, all more advanced phyla of multicellular animals with complex body plans (Metazoans) appeared, including annelids (worms), arthropods (animals with segmented bodies, including, among many others, arachnids and insects), brachiopods (shelled animals), chordates (vertebrates and other species with a predecessor that had a spinal column, such as amphibians, reptiles, birds, fish, mammals), echinoderms (radially symmetrical marine animals like starfish, sea-cucumbers and sea-lilies) and molluscs (invertebrates without segments or limbs, usually secreting a shell).

The complex body plans involved many cell types. The so-called Hox gene (Section 4.4), containing an ancient DNA sequence involved in anatomical development regulation known as a 'homeobox', is essential in this context. It controls the body plan through advanced hierarchical genetic control, i.e. genes controlling other genes. This explains why few genetic changes can produce considerable variety of phenotypes, as happened during the Cambrian period. Other factors played a role as well. Life in the period before the Cambrian explosion had created the right conditions, notably a higher concentration of oxygen in the atmosphere and subsequently in the oceans, allowing for larger animals. In addition, the concentration of calcium in oceans had increased due to marine volcanoes and enhanced weathering of rocks on land under the influence of mosses. This allowed multicellular marine organisms to evolve skeletons, hard body part and shells.

The first eyes appeared about 540 million years ago in *Redlichia* (a type of trilobite), early on during the Cambrian explosion. These eyes were similar to those of modern insects, i.e. with a compound design. Eyes ultimately changed the living world: movement became easier and quicker, causing competition and fights to become fiercer. This contributed to an evolutionary arms race allowing evolution to explore new morphological and behavioural adaptations. Possibly, the evolution of eyes was another factor contributing to the Cambrian explosion of new life forms. But vision through eyes is not a universal feature of multicellular animals. Only 6 of a total of 37 phyla have evolved it. They are, nevertheless, the most abundant and widespread animals on Earth, accounting for about 96 per cent of all living species (see Box 2.2).

Some 500 million years ago, vertebrates arrived and 100 million years later the first amphibians and insects. About 360 million years ago, reptiles, seed-carrying and vascular plants appeared, while extensive forests became common. About 245 million years ago, dinosaurs could be found, 200 million years ago birds and mammals (insect eaters) appeared.

About 470 million years ago, land plants appeared in greater numbers. They evolved from green algae, probably at the edges of shallow waters. The dating is uncertain – estimates range from 500 to 600 million years ago. This transition was difficult as the conditions on land were harsh in comparison with those in waters. Some 100 million years later, currently standard elements of plants such as roots, leaves and, somewhat later, seeds evolved. About 140 million years ago, flowering plants appeared. Once vascular systems had emerged, flowering plants and especially conifers (gymnosperms), unlike mosses and ferns, could resist gravity and grow higher, which allowed the evolution of terrestrial forest ecosystems.

Another major transition occurred 90 million years ago with the first mammals making their appearance. This was a consequence of a virus infection which entered genetic information in the animal genome allowing the formation of the placenta.

Some 65 million years ago, many groups of organisms become extinct, including dinosaurs, owing to a major environmental disturbance (see Section 4.4). Only small animals were able to adapt to the strongly altered environmental conditions. This explains major radiations of mammals and birds. Indeed, among the dinosaurs, only the feathered ones that were small, a necessary condition for flying, survived (Balter, 2014). In addition, owing to the evolution of flowering plants, pollinating insects radiated.

So, we see about 65 million years ago a whole range of archaic primates and mouse-sized animals evolving out of certain insectivores, distantly related to bats and rodents. A main new feature was a set of teeth that were suitable for eating leaves and fruits. Only 50 million years ago, the earliest true primates appeared on the scene, followed 5 million years ago by ape-like ancestors of humans, and in the last few hundred thousand years modern humans. Section 11.2 provides more detail on the primate–human lineage.

Main Evolutionary Innovations

The previous chronological account focusing on species can be complemented by listing major innovations, regardless of the species or time order in which they occurred:

- *Genetic code*: General structure suitable for specifying 20 amino acids. Fixed in the ancestral line. Largely defines phenotype and allows for accurate replication.
- *Cell*: Controlled environment for complex chemical reactions and energy accumulation; prolongs existence, allows increasing complexity and makes replication easier. The organelles in a cell are the functional analogue of organs in a multicellular body.
- *Photosynthesis*: Taps into a new energy source and produces oxygen.
- *Sexual organisms*: Allows for exchange and combination of genetic material between two organisms. It has contributed to increased variability and hence more rapid evolutionary change.
- *Cell differentiation and multicellular life (Metazoans)*: Allows for complex structures and labour division (organs). Its advantage is shown by the fact that it has developed independently in unrelated groups of unicellular species (bacteria, algae, amoebae, etc.), in the kingdoms of plants, animals and fungi. Note that fungi are more closely related to animals than to plants. Indeed, the two are classified as opisthokonts, characterised by flagellate cells, such as certain animal sperm and fungal spores, moving via a posterior flagellum.
- *Organs*: An organ or viscus is a structured collection of cells or tissues that delivers a useful function for the body of which it is part. Not all multicellular organisms have organs. For example, some sponges do not show any specialisation at the level of cells or groups of cells. Most fish, reptiles, amphibians, birds and mammals have about 80 types of organs divided into more than 10 organ systems, such as digestive, urinary, circulatory, nervous, endocrine, immune, respiratory, muscular, skeletal, skin and reproductive systems. These same organ

systems are even found in insects, although their specific organs are quite different. Plants have few organ systems, namely roots, stems, leaves and reproductive parts. Some animals have unique organs, such as silk-producing glands in spiders. Organs allow specialisation and labour division within the body, which means more efficiency in performing associated processes or developing new capabilities.

- *Endocrine system*: Chemical communication for internal regulation and homeostasis. This involves special organs that produce hormones which then enter the bloodstream and contact specific cell receptors.
- *Vertebrates*: Development of a complex neural system linking all parts of the body with the brain as the coordinating unit. This in turn allowed for the evolution of more specialised organs.
- *Animals*: Although there are many differences between animals, they are all heterotrophic, meaning they take their energy not from light but from organic and possibly nonorganic molecules, many of which are being produced by autotrophs (plants). In higher animals other distinctive features have developed, such as subtle ways of movement and complex nerve systems. The capacity for mobility exists in various ways in various organisms, in water, on land, in the air, involving moving hairs, peristaltic movements, walking, flying, etc. This allows a wider array of food opportunities, fleeing from danger (environmental threats or predators), and a broader range of potential mating partners.
- *Transition from water to land*: Tetrapods (four-limbed vertebrates), covering amphibians, reptiles, birds and mammals, evolved fewer than 400 million years ago from ancient fish-type animals, through lungfish to the first amphibians. This transition was preceded by the emergence of early lungs. This has been attributed to two factors. First, heavy fish, due to bones and scaled skins, needed uplifting power (buoyancy) in the form of a lung-cum-swim bladder. Second, as the bladder was open to the gut, it could help to provide oxygen intake to satisfy the energy needs of active and large fish. The transition likely happened in shallow wetlands, as here oxygen levels in the water were sometimes deficient, while those in the air were relatively high and stable. This might have been stimulated by the documented rise in the general concentration of atmospheric oxygen in the preceding period. Overpopulation of waters and local droughts have also been suggested as relevant factors. Another factor has been formation of the ozone layer due to oxygen (O_2) molecules splitting under influence of solar radiation into single O molecules and then combining with O_2 to form ozone (O_3). As the latter absorbs ultraviolet rays, which cause a high rate of mutation in genetic material, the thin layer of ozone at high altitude in the Earth's atmosphere acts as a protecting shield against life-threatening radiation. The formation of this shield is believed to have occurred about 600 million years ago. It not only enabled organisms to live on the land, but also may earlier have created favourable conditions for the above-mentioned Cambrian explosion of life forms in the seas.

- *Transition from land to water*: Some species evolved back from land to water, giving rise to cetaceans – whales, dolphins and porpoises. These are the outcome of convergent evolution with fish and water reptiles. While fish swim by moving their body sideways, cetaceans do it by moving it in vertical directions, just like the land animals from which they are descended.

- *Development of senses*: Senses like seeing, smelling, hearing, tasting and touching allow the identification of food, potential partners and danger. The ability to see, beginning with the more primitive capability of identifying light and shadow, has an enormous selective advantage (see Box 2.2).

- *Cellulose in plant cell walls*: Creates stable, more protected structures and allows for larger organisms (trees).

- *Vascular plants*: Better access to water and nutrients; allowing the colonisation of dry and nutrient-poor areas. Vascular plants ultimately gave rise to complex ecosystems, with many types of vegetation and large animals, in many different climate zones.

- *Seeds, flowers and fruits*: The evolution of seeds in gymnosperms meant an improvement over spores as produced by mosses and ferns. Seeds allow for sexual recombination of the genetic material of the two parents, are protected by a hard coat and contain nutrition for the embryonic plant. The evolution of flowering plants meant a change from more primitive plants – gymnosperm with naked seeds – to angiosperm with seeds inside a protective space, as in fruits produced by flowering plants. Fruits could evolve as an adaptation because they served well the functions of not only protecting but also stimulating the dispersal of seeds. Animals would eat the fruit and when defecating expel the seeds without any damage, as well as automatically fertilising them. A group of species producing a wide variety of fruits is the rose family (Rosaceae), yielding fruits such as almonds, apples, apricots, blackberries, cherries, peaches, pears, plums, raspberries, rosehips and strawberries.

- *Diversity of reproduction mechanisms in plants*: Groups of species such as mosses, ferns, conifers and flowering plants show distinct reproduction mechanisms. The flowering plants came last but are now dominant, largely owing to their more efficient reproduction relative to other plants. This is largely because of their close ecological relationship and resulting coevolution with pollinating insects. Flowering plants provide nearly all food, natural fibres and medication for human purposes.

- *Placental and marsupial mammals*: Care for young led to more advanced behaviour and ultimately gave rise to complex social behaviour.

- *Warm-bloodedness (homeothermy)*: This allows mammals and birds to maintain a body temperature higher than their environment, through regulating metabolic processes and insulation (fat, fur, feathers). Warm-blooded animals can live in colder regions. As amphibians, reptiles and insects are highly susceptible to fungal infections, and few fungi survive the body temperatures of homeothermic animals, this has been speculated to be a main advantage explaining the evolution of homeothermy.

- *Regulatory genes for controlling development*: Allow for control of rate, timing and three-dimensional pattern of changes in the form of an organism. The Hox gene (Section 4.4) is perhaps the best-known example: it controls the body plan of an embryo along the head–tail axis, resulting in particular segment form and function, such as antennae, legs and arms, or wings.
- *Social groups*: This occurred only in certain species, especially in ants, fish, birds and mammals (E.O. Wilson 1975, Chapter 20). It involves fighting for territory, cooperation, sharing, division of labour among organisms, and vertical and horizontal learning.

Fundamental Changes in Natural History

At an abstract level, one can identify four fundamental changes over the course of natural history (the first three are mentioned also by Maynard Smith and Szathmáry, 1995, Section 15.2):

1. *Division of labour*: A process that breaks down a function into a set of specialised tasks, leading to an increase of efficiency in performing the function. Efficiency usually relates to energy use, but can generally relate to anything that is scarce or limited. Important examples of an increased division of labour in the living world over evolutionary history are: cells with many specialised organelles, multicellular organisms with dedicated body parts or organs, sexual organisms with labour division between the sexes, ecosystem and complementary multispecies relationships (e.g. mutualism), and social groups with internal task division. Note that labour division reflects the use of an economic metaphor, as recognised already by Darwin (see Section 2.2).

2. *New ways of transmitting information, both phenotypic and genotypic (replicative)*: This involves the development of interaction systems among cells in multicellular organisms (hormones and nerve systems), the development of the various senses (sound, smell, vision), the development of social interaction, and ultimately the development of symbols and language by humans. Information transmission is, of course, related to division of labour since the latter requires communication. Indeed, advanced labour division requires an appropriately advanced, associated communication system.

3. *Emergence of new levels of meaning and selection*: From the unit of heredity, the gene (a sequence of base pairs in a DNA molecule), to cell to multicellular (organisms) to group (social systems). Individual selection means selection at the level of the gene, cell or multicellular organism. Multicellular organisms generally do not show a conflict between selection at the cell and organism levels, since all cells derive from a single cell so that the cells are genetically identical (apart from somatic mutations). Group and individual selection can conflict, though, which has given rise to an ongoing debate about the relevance of group and associated multilevel selection (Chapter 6).

4.   *Access to new energy sources (heat, chemical energy, solar light or wind), and new ways of capturing and transforming energy*: Evolution is in essence the harvesting of energy by life forms to maintain metabolic processes and reproduce. The main event in this respect was the 'oxygen revolution' triggered by cyanobacteria, which allowed for a large-scale shift to energy-rich life processes. The basis of this was photosynthesis. In addition, integration of mitochondria, the energy factories, in eukaryotic cells represented a major transition in terms of energy availability. Smaller 'energy transitions' involve the use of wind energy through gliding and flying, or cold-blooded reptiles' control of skin colour to capture more solar heat.[8] In addition, evolution of new predating or parasitic strategies can be seen as improving access to energy for the respective organisms.

An intriguing question is 'What can explain the major transitions?' Is one mechanism sufficient or are multiple mechanisms needed? Gradual microevolution and non-gradual macroevolution both played their role (Section 4.4). Maynard Smith and Szathmáry (1995) argue that there are three ways in which genetic complexity may increase. First, duplication and divergence, which is the way eukaryotes replicate. This probably has had no direct impact on the major transitions. A second way is symbiosis, which may have been crucial for the evolution of chromosomes out of separate genes and the eukaryotic cells with organelles. A third way is epigenesist, that is to say, inheritance beyond genes or DNA (Section 4.2). This denotes that entire chromosomes, including DNA and proteins that control gene activity, can be inherited by daughter cells. Knowledge about the latter is needed to fully understand cell development and specialisation.

In addition to the above information communication issues, two other information-theoretical aspects of evolution are interesting. One is that evolution relates to the ability to reproduce, at any level: molecule, chromosomes, cell, organism and ecosystem. A set of minimum elements of a self-reproducing system is as follows (Ayres, 1994, p. 93):

- A blueprint: storing information; in living entities this is DNA (deoxyribonucleic acid).
- A factory: manufacturing products; in living entities this operates through DNA transferring some information to messenger RNA, which is read by a ribosome using amino acids as inputs.
- A controller: supervising the factory's operations; certain enzymes fulfil this function in living entities.
- A duplicating machine: to copy the blueprint; in living entities this is DNA replication.

According to Ayres (1994, pp. 93–94), biological evolution as we understand it is based on information flowing from DNA to proteins. Information flowing in the opposite

---

[8] For more discussion, see this special journal issue of *Philosophical Transactions of the Royal Society B*: http://rstb.royalsocietypublishing.org/content/368/1622/20120253.

direction, from proteins back to DNA, is not possible. This would be an instance of Lamarckian type of inheriting acquired characteristics, i.e. a learning machine, which translates phenotype adaptations, reflected by the presence of `particular proteins, to genotype changes. However, as we will see in the next chapter (Section 4.2), some studies suggest there are exceptions to this rule.

A second general information-theoretical aspect of genetic evolution is dominance of vertical transmission (within species), such as transferring genetic material from parents to offspring. As we will see later, horizontal and oblique transmission are relevant to social-economic evolution, distinguishing them from genetic biological evolution. In this respect, Lovelock (2014) argues that while biological evolution is mainly about harvesting energy, the emergence of human culture meant a transition to a focusing on harvesting information, which was strengthened by the Industrial Revolution. This is not to say that horizontal information transmission is completely absent in natural systems. Indeed, bacteria are sometimes involved in a process of genetic exchange and recombination.

4 Advanced Ideas in Evolutionary Biology and Genetics

4.1 Beyond Neo-Darwinism

After the brief historical context and basic introduction, it is time to delve a little deeper into the fascinating field of evolutionary biology. This involves elaborating certain aspects of evolution, such as molecular processes in the theory of neutral genetic drift, and broadening the set of evolutionary mechanisms, by characterising symbiotic evolution and epigenetic inheritance. Some other ideas partly contrast with standard theory, notably non-random mutations, and, to a lesser extent, group selection and macroevolution. A rudimentary understanding of all these theories creates a fertile basis for transposing ideas and insights to the social sciences.

In addition, we consider here early chemical evolution and the origin of life in some detail, as they complete the picture of evolving organic life. Some would claim that the origin discussion does not belong to evolutionary biology. Strictly seen, it indeed precedes organic evolution. On the other hand, the origin of life is about chemical evolution which is subject to the same generalised evolutionary mechanisms of diversity, selection, innovation and inheritance, operating at the level of complex chemical molecules and cycles. Moreover, the separation between chemical proto-life and organic life is not sharp but fuzzy. Both are reasons to address the origin of life here.

Another topic dealt with is evolutionary ecology. Ecology represents a founding stone of evolutionary thinking, as it clarifies the elementary features of population and multispecies dynamics. In addition, ecology shares a focus on populations and communities with evolutionary social sciences, while spatial scale and extent tend to be comparable among studies in the two areas. The central notion connecting ecology and evolutionary biology is coevolution. It will be argued that all evolution is actually coevolution, in biology as well as in socioeconomic-technological settings.

A final theme considered is the meaning of economic processes and reasoning for a full evolutionary explanation of natural history. Touching upon issues like specialisation, labour division, cooking and mercantile (as opposed to kin or nepotistic) exchange, it provides a connection to later chapters (7 and 8) on cultural and economic evolution. Table 4.1 provides an overview of the various concepts and theories addressed to make it easier for readers to identify topics or keywords that may be to their interest.

Building on Figure 3.2 in the previous chapter, Figure 4.1 is an attempt to provide a more complete picture of evolution, by integrating essential elements of the theories listed in the table. While this doubtlessly still represents a simplified picture, it shows

Table 4.1 Advanced biological concepts and theories addressed in this and later chapters

Theory	Discussed in	Core concepts	Degree of debate among biologists
Neutral genetic drift	Section 4.2	Molecular evolution, molecular clock, deleterious versus advantageous versus neutral mutations, population size, isolation, speciation	Low
Epigenetic inheritance	Section 4.2	Non-genetic inheritance, chromosomes, histones, DNA methyl markers, gene regulation (switching genes on/off), architectural memories, prions, RNA interference, reversible change, adaptation, environmental pressure and change	Medium
Directed and adaptive mutations	Section 4.2	Phenotype affects genotype, non-random variation, rate and direction of mutation, environmental stressors, sexual versus asexual forms, immune system, extreme pressure (e.g. starvation)	Medium to high
Non-gradual, modular evolution	Section 4.3	Compositional evolution, pre-adapted module, jumps, macromutation, recombination, sexuality, symbiosis, evolvability, transitions, regulatory genes, group selection	Low
Symbiotic evolution	Section 4.3	Symbiosis, biophysical integration, organelles, prokaryotes, eukaryotic cells, mobility of cells, ecology, multispecies relationships, parasitism, mutualism, coevolution, evolutionary transition	Low
Macroevolution and downward causation	Section 4.4	Speciation, multilevel selection, upward and downward causation, regulatory (Hox) genes, macromutation	None
Convergence and inevitability of evolution	Section 4.4	Evolutionary routes, cumulative effects, eye, homeothermy (warm-bloodedness), primates, path dependence, micro- and macroevolution	Medium
Macroevolution and environmental change	Section 4.4	Major environmental change, mass extinctions, extra-terrestrial causes, continental drift and plate tectonics, volcano eruptions, geographical spread of humans	Low
Punctuated equilibria, sorting and 'exaptation'	Section 4.4	Palaeontology, geological time, speciation, Cambrian explosion, stasis (delay of species diversification), high rate of cumulative changes, sorting instead of selection, exaptation instead of adaptation	Medium

Table 4.1 (*cont.*)

Theory	Discussed in	Core concepts	Degree of debate among biologists
Origin of life and early chemical evolution	Section 4.5	Panspermia, RNA world, protocells, feedback, autocatalytic cycles, hyper-cycles, coevolution, building blocks of life, polymers, primal soup, geological change, Gaia, atmosphere	Low
Evolutionary ecology, niche construction and coevolution	Section 4.6	Population, behavioural and systems ecology, sociobiology, ecosystem, food chains/webs, trophic levels, primary producers, succession, logistic growth, species interactions, herbivore, plant, predator, prey, competition, mutualism, r- vs K-strategy, foraging, group size and structure, reproduction, coevolution, co-adaptation	Low
Sexual selection	Section 4.7	Competition for mates, Red Queen effect, intra- versus intersexual selection, male–female differences, coevolution and indirect sexual selection, parental investment, promiscuity, monogamy, polygyny, parental bonding, great apes, large human brain, menopause, cultural effects, marriage, endosomatic versus exosomatic adaptations	Low
The role of economics in biological evolution	Section 4.8	Efficiency, labour division, market metaphor, investment, advertisement, evolutionary game theory, exchange and trade, orientation in space and time, trust, mercantile (vs. kin or nepotistic) exchange, brain size, learning	Low
Group and multilevel selection	Chapter 6	Population structure, conflict, non-additive interaction, cooperation, free-riding, isolation, group stability, propagule pool, migration, assorting, spatial clustering, endogenous splitting, group institutions (norms, punish, reward, sharing), groupish behaviour, cultural drift	High
Self-organisation in evolution	Sections 9.5 and 9.6	Ontogeny, development, stability, edge-of-chaos, *NK* model, nonlinearity, attractor, bifurcation, chaos, emergence, local interactions, macrostructure, hierarchy, distributed control, feedback, recursiveness, entropy, exergy, far-from-thermodynamic equilibrium, resilience	Medium

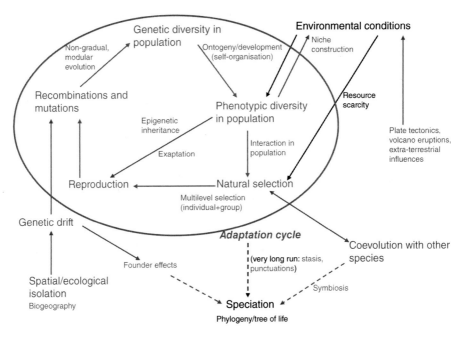

Figure 4.1 An extended dynamic systems view of evolution.
Notes: Extension of Figure 3.2 (blue and black objects) shown in purple. Broken arrows indicate effects over longer time periods.
(*A black and white version of this figure will appear in some formats. For the colour version, please see the plate section.*)

that the common impression given of biological evolution as a fairly simple process leading to complex outcomes is inaccurate. It is more correct to characterise biological evolution as a complex process spawning organic complexity.

4.2   Drift, Epigenetics and Directed Mutations

Here we expand the introductory discussion of genetics in Section 3.3 by considering various advanced topics.

Neutral Genetic Drift

The first 'advanced theory' is known under a variety of names: Sewall Wright effect, molecular drift, genetic drift, random drift, neutral evolution and neutral theory of molecular evolution. The broader area of research of which it is part is molecular biology or molecular evolution. At the basis of genetics are three simple ideas, namely mutations occur at a quite regular, almost constant, rate over time; the nature of a mutation and its impact are random; and the majority of mutations have no phenotypic consequences and, if they have, they often are selectively neutral, meaning they do not affect the survival, or fitness, of organisms. As a result, such mutations are subject to random drift. This is particularly prevalent in small biological populations, since the chance that these mutations become fixed is negatively related to population size. Drift, therefore, is a main factor behind observed genetic variation in small populations. The regular pace of mutations suggests the existence of an 'evolutionary clock' at the molecular level guided by physical-chemical constraints and regularities. This clock possibly has a pace that differs among groups of species and parts of the genetic material. For instance, as higher species have improved DNA repair systems, they tend to show a slower rate of genetic mutation than lower species.

Motoo Kimura first proposed 'neutral evolution' as a coherent theory in *Population Genetics, Molecular Evolution, and the Neutral Theory* (1968), and elaborated it in *The Neutral Theory of Molecular Evolution* (1983). Kimura's theory became well-known as he argued that molecular evolution is dominated by selectively neutral mutations, giving rise to the so-called 'neutralist-selectionist' debate about how common neutral mutations and resulting neutral alleles are in genomes. Now it is widely agreed that while so-called 'purifying selection' eliminates deleterious mutations and positive, directive selection makes advantageous mutations more common, neutral mutations are by definition insensitive to selection pressure. In other words, from a selection angle, random drift is non-directional. This means, more concretely, that it has no or hardly any impact on the structure of cells, organisms and societies (Wilson 1998, p. 130), for which reason it is also known as the 'nearly neutral theory' (Ohta, 1992).

Since drift is more common in smaller populations they tend to evolve faster than large populations. In line with this, it is now widely believed that isolation of sub-populations, allowing neutral drift to make them different, has been an important mechanism behind speciation (Li, 2006). This illustrates that, while some mutations are initially neutral, on a later date they can become advantageous or adaptive due to particular changes in the genome or the selective environment. In other words, such mutations indirectly contribute to 'non-neutral evolution'.

Epigenetic Inheritance

There is increasing support for the view that biological inheritance or heritability concerns more than DNA. It was noted in Section 3.3 that not just DNA with genes on it is replicated, but that in fact entire chromosomes (or the lower order equivalent known as chromatin)[1] are subject to the process of inheritance. A chromosome includes the DNA, methyl markers and proteins such as histones. The latter controls the spatial form and winding of the DNA, which is an extremely long molecule. The structure of chromosomes represents a chief case of so-called epigenetic information, which can be regarded as switching on and off specific genes on the DNA, known as 'gene expression'. Switches take the form of changes in the methyl markers or histones.

---

[1] DNA mostly adopts the lower order form of chromatin because it then can be read to generate proteins. It is condensed during mitosis (cell division), resulting in the higher order chromosome form, which allows for accurate replication and division between the nuclei in the two daughter cells.

These then alter the chromosome, without affecting the DNA structure itself – hence the term 'epigenetic' to distinguish from 'genetic', i.e. pertaining strictly to the DNA.

Such epigenetic control of gene activity can – in some cases – be inherited by daughter cells through copying the relevant epigenetic information along with the DNA. While the practical significance of this insight has been accepted in medicine, its role in evolution is less clear. If epigenetics is true, it means that environmental factors can trigger epigenetic tags, activating or switching on particular genes, which can be transferred to the next generation.[2]

Other types of epigenetic inheritance have been discovered. One is inheritance in some species of cell structures, what Jablonka and Lamb (2006) call 'architectural or cell memory'. For example, mother and daughter cells share a membrane during a division process which has been argued to be necessary for the daughter cell to develop. This means not all information about cell development is contained in the DNA (Cavalier-Smith, 2000). Another case is disease-causing prions, which, if present in one organism that is eaten by another, may cause fundamental changes at the protein level in the receiver. This will create disease features, such as BSE, or mad cow disease, and, in humans, Creutzfeldt–Jakob disease. These are easily misinterpreted to be the result of genetic deficiencies. A fourth type of epigenetic inheritance system is 'RNA interference', discovered in the late 1990s and still imperfectly understood. It involves deactivating ('silencing') genes with one molecule. Since such a molecule can multiply and migrate in the body, phenotypic information transfer – effectively non-genetic inheritance between cells – can occur.

Epigenetic inheritance has a number of advantages over mutations in the genetic material. First, changes in values of 'epigenetic switches' are reversible. This allows reversion to the 'normal state', for instance, after an extreme environmental pressure has disappeared or in response to temporal – daily or seasonal – cycles. Castro Dopico et al. (2015) find that a quarter of human genes differ in activity throughout the year, notably between summer and winter, affecting the immune system and other physiological features of the body. Second, epigenetically triggered changes can occur at a higher rate than mutations. Third, changes can take the form of partly non-random responses to environmental pressures. All three allow for a better adaptation of an organism to environmental change. Using a hypothetical example, Jablonka and Lamb (2006, pp. 114–119) show that even in the absence of genetic variation, heritable epigenetic variation allows a process of evolution to occur. They further note it is likely that epigenetic inheritance is ancient, notably the type based on chromatin marking and RNA interference. It could have allowed early cells to quickly adapt to constantly changing environmental circumstances or to defend against invading DNA or RNA viruses.

---

[2] Chopra and Tanzi's 'supergenes' (2015) takes this idea to the extreme. It suggests that we can positively affect activity of genes in our body and even in that of our offspring, through various strategies: eat less processed food, avoid television before going to sleep, achieve positive emotions, undertake exercise, meditate regularly, take the right personal 'epigenetic medicine' at the right time and possibly use stress-relieving gadgets. Although most of these strategies sound familiar, we still await evidence of their systematic connection with genes.

How important has epigenetic inheritance been for evolutionary history? Most biologists would probably say 'little', pointing out that evolution is mainly about adaptations due to selection working upon genetic variation. Rutherford (2015) argues that the large majority of epigenetic modifications in the DNA tend to get reset each generation, and refers to evidence in mice suggesting that the ones that do get passed down from parent to child disappear after several generations. Research on epigenetics is complex and fascinating, but still in its infancy, meaning that we cannot draw far-reaching conclusions yet.

Directed and Adaptive Mutations

Epigenetic inheritance involves a process through which an organism during its life can achieve adaptations and transmit these to offspring. This suggests a Lamarckian type of evolutionary mechanism as acquired information is inherited (see Sections 1.4 and 7.5 for various interpretations of Lamarckism). Here we present another set of Lamarckian-type mechanisms, taking the form of the phenotype affecting its own genotype.

Our immune system is often mentioned as a case of the phenotype providing feedback to the genotype. DNA in a white blood cell (lymphocyte), which produces antibodies to fight bacteria, is restructured and becomes different from that of other lymphocytes and cells in the body. Diseases are so abundant and evolve so rapidly that only an evolutionary process like the immune system is able to resist them. Not surprisingly, the immune system has been called a 'Darwin machine' (Calvin, 1987) or an example of 'clonal selection' (Burnet, 1976). One explanation is that mutations occur simply in this case because stressed cells no longer function well at the molecular level or may have turned off DNA repair systems to save energy or other resources, making DNA change more likely. Edward Steele (1981) has argued that similar rapid mutation processes may occur in germ-line cells and thus can affect evolutionary change. He and co-authors have undertaken experimental work to support this view (Steele et al., 1998).

While it is generally assumed that mutations are random or undirected, it is impossible to prove that this is really always true. 'Random' is often used interchangeably with 'undirected'. But the latter is really more adequate, since mutations are not really random with respect to their position in the DNA or their frequency as a function of environmental conditions. Some studies suggest that not all mutations are undirected, as mutations may arise in reaction to a need by an organism (Cairns et al., 1988). It is established that the rate of mutation can respond to environmental factors, such as ultraviolet radiation damaging DNA or a lack of food or essential molecules nearly causing starvation. A recent empirical study suggests that the rate of mutation is evolutionarily optimised so as to reduce a risk of mutations harmful to an organism (Martincorena et al., 2012). The general opinion among biologists, however, seems to be that mutations are undirected under normal, i.e. non-extreme (like starvation), conditions.

Definite proof of semi-directed mutation through experimentation has turned out to be extremely difficult to find. Testing has been limited to a few microorganisms, notably: repair of DNA damage (the so-called 'SOS response') and adaptive responses to starvation in the bacterium *Escherichia coli*; and adaptive responses to a lack of intake of the amino acid tryptophan by the yeast *Saccharomyces cerevisiae* (McKenzie et al., 2000). However, the results of these studies are inconclusive.

In a review of the literature on adaptive or directed mutation, Sniegowski and Lenski (1995) note that many examples of apparently adaptive mutation were later rejected. Ascribing adaptive significance to mutation mechanisms is conceptually problematic, while there are many experimental and conceptual problems associated with adaptive mutation. The authors argue that the most plausible molecular models explaining the phenomenon of adaptive mutation are consistent with, and so do not undercut, natural selection working on randomly generated variation.

According to Jablonka and Lamb (2006), it would be surprising if evolution, having tried out so many chemical processes in cells and organisms, had not developed mechanisms to control the rate and direction of mutations under certain extreme circumstances. They note that in species appearing in both sexual and asexual forms, asexual reproduction dominates as this is cheaper (no need for energy-wasting males) and preserves well-functioning genes. On the other hand, under stressful conditions sexual reproduction takes over to create more variation, thus contributing to survival of individuals and adaptation of the species to altered circumstances. This arguably suggests that natural selection indirectly adjusts the rate of recombination and mutation. The latter does not imply directed mutations, but simply more frequent ones that possibly speed up evolutionary change.

The presence of adaptive mutations suggests a vague boundary between development and evolution. It implies that the classic distinctions 'instruction versus selection' and 'proximate versus ultimate causes' become less sharp. Others would claim that adaptive mutations are perhaps not impossible but are likely to have played a minor role in evolutionary history. Section 7.5 offers a discussion of the related Baldwin effect and genetic assimilation.

4.3   Modular and Symbiotic Evolution

Modular Evolution

Darwinism has been called a gradualist framework of evolution, stressing accumulation of small changes. Darwin himself noted that evolution does not make jumps or 'saltations'. Goldschmidt (1940) proposed the idea of 'macromutations' which the then dominant evolutionary biologist Ernst Mayr coined 'hopeful monstrosities', arguing that macromutations cannot have been important in natural history as they would have destroyed the genetic adaptive capacity, possibly even undercut the viability, of organisms. This does not deny, though, the relevance of other forces that contribute to non-gradual patterns of evolution. These include phenomena such as recombination (through sexuality or lateral gene transfer as in bacteria), symbiosis and regulatory genes (Section 4.4). The similarity of these is that they all combine 'pre-adapted modules' of some type to form new entities. One could speak of modular or compositional evolution (Schlosser and Wagner, 2004). Small changes in genes affecting how modules are composed can have large effects on the phenotype, and thus be perceived as macromutations.

Because of the nature of their replication mechanism, endogenous retroviruses have provided many pre-adapted genetic modules in the evolutionary history of life. For example, the placenta in mammals arose due to retroviruses. Retroviruses of all kinds frequently appear in the genomes of jawed vertebrates, including humans. Although at the original time of entering the genome of their host they must have caused it to become sick, their fixation in the genome is ancient as well as often partial. As a result, it generally does not create disease effects in current carriers, as this indeed would select against the hosts, which would stimulate an appropriate adaptation. Nevertheless, it is believed that some retrovirus gene sequences in humans may be behind certain cancers and multiple sclerosis.

Algorithmic approaches in computer science and mathematical modelling have shown that modular evolution can span a space much faster and more completely than gradual evolution is capable of (Watson, 2006). In addition, constellations can be reached that are 'unevolvable' in terms of gradual evolution. A modular evolutionary approach may therefore help to fully explain rapid evolution and the major transitions in the history of life. Incidentally, Chapter 10 will show that non-gradual evolution based on pre-adapted modules is equally relevant to understanding technological evolution. According to Clune et al. (2013), modularity in networks, underlying a wide range of systems from gene regulation to brain processes, contributes considerably to evolvability. Modular networks are characterised by highly linked clusters of nodes that are sparsely connected to nodes in other clusters. One cluster can then change quickly under evolutionary pressure, without affecting the remainder of the network. Hence, existing functions supported by the network that do not require adaptation to environmental change remain undisturbed.

Although modular evolution is implicit in many aspects of evolutionary biology, such as sexuality, recombination, symbiotic evolution and major transitions, most explanations of evolutionary theory tend to emphasise its marginal, gradualist nature. But the one does not exclude the other: both mechanisms, gradual and non-gradual evolution can be present. Jason Potts (2000) has argued that such evolution is not so easily captured in mathematics such as differential equations, unlike gradual evolution, but instead requires combinatorial mathematics, such as in the form of (multi-)agent-based or network (graph theory) models. This may explain why gradual evolution dominates in the thinking and teaching about evolution and only sparse attention is given to the role of modular evolution.

Symbiotic Evolution

A special case of modular evolution is symbiotic evolution. This refers to a theory suggesting that various transitions in evolutionary history (Section 3.6) result from symbiosis, a close and long-term relationship between organisms. The evolutionary step here is that in some cases symbiosis ultimately gave rise to a physical integration between the organisms involved, resulting in their subsequent continuation in the form of a single species. This is also known as 'conjunctive symbiosis'. Symbiosis denotes all intermediate relationships between interactive species, including temporary as well as incomplete physical integration. This is common in nature, as it increases adaptive flexibility in the presence of changing environmental circumstances. Symbiotic organisms have properties that may differ considerably from those of its component organisms. In fact, some could not live outside the symbiotic relation.

Without symbiosis the natural history of Earth would look quite different. Among others, symbiotic species have been critical for mineral take-up from soils and so have aided early colonisation of land. In many cases, symbiosis occurs between unicellular organisms and multicellular organisms. Examples are dinoflagellates and green algae living in marine organisms (e.g. *Hexacorallia* and *Octocorallia* forming coral reefs), lichens (a composite of a fungus and photosynthetic algae or cyanobacteria or both). Another case is mycorrhiza, a symbiotic relationship between mycelia, i.e. a branching colony of fungus and the roots of various tree species. For example, truffles tend to enter into this type of symbiosis with beech, birch, hazel, oak, pine and poplar trees.

One can regard symbiotic evolution as a type of coevolution (Section 4.6). It denotes an increasingly intensive relationship ending in permanent biophysical assimilation or merger. It thus can be understood from the perspective of the evolutionary ecology of multispecies relationships. Mutualism and parasitism are the most obvious types that can precede symbiosis as they can transform in closer relationships (Section 4.6). In a spatial sense, this requires being near to each other continuously. Endosymbiosis denotes that individuals of one species may even be continuously within the organism of individuals of the other species, due to a process that has been referred to as 'melting' or 'cannibalism'.

Symbiotic evolution can produce new organisms that combine features of the integrating original organisms. In other words, it serves as a mechanism for fundamental evolutionary innovations. This idea of evolution through symbiosis, first suggested in the 1910s and 1920s by Russian biologists, was elaborated by Lynn Margulis (1970). Among the best-known examples are organelles like mitochondria and chloroplasts in eukaryotic cells. They descend from free-living prokaryotes, namely purple non-sulfur bacteria and cyanobacteria, respectively. A symbiosis between these, probably following a parasitic or slavery relationship among prokaryotes (Maynard Smith and Szathmáry, 1995, Section 8.6), around 1 billion years ago, has resulted in the present structure of eukaryotic cells. This explanation has been confirmed by the fact that both types of organelles contain DNA. Moreover, this DNA is circular, as in prokaryotes, while their ribosomes resemble those found in bacteria. The fusion of mitochondria with other cells has probably occurred only once. The same holds for chloroplasts. Another type of symbiosis is still debated, namely as the origin of the mobility of cells through cilia and flagella, i.e. flexible hair- or whip-like affixes. Maynard Smith and Szathmáry (1995, p. 35) draw attention to the notion of 'chemical symbiosis', where the waste of one chemical cycle serves as a nutrient for another and vice versa, which was probably critical to starting life on Earth (Section 4.5). It has even been argued that sex is a type of symbiosis, more particularly cannibalism of the male germ cell by the egg cell (Margulis and Sagan, 1986). In addition, superorganisms such as ant colonies, and even the close relationship between ants and fungi (a sort of farming), have been suggested as cases of symbiotic evolution (Hölldobler and Wilson, 1994, Chapter 8). The idea is, therefore, that they are as a whole subject to selection.

4.4   Varieties of Macroevolution

Some argue that microevolutionary theories need to be complemented by macro-evolutionary ones to achieve a complete explanation of natural history. The micro level is traditionally dominated by geneticists and ecologists and the macro level by palaeontologists. According to Matt Ridley (1997, p. 226) speciation is the traditional dividing line between micro- and macroevolution. The notion of macroevolution has many meanings. One can see it as operating on a time scale of millions of years that allows it to explain major changes in body plans, speciation and mass extinctions. One particular view is expressed by so-called punctuated equilibrium theory, which suggests that unlike microevolution, which is largely gradual, macroevolution often has a non-gradual character. This relates to an old discussion about macromutations (Section 4.3 and below). Another view relates to long-standing debate about the levels and units of selection and the upscaling of microevolution to a macro level.

Downward Causation, Regulatory Genes and Macromutations

It is now widely accepted that selection occurs at multiple levels, from genes to organisms. Some defend even selection of groups and ecosystems (Chapter 6). Suggested candidates for selection levels or units thus include genes, combination of genes (genomes), regulatory genes, chromosomes, organisms, colonies, groups, populations, species and ecosystems. So natural selection works on many units simultaneously. Only if there were a unique gene for each feature or behaviour of an organism could everything be summarised as, or reduced to, genetic selection. Instead, however, non-linear and non-additive interaction among genes determines the characteristics on the level of the organism, which makes the link between selection pressure and genetic response rather indirect and complex.

The combination of various types of micro- and macroevolution leads to a framework with multilevel causation (van den Bergh and Gowdy, 2003):

- *Downward causation*: effects of individual selection on gene frequencies; the effect of regulatory genes on the impact of other genes; or the influence of speciation on individual organisms.
- *Upward causation*: microlevel evolution affecting higher levels, such as genes affecting organisms, or random genetic drift having long-term selective and adaptive consequences for small genetically isolated populations.

An important case of downward causation is associated with a class of genes known as 'Hox genes'. They are responsible for a complex, hierarchical system of genetic control, through genes activating or deactivating other genes during development (Carroll, 2000). Such regulatory genes explain the existence of different cell types in multicellular organisms, as each type is associated with the activation of particular genes. The existence of these genes was not recognised in classical population genetics. Changes in them, sometimes referred to as 'macromutations' (Section 4.3), could be responsible for many non-gradual or discrete changes ('saltations'). This can explain certain sudden, major changes in phenotypes and subsequent new evolutionary directions, which show up as 'punctuations' in the fossil record. Changes in regulatory genes have been less common in the evolution of frogs than mammals, explaining why the more than 3000 species of frog are similar in appearance and behaviour, unlike the 3800 species of placental mammals (Strickberger, 1996, p. 273).

Convergent and Recurrent Evolution

The next issue is not so much a theory as an idea, though a fundamental one with huge implications for the macro-level picture of biological evolution. Both micro- and macroevolution lend support to the view that evolution is path dependent, meaning it is irreversible and greatly influenced by historical micro- and macro-level accidents. This suggests that the specific direction evolution takes is determined by what happened during early natural history. Another view is proposed by Conway Morris (1998, 2003), namely that many facets of evolution are directional and thus predictable in general terms. He argues that the number of distinct feasible pathways from DNA to functional, behavioural or physiological characteristics of organisms is limited. Moreover, good evolutionary solutions for certain environmental challenges often look the same. Taken together, this would imply that the number of alternative and fundamentally different evolutionary lineages is limited. Hence, convergent evolution is, in Conway Morris' view, more common than traditionally believed. The evolution of the eye illustrates this. Even eyes with an independent evolutionary history often share many physical and chemical characteristics among species, supported by similar genes. Another example is the independent evolution of the capacity to fly in birds, insects and bats.

Convergence and inevitability, also known as recurrent evolution, are connected in the sense that widespread convergence provides strong support for the inevitability of certain characteristics and mechanisms of evolutionary change. Inevitability is due to the fact that certain features create such a clear advantage in species over ones that lack them that they must evolve at some point in time: this holds for general features such as productivity, speed, vision and intelligence (Vermeij, 2010). Conway Morris also regards the evolution of warm-bloodedness and primate-like species as inevitable. Regarding primates, he suggests that if human evolution, for some reason, could not have occurred in Africa, it would ultimately have happened in South America. Inevitability should not be taken literally, though. Due to limited evolutionary time, the number of evolutionary end points is restricted.

Macroevolution and Environmental Change

An important driver of macro-evolutionary patterns, notably major trend-breaks, is large-scale environmental change. This may 'unlock' existing constellations by disturbing gradual evolution in a stable environment. This usually involves mass extinctions, which promote a major evolutionary innovation. Probably the best-known example is the last mass extinction, arguably caused by a large meteorite hitting the Earth about 66 million years ago and causing extreme climatic change at a global scale. This marked the end of the era of large dinosaurs and paved the way for expansion of the mammals. The collision abruptly created an extremely harsh environment[3] characterised by fires, dust, extreme temperatures and collapse of food chains, which particularly threatened the survival of large specialised animals living in relatively small populations and often positioned at the top of a food chain, such as dinosaurs. Species with smaller bodies, like mammals, could subsequently occupy the wide variety of ecological niches previously occupied by dinosaurs, and evolve into larger animals. Another hypothesis is that many dinosaurs had already been extinct before the meteorite collision with the Earth, as a result of changes in the environment triggered by a high frequency of volcanism and small meteorite impacts.

Other environmental factors have been documented to have altered the course of evolution at some point in natural history (Maynard Smith and Szathmáry, 1995; Ridley, 1997, Section G; Ruse, 1999, Chapter 11):

- Alteration of climatic and biogeochemical conditions: Periods of continental glaciation followed by warm periods with high sea levels may have affected both land and marine life.
- An increase in the concentration of oxygen in the atmosphere: This was the result of population growth of species undertaking photosynthetic processes. In turn, it allowed more biomass production, longer food chains, more complex food webs and new habitats offering niche opportunities for new, evolving species to specialise and diversify.
- Continental drift and plate tectonics: This changed climate conditions and isolation patterns of large pieces of land, with enormous consequences for species that had developed during an extremely long period of isolation. In addition, movement of the continents transformed the shape and extent of shorelines, in turn affecting many life forms.
- Extra-terrestrial causes of mass extinctions: These involve mainly meteorites hitting the Earth. In only the last 440 million years, after the Cambrian explosion, five major extinction periods have been identified, during which a significant proportion of all species disappeared. According to Raup and Sepkoski (1984), palaeontological data indicate periodicity of mass extinctions. The authors suggest as an explanation the periodic occurrence of a comet shower, possibly related to a distant companion star to the Sun. When it periodically closes in on the solar

---

[3] See for a brief and clear explanation: http://www.astronomynotes.com/solfluf/s5.htm.

system, it creates gravitational disturbances which generate a comet cloud. Some studies, however, debate that estimates of extinction rates are reliable, based on the argument that sedimentation of earth layers into rock material – later available for sampling of fossils – is not constant over time (Peters and Foote, 2002).

- Large volcanic eruptions : These have likely influenced both the average global temperature and the composition of the atmosphere, thus stimulating the extinction of certain species. A major geological 'event' in this respect was the formation of the Deccan Traps 65 million years ago in West India. This took at least several tens of thousands of years, during which millions of cubic kilometres of lava reached the surface of the Earth, affecting an area of about 500 000 km^2. This has been suggested as an alternative cause for the extinction of the dinosaurs. Possibly, the combination of this phenomenon, the meteorite impact and competition with other animals created overall harsher living conditions for the dinosaurs.

- Spread of humans: More recently, the geographical spread of humans is supposed to have contributed significantly to the extinction of certain species, notably megafauna. A well-known example is the arrival of humans in Australia some 56 000 years ago. Before that time, 24 larger mammalian, bird and reptilian species lived in this continent. Except for one type of kangaroo, the species have all become extinct over the course of an evolutionarily brief period of time, in the order of tens of thousands of years. One explanation is that these species had evolved in the absence of a human or other predator, so that they were not adapted to predation pressure, which increased due to growing human populations. Another explanation is that these species' food and habitats were being destroyed by humans hunting smaller animals and burning large areas of vegetation. Another factor may have been a changing climate. The importance of each of these factors likely differed among species.

- A similar history holds for North America, somewhat later. When humans entered this continent over ice plains between Asia and America some 14 000 years ago, it hosted 41 larger animal species of which only a few remain now. The buffalo (bison) and moose probably survived because, as well as living on the plains, they were able to live in cold climates in the far north, with sparse human populations and therefore fewer predators. Note that animal extinctions caused by hunting were possible because the respective species had evolved in the absence of, and so had not adapted to, intelligent hunters like the primate ancestors of humans (Diamond, 1997a). Adaptation might have involved being nervous or jumpy and showing fast fleeing behaviour, as is observed in African species such as antelope, gazelle and zebra. As these evolved in the presence of human ancestors, they were able to survive ancient primate hunting pressure through coevolution.

Punctuated Equilibria, Sorting and Exaptation

Macroevolution is often equated to the 'punctuated equilibria' theory. But it is much broader, as clarified by the range of themes addressed in this section. Niles Eldredge and Stephen Jay Gould (1972) proposed the notion of punctuated equilibrium, which since then has been debated and remains controversial. Eldredge and Gould claimed that the fossil record of 'morphospecies', i.e. species distinguished on the basis of their morphology, showed that evolution did not have a constant pace, but instead is characterised by long periods of stasis interrupted by bursts of rapid change. Their term 'punctuationism' suggests an opposition to gradualism, namely speciation as the result of an accumulation of gradual adaptations due to repeated selection. Hence, many observers have – incorrectly – concluded that Eldredge and Gould shot holes in neo-Darwinian evolutionary biology.[4] Anyway, rather than a new theory, their work is best regarded as a novel interpretation of palaeontological data.

Nevertheless, as clarified in previous (sub-)sections, it is now widely accepted that evolution, while having gradualism at its core, also involves non-gradual changes – notably owing to modularity and regulatory genes. This means punctuated equilibria are not inconsistent with neo-Darwinism at all. Moreover, if one looks precisely, such equilibria are not at all about abrupt changes but merely reflect a relatively high rate of cumulative changes over geological time. Seen from a distance, or at an aggregate time scale, this gives the impression of a sudden and large change, but it can easily involve thousands of generations over the course of tens of thousands of years. 'Punctuationism' can thus be completely consistent with gradual evolution.

As already suggested by Ernst Mayr (1959), speciation is most likely to occur in small, peripheral (isolated) populations at the edge of the main species range. Therefore, fossils of the speciating population will be rare and unlikely to be found in the same location as the ancestral lineage, which can contribute to punctuation patterns in the fossil record. According to George G. Simpson (1944) there is no constant rate of evolution but its rate varies significantly among species and periods of time. He used the terms 'bradytelic' for low, 'horotelic' for medium, and 'tachytelic' for high rates and the term 'quantum evolution' to indicate that crossing a certain rate threshold results in a path to extinction or major evolutionary innovations in the respective species. As examples, he noted that higher species generally have evolved more rapidly than lower ones and mammals more rapidly than reptiles. The latter is currently explained on the basis of regulatory genes. The idea that the pace of evolutionary change might be rapid confirms certain aspects of, rather than overturns, traditional theory. Richard Dawkins (1986) has suggested that against this background Eldredge and Gould deliberately played down the relevance of rapid gradualism in order to give more importance to their own idea of punctuations or macromutations.

According to Gould and co-authors, the radical aspect of punctuated equilibrium lies in challenging natural selection as the only important mechanism of evolutionary change. Gould and Eldredge (1993) present evolution as a hierarchical process with natural

---

[4] Darwin had stressed the gradual aspect of evolution – expressed by the statement 'natura non facit saltum [nature does not make jumps]' – as illustrated by the summary of his *Origins* Chapter 6 in Table 3.2. It has been argued that Darwin held this view because a convincible alternative was needed to counter the biblical story of sudden creation (a special case of 'saltation') by a god of the Earth and species (Hannan and Freeman, 1989, p. 37). Darwin seems to have been fearful of receiving extreme criticism on his work because the same had happened to most previous works offering an alternative to the creationist explanation of life and nature.

selection operating at the individual level and other, biological, climatic and biogeochemical changes, being co-responsible for the array of species and ecosystems present at any given historical moment. According to them, it is the notion of hierarchical selection that really embodies the radical content of punctuated equilibrium thinking because it challenges the idea that what now exists has won the struggle for survival at the micro level. Instead, they promote unique, historical causes or 'contingencies' as critical to certain speciations and extinctions. This is consistent with the earlier discussed relevance of macroenvironmental factors behind extinctions as a source of 'evolutionary innovation'. However, to suggest that this adds another, independent mechanism to natural selection goes too far. It would be more correct to say that large-scale environmental changes affect the course of evolution as they tend to considerably alter the selection environment for most species. This is done through abiotic changes – less oxygen, higher sea level, higher atmospheric temperature, etc. – as well as biotic changes – e.g. species getting extinct, affecting competition and predation pressures for other species. The earlier discussed case of extinction of the large dinosaurs illustrates this: small mammals faced less competition and could occupy the freed niches.

On balance, the most meaningful aspect of the idea of punctuated equilibria is not so much the sudden changes – that is to say, in a geological time frame – but the extended equilibria or periods of stasis, during which relatively little or no observable change occurs. To complicate things, stasis is not necessarily identical to lack of change; it may just reflect unidirectional change, resulting in relatively small variations around a mean. Erwin et al. (2011) suggest that the period of stasis involves molecular rearrangement and accumulating mutations for development in the genome, which only much later contribute to potential ecological success.

To capture the difference between micro- and macroevolution, Gould and Vrba (1982) proposed, consistent with Gould's idea of evolution as a hierarchical process, the notion of 'sorting'. This can be seen as a generalised selection concept, meaning something like differential survival rates without a specific cause. The following examples of sorting were often given: non-Mendelian sorting of genetic elements ('hitch-hiking genes'), extrinsic control as a result of different climatic and geological histories, and genetic drift (Section 4.2). The question is, of course, what one gains by adding such a broad umbrella term. Moreover, as Maynard Smith (1995) has noted, such downplaying of adaptation as the core causal mechanism of evolution implies that in evolutionary analysis we should be less focused on asking about the adaptive function of observed features of organisms; but this would hardly be an effective strategy to advance evolutionary biology.

In line with the notion of sorting, Gould and Lewontin (1979) have suggested that the term 'adaptation' is regarded too broadly and uncritically as a description of all consequences of evolution. They illustrated this viewpoint by drawing attention to the spandrels found between the arches in the San Marco Cathedral in Venice. Whereas the arches perform the function of holding up the ceiling, the spandrels have no clear function but just fill up the empty spaces between the arches (whether this is true has been debated though). What they initially called 'spandrels' were later termed 'exaptations' by Gould and Vbra (1982), referring to features that 'evolved for other uses (or for no function at all) and later co-opted for their current role' (p. 6). Exaptations are defined as fit for their current role but not designed for it. Their current utility should not be confused with reasons for origin. A main problem with this notion, though, is that evolution does not have foresight nor does it anticipate. Moreover, many adaptations are based on past genetic drift or pre-adapted modules. As a result, the distinction between adaptation and exaptation is unclear. Darwin recognised that the function of a trait can change over time and that this can even happen frequently. Gould and his co-authors would, nevertheless, prefer to use the term exaptation for this case. But exaptation is really nothing more than pre-adaptation, a common notion in evolution and an integral part of adaptation. It does not in any way undercut Darwinian evolution.

Gould's strongest critic, philosopher Daniel Dennett (1995, Chapter 10), argues that Gould and his co-authors contribute to a false, biased view of evolutionary theory, and downplay the role of adaptation, selection and gradualism without good evidence or a convincing theoretical argumentation. He thinks they have exaggerated the importance of their own-named mechanisms 'punctuated equilibria', 'sorting' and 'exaptation' to suggest a paradigm shift in evolutionary biology. Dennett's judgement is that Gould's 'declarations of revolution have all been false alarms'. In response, Gould accuses Dennett of defending 'ultra-Darwinism' and 'Darwinian fundamentalism'[5] (Gould, 1997a) and showing 'little understanding of evolutionary theory beyond natural selection' (Gould, 1997b). But at the same time he admits that he did not contribute to any revolution or paradigm shift.[6]

Many will claim that Gould and co-authors were attacking straw men. Few serious evolutionary biologists, even before 1979, considered selection as the only force of evolution, or all change as adaptive. In fact, Darwin himself recognised these issues – as admitted by Gould – and Dennett, Maynard Smith and Dawkins certainly do so too. This then makes the accusation of 'ultra-Darwinism' odd if not empty. To the credit of Gould, he incessantly reminded us of the caveat of extreme adaptationism, suggesting instead that not each physiological, mental and behavioural trait of an organism can be attributed to selection for functionality and utility (Lynch, 2007).

4.5   Early Chemical Evolution and Origin of Life

One fundamental aspect of natural history still awaits our attention, namely the beginning of organic life. Prevailing hypotheses have been a popular target of criticism by creationists. Some claim, though, that the origin of life is not part of evolutionary

---

[5] In disagreeing with others, Gould frequently used terms like 'ultra-Darwinian', 'Darwinian fundamentalist' and 'hyper-Darwinian'. Ironically, Gould claims immense respect for Darwin, but uses the latter's name to create such intentionally denigrating terms, hence suggesting that something is fundamentally wrong with Darwin's original theory – a view shared by very few evolutionary biologists.

[6] Maynard Smith (1995), a widely respected evolutionary biologist and not known for sharp statements, has said about Gould: 'Because of the excellence of his essays, he has come to be seen by non-biologists as the preeminent evolutionary theorist. In contrast, the evolutionary biologists with whom I have discussed his work tend to see him as a man whose ideas are so confused as to be hardly worth bothering with, but as one who should not be publicly criticized because he is at least on our side against the creationists.'

biology, as the latter is confined to organic evolution once life had emerged. The problem with this separation is that the boundary between replication of molecules and life is fuzzy. Moreover, as this section will show, evolutionary mechanisms are at the heart of explanations of life's chemical origin. Nevertheless, a separate field of research, namely abiogenesis, is devoted to the scientific quest of the origin of life.

What is life exactly? A variety of definitions has been proposed: being machine-like, possessing a metabolism, dynamically interacting with the environment, being active over an extended period of time, having the potential to reproduce, and combinations of the preceding features. An evolutionary view on life might highlight descent with modification and heritable variation. According to a strictly phenotypic definition, metabolism and development are critical factors. A genotypic definition would draw attention to replication and variation of chromosomes. Maynard Smith and Szathmáry (1995) argue that the combination of metabolism and replication is essential to defining life: metabolism supplies the resources for replication, while the latter, influenced by mutation and recombination, slightly alters the details of this metabolism in subsequent generations of organisms.

It is believed that the first life forms consisted of merely replicating genes, like modern-day viruses, without a well-bounded cell space to accommodate and protect any metabolic processes. To imagine how difficult it was for evolution to transform such virus-like life into cell-based life, one should imagine that an average cell, despite being extremely small – with a mass in the order of magnitude of 1/1 000 000 mg – contains up to 100 000 different types of proteins, and in the order of magnitude of 600 million proteins in total, to keep its metabolism going.

Given that for a long time most people resisted the idea of a spontaneous origin of organic life, it is somewhat at odds that in human intellectual history various people have proposed a regular spontaneous generation of life. The experimental work by Louis Pasteur (1822–1895) and John Tyndall (1820–1893) was critical in getting rid of this false belief. Now we know this belief was due to the omnipresence of extremely small, virtually invisible bacteria, fungi and eggs of insects, which could multiply or grow into larger, visible beings.

There remain now two main hypotheses about the origin of all life on Earth: spontaneous origin on Earth and Panspermia. As the first receives most support nowadays, we start with a brief exposition of the second.

Panspermia

The Panspermia hypothesis suggests an extra-terrestrial origin of life. One idea is that life spread spontaneously through the universe in the form of microorganisms or spores. Alternatively, life is intentionally propelled by an intelligent species elsewhere in the universe. One line of empirical support for Panspermia is identification of organic compounds in meteorites and comets, motivated by the belief that life might have arrived through these to Earth. Panspermia has been supported, among others, by renowned astrophysicist Fred Hoyle and by Nobel laureate Francis Crick (1981), who with James Watson unveiled the structure of DNA.

Estimates by astronomers of the number of planets similar to the Earth in terms of size, distance from a star, and several other features required for creating favourable conditions for life (as discussed below), range from several million to several billion (note that there are an estimated 150 billion stars in the universe). This suggests that if life could have originated on an Earth-like planet it likely originated also on another planet.

Evolution of life forms on other planets should follow the same logic as on Earth. The main difference might – depending on the initial atmospheric composition and availability of nutrients – pertain to the structure of the basic building blocks of life, notably amino acids and associated genetic codes and proteins (Ayres 1994, p. 87). Different building blocks of this kind are likely to have major implications for the type of species that can evolve on the basis of them. This means that life forms on other planets, if they exist, may be expected to considerably differ from those on Earth. Nevertheless, general characteristics will likely be the same. That is, if life had had enough time and space to evolve elsewhere, it would almost certainly have given rise to multicellular organisms, plant (static) and animal (moving) species, predators and preys, social species, eyesight in many species and a form of sex (i.e. exchange and recombination of genetic information). This is in line with the idea of convergence as discussed in Section 4.4.

The existence of differences between environments and hence building blocks of life on distinct planets lowers the probability of Panspermia. Indeed, life that evolved under conditions elsewhere would have building blocks that might not be adapted to Earthly abiotic conditions at the time when a supposed Panspermia occurred. Additional reasons to cast Panspermia into doubt are associated with long-distance travel through space: it would involve an enormous amount of time and at best allow for microorganisms, which would have to survive high-energy radiation in outer space.

More importantly, Panspermia does not offer an ultimate explanation for life as it still requires an explanation for life originating on a supposed 'mother planet'. Therefore, we move now to a fundamental explanation of the origin of life.

Spontaneous Origin of Life on Earth

A number of conditions satisfied by the early Earth made it suitable for a spontaneous origin of life. These conditions include an appropriate distance from the Sun – not too hot, not too cold – and an adequate gravity force – allowing for a stable atmosphere and water bodies. In addition, the presence of a moon was critical as it affected the magnetic field of the Earth, which makes high-energy particles circle around the Earth and enter at the poles – where they cause the auroral lights. The combination of the atmosphere and magnetic field functions as a shield against particles and radiation that would have damaged any early organic life.[7]

---

[7] Beyond these conditions for the Earth's favourable conditions for life, there are critical values of parameters associated with the fundamental physical conditions that make the universe friendly to life – the so-called 'anthropic principle'. Among others, if the strengths of the strong force (that binds nuclear elements of atoms), the electromagnetic force and gravity would have been different, the universe and the atoms of which it is made up would not exist in their current forms or at all. Or if the amount of mass in the universe

Robert Hazen (2005) distinguishes between three scenarios of the emergence of life: first, beginning with metabolism, with genetic replication following later; second, beginning with genetic replication, and metabolism following later. The first seems more likely than the second, as creating a stable gene-like molecule out of pre-biotic soup without any metabolism involved is difficult if not impossible. A third scenario would be a ping-pong between these two scenarios, namely simple proto-metabolism arising simultaneously with proto-replication, developing in interaction and in complexity due to mutations and selection.

Anyhow, the beginning of life took the form of chemical evolution. This involved the emergence of chemical cycles, which slowly became more complex owing to repeated variety creation and selection of molecules and cycles. According to one theory, spontaneous chemical and physical factors led to chemical networks involving amino acids, which were capable of copying and multiplying themselves. An alternative theory suggests that self-replicating RNA molecules are precursors to current life, as RNA can store genetic information like DNA and catalyse chemical reactions involving enzyme proteins. The early transition from a hypothesised RNA world to the co-existence of DNA, RNA and proteins had the advantage that DNA is a more stable storage form of data and can program for a greater variety of amino acids. On the other hand, DNA needed a more developed supporting chemical network. Richard Dawkins (2007) has called this a Catch 22: DNA cannot come into existence without proteins, and proteins cannot exist without DNA. RNA can be regarded as a workable combination of replicability and catalytic or enzyme-type features. Decisive evidence for this 'RNA world hypothesis' is that the composition of many critical and stable, and thus likely old, components of cells is dominated by RNA.

In the next phase of early natural history, localised organisations evolved, referred to as chemical microspheres or protocells. The process leading to this has been called 'chemical symbiosis', as the waste of one chemical cycle became the nutrient input of another (Maynard Smith and Szathmáry 1995, p. 35). As explained in Section 4.3, symbiosis means a physical integration following an ecological relationship, in this case among what have been called 'chemical or molecular species'. The crucial step in chemical evolution towards life was the interaction among complex molecules through positive feedback cycles. Two different views on this have been exposed. A first one is that chemical evolution was an accumulation of small incremental improvements. The work on 'hyper-cycles' by Manfred Eigen provides a basis for this view. According to a second view, systems move suddenly to another stable structure which can replicate. Such a self-organising viewpoint originates from the work of Ilya Prigogine.

For a long time, it was thought that the probability of life spontaneously coming into existence on the young Earth must have been extremely small. Stuart Kauffman (1995), however, argues that this probability may have been quite large. He stresses self-organising positive feedback cycles of increasingly complex molecules in an initially extremely diverse atmospheric soup of molecules. These are known as autocatalytic

---

would have been larger, the universe might, under the continuous force of gravity, have collapsed too early for life to appear (Rees, 1999).

cycles. They were first presented in Tibor Gánti's 'chemoton model' (Maynard Smith and Szathmáry 1995, Section 2.3). The first instance of evolution in terms of diversity, competition and selection is found here. Namely, competition between different cycles meant that the one operating more quickly would ultimately dominate. Alterations over time (mutations) in the cycle would then make for a complete early evolutionary process. The model of chemical evolution is extended with the notion of hypercycle (Eigen and Winkler, 1983): a cyclic connection among self-reproducing catalytic, singular cycles. This competition between cycles also allows cooperation, creating positive feedback or nonlinear, even exponential, growth. Some believe such chemical evolutionary processes were stimulated by energy abundance – in the form of direct solar radiation and heat in the Earth's waters and atmosphere – thus permitting energy dissipation to occur more effectively and in higher quantities. Similarly, energy abundance in ecosystems will soon result in an existing species inhabiting the associated niche to benefit from the unused energy, or for evolution to sooner or later alter species so as to occupy that niche.

Competition between different hyper-cycles ultimately resulted in the five main categories of building blocks of life (Ayres, 1994, Section 4.2):

- Bio-molecules like OH, CO, CH_2, CH_3, HCN, NH_2 and NH_3 form the basis for most other building blocks. Oxygen (O), carbon (C), hydrogen (H) and nitrogen (N) are the core elements of these. This is reflected, among others, by their share in the human body weight, respectively: 65, 18.5, 9.5 and 3.5 per cent. In addition, calcium and phosphorus make up 1.5 and 1.0 per cent of the body weight, respectively. All other elements contribute less than 1 per cent (Campbell, 1996, p. 27).
- Amino acids, of which 20 types are found in nature, although many more are chemically feasible. Chains of amino acids form proteins, which serve as chemical catalysts in primitive cells and fulfil a large number of functions in higher organisms.
- Fatty acids, which are carboxylic acids with a long chain of carbon and hydrogen atoms, with saturated or unsaturated bonds between them. They form the basis of lipids ('fats'), which serve as the major energy storage medium of plants and animals, and as the semi-permeable membranes that define the boundary of cells.
- Sugars, which include lactose, glucose and ribose. These act as energy sources for metabolic processes. Chains of sugars form polysaccharides like cellulose, essential to cell walls of plants and other organisms; starch and glycogen are the main energy-carrying polysaccharides in plants and animals, respectively.
- Nucleotide bases connected to chains of phosphates and sugars form DNA and RNA.

The possibility of a spontaneous origin of several of these molecules has been confirmed in lab experiments. This started in the 1950s with the famous Urey–Miller experiments which tried to approximate the conditions of the early Earth's atmosphere (the 'primal soup').

Of course, the origin of life puzzle can never be completely solved in all its detail, as it happened so extremely long ago, without any observers being around, and did not leave any traces in the fossil record. It is uncertain, for example, what provided the most favourable conditions: the atmosphere (clouds, in particular the surface of water droplets), the oceans, tidal pools, lakes or even arctic areas. The atmosphere is less likely because it lacked a sufficiently high concentration of molecules. Oceans, particularly near volcanoes, small tidal pools and geysers make the most sense, because of relatively high temperatures and concentrations of many molecules, which jointly can stimulate relevant chemical reactions. In any case, one type of early life form survived and subsequently was kept alive through an enormously long chain of replications. Life in this sense is analogous to a flame or fire that has to be transmitted in order to survive.

Geophysiology and Gaia

Geological change has interacted with biological evolution. Before life existed on Earth, collisions with meteors and volcanic emissions dominated the Earth's crust dynamics. Weathering processes and organic activity have caused the results of most collisions to be no longer visible. The internal heat of the early Earth was three times as large as the present-day level, causing more intensive volcanic activity. In addition, hydrogen disappeared from the atmosphere into space. Some results of these and other processes are that the oxygen concentration of the atmosphere increased, which was reinforced by organic life later on. Moreover, the atmosphere and hydrosphere gradually changed from being alkaline to being acid. Ayres (1994, Chapter 4) notes two other interesting consequences of the presence of non-biological factors. First, gaseous carbon dioxide (CO_2) is continuously removed from the atmosphere by weathering of basaltic rock. Owing to the fact that this process has been operative for over 4 billion years the amount of organic carbon in sedimentary rocks (shale) is about 20 000 times that in the world's coal reserves (which arose due to, and thus after, organic life). Second, water is plentiful on Earth, for which no fully satisfactory explanation seems to exist. Water not only is essential for weathering and sedimentation on Earth, but also dominates life as a medium for chemical solutions and reactions. As a result, water is a major substance of living organisms. According to Ayres (p. 74) any initial water present on Earth would have 'boiled off' or its hydrogen parts would have disappeared to space. Moreover, the leaching and weathering of crust rocks since the formation of the Earth cannot explain the amount of current water. Instead, it is now widely accepted that the main source of water was provided by relatively small comets (100 tons each) that consisted mainly of ice. During an early period following its formation, the Earth was hit by a regular pattern (in the order of 20 every minute) of these.

The interaction of life with geological and geophysical processes has several dimensions, studied by geophysiology, the close cooperation of geology and biology. One is the balancing of the gaseous composition of the Earth's atmosphere. This is associated with global functions like climate (temperature) regulation, control of the likelihood of fires, oxygen and ozone regulation, pH control, etc. The Gaia hypothesis, formulated by James Lovelock (1979) and elaborated also by Lynn Margulis (Lovelock and Margulis, 1974), states that Gaia is the complex cybernetic or feedback system encompassing the biosphere, the atmosphere, the oceans and the crust of the Earth, which strives to create an optimal physical and chemical environment (homeostasis) for life on Earth. Gaia represents the highest level of ecosystem thinking which considers the abiotic environment as an integral part of the ecosystem. This conceptualisation is only correct at the global level and not on any lower level as ecosystems are then typically open towards an environment, which implies that they neither completely control their environment nor are completely controlled by it. The relevance of Gaia is that it helps us to understand the processes and functioning of the biosphere as well as the logic of its evolutionary history. The hypothesis was criticised by biologists for lacking clear mechanisms and suggesting evolutionary teleology. To overcome resistance to it, Lovelock (1992) tried to capture the biosphere's self-regulating capacity formally in an evolutionary model, called 'Daisyworld'. It described global biodiversity and climate change at a high level of abstraction (see Section 14.2).

The present gaseous composition of the atmosphere is shown in Table 4.2, along with some explanations in terms of related functions of each gas. As indicated by the fourth column, the atmospheric composition is far removed from a chemical equilibrium, i.e. without organic life. From the Gaia perspective, this is explained as the Earth, including its atmosphere, being a sort of living organism rather than a purely chemical system. The balance of concentrations of gases in the atmosphere is delicate. Ayres (1994, p. 76) illustrates this by noting that if the oxygen concentration were 4 per cent higher, forests could not exist as forest fires would be too frequent and hot.

The capture of carbon dioxide and production of oxygen through photosynthesis, as well as the use of oxygen in respiration by organisms, play chief roles in the composition and stability of the atmosphere. As discussed in Chapters 13 and 14, some of these

Table 4.2. Important chemically reactive gases of the air

Gas	Abundance (%)	Flux in megatons/yr	Extent of disequilibrium	Possible function under Gaia hypothesis
Nitrogen	79	300	10^{10}	Pressure builder Fire extinguisher Alternative to nitrate in the sea
Oxygen	21	100,000	None, taken as reference	Energy reference gas
Carbon dioxide	0.03	140,000	10^3	Photosynthesis Climate control
Methane	10^{-4}	1,000	Infinite	Oxygen regulation Ozone regulation
Nitrous oxide	10^{-5}	100	10^{13}	pH control
Ammonia	10^{-6}	300	Infinite	Climate control
Sulfur gases	10^{-8}	100	Infinite	Transport of sulfur cycle gases
Methyl chloride	10^{-7}	10	Infinite	Ozone regulation
Methyl iodide	10^{-10}	1	Infinite	Transport of iodine

Source: Lovelock (1979, Table 3, p. 68). By permission of Oxford University Press.

processes are currently being affected by human activities, notably through the emission of greenhouse gases. In addition, natural cycles of the basic nutrients nitrogen (N), sulfur (S), phosphorus (P) and potassium (K) are disturbed as a result of industrialised agriculture and motorised transport. Finally, for another dimension of interactions between Earthly abiotic and biotic domains, see Box 2.1 on pseudo-coevolution of minerals and life.

4.6   Ecology, Coevolution and Niche Construction

Ecology and evolutionary biology have many connections. Both are in essence about the relationship between living organisms and their environment. Notions like competition, scarcity and a population of organisms occupy a central position in either field, which points at the crucial interface role played by population ecology. But behavioural ecology and ecophysiology – functional diversity in relation to nutrient and hydrological conditions – are essential as well to arrive at a complete understanding of evolutionary biology. The combination of ecology and evolution leads to the notion of coevolution. This will prove to be relevant for later conceptualisations of social-economic evolution as well as its interaction with biological evolution.

From Individual to Ecosystem

Ecology studies the relationships of organisms with their biotic and abiotic environment at four levels: individual, population, community and ecosystem. A population relates to a single species. A biotic community is the assembly of different populations, possibly of distinct species, in a given area. An ecosystem combines various communities of plants and animals that interact with the abiotic environment.

Population ecology studies the environmental factors that determine the size of a population and its distribution. The simplest model of density independent growth is based on compounded growth of population size N: $dN/dt = r \cdot N$, with the explicit solution $N(t) = N(0) \cdot e^{rt}$; here $r > 0$ denotes the birth minus death rate. This describes the well-known phenomenon of unbounded, exponential population growth, as shown in Figure 4.2. It requires net production of biomass, which occurs when the amount of energy fixed is larger than that lost in respiration, that is, the energy needed for maintenance. Net production may take the form of body growth or reproduction.

Unbounded growth has only been found to occur in nature over limited periods of time and up to a certain threshold level of population. The simplest extension that explains this is density-dependent growth taking into account growth-limiting factors. The basic model here is logistic population growth, also shown in Figure 4.2. Its mathematical form is $dN/dt = r \cdot N \cdot (1 - N/K)$, which has the explicit solution $N(t) = K / [1 + (K/N(0) - 1) \cdot e^{-rt})]$. Here r is the intrinsic or instantaneous growth rate, free from environmental constraints, while K denotes the environmental carrying capacity of the population considered, reflecting limiting environmental factors like food (resources), space or predators. The model generates an equilibrium population level K; the net

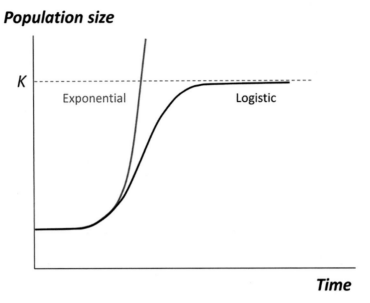

Figure 4.2 Exponential versus logistic population growth.

growth rate for small N approximates r, implying close to exponential growth, while for N close to K the growth rate becomes almost zero. So the population magnitude is dominated by K, while its dynamics are controlled by r. Note that this model is formulated in continuous time; a discrete time, or difference equation, formulation, motivated by sequential generations of a population, can generate different and sometimes strange nonlinear dynamic behaviour, such as bifurcations (May, 1976).

An extension of population dynamics to multiple, interacting species leads to interspecific models, describing relationships between different or similar trophic levels. Table 4.3 shows the main possibilities. One might say that these more complicated models elaborate and endogenise the carrying capacity K in the previous model by a dynamic equation for another species – like a competitor, predator or parasite.

Table 4.3. Types of multispecies interactions

	Type of interaction	Species 1	Species 2
1	Competition	−	−
2	Predator–prey	+	−
	Parasite–host	+	−
	Herbivore–plant	+	−
3	Mutualism	+	+
4	Commensalism	+	0
	Amensalism	−	0

Note: +/−/0 indicates that the population size of one species is positively/negatively/not affected by that of the other.

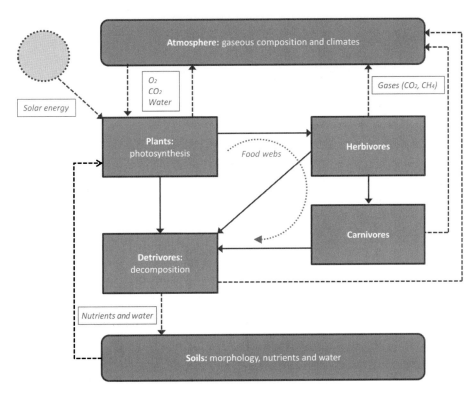

Figure 4.3 Main relationships between biotic and abiotic factors in a terrestrial ecosystem.
(A black and white version of this figure will appear in some formats. For the colour version, please see the plate section.)

To further complicate the analysis, and describe an ecosystem, the concepts of trophic levels, food chain (connecting trophic levels), food web (connecting food chains), and key species are needed. These will describe the energy and material flow through the system and relate the species to each other. Trophic levels are groups of species equally far away from the primary producers, i.e. the green plants, in terms of 'food steps'. Figure 4.3 illustrates the general, aggregate structure of a terrestrial ecosystem.

The dynamics of ecosystems are complex and can be regarded to involve at least four levels:

1. *Reversible, mechanistic population growth*: This involves the discussed logistic pattern, to reflect some carrying capacity and positive or negative interactions with other populations, of predators, prey, competitors, plants, herbivores, parasites, etc.

2. *Succession*: A reversible pattern of successive ecosystem structures towards a climax system. This involves a process of change in the community structure of an ecosystem. A climax can be considered as a sort of maximally efficient level of biomass production, (solar)energy binding and recycling of nutrients. The climax can be a complex or a relatively simple ecosystem, depending on the prevailing abiotic environment – notably its soil, climate and water conditions and whether stable or not. Whereas stable environments allow for complex, fragile ecosystems,

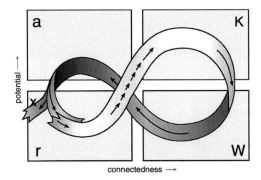

Figure 4.4 Cyclic movement of an ecosystem.
Source: Gunderson and Holling (2002, p. 34). Copyright © 2002 Island Press. Reproduced by permission of Island Press, Washington, DC.

in variable environments simple ecosystems tend to be most robust. The climax system can be seen as reaching maximum stability under the given environmental conditions.

3.   *Reversible or cyclic movement around an ecosystem equilibrium where extended notions of stability are relevant*: A steady state in terms of structure and functions; homeostasis as the tendency to resist changes and remain in an equilibrium; resilience as the ability of a system to maintain its structure and patterns of behaviour in the face of disturbance, or as the speed with which the ecological variables return to their equilibrium values after a perturbation (Holling, 1986). Such a steady state may be disrupted by natural or human-induced catastrophes, such as forest fires, storms, floods, droughts, and animal or plant diseases. As a result, the linear, progressive succession process to a climax is replaced by a cyclic type of mechanism, as illustrated by the so-called 'lazy-eight adaptive cycle' model by C.S. Holling in Figure 4.4. Note that the tail labelled 'x' in the figure indicates the potential for the ecosystem to undergo a regime shift.

4.   *Alteration of the ecosystem*: Evolution alters ecosystem components and in the longer run may even change the fundamental structure of an ecosystem. It represents a slow but irreversible process.

Many ecosystems nowadays are controlled or entirely artificial. Examples of the latter are agroecosystems. These can be regarded as young ecological systems in which organic matter and biomass are not allowed to accumulate, so that a high productivity of desired domesticated plant or animal species is guaranteed. Because natural conditions are insufficient to achieve or maintain such a state, certain artificial conditions are imposed through fertilising, irrigation, drainage and pest control.

Evolutionary Ecology

Two basic strategies associated with the r and K parameters in the above logistic population growth model affect population size and, indirectly, the evolutionary success of species. The *r*-strategy comes down to producing many offspring and applying little parental care. This is typical for insects and amphibians, which are small, mature rapidly and have a relatively short life span. Annual plants also belong to this type. On the other hand, the *K*-strategy entails producing few offspring and applying much parental care. This is typical for birds and especially mammals, which are larger, mature slower and have a longer life span. Forest trees are also often considered as belong to this type.

Because of their features, the growth curves of *r*- and *K*-strategy species differ. Population of *r*-strategists can overshoot and then collapse, thus creating cycles around the equilibrium, partly triggered by unstable environmental conditions. *K*-strategists instead follow the logistic growth curve, with the difference that for low viable levels extinction results. This tends to generate stable populations. Note that selection can change the values of the *r* and *K* parameters in species, which causes any equilibrium to shift.

Young ecosystems in early succession stages are dominated by *r*-strategists. Owing to their features, they are perfect colonisers of ecosystems in which the density of organisms (life) is relatively low so that there is room for (close to) exponential growth. Larger *K*-strategists with a stable population size are typically found in older climax ecosystems which have a more stable environment, a higher density of organisms and more complex food-web structures. Typically, *K*-strategists have evolved later in natural history, as they required a complex ecosystem for ecological support and a larger brain for performing care functions. Although the *r*- and *K*-strategy distinction is somewhat simplified, evolution nevertheless seems to have promoted a movement towards one or the other extreme, depending on whether initial selection pressure pushed a species in one or the other direction – an example of path dependence – through positive feedback or reinforcement.

Evolution, in an ecological context, essentially means ecological adaptation. This involves foraging and reproduction strategies as well as size and social structure of groups (Putman and Wratten, 1984). With regard to foraging strategy, it is evident that efficiency of food gathering, in terms of intake/effort ratio, is essential to survival. It is the net result of locating, gathering or catching, and eating food. Depending on the species, one even has to account for processing, transporting and storing food. The energy costs associated with each of these activities differ between animal types. For predators, the hunting strategy is important: sit-and-wait, search, pursue prey, etc. These strategies go along with physical and mental features of the associated predators. The different options explain the wide diversity of foraging behaviour among similar species that can be regarded as specialisation.

Reproduction strategy or sexual behaviour involves various aspects, such as a socio-sexual system (monogamy, polygyny, etc.) and investment in offspring, from producing sex cells, through supplying food, to teaching of young. One extreme strategy is deserting the partner after copulation, to avoid costs of care for offspring. Higher mammals invest much in offspring and females more than males. In particular, pregnancy implies a considerably larger investment by females than males. The male–female investment differences are the largest for mammals. As males are 'free' after a minor, successful coitus investment and females are tied because of their pregnancy, in mammals one therefore finds considerable male promiscuity and polygyny (the habit of mating with more than one female). In humans, the pressure of social norms – probably triggered by the higher survival of offspring in two-parent than one-parent families – has reduced promiscuity significantly. Possibly this had a selective advantage given the extremely long period during which human children depend on, and learn from, both parents. This topic is discussed more extensively in Section 4.7 on sexual selection.

It has been argued that both group size and social structure adapt to the environment. This occurs, among others, through the previous strategies, i.e. foraging and reproduction. Advantages of certain group size and structure are related to sensitivity to predating (detection, avoidance, defence), efficiency of foraging, access to the opposite sex and need for social learning under environmental conditions. Evidence for a wide range of species shows that particular environments enhance specific group structures: inside dense forests solitary behaviour is found; at the forest edge small family groups dominate; and in open grassland or arid zones individuals tend to be part of large groups.

Coevolution

Species are shaped by evolution in interaction with other evolving species. This insight has given rise to the notion of coevolution. It was originally proposed in ecology to refer to the joint evolution of butterflies and flowering plants (Ehrlich and Raven 1964). But examples of coevolution go far beyond this. Early evolution already depended on coevolutionary processes. Large animals and longer food chains could not have developed in an anaerobic environment as characterised the early atmosphere, but had to await an aerobic environment enhanced by oxygen-producing species. Moreover, symbiosis evolved from multispecies ecological relationships, notably parasitism and mutualism. In addition, evolution of ecosystems permitted interaction among autotrophs (green plants and cyanobacteria), using light as their energy source, and heterotrophs, using organic compounds as their energy source.

Coevolution explains the evolution of communities involved in multispecies relationships. Since these communities make up the living part of the ecosystem, coevolution really is a theory of ecosystem evolution. It is not simply Darwinian individual selection scaled up but, because of interaction among species, introducing a highly nonlinear effect in evolution. Coevolution means that multispecies categories are not fixed forever. For example, coevolution of parasites and hosts can lead to commensalism, where the parasite inflicts minimal damage to, or even creates benefits for, the host. A stable predator–prey relationship can be regarded as commensalism at the population level. Namely, the predator population does not 'kill' the prey population, even though individual predators kill individual prey.

Due to repeated mutual selection, prey have become better adapted to predators and vice versa. For instance, both run faster or prey hear better, or predators work together. The 'arms race' has been used as analogy for the reinforcement mechanism or positive feedback involved: predation will select prey that run faster, which in turn will select predators that run faster, and so on. Bacteria and viruses that cause diseases in hosts, leading to rapid death, and animals carrying them, tend to coevolve in a way that the pathogen and host can co-exist for longer and the disease is more easily spread.

Evolution may even turn the relationship from parasite–host into mutualistic.[8] In addition, under the pressure of selection, competing species specialise such that they do not occupy the same niches. For instance, certain insect-eating species in the same locus may focus on different parts of plants (leaves, branches, trunks).

Plants evolved to develop characteristics like colourful flowers to attract insects that spread the pollen. The insects develop certain body shapes to be able to get to it. The changes in the flower features can create a barrier to certain pollinating insects such that some kind of ecological isolation of subpopulations of flowers is realised and, in the long run, speciation is promoted. Driven by selection pressure, plants also developed chemical and physical characteristics to resist insects or mammals that eat them. In response, the latter evolve chemical or physiological features that warn about or allow circumventing the plant defence system. Examples are taste, to identify bitter and thus potentially dangerous foods, and a liver to detoxify poisonous substances.

The response in pregnant women to morning sickness, that is, aversion against common foods or even vomit them, can also be understood in this context (Profet, 1988). Such sickness stimulates the mother to be careful in eating bitter, flavoured or novel foods, which protects the foetus against potentially harmful substances, even when these might not harm the mother herself. The cost of such behaviour may be reduced intake of nutrients. Pregnancy sickness typically starts when the embryo's organs begin to develop, at which point the wrong chemicals might cause serious birth defects, while it fades as soon as the organs are in place and the stage of rapid growth, requiring many nutrients, takes off. This suggests that evolution has been capable of optimising the mother's eating behaviour to the specific needs of the foetus. Likewise, there is evidence that people with allergies are at much lower risk of some types of cancers, arguably because they have a lower intake of carcinogenic substances (Sherman, 2008).

Coevolution not only pertains to visible animals and plants. Even though such life forms have become immensely diverse after the appearance of eukaryotes, prokaryotes (including all the bacteria and Archaea) are still critical to ecological conditions and processes on Earth. As an indication, their collective worldwide biomass is being estimated at about equal to that of eukaryotes. Bacteria are often associated with diseases, but the ecological relationships between bacteria and other species are much more diverse. Many prokaryotes, along with fungi, are decomposers of organic and nonorganic waste. Others are involved in symbiotic relationships with other organisms in the form of mutualism, commensalism or parasitism (Table 4.3). Many of them are nowadays even used in human-controlled, economically productive processes, such as sewage treatment, food processing, production of medicines and genetic modification.

Taking coevolution to a different level, William Durham (1991) has elaborated the interdependence of human genetic and cultural evolution, identifying five categories of coevolution. This is discussed in detail in Section 7.4. Human invention of agriculture,

---

[8] There is much now known about the kinship and family tree of viruses affecting algae, plants and animals, including humans. This is especially accurate for larger viruses as more genes can be compared with other viruses. Such a tree of viruses has multiple origins, unlike the tree of life, and is not part of the latter tree as viruses are not life forms. For more information see the Tree of Life Web project (http://tolweb.org/tree).

involving domestication of animals and especially plants, and subsequent cultural-economic developments can be regarded as a special case of coevolution among animals and plants. Humans depend on the cultivated and selectively bred plants, and the plants depend on human control. This line of research still continues, paying increasing attention to how current small-scale societies evolve under the pressure of rapid transformation of their natural environment, including how they adapt their practices of environmental management (Reyes et al., 2017).

A warning is justified. Coevolution is often used in a loose, imprecise way, namely to denote mere mechanistic interactions between variables or subsystems within a larger system. This misuse of the concept is especially common in the social and environmental sciences. Coevolution in a strict sense is a much more complicated process, involving interaction between evolving populations with internal diversity, resulting in mutual selection. One would best avoid the use of 'coevolution', and instead opt for the term 'co-dynamics', to refer to any other type of interaction among subsystems in a larger complex system (van den Bergh and Stagl, 2003; Winder et al., 2005). Another incorrect use of coevolution is to denote that multiple characteristics of an individual jointly undergo evolutionary change in response to the same selective pressures. A more precise term for this is 'co-adaptation'.

Niche Construction

The notion 'ecological niche' denotes an organism's or species' place in the living environment, in terms of its dealings with food, competitors and predators. Niche construction theory introduces additional feedbacks in the already complex process of evolution we have arrived at so far. It denotes that organisms evolve in a niche that is not completely out of their control. Instead, they modify it and hence alter natural selection pressures for themselves, and by default for other species (Laland et al., 1996; Odling-Smee et al., 2003). Richard Dawkins (1982) expressed a similar idea in his notion of 'extended phenotype'.

Niche constructing theory adds a distinct ecological dimension to evolution by recognising that many organisms influence the features of the habitat they and other species use. They are practising what has been called 'ecological engineering', a term common in human management of ecosystems (Mitsch and Jørgensen, 2003). The immediate reason of such engineering is to redirect or control energy and material flows so as to benefit from these in terms of food intake, creating a local climate or protect oneself from predators.

The evolutionary significance of niche construction is that it changes the selection environment, in turn affecting future evolutionary dynamics. Niche construction can generate both evolutionary innovation and stability and can contribute to rapid change or delayed responses to selection. For example, beaver dams may, after beaver activity has stopped, give rise to meadows that can last for decades without being taken over by the original vegetation. An early example of niche construction in natural history is the case of cyanobacteria altering *c.* 2.3 billion years ago their niche by releasing oxygen through photosynthesis. This ultimately gave rise to the oxygen-rich atmosphere of the Earth, which was a crucial factor of the direction taken by subsequent biological evolution.

Like coevolution, niche construction challenges the old idea that organisms are passive objects subject to selection by their abiotic and biotic environment and that organisms adapt to their environment and not vice versa. One difference is that while coevolution is rather symmetric regarding the species involved, in niche construction one species is asymmetrically changing the environment for itself or others. Another difference is that niche construction further involves not just biotic but also abiotic elements, as the environment affected includes both of these. To complicate things, niche construction, if effective, is likely to influence the existing coevolution with other species. Niche construction is not a rare exception: all living organisms alter elements of their environment, to a greater or lesser degree. Humans have gone extremely far in altering their niches over time, through cultural, economic and technological mechanisms (Kendal et al., 2011). In line with this, niche construction theory has been argued to be useful in thinking about strategies to foster a transition to a sustainable economy (Laland et al., 2014).

In an interesting interaction between Laland et al. and Wray et al. (2014), the first team argues that an extended evolutionary synthesis is needed around the relatively new concepts of phenotypic plasticity (Fusco and Minelli, 2010), inclusive inheritance, niche construction and organism development. The second team argues, however, that these phenomena are just 'add-ons' to the basic evolutionary processes of natural selection, drift, mutation, recombination and gene flow, and that Laland et al. suggest divisions which do not exist.

4.7   Sexual Selection and Cultural Effects

One category of natural selection often receives separate attention. We continue this tradition here. It concerns sexual selection, which is at stake when individuals of one sex are involved in the competition for mates of the other sex. The relevance of this competition for evolution will not come as a surprise given that the mating of sexual partners is the process by which offspring is generated. Key to sexual selection is that fitter individuals are better able to find fit mates, and so will have more reproductive success in terms of more or better quality offspring. Darwin and Wallace equally recognised the importance of sexual selection. Darwin considered it to be so significant that he wrote an entire book on it, *The Descent of Man, and Selection in Relation to Sex* (1871). Some think that this represents his most ingenious work (Van Rhijn, 2013). Anyhow, he was far ahead of his time, as sexual selection remained a neglected topic of research until almost a century later.

Sexual selection has been particularly apt in creating unusual or extreme physiological or behavioural characters in certain animals, such as the peacock's tail, and – arguably – the human brain (Section 11.3). This has involved what is known as a 'Red Queen effect', a kind of evolutionary arms race, predominantly among males competing for females (Ridley, 1995). The notion of 'Red Queen' comes from Lewis

Carroll's book *Through the Looking Glass* in which Alice is told by the Red Queen that "in this place it takes all the running you can do, to keep in the same place".

Something similar to sexual selection is relevant to flowering plants. Their interaction with insects has given rise to interactive evolution with mutual selection exerted by the plants and insects. Such coevolution generates, like sexual selection between the sexes, an evolutionary arms race, but then between distinct species instead of between males and females of a single species, as in sexual selection. This explains why one can find extreme adaptations resulting from either process. One might perhaps say that coevolution of flowering plants and insects represents a kind of 'indirect sexual selection'.

Sex can be loosely defined as exchanging and combining genetic material among different living organisms. This involves the process of meiosis, causing a transformation from diploid to haploid phases in the life cycle of an organism. Sex has the advantage of increasing genetic, and thus phenotypic, variability, in turn accelerating evolutionary change. It can occur in both prokaryotes and eukaryotes, but is more frequent in the latter, explaining the faster pace of evolution and major changes in eukaryotes. Sex was probably preceded by recombinant repair in cells. Indeed, a similar chemical mechanism is involved, involving crossover of chromosomes, requiring some sort of fracture and repair that gives rise to limitless variations (Maynard Smith and Szathmáry 1995, p. 149).

Sex augments the diversity space as genotypic information of any male in a population can be mixed with that of any female. Another benefit of sex that may have contributed to its emergence is that it makes organisms more resistant to microorganisms that cause diseases (pathogens) or remove energy and materials (e.g. through parasitism). Through sex and the associated major genetic recombination, protection against such microorganisms is more easily altered or renewed from generation to generation, unlike in asexual organisms. This is, though, still debated (Otto and Nuismer, 2004).

One can regard sex as a specialisation which has a certain cost, relative to parthenogenesis, a form of (animal) reproduction in which the egg develops into a new individual without fertilisation. However, at the population or species level it enlarges (potential) diversity enormously, which is crucial for species survival in changing environments. While most plants and many lower animals are hermaphroditic – capable of producing both types of sex cells – animal species predominantly consist of strictly male and female individuals. Sex is asymmetric in many ways, not just because of differences in male and female brains, genitals and many other bodily aspects, but also in the sexual strategies undertaken, including the gametes produced. The relatively large female egg cell (ovum) incorporates the contents of the sperm – which can be seen as a type of symbiosis (Section 4.3). The sperm is just half a set of male genes, while the egg is a complete cell, with all its metabolic equipment needed to survive and divide. The reason for this particular asymmetry is arguably that merging two complete cells would represent a much more difficult challenge, physically and chemically. Moreover, it would lead to severe conflicts between virtually different organisms. Notably, the mitochondria of the two parents would end up in a battle. Ironically, whereas the male is generally stronger than the female, his gamete is defenceless against the egg cell. Incidentally, conflict between the sexes is a common phenomenon in the history of sexual selection.

Adaptations that benefit one sex may go at the cost of the other, the extreme case being some female spiders eating their partner after copulation. This has been explained as not only providing nutrition for egg formation but also as selecting against older or otherwise unfit males who might through repetitive copulation weaken the species' gene pool. Fit males would be able to escape a cannibalistic fate.

More than two sexes (or genders) are uncommon in nature: it increases the possible interactions and conflicts, without adding many benefits. Not surprisingly then, species with more than two genders do not create offspring that integrate more than two gametes. Notice that in the science fiction literature one can find examples of three-gendered species, as in Isaac Asimov's *The Gods Themselves* and in Orson Scott Card's *Speaker for the Dead* (from the celebrated *Ender Saga* series). For more discussion of it in the real world, see Joan Roughgarden (2004, notably Chapter 6). This book presents an unorthodox viewpoint, arguing that sexual selection theory cannot explain all the diversity in observed sexual behaviours and characters. To support this view, it mentions more than 300 vertebrate species in which individuals have both male and female organs, change their sex during life, show homosexual behaviour or have other uncommon sex-related features. Not surprisingly, this book received mixed responses. One critic claims that 300 out of 40 000 vertebrate species means less than 1 per cent of species show special features or behaviours, and suggests that she wrote the book merely to enhance equality for gay and transgendered people (Allen, 2005).

Darwin already noted that males and females differ, especially in polygamous species, and are similar in monogamous species. A relatively large, manifest or ornamented body can be explained by creating force and power to compete with other males for mates – known as intrasexual selection – and exerting sex-appeal towards females – intersexual selection – allowing for sexual relations with multiple females. Males usually invest less of their energy in parental care of offspring than females. The evolutionary theory of parental investment explains that females are generally pickier and less promiscuous than males because of the gestation period, which means they invest more in the offspring (both before and after birth) and can produce fewer offspring than males. One might be tempted to conclude that women are the selective sex exerting strong selection pressure on men, but men's taste exerts selection pressure on women as well. Women and men tend to use different criteria in looking for partners – though perhaps not so much when it comes to lovers. Pinker (1997, p. 482) summarises this as 'In our society, the best predictor of a man's wealth is his wife's looks, and the best predictor of a woman's looks is her husband's wealth.' Indeed, women are generally more preoccupied than men about looking attractive through appropriate clothes and make-up, reflecting that men are more interested in women's looks than the reverse.

In most animal species, males are relatively promiscuous, aided by the small invest-ment needed – a few minutes, or even less time – in copulation, with no strings attached afterwards. This has caused males to compete for access to females, generally giving rise to strong and large males. Only in a few species are females bigger, stronger or better armed, which is due to their polygamy. While these are all endosomatic adaptations, in humans with their technology-driven societies males' desire for sexual variety has given rise to an exosomatic adaptation, namely a large pornography industry.

Sexual selection has given rise to an evolutionary arms race and excessive phenotype features, leading to sexual dimorphisms, i.e. distinctive secondary characteristics relevant to courtship. This is typical for most animals, including for humans, and the more pronounced, the more competitive males are for access to females. When such competition is ineffective – e.g. when females live in large groups or are all fertile at the same moment – sperm competition is more effective. This takes many subtle forms and is especially relevant in some species where females have intercourse with various males, leading to competition between sperm from different males in the vagina. This adds to the already existing competition between the millions of sperm cells of a single male that enter the female body during copulation. Sperm competition guarantees quality, as only the healthiest and fastest sperm have a serious chance to make it to the egg cell, while weak and defective ones will be selected out. Not surprisingly, it has given rise to many specific adaptations in insects, reptiles and mammals to assist in winning the sexual competition. Examples are: assuring sperm can survive for a sufficient time inside the vagina, insertion or spontaneous emergence of copulatory plugs after copulation, release of fluids by the male during copulation that work as an anti-aphrodisiac in the female or destroy the sperm of males with whom she copulated recently, guarding the female against other males generally or during fertile periods and producing large testicles so that more sperm can be ejaculated during coitus.

Typically, monogamous species or species with dominant males and harems, such as gorillas, have relatively small testicles as they do not have to compete for females. But males who cannot dominate the large groups in which they live and have to compete for females, such as chimpanzees, have relatively large testicles. Human males are intermediate in this respect. They are about 1.15 as large as women, suggesting that men have competed for women during human evolutionary history. As humans live in large groups, there is a risk of promiscuity, which has been limited by the emergence of monogamy-encouraging marriage. One can speculate about why this has become such a widespread institution across human societies. A possible reason is that it reduced violent conflict between men over women in early societies, which not only played in the cards of the leaders – who could then use scarce manpower to fight external enemies – but also contributed to more equality. Pinker (1997, p. 478) suggests that early Christianity appealed to poor men as its requirement of monogamy offered them a fair, egalitarian chance of marriage. Regarding monogamy, though, there is considerable diversity among (current and past) human groups. In fact, of the 1231 cultures documented in the *Ethnographic Atlas Codebook*, 186 were monogamous, 453 had occasional polygyny (one husband with multiple wives), 588 had frequent polygyny, and 4 had polyandry (one wife with multiple husbands, mostly brothers).[9]

Human sexual dimorphism expresses itself through many differences between men and women, in terms of body shape and size, muscle force, brain, breasts, voice and body hair. Voices are an essential part of sexual attraction and differ as they signal body size: women like deep voices as they are typical of large, firm-bodied men, while men

---

[9] See http://www.psychologytoday.com/blog/darwin-eternity/201108/are-people-naturally-polygamous-0. And http://eclectic.ss.uci.edu/~drwhite/worldcul/atlas.htm.

prefer small bodies and high-pitched voices (Xu et al., 2013). Breasts are another notable difference for which different evolutionary explanations have been offered: one is that permanent breasts hide pregnancy, keeping males continuously interested in sex; another is that bipedalism stimulated a missionary sex position in which breasts substituted for the sexual stimulation function played by buttocks during coitus from behind (doggy-style position). Both explanations are speculative.

Why do humans have sex even when the woman is not fertile? According to Jared Diamond (1997b), it creates a strong, long-term pair bond and causes men to be frequently at home to contribute to taking care of their children – contributing to a high survival rate of human babies and young children. Survival is at stake without good parental control, since humans are immature and even helpless for a long time, having been born relatively early to avoid birth complications in view of the large human brain having to go through a narrow birth channel. For this reason women have also evolved a wider pelvis than men. Since further widening would have hampered them in walking upright, evolution has arranged a compromise between the two crucial functions of the female anatomy. Indeed, stability of walking until old age – while carrying food, children or artefacts – may have been so important in evolutionary human history that under the influence of hormonal changes during the menopause women's hips tend to become narrower (Huseynov et al., 2016). Women undergo this menopause at an older age, generally between 45 and 55, implying the end of fertility. An evolutionary explanation is that it allows a woman to be free of the risk of bearing another child to fully concentrate on assisting her adult children to raise their offspring. Menopause is unique to humans, i.e. not found in other mammals. It may have been a crucial condition for the evolution of human intelligence, as grandmothers improved the survival rate of their large-brained grandchildren (Croft et al., 2015).

Diamond further argues that a man gains from a long-term parental bonding as he can be more confident that the woman's offspring is his than in the case when he would be sporadically or never at home. In contrast to many other species where males do not provide parental care, human males have no clue about when a woman is fertile. This contributes to men guarding their women against sexual advances made by other men as well as being prepared for sex at any time, even on days when fertilisation is not possible. It explains why most human sex is mainly fun and not functional in terms of immediate fertilisation or reproduction, whereas for most other (mammal) species – exceptions being bonobos and dolphins – sex is directly functional. Here females tend to advertise the phase of ovulation by sending out chemical or visual signals that arouse males.

Why do humans generally practise sex in isolation, invisible to others? Is it because of shame feelings? We also wear clothes and most of us feel uncomfortable showing our naked body to anyone. Or is it, as van Hooff (2002) has suggested, because it reduces sexual competition between males? He thinks it is not part of a cultural adaptation but is a much older habit. In any case, if one considers sexual dimorphism, testicle size relative to body size and socio-sexual systems (monogamy, polygyny, etc.), then humans do not resemble any of the other primates. Surprisingly, as one can see from Table 4.4, we are less similar to chimpanzees and bonobos than to gorillas and

Table 4.4. Socio-sexual features of the great apes and humans compared

Hominid	Sexual dimorphism in terms of weight male/weight female	Size testicle as % of body weight	Socio-sexual system
Gibbon	1.0	0.6	Monogamy ♂♀
Orangutan	1.7	0.4	Polygyny ♂♀♀♀
Gorilla	1.9	0.2	Polygyny ♂♀♀♀
Chimpanzee	1.2	2.7	Polygynandry ♂♂♂♀♀♀
Bonobo	1.1	3.1	Polygynandry ♂♂♂♀♀♀
Human	1.2	0.3	Monogamy and polygyny* ♂♀(♀♀)

Source: van Hooff (2002, Figure 4, p. 41), slightly adapted and simplified.
With permission of J.A.R.A.M. van Hooff.
Note: The term 'polygamy' is more commonly used for humans, but is broader as it refers to the practice of having more than one spouse, whether this applies to a man (polygyny) or a woman (polyandry). An even broader term is polyamory, the practice of amorous relationships involving multiple people.

orangutans, but perhaps we are most similar to the monogamous gibbon. As discussed, the large majority of historical human societies were polygynous, and currently widespread monogamy is likely due to cultural-religious factors. Illustrative of this is that human monogamy involves frequent promiscuity and infidelity. So, at best, we are what Engber (2012) calls 'monogam*ish* '. This is perhaps not surprising if one takes into account that perfect monogamy is exceptional anyway among animals in general and the great apes in particular.

Sexual selection has been suggested to be co-responsible for our large brain, as elaborated in Section 11.3. It has further been hypothesised that some human behaviours, such as the creation of art and music and verbal creativity, including humour, are adaptations to make courtship more effective, having been favoured through a process of sexual selection. Regarding music, biologists tend to give more credit to it being a by-product of brain evolution (Section 7.8). There is no doubt that sexual selection has influenced physiological, brain, mental and thus behavioural differences between men and women. An extreme feminist perspective might suggest that human culture resembles a patriarchy or even male conspiracy that has exploited women. In a very informative and well-documented article, Barbara Smuts (1995) contends that human societies show more male dominance than most other primates. She offers various hypotheses to explain this: a reduction of kin dominance in social relationships leading to fewer female–female coalitions; elaborated male hierarchy formation and alliances in cultural 'network societies'; traditional male control over resources; female sexual strategies that reinforce male control over females (such as preference for resource-rich men); and the evolution of language which has tended to reinforced male power. She ends on a positive note, suggesting that while it may be hard to escape the current situation, it should not be judged as impossible, since humans show a tremendous plasticity and variation of behaviour.

Another view, employing evolutionary thinking, is that culture is a system which has competed with other systems – including no culture and distinct cultures – and used both men and women for this purpose. Following closely the arguments of social psychologist Roy Baumeister (2007, 2010), Box 4.1 discusses how this may have affected gender roles in a cultural setting, arriving at rather similar conclusions to Smuts. A still different take is offered by Konner (2016) who suggests an 'X-chromosome deficiency syndrome', that is, males and masculinity being the outcome of some birth defect. He emphasises that this 'syndrome' causes a series of deficiencies in its carriers, including serious ones such as being violent, having a short life span, ruthlessly aiming for power and suffering from various brain defects, as well as less serious ones such as premature hair loss and attention deficit.

4.8   Economic Phenomena in Biological Evolution

Many writers have noted the similarity between biological and economic systems. It inspired Vermeij (2006) to write a fascinating book entitled *Nature: An Economic History*. In it, he argues that both systems are characterised by growth, decline, diversity and inequality, and are driven as much by competition, cooperation, specialisation, exchange and innovation as by energy availability and efficiency, distribution of material wealth and power, and external disturbances. Vermeij thinks that biology is equally well described by generalised economic concepts and processes as economics is by evolutionary ones.

Darwin recognised some of these similarities. In Section 2.2, we discussed the well-known influence of early economists, notably Malthus, on Darwin. But there is much more to say about this. To illustrate, in Chapter V on the 'Laws of Variation' (p. 114) of his *On the Origin of Species* (1859), Darwin cites Goethe's 'law of compensation or balancement of growth [. . .] in order to spend on one side, nature is forced to economise on the other side'. In the ensuing discussion, Darwin argues in favour of a related but somewhat less strict principle, that 'natural selection is continually trying to economise in every part of the organisation. If under changed conditions of life a structure before useful becomes less useful, any diminution, however slight, in its development, will be seized on by natural selection, for it will profit the individual not to have its nutriment wasted in building up a useless structure.'.

Darwin thus introduced economic reasoning into evolutionary thought. He also made use of the concept of labour division, which relates to specialisation, both in and between organisms (e.g. in groups or sexual mates). Subsequently, many other authors in evolutionary biology have employed economic mechanisms and reasoning. For example, it is now quite popular to interpret interactions among individuals from one

---

Box 4.1 An evolutionary view of gender roles and successes in cultures

With the rise of culture, the world of men changed more than that of women. In early societies, women continued their care-taking tasks as before. Men, however, used their evolved abilities by participating in large networks of relationships that make up economies, religion, science, trade, military organisations, etc. Here a gender distinction in social behaviour plays a role: women are generally seen as more social than men, and indeed they more often show close, intimate relationships with friends and family in which they are not afraid to show their emotions. Men are also social but in a different way: they tend to have, already in childhood, more superficial contacts, often hiding their genuine feelings; this is functional in larger networks of social interaction characterised by competition, formal relations or hierarchy. As a result, men were more successful in such network interactions, which in turn delivered them power, knowledge and wealth. This has resulted in increasing inequality between the sexes. To this should be added that all cultures tend to protect women while they employ men for dangerous, risky activities. This provides an additional explanation for why men disproportionately achieve wealth, power, success and fame – both in history and modern times.

But inequality also exists among males. The reason is that men who are more talented or more motivated to undertake risky tasks are more likely to become successful. Talent and motivation, though, are not equally distributed. In fact, data indicate that while men and women are on average equally intelligent, among men there are more extremes. Indeed, men are overrepresented among both the most and least intelligent individuals. One might say that nature gambles more with men than with women. This greater phenotypic variability among men also holds for other features, such as height, weight and physical performance (Lehre et al., 2009). These differences easily translate in higher average salaries for men than women, even in the absence of a significant difference in average performance. Cultural institutions and reward systems are unable to correct for this. For example, a minimum wage as is common in many countries does not compensate for any excessive salaries received by high-performing males.

Another factor is motivation, which is partly genetic and partly cultural. Most people, both men and women, do not like mathematics. But among those who do like it, men are overrepresented. Combining the distribution of talent and motivation can explain why men are more found in activities associated with mathematics. In many countries, current public policy is based on the premise that motivation is scarcer than talent, resulting in efforts to improve the motivation of young women to study mathematics, engineering and physics, fields traditionally dominated by men. Likewise, policies can be designed to stimulate men to adopt more care-taking tasks.

To understand gender roles in culture, from an evolutionary angle one must consider differences in sexuality as well as in reproductive motivations and strategies between men and women. Whereas women can give birth to a limited number

Box 4.1 (*cont.*)

of offspring during their life, men can produce many more. This is consistent with the insight from DNA studies that current humans descend from twice as many women as men. Possibly, roughly 80 per cent of women and 40 per cent of men in the past have reproduced. In line with this, we descend more from socially and biologically successful men – the extreme examples being the many kings and sultans in history who had hundreds of children.

Women's reproductive strategy traditionally has involved avoiding risks and being attractive to men. Men's strategy traditionally was to take risks and compete economically and sexually with other men. This might also explain why in many cultures the life of women is valued and protected more than that of men. Although not everyone will appreciate this explanation, and some will quickly judge it as being politically incorrect, we should worry more about evolutionary and psychological incorrectness. It is undeniable that the fundamental differences between how men and women on average operate in cultures have biological origins. In fact, one of the first classifications humans make regarding other humans is 'male/female'. Children become aware of the gender difference around the age of two, and start developing their own gender identity not much later. Of course, this is affected by the extent to which gender stereotypes are offered in the relevant cultural setting of each child. In this respect, one can observe differences among cultures and changes over time, guided by public policies. Something as simple as a beard is telling is this regard: in many Western countries it has become an exception among men, causing an obvious distinction between the sexes to be less prevailing.

Many cultures have adapted to the widespread desire to reduce differences in opportunities and successes between men and women. It would be wise to formulate policies that aim to reduce gender inequality consistent with the previous insights. This includes recognising why men and women have historically played a complementary role in cultural processes and understanding the challenges associated with breaking away from it. A well-informed scientific approach would contribute to the effectiveness of gender equalisation and women's empowerment policies.

Inspired by Baumeister (2007, 2010).

or multiple species as markets, including mate choice in terms of a sexual market (Werner et al., 2014). The use of a market metaphor inevitably leads to identifying price variables. The evolutionary angle is that market demand can select certain characteristics of supply and vice versa, giving rise to gene-based adaptations over time. Coevolution and symbiosis are concepts often used in this context. Likewise, in evolutionary economics, we can find many agent-based models that take the form of coevolution of populations of suppliers and consumers (Box 8.1).

Applications of market metaphors in biology include grooming among primates for favours in return, mate choice driving the evolution of secondary sexual characters, cleaning activities by one fish species of another, and mutualism of fungi or bacteria and plants. A relevant indicator for evaluating and predicting such interactions is the supply/demand ratio. If supply is small relative to demand, a supplier can choose among many alternatives, which will work to his benefit. Either he can devote less energy (effort) to interactions, i.e. wait for demand because scarcity works to his advantage, or select the best option. In evolutionary game analysis of such issues, economic notions are used, such as least-cost solutions, competition by contest or outbidding and preference for best value services. In the context of pollination markets, metaphors like 'investment in nectar', and 'advertisement through attractive flowers' are used. Such investments and advertisements are found to be stimulated by a high ratio of flowers to pollinating insects (Noë and Hammerstein, 1995; Noë et al., 2001).

In an original book, Haim Ofek (2001) offers arguments, evidence and speculation to convince his readers that important aspects of human evolutionary history can only be fully understood if economic arguments are brought in. In particular, he suggests that exchange, or trading one thing for another, had arisen before humans had evolved to the current modern form, and, in turn, affected their biological-genetic evolution. In particular the capacity for 'mercantile' — or non-kin or non-nepotistic — exchange is unique to humans and depends on skills like communication, negotiation, quantification, abstraction and orientation in space and time. In turn, these require appropriate linguistic, facial expression, gesture, memory and mathematical capacities. In addition, mercantile exchange goes along with mutual trust, which can reinforce one another.

An unresolved question in human evolutionary biology, according to Ofek, is what function was served by the higher faculties of the human mind during the early stages of the human species. Candidates suggested include care-taking, sexual competition and, above all, interaction and cooperation in large groups of humans, which all set demands on such faculties. The capacity for mercantile trade as an additional explanation for the larger brain has not received much attention, even though such trade contributes positively to survival and reproduction of individuals or groups involved in it.

Such success derives from the fact that exchange between humans, or even human groups, allows for diversity in consumption, specialisation and labour division in production, redistribution of wealth, cooperation and maintaining inventories. These contribute to an increasing efficiency of certain productive activities, such as hunting, food gathering and child rearing. This labour division moreover allowed for a great capacity to learn, notably through the task-oriented plasticity of the hand, including the complementary pair of hands and hand–eye control. This is confirmed by the fact that the largest part of the motor cortex is devoted to control of hand-related functions.

Ofek spends a great deal of time on deliberating how market or mercantile exchange, 'by treaty, by barter, and by purchase' in the words of Adam Smith, is similar or analogous to exchanges observed in nature, of which he identifies two cases: (1) nepotism depending on kin selection and involving physical interactions (as in social insects) or 'emotional currency' (like begging and submission in wolves); (2) multispecies symbiosis that is reciprocal, i.e. with mutual benefits, and depends on coevolution involving instincts or reflexes (e.g. flower–insect relations). Market exchange is not restricted to kin but extends to non-kin. Whereas symbiosis is restricted to a particular resource in a specific environmental setting, economic market exchange involves multiple and potentially novel resources. More importantly, while symbiotic exchange is between individuals of different species, nepotistic and market exchanges are intraspecific, which can stimulate labour division and specialisation. Another aspect of exchange is asymmetry. Ofek notes similarity between asymmetry in sexual markets, where males compete and females choose, and in economic markets, where vendors compete and customers choose. The similarity is due to the fact that both females and customers suffer the long-term consequences of any wrong decision made and hence are stimulated to be more careful than males and vendors.

Labour division is a theme that runs through both evolutionary biology and economics, being responsible for productivity and efficiency improvements in both fields. Labour division underlies cell differentiation and complementary organs within multicellular organisms as well as sector and product diversity in economies. One might add that labour division is also observed between the sexes, a third category of exchange of the interspecific type, which Ofek puts under nepotistic exchange. Sexual labour division was crucial to hunter-gatherer societies: men were mainly involved in hunting and women in gathering. Ofek argues that it became less important after the transition to agriculture, even though some tasks, such as ploughing, were less suitable for women than men as they required greater body size and force. Early industrialisation with heavy machinery, such as steam engines, relied relatively much more on male inputs, while the emergence of electricity and health care prolonging life beyond child-bearing age, allowed a steady increase of women participating more equally in labour markets – a trend that is still ongoing. This illustrates a general phenomenon, namely that sexual patterns of division of labour in human societies alter in response to technological and environmental changes.

Labour division can only be driven by evolution if specialisation and labour division occur at a level of organisation below the unit of selection. The reason is that labour division involves interaction or cooperation between entities that need to share and redistribute the benefits in a way that they all come out well. This is achieved by selection, or competition, among units characterised by distinct schemes of internal labour division. If a certain labour division then works out well for the unit, it can be selected and become more common, i.e. after replication and diffusion.

For selection to achieve labour division, entities within an organism or group need to have the phenotypic plasticity that allows them to specialise in a particular task. For example, labour division in human societies is so extensive because the human mind is flexible and, along with it, the human hand (controlled by the brain–mind), allowing for endless task performance by humans. Ofek notes that perhaps only the trunk of an elephant, also a highly social animal, can be compared to the human hand in terms of task flexibility. Social insects also show great phenotypic plasticity, not through individual brains but through morphological phenotypic diversity resulting in castes performing distinct, complementary tasks – workers, soldiers, gardeners, etc. Social mammals like canids and primates are more similar to humans, as their plasticity depends on their brain. But human labour division, and associated mercantile exchange, is the only one that goes far beyond kin relationships – owing to the human capacity for consciousness about mutual benefits, indirect reciprocity and language to build up trust.

The previous arguments illustrate that interaction between biologists, economists and historians may contribute to a better understanding of genetic biological evolution. Specific themes along these lines are elaborated in Chapters 6 and 7 on the evolution of groups and culture, and in Chapter 11 on ancient human socioeconomic history.

Part III

Bridging Natural and Social Sciences

5 Evolution of Social Behaviour in Animals and Humans

5.1 Sociobiology

The emergence of social evolution in various types of animals meant the appearance of a new level of reality, characterised by advanced communication, cooperation in large groups, complex social organisation, and institutions to stabilise society or enhance learning. Social-cultural evolution can involve genetic and non-genetic changes. Social organisation in relatively simple creatures, such as ants, has been dominated by genetic factors, through the mechanism of so-called 'kin selection'. Selection of higher animals with a more developed memory has further been affected by 'reciprocal altruism'. In even more intellectual animals, such as primates, not only kin and reciprocal altruism but also indirect reciprocity and associated reputation effects have left their imprint. The evolution of human culture has most likely been shaped by a combination of all these types of selection, as well as by group selection (Chapter 6) and non-genetic cultural evolution (Chapter 7).

Sociobiology denotes the body of literature that originally introduced an evolutionary angle to the analysis of social behaviour and strategies of animals. This meant an integration of evolutionary theory and ethology, the study of animal behaviour. Socio-biology was intended to clarify group composition and size, social organisation, communication and the division of labour within populations, colonies and other groups of animals. A main accomplishment was explaining the balance between competition and cooperation, as being driven by combinations of selfishness and certain types of altruism.

Foundational insights of sociobiology were derived by William (Bill) Hamilton (1964), Robert Trivers (1971) and John Maynard Smith and George R. Price (1973). An early synthesis of the literature is found in a much-debated book *Sociobiology: The New Synthesis* by myrmecologist (ant-expert) E.O. Wilson (1975), who, in a final chapter, touched upon the most sensitive theme, namely human sociobiology. In an effort to counter strong criticism, this theme was continued in Wilson (1978) and Lumsden and Wilson (1981, 1983), where a connection with human psychology was established.

Social evolution mainly applies to animals, because it requires activities and movement. These features are found in most animal categories, from insects to mammals. In addition, corals and other colonial invertebrates represent societies of individuals, namely zooids. Along with social insects, such as ants and termites, they resemble multicellular organisms in terms of both physical structure and labour division of elementary units – i.e. cells in organisms are analogous to individual members in colonies. The social strategies found in all these groups of species cover a wide array of particular actions, including aggression, territorial behaviour, dominance, sexual behaviour, breeding, parental care and play. At a fundamental level these can be categorised as expressing some blend of selfishness, altruism and spite.

Sociobiology has generated clear and sometimes surprising insights, which understandably have raised much debate (Ruse, 1979). The argumentation used in sociobiology is mainly theoretical (genetics, logic), while definite empirical genetic statements have been provided mainly for social insects. For higher animals, and certainly mammals, including primates and humans, such irrefutable empirical evidence is missing (de Waal 1996). Nevertheless, many writers now agree that the early criticism of sociobiology was unfair, as it is not so much a theory as a field of research.

According to Durham (1991), a coevolutionary view in which the genetic dimension is matched by a cultural component undercuts the criticism on human sociobiology (e.g. Lewontin et al., 1984). This type of approach, matching anthropological and social-psychological factors and insights, gained much attention and respect through the work of Robert Boyd and Peter J. Richerson (1985). Sociobiology was subsequently integrated into the area of behavioural ecology, while human sociobiology has continued under the name of evolutionary psychology. The latter employs the same concepts and methods as sociobiology, but with the advantage of tapping into a broader set of methods and insights within psychology at large.

The most important contribution of sociobiology has been to offer various explanations for a range of seemingly altruistic behaviours. Core notions in this respect are 'kin selection' and 'inclusive fitness'. The first denotes that altruists are protecting their own genes by helping close relatives survive. This is most evident in ants, mammals and birds, which share food with their offspring and defend them against all kinds of threats – from the environment, predators and aggression by other members of their species – thereby often putting their own life and well-being at risk. Effectively, such individuals are maximising the inclusive fitness of a relevant gene, which accounts for the performance – survival and replication – of both the individual and its relatives who carry the same gene. Inclusive fitness was mathematically clarified by Ronald Fisher (an eminent statistician and biologist) and J.B.S. Haldane (a polymath and biologist) in the early 1930s. Haldane made the famous joke that he would be willing to die for two brothers or eight cousins as in genetic terms this would mean no loss (or gain). Considering inclusive fitness is especially relevant when assessing the evolutionary consequences of strategies that involve social interaction – such as altruism, cooperation, spiteful actions or status-seeking – as these will positively or negatively affect the survival and reproductive chances of others. Note that although spouses are not kin, they may treat each other as 'semi-kin' if they have offspring. The reason is that their genes are bundled in their descendants, causing their 'genetic interests' to overlap.

Reciprocity is a second important force underlying altruism. It refers to behaviour based on the expectation that favours will be returned, which implies that short-run costs are compensated by long-run benefits. For this reason it is sometimes referred to as delayed self-interest. It can be associated with such different acts as helping, trade, revenge and retaliation (tit-for-tat), and it involves emotions like gratitude, sympathy, guilt, shame, trust and distrust. Since there is a time lag between giving and receiving, individuals need to meet, probably regularly, and keep track of who has done them a favour (or harm). If a favour is not returned, trust is damaged and contact may end, or even punishment may follow. Reciprocity of this type obviously requires memory and a minimum level of intelligence. Not surprisingly, it is not found in organisms with relatively little brain complexity. On the other hand, in humans it is a key force behind social relationships. Not only does it play a role in pairwise bonds and friendships, but also it aids in buttressing less frequent relationships with strangers.

Sometimes indirect reciprocity is used to denote a more complex, transitive inter-action, involving more than two individuals, such as giving to 'givers' or punishing violators of reciprocity. This, however, seems only applicable to certain primates with considerable cognitive capacity, as it requires an advanced understanding and recollec-tion of complex social relationships. In this setting, the related notions of reputation and gossip are vital as they provide indications of how agreeable and altruistic past behav-iour of an individual generally has been (Nowak and Sigmund, 2005). This has been suggested to explain why intelligence (neocortex size) and group size are found to be positively correlated in primates. Humans tend have a social network of some 150 people, known as 'Dunbar's number', consistent with the typical size of groups in which humans evolved (Dunbar, 1996).[1]

In line with these previous notions, Richard Dawkins (1976) developed the metaphor of the 'selfish gene' in his book of the same title. His aim was to clarify to a broad audience that selection in social interactions can be understood through genetic account-ing. The individual helping its relatives or others is acting in the interest of overlapping genes, even though s/he is not aware of this. The selfish gene should not be misinter-preted to mean that there is a gene for selfish behaviour or that altruism is faked by individuals. The selfish gene idea is a metaphor for saying that selection and adaptation outcomes are guided by the net effects on all identical genes in the population. This holds equally for kin and reciprocal selection. The first, meaning that own genes benefit from seemingly altruistic acts, has been shown to be a sufficient explanation of social organisation of simple organisms, such as ants (Box 5.1). The second means that the own genes ultimately benefit in terms of the sum of direct altruistic costs and indirect, reciprocal benefits. If such benefits fall short, the gene is genuinely unselfish and will not replicate well. It has been said that the selfish gene metaphor does not sufficiently reflect that genes, through the proteins they program for, interact in complex ways – sometimes in a sort of hierarchical manner, such as through Hox genes (Section 4.4). In addition, epigenetic inheritance associated with activating specific genes on the DNA (Section 4.2) suggests that not all information inherited by organisms runs through genes. These complications mean that generally there is not a one-to-one connection

---

[1] Dunbar (2016) affirms that this number also applies to online social media. It suggests that the cognitive constraint on the size of social networks cannot be relaxed by new communication technologies, arguably as stable online relationships require frequent maintenance.

Box 5.1 Social insects – the eusocial colony or superorganism

All species of ants and termites, some bees and wasps, several spiders and the naked mole rat are 'eusocial', meaning they have the following features: breeding is primarily restricted to a single female (the queen); they cooperate in caring for the young; they show reproductive division of labour, with workers being sterile; and overlap exists between at least two generations so that offspring can help parents during a part of their life. These features are the reason for eusocial species generating among the most complex animal societies. Not all the three conditions are always satisfied. If one or two apply, the species is considered 'pre-social'.

Sociality of termites has a single evolutionary origin, while it has evolved several times in ants, wasps, bees and spiders. There are indications that ants, which came into existence some 120 million years ago, were already social some 100 million years ago. Key to eusociality is so-called 'haplo-diploidy', meaning that unfertilised eggs develop into haploid males (with one copy of each chromosome), and fertilised ones into diploid females (with two copies of each chromosome). This distinction allows the queen to control the sex of her offspring. As the haplo-diploid character of genetic relatedness is higher between sisters (0.75) than with offspring (0.5), workers helping the queen (their sister) to raise her offspring, rather than producing own offspring, is a strategy that genetically pays off. This explains why evolution occurred through inclusive fitness (Section 5.1)

Labour division or caste specialisation in eusocial societies can include the following tasks: caring for the queen, cleaning, caring for larvae, processing food and collection of food. In ants and termites, the roles are for life, associated with anatomical differences, whereas in some bees tasks shift depending on the age of a worker, stimulated by changes in hormones. Selection leads to adaptation of the labour or caste structure over different generations of colonies. This can lead to new, specialised castes and to improvements in the functioning of existing castes. The basic labour division comprises the queen, workers and soldiers, which are all females. In some species the number of specialised tasks (functional types) exceeds ten. As in many mammalian species, in many ant species males are either not part of the colony/group or do not perform a specific task, but live merely for the purpose of mating. In termites, males do form part of the colony structure though.

Communication and its evolution are crucial to the complex organisation of insect societies. This involves tapping, stroking, grasping, contacting of antennae, tasting and secretion of chemicals. Chemical signals seem to be more important than vision or sound. Signals trigger a range of social responses, such as recognition, alarm, grooming, exchange of oral/anal liquids or food particles. Richness of communication tends to correlate with complexity of social organisation.

The success of a colony is due to efficient food gathering and breeding, as well as effective defence against enemies, predators and diseases (through cleaning). The solitary lifestyle does not exist any more in ants and termites, only in distantly related species such as the cockroach. Colonies can be extremely large, in numbers of individuals and biomass. This usually goes along with control of a large territory.

Box 5.1 (*cont.*)

Ants consist of more than 10 000 species, with about 10^{15} living individuals. They are found in all areas, except for the polar regions. They are generally more successful than termites because they are predators, nest in energy-rich soils, and have an acid secretion that inhibits growth of microorganisms. Finally, some ant and termite species grow and control fungus species. In fact, they made a transition to agriculture long before this was achieved by humans.

A colony of ants or termites can be interpreted as a superorganism, to be compared with a multicellular organism,[2] for several reasons.

- The superorganism is the phenotype of the ant genotype that adapts to the environment.
- The ant colony is selected as a whole.
- It creates a microclimate by constructing a nest (cf. temperature control in mammal bodies).
- It is characterised by labour division (cf. cell specialisation in multicellular organisms). Chemical substances cause individuals to specialise, as do cells in the body due to hormones.
- Only one part of the colony (the queen) specialises in reproduction, analogous to the genitals in an organism. Like somatic cells in a body, workers in a colony cannot reproduce their genes.
- A new generation comes into existence when one of the daughters flies out of the colony, copulates with one or more males from another colony, and subsequently starts a new colony. Only a tiny proportion of those who try will succeed, due to natural selection.
- Both bodies and ant colonies come forth out of a single cell.
- Like cells in a body, ants are strongly tied to the colony and cannot survive without it.

Sources: Maynard Smith and Szathmáry (1995, Chapter 16), Hölldobler and Wilson (1994) and Schowalter (2011).

---

between single genes and reproductive success of physiological features or behavioural strategies. This adds some nuance to the metaphor, but does not completely undercut it, as it can still apply to sets of activated genes.[3]

Additional explanations offered for altruism, next to kin relations, reciprocity and indirect reciprocity (reputation effects), are genetic and cultural group selection,

---

[2] Hofstadter (1979, pp. 311–336) elaborates on an analogy between the brain and an ant colony, noting that neither individual ants nor neurones are aware of the powerful structure they contribute to.

[3] Not everyone is of this opinion: David Dobbs argues in 'Die, selfish gene, die' that the selfish gene view makes us blind to real-world complexity:
https://aeon.co/essays/the-selfish-gene-is-a-great-meme-too-bad-its-so-wrong.

discussed extensively in Chapter 6. Related suggestions by Koopmans (2006) are so-called 'ideological altruism' and 'missionary altruism'. The first refers to cultural group selection being specifically associated with religions and political ideologies. The second is an original concept reflecting that some people sacrifice themselves for others with whom they have nothing in common – no ideology, no religion. It means that altruism increases the probability that the receiver will copy the altruism, along with the culture and religion, of the provider. Christian missionaries illustrate this well, as they helped unrelated people from other countries, resulting in whole tribes and populations converting to their religion without violence, through mere example and imitation. The reason is that a missionary represents a role model, appearing richer, more powerful and more educated than the people he is helping. This radiates success and will be seen by others as a good example to be imitated.

Another explanation offered for seemingly altruistic cooperation is signalling, which can occur in sexual interaction, group cooperation and even to indicate to predators that a potential prey is healthy and thus a hard catch. Signals can be honest or deceiving, and they may involve a cost or risk. The Zahavi handicap principle suggests that costly signals provide reliable information about the quality of the signaller, in terms of sexual ability, health, strength, etc. (Zahavi and Zahavi, 1997). Weak individuals are not able to send the signal: for example, they will be quickly killed by a predator. The problem with signalling as a trigger for cooperation, however, is that it shifts the explanation of altruism to the act of signalling. Why would an animal signal if there is a cost involved and not a guaranteed benefit? So then we have to look again for an explanation such as kin selection, reciprocity or group selection to explain the signalling itself.

By combining evolutionary game theory with stylised empirical facts from archaeological evidence on causes of death of hunter-gatherers during the transition from Pleistocene to Holocene, Samuel Bowles (2009) suggests that altruism in humans has been strengthened through a long history of conflict and war between human groups. This sounds paradoxical, but the argument behind what has been called the 'bloody roots of altruism' is strong: altruism makes group members more cooperative and willing to make sacrifices for the benefit of the group. This in turn causes such groups to be stronger, which can help them to win conflicts with groups that lack the same altruistic features. Peter Turchin (2015) provides a historical account with many telling examples. Bauer et al. (2016) consider a broader set of insights to support the idea that war can foster cooperation. We will scrutinise this line of thought in the next chapter on group selection.

Why are not all animals social, and why are there degrees of sociality? The reason is that the solitary and social strategies have different costs and benefits, which can be translated in fitness, i.e. representation in the next generation. An important general benefit of social groups is that they may allow for higher numbers of offspring, which increases the proportion of socially behaving individuals in the total population. Additional advantages of sociality include cooperative benefits of hunting, protection against predators and guarding against out-group members of the same species who are looking for sexual mates. However, groups also may have disadvantages for individuals, in terms of a higher level of stress, owing to frequent social interactions and intense intragroup competition for food and mates.

Traditional ethology continued alongside sociobiology, later called behavioural ecology to avoid the negative responses that sociobiology received. To reflect this further development, Frans de Waal (1996, pp. 13–20) proposed a distinction between 'Calvinist or Classical' and 'New and Improved' sociobiology. The first is gene centred, argues that all social behaviour springs from selfishness, and regards all social behaviour as entirely the result of genetic or individual selection. A well-known standpoint is the 'selfish gene' interpretation, according to which the gene is the only relevant selection level, and organisms are merely carriers of genes. A popular expression of this is: 'a chicken is an egg's way of making other eggs'. According to de Waal (1996, p. 14), extrapolating this to all human social behaviour is overly cynical, as it would imply that Ghandi and Mother Teresa have similar motives to any thief or drug dealer. But selfishness is merely a metaphorical notion applied to genes, meaning that it is not used here in the normal, psychological sense of denoting intentional, conscious acts. Moreover, such metaphorical selfishness of genes does not automatically translate to selfish motivations or acts of individuals carrying such genes.

In an effort to avoid misinterpretations, Elliott Sober and David Sloan Wilson (1998) propose a distinction between evolutionary and psychological (or vernacular) altruism and egoism. Evolutionary altruism means that an individual's behaviour is beneficial to other organisms while being detrimental to the individual itself. Psychological altruism denotes that an altruistic act results from a purposeful, conscious choice that takes into account the beneficial effect on the other and the detrimental effect on oneself. These points solve the apparent paradox: psychological altruism can be consistent with evolutionary egoism. This distinction is essential in order to discuss the evolution of morality and norms, which integrates insights from psychology, with a focus on intentions, and sociobiology, with a focus on effects. The distinction is consistent with the idea that, whereas some animals morally reason, evolution or natural selection has no moral dimension.

5.2   Social Behaviour and Organisation in Animal Species

It is instructive to consider the main categories of social animals as it allows identification of the core features of social behaviour and organisation (E.O. Wilson, 1975). Colonial invertebrates are physically integrated to such an extent that it is difficult to distinguish between a (super)organism and a society. This assures protection against physical conditions (flowing water), better dispersal of propagules (spores or seeds) and defence against predators. Colonialism is found in corals, which are small colonies of thousands of identical polyps, and in the social insects (Box 5.1). Cold-blooded vertebrates are a broad group in which one can frequently encounter social behaviour. A typical example is fish schools, consisting of tens to millions of individual fish. They exist temporarily, and are self-organised, i.e. without any leadership. The main advantage is protection against predators. Another advantage is energy conservation, resulting from more efficient movement and maintenance of heat. Some dinosaur species, like hadrosaurs, are thought to have also lived in groups. While amphibians and reptiles are solitary, in some species females share nests or participate in communal egg laying (Doody et al., 2009). Social behaviour is common in birds and takes various forms, such as holding communal nests, flocking or cooperation in feeding. Many such birds have *K*-strategy features (Section 4.6), notably a strong parent–offspring relationship, reflected by parents feeding the young.

Of course, the most impressive examples of social behaviour are found in mammals. The change from solitary to social mammals has been suggested to have been stimulated by the movement out of forests to grasslands and savannahs (areas combining grasses and woodland characterised by open canopy). Indeed, most forest mammal species are solitary, while most mammal species living in open habitats are social. This holds for herbivores as well as their predators, the carnivores. The social features of both have most likely been shaped by evolving predator–prey interactions (Section 4.6). The essential relationship for social organisation of mammals is the strong bond between mother and offspring, created during the early life of the offspring when milk is provided as food. Without exception females take care of the young. Social features differ among species, such as male–female bonding and group formation beyond the family.

Two extremely social mammal species are dolphins and elephants. The first are sea mammals that live in groups of 10 to 100 members, have a relatively large brain, have a great capacity for imitation of sound and movement, and show cooperative behaviour in helping disabled group members. The second have strong bonds among females, their offspring and their grandchildren, which can last for decades. Group size is about 20 individuals. Adult males live alone or in loose groups of males and show more simple social behaviour. Hoofed mammals (ungulates) tend to live in herds, and include more than 180 species of horses, pigs, antelopes, cattle, sheep, etc. Carnivores include more than 250 species. They are relatively social compared with other mammals and, in general, show the most subtle and varied social behaviour aside from primates. The majority, however, are solitary. Among cats, lions are socially the most advanced. They form groups of various males and females that are all potential breeders. Such groups are based on a bond among females, sisters or cousins, which show considerable cooperation in both childcare and hunting. The efficiency of joint hunting is evident, as single lions catch about half as much prey. Males are sort of parasites, enforcing priority in eating killed prey. Canids are characterised by pair bonding (e.g. foxes) or families headed by a breeding pair (e.g. wolves). Unlike lions, they develop around a dominant male. Cooperation in hunting is extremely subtle. Some species hunt in packs, such as the wolf, the African wild dog and the dhole of Asia. This allows them to take down much larger animals than could single predators. Among canids, African wild dogs are unique for their altruistic behaviour, showing unusual cooperation, subtle communication and rituals, playful interaction and egalitarianism in terms of sharing food.

Boxes 5.1 and 5.2 illustrate complex social behaviour and organisation in very different species, namely social insects and primates. Ants are relatively simple creatures, which, according to Bert Hölldobler and Edward O. Wilson (1994), show behaviour that is almost deterministically, namely chemically, related to the information in their genes. They are unconscious and have little or no memory, only instinct. In line with this, their social behaviour and organisation are completely determined by genetic

---

Box 5.2 Nonhuman primates: the social ancestor of humans

Nonhuman primates are characterised by groups of various males and females. They include tarsiers, lorises, lemurs, New and Old World monkeys and apes. Like in many other mammalian species, to which primates are phylogenetically connected, males tend to be polygynous and aggressive to each other. Females – or more specifically, mothers – form the centre of the society, fostered by a lasting bond with their offspring. Compared with other forest mammals, primates are large, which was possible due to their intelligence, allowing effective foraging and defence against predators. Compared with other mammals, primates generally have good eyesight and reasonable hearing, but less-developed smelling capacities.

Social strategies of primates involve alliances, which may be one origin of social equity. They have the ability to rise in the social hierarchy, by opposing higher ranked individuals, or even becoming the dominant individual of a troop. Two organisational structures are found: centripetal, in which members observe and react to a dominant male; and acentric, in which females and the young live mostly separate from males who tend to live at the edge of the troop, thus protecting it against predators. Social organisation and behaviour tend to be influenced by the environment (ranging from dense forest to open savannah) and food habits (leaves, fruits, insects or meat), the combination of which influences food distribution among group members. To illustrate, most insectivores are solitary, while most pair-bonding species are vegetarian.

There is a wide diversity of primate species. Some lemurs live in groups with males and females, ranging from several to more than 20 individuals. Aggressive behaviour is common, and shapes the social structure. Unusual for primates, females dominate males. Aggressive interactions involve not only visual and vocal, but also chemical expressions (bad smells in fights). Many baboons live as males with a harem (1–10 females), which can be integrated into bands ranging from 12 to 750 individuals. The latter cooperate in foraging and defence against other bands. Males show jealous behaviour in the form of possessiveness of females. Aggressive interactions among males involve much bluffing. Single males, i.e. without a harem, live in small bands. Tailless gibbons, the 'lesser apes', live in small groups of 2–6 members, composed of parents and offspring. Both parents have a close relation with the young. The father is protective, intervening when young play too aggressively.

Among orangutans, arboreal great apes of Indonesia, females and young live in small groups of four or fewer individuals. Adult males are solitary, often moving from one to another female group, maintaining a distance, which keeps aggressive interactions at a minimum. Gorillas are the largest primate, being up to 2 m tall, while some weigh 180 kg or more. They are vegetarian, non-aggressive and live in groups of 2–30 individuals, with a dominant male. Individuals in the troop stay close together. Dominance depends strongly on size and age, and is effected through subtle communication rather than aggressive behaviour. In fact, gorillas are the most relaxed and slow-living primates. This obviously relates to their large size, which requires much energy and provides easy

> **Box 5.2** (*cont.*)
>
> protection against predators. Audio-visual communication involves some 17 signals and is closer to monkeys than to other apes. Facial expressions are highly developed and used to recognise individuals.
>
> Without any doubt, chimpanzees and bonobos (formerly called pygmy chimpanzees) are the most socially advanced nonhuman primates. They live in large societies of 30–80 individuals, which easily break-up and reform. These are organised in a complex hierarchy around a dominant male, who is not necessarily the largest but the most socially capable. Mixing of separate groups can occur through troops exchanging adult females, something not seen in other primates. Males show unusual cooperation in hunting. Both species demonstrate human-like social behaviour: resolving conflict non-aggressively; arbitrating and mediating in conflicts, including lower-ranked individuals; consolation of hurt or otherwise sad mates; and deliberately misleading or fooling others. Chimpanzees are omnivorous and eat fruits, leaves, bark, seeds, insects and small mammals. They share food and beg or negotiate for it. Individuals who do not share regularly do not receive much from others. Communication is the most advanced of all nonhuman species. It involves a large number of vocal signals, facial expressions, touching (grooming) and body postures and movements. Chimpanzees have a great capacity to recognise and memorise others, including humans.
>
> *Based on*: Wilson (1975, 1978) and de Waal (1996).

or kin relationships. Surprisingly, their social organisation is rather complex, even when compared with higher animals. Whereas ants are unconscious of their role in the colony, the latter shows quasi-intelligent behaviour.

Social behaviour of primates is of particular interest as it represents a transition from animals to humans (Box 5.2). Primates show the most complex and subtle social behaviour of all animals because of their memory, degree of self-awareness and capacity to feel empathy or even sympathy for others. Associated with these features, reciprocity and indirect reciprocity play an important role in the social behaviour of primate species. Their social organisation involves such diverse categories as cooperation, fighting, coalition formation, personal relationships (grooming, greeting), retaliation, food sharing among non-kin and recognition of facial expressions.

There are several commonalities between ants, primates and other animal species. One is that in most social groups, the mother plays a central role. Early socialisation in mammals happens through the intense mother–child bond, which in primates and some other species, such as elephants, lasts for a long time. This bond acts then as the cement of social organisation. Social organisation further enhances two features which evolution tends to promote generally, namely a division of labour (within organisms as well as groups) and more subtle ways of communication. The two obviously go hand in hand as labour division can only function with adequate communication among workers or their groups.

5.3   Evolution of Empathy, Morality and Altruism in Primates

The examples of primate species mentioned in Box 5.2 raise the question of whether human social behaviour and organisation can be seen as an extrapolation of an evolutionary trend in primates. According to de Waal (1996, p. 1), biologists would expect a degree of continuity between the behaviour of primates and humans and finds support for this in the moral behaviour, empathy, sympathy and intentional altruism encountered in many primates.

Key philosophical writings on moral behaviour and ethics tend, though, to neglect the evolutionary history of humans. Philosophers such as Aristotle, Plato, Confucius, Aquinas, Hobbes, Spinoza, Hume, Rousseau and Kant may be excused as they were writing before the rise of modern evolutionary biology. For those writing after Darwin, however, matters are somewhat different. Whereas an important thinker like John Stuart Mill did not take Darwin's theory seriously, Friedrich Nietzsche rejected it categorically. For modern writers on ethics, there is no excuse at all to ignore insights about evolution of the brain and associated mental faculties, or similarities in behaviour between humans and primates.

Evolutionary ethics, or 'biologising' ethics, has two advantages over ethics as traditionally studied: it offers a historical account of the roots and development of human ethics as part of the general natural history of the human species; and it can identify the possible limits to changing our ethics or behaviour. This is of utmost relevance to the social, environmental and policy sciences. De Waal (1996, p. 10) offers clear arguments for why philosophers should take evolutionary biology more seriously when writing about morality and ethics. He stresses that ethics starts with the gradual development of moral reasoning in children, which is constructed upon fear of punishment and a desire to conform to other humans.

Figure 5.1 summarises core concepts and mechanisms associated with the evolution of social behaviour in animals and humans, drawing upon discussions in this and previous sections. On a psychological level, the crucial change and stepping stone from non-moral animals to moral humans is the evolution of perceptions and empathy. These lead to a moral community in which direct and indirect, one-to-one interactions matter. Individuals care about good relationships between others, and even undertake mediation and arbitration to deal with conflicts, what de Waal (1996, p. 34) refers to as 'community concern'. Empathy is the ability to be affected by feelings of another organism or the situation in which it finds itself. It should be distinguished from sympathy, which denotes compassion or sorrow, going one step further. In other words, where empathy is related to understanding, sympathy connects to well-being. Empathy follows upon self-awareness, distinguishing oneself from the environment or the rest of the world. It involves perception, picturing oneself in the position or situation of the other. Therefore, empathy requires a minimum level of intelligence. It builds on memories of one's own, similar experiences, which are triggered when seen in others. Apes can imitate in a subtle way ('aping') and deliberately mislead or fool others, both following from their empathetic capacity. Empathy can thus be seen as

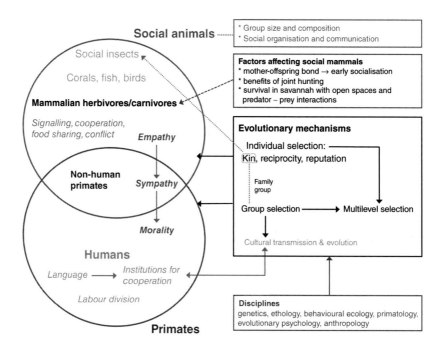

Figure 5.1 Evolution of social behaviour.
(A black and white version of this figure will appear in some formats. For the colour version, please see the plate section.)

the stepping stone from apes to humans, though it is not limited to apes. It is found in several other intelligent mammals. Dolphins seem to understand that they see their own image when looking into a mirror, as do some elephants. Sympathy in the form of comforting behaviour has been observed in both these species as well as in wolves, dogs, pigs, certain birds (e.g. ravens) and even some rodents (certain rats, mice and voles).

Humans have a more developed empathy, internalisation of rules, sense of justice and community concern than other primates. This is associated with evolution having planted more refined psychological features in humans. Without trying to be complete, examples of unique human features are: blushing, which relates to one's reputation as a consistent member of society; laughter at someone else's bad luck, following from feelings of self-esteem and fairness; crying, so as to attract attention and consolation; and concern for other species, a form of extended empathy. Human culture has developed many rules to avoid conflict, such as 'ladies first', which can be understood as a solution to a coordination game (Buskes, 2006, Chapter 12). Other rules have been clarified by Robert Axelrod (1984), such as avoiding conflicts through cooperating as long as others do so, retaliating if others free ride, forgiving and cooperating again if others, in response, alter their behaviour, and behaving predictably and transparently so that others can easily adapt to it.

Sarah Hrdy, a researcher who has written much on the role of women in human evolution, has argued that empathy develops better in extended families where children have many surrogate mothers, notably aunts and grandmothers (Hrdy, 2009).[4] This may mean that the development of empathy in children is at risk in current societies with generally small families. This could increase the propensity of social handicaps among children and subsequently adults. On the other hand, formalised childcare, where children are confronted with many different adult caretakers, may compensate. Hrdy, nevertheless, thinks that recent trends in cultural evolution may do damage to the typical, unique psychological features of humans that were formed through a lengthy course of primate evolution.

5.4 Human Sociobiology and Evolutionary Psychology

The suggestion of sociobiology that the study of social behaviour of humans should be placed in the wider context of animal social behaviour created much opposition within the social sciences, notably sociology and anthropology (Ruse, 1979). Social science's focus on nurture of social relationships is often regarded as the antithesis of sociobiology with a focus on nature or genetic constraints. Both approaches support the view that human behaviour can only be understood in a social context. But they differ in an essential way: the sociobiological approach emphasises the commonality of social phenomena between different human cultures, groups or states, whereas the sociological approach tends to stress their diversity. Hence, the two approaches are complementary and contrary to some extent.

The relevance of sociobiology for the social sciences can be compared to that of physics and chemistry for biology: as physical and chemical processes constrain living processes, biological processes constrain cultural ones. Obviously, this should not be interpreted as biological determinism, if only because the social organisation has its own dynamics and is characterised by considerable plasticity. This means it cannot be regarded as fully explainable on the basis of biological regularities and principles applying to the living organisms that make up social organisations. This is consistent with the notion of emergent properties according to which a new level of reality has features that were not foreseeable at the lower level from which it arose (Section 9.5). These subtleties, though, did not prevent early opponents of applying sociobiology to humans to – inappropriately – hint at biological or genetic determinism. Later, the antagonism to sociobiology became weaker, as indicated by broad acceptance of the scientific approach of evolutionary psychology. This went along with a better general understanding that genes do not act as blueprints but that behaviour is codetermined by genes and the social environment.

Sociobiology draws attention to similarities of behaviour among groups within a certain species, as well as between species, accounting for overlapping phylogenetic history and genes. Behavioural differences among individuals and groups within a

---

[4] With her work, Hrdy rehabilitates evolutionary behaviourist Niko Tinbergen, who in his acceptance speech for the 1973 Nobel Prize in physiology/medicine hinted at such a relation. This was criticised, though, for his focus on autism. Hrdy instead addresses social handicaps more broadly.

single species are then regarded mainly due to the effects of distinct environments. The observed similarity of human behaviour and cultures in different parts of the world is consistent with the fact that the genetic differences between human groups are miniscule. According to Jared Diamond (1997a), the ultimate causes of differences among current human cultures are mainly the result of cultural evolution and historical accidents, associated with geographical factors and diversity of domesticable plants and animals (see Section 11.6).

Evolutionary psychology studies human psychology on the basis of evolutionary insights about human individuals, societies and their environments (Barkow et al., 1992; Buss, 1995; Crawford and Krebs, 1998). In particular, it aims to identify constraints to human behaviour, including human instincts, on the basis of understanding the social and environmental conditions that prevailed during human evolution as hunters and gatherers in the Pleistocene era. Cosmides and Tooby (1997) have stressed that as the human species has lived most of its time without modern technology and industry, and as natural selection is a slow process, there have not been enough generations for human minds to evolve away from their Stone Age roots so as to adapt to post-industrial conditions. Put another way, biological evolution of the human brain has not kept pace with the evolution of human culture, causing humans to be not particularly well adapted or even maladapted to certain aspects of modern life.

By focusing on evolutionary roots, ultimate causes of behaviour can be traced. Moreover, it gives an insight into human instincts, which many social scientists falsely think have been replaced by rational thinking or culturally formed habits and imitation. In this context, evolutionary psychologists often refer to the 'standard social science model' (SSSM) which depicts the human mind as a blank slate or '*tabula rasa* ', to be filled with content mainly through individual experience, including social interaction (Pinker, 2002). The SSSM also tends to see the brain as a general purpose tool. However, evolutionary psychology instead stresses that different neural circuits resulted from solving distinct adaptive problems over human evolutionary history. So the mind is not capable of fitness maximisation as a general goal, but instead has developed through evolution to respond to many problems through domain-specific modules. This does not deny that the complete brain has a multipurpose flexibility. Indeed, this explains why by-products of evolution appear in relation to the oversized human brain, an example being arguably our interest for, and enjoyment of, music (see Section 7.8).

Evolutionary psychology instead tries to identify the instincts and human biases by reverse engineering the brain, thus figuring out the functions of its mental modules. This involves understanding which neural circuits or modules were designed by natural selection to solve which particular problems faced by our ancestors.[5] This idea goes

---

[5] Some argue that the 'modular brain', consisting of neural circuits, is underpinned by the modern understanding of neural and chemical proximate mechanisms in the brain. Others reject this view. Taking the middle ground, one might accept that there are neither perfectly isolated and well-bounded modules nor seamless general domain phenomena, but merely proxies of either. One should also note that it is common to divide the brain into anatomical regions. Moreover, brain modules have been defined in various (other) ways. Glimcher (2016) provides an accessible introduction to supposed modular mechanisms and explains their relevance for the social sciences. He identifies three types of basic processes in the mammalian brain:

back to Darwin who noted at the end of *On the Origin of Species* (1859) that 'Psychology will be based on a new foundation', an idea elaborated in his later works *The Descent of Man, and Selection in Relation to Sex* and *The Expression of the Emotions in Man and Animals* (1871). According to Pinker (1997, p. 22), Darwin's prediction is still underway as many in the social sciences are resistant, associating evolutionary thinking, erroneously, with determinism and justification of inequality. Chris Buskes thinks resistance, notably in the 1960s and '70s, had a lot to do with several widespread beliefs in the social sciences: that it is an autonomous area of research; that humans, because of their unique intelligence, have largely escaped biological influences and limits – i.e. that nurture (culture) dominates nature; that society can be managed to create human progress; and that egalitarianism is at odds with evolutionary thinking about humans. A problem here is that explanatory and normative issues are often not well separated by critics or disbelievers (see Section 1.4 and Chapter 15).

Evolutionary psychology has gone beyond older sociobiology in two ways. First, it focuses on the brain as one organ, consistent with how many evolutionary biologists have specialised in the evolutionary analysis of other organs. Second, it combines cognitive science with evolutionary thinking. Both advances mean that evolutionary psychology offers a broad basis of support for evolutionary reasoning. This involves identifying interactions between four levels of explanation in evolutionary psychology, namely adaptive evolutionary, cognitive psychological, neurophysiological and phylogenetic. The latter represents looking for similarities, connections and consistencies in behaviours, emotions and social systems through comparative analysis of current and ancestral forms of humans, notably within the line of primates.

The topics addressed by evolutionary psychology include general ones like information processing, the role of emotions, time allocation, problem-solving, cooperation and language. In addition, it treats issues associated with social behaviour, such as aggression and violence, sexual behaviour, formation of relationships, jealousy, parental strategies and preference for status and power. One main insight is that human behaviour is predominantly automatic and characterised by heuristics and biases instead of being rational. Among others, evolutionary psychology claims that the human mind has modules for specific tasks, such as language, learning, choosing mates, parental love, recognising people, interactions with non-kin and moral decisions. The latter may be related to the human propensity to conformity, docility and detection as well as punishment of dishonesty or cheating. This view moves away from the common idea in social sciences of a general purpose mind. In addition, it helps to clarify, from an evolutionary

areas constructing an estimate of the value of each option considered, areas that compare these values and transfer the outcome to movement control, and learning areas that update option values if the quality of the chosen option deviates from expectations. The latter involves an important role of the neurotransmitter dopamine. Preferences are not constant as they depend on context and associated motivation. A change of context can shift decision-making to a distinct brain area, in turn altering preferences. For instance, people who get hungry start behaving differently in social or work situations due to changes in the state of their hypothalamus. It has been found that in particular their risk attitudes alter, which may even affect decisions such as investment in stock markets (Levy et al., 2013).

angle, persistent gender differences observed in many cultures (see Box 4.1). It further provides insight into evolution of early public persona and politics (Box 15.2).

To illustrate that our mind reflects our ancient environment, note that Western people still will cite certain animals, notably lions, tigers and snakes, as among the most frightening things they can think of, even though such animals are unlikely to be encountered in modern life. However, from an evolutionary psychology angle, this makes sense as we have been selected to fear such animals given that they are extremely ferocious or venomous. More generally, the divide between passion and reason has been created through evolution to assure that we have the best solution for dangerous or otherwise urgent situations, namely react fast, run away, or attack, whatever best serves our survival. In such instances, reason would be a second-best option, as it is too slow.

As can be expected, evolutionary psychology, like human sociobiology, has been subject to fierce criticism, partly motivated by political ideology (Lewontin et al., 1984; Rose and Rose, 2000). Opponents tend to emphasise that evolutionary logic, in terms of adaptation to environmental circumstances, often sounds convincing, but can lead to theoretical 'just so stories' without any factual evidence. They would instead explain many psychological phenomena as by-products of the complex brain. Pinker argues, however, that complex organs and behaviours that serve a clearly useful purpose in terms of survival must be an adaptation due to natural selection, and only less complex aspects of behaviour, such as taste for music and art, which do not have a clear adaptive role, can be a by-product of some adaptation or result from genetic drift. On the other hand, the distinction complex/simple and non-useful are not entirely straightforward (Mitchell, 1999).

It is now widely accepted that as long as results of sociobiology and evolutionary psychology are judged critically, they can provide interesting hypotheses for empirical testing and provide a useful starting point for thinking about social behaviour (Cosmides and Tooby, 1994; Ben-Ner and Putterman, 2000; Jackson, 2000). Psychological research comparing identical and non-identical twins has provided support for the idea that genes have a significant and identifiable influence on human behaviour. To repeat, this should of course not be confused with, or simplified to, genetic determinism. Adaptation as an explanation should be built on firm ground, involving multiple sources of evidence: theory, history, experiments and empirical facts. This implies an integration of insights from evolutionary and social psychology. De Waal (2002) emphasises in this context the 'dilemma of the rarely exercised option': an evolutionary, adaptive explanation of atypical behaviour should be consistent with explanations of the typical, dominant behaviour found in reality. In exceptional acts, such as rape of women, and child abuse by step-parents, this is not the case (the dominant behaviour is no rape and no abuse), which causes related adaptive stories to be suspect. Moreover, as de Waal notes, rape would require its own genetic basis separate from other sexual tendencies to count as an adaptation. Partly in response to criticisms, one can see different approaches now in evolutionary psychology, ranging from those linking closely to the original sociobiology and modern counterpart behavioural ecology, through those connecting with other areas of psychology (developmental psychology, social psychology or neural sciences), to adopting a gene–culture coevolutionary view (Section 7.4). See Griffiths (2008) for more discussion.

Evolutionary psychology differs from another evolutionary approach within psychology or broader cognitive science, namely invoking evolution to explain human thinking and selection in brain processes (Calvin, 1987). This goes back to early ideas that human creativity and learning can be seen as following evolutionary processes of variety creation and selection (Ashby, 1960; Campbell, 1960, 1969; Skinner, 1953; Young, 1965). A theory of 'neural group selection' was proposed by Edelman (1987), depicting the development of neurones (nerve cells) and their synapses as the result of environmental selection. Finally, next to evolutionary psychology, human behavioural ecology studies social evolution in humans from a predominantly biological angle, with close links to biological and physical anthropology (Smith et al., 2001). This comes down to applying the same theoretical angle and empirical methods as used in animal behavioural ecology to human populations. This focuses on foraging strategies, cooperation, sharing food, reproductive strategies and parental care in response to varying ecological conditions. The results of this can complement arguments and insights from evolutionary psychology.

5.5   Evolution of Human Language

Social evolution in early humans critically depended on subtle communication, initially and for a long time only orally, and later also through writing. One can see language as a symbolic representation of one's inner thoughts and the world. It has the capacity to create emotions, images, understanding and willingness to act in other human beings. Language has served as a catalyst in creating strong relationships and social organisation within as well as beyond kin groups, ultimately giving rise to tribes, villages, empires and states. Through language, humans are able to think precisely and expansively, as captured by the term 'linguistic thought'. The human capacity for individual, social and even intergenerational learning depends strongly on language capabilities. For all these reasons, language has been essential to the development of human ultra-sociality and complex human cultures.

Language allowed humans to go beyond symbolic communication, as mastered by apes, towards a subtler, richer and more substantial way of communication. This evidently has an evolutionary advantage since it allows effective and complex cooperation as well as planning ahead in time. In addition, language has helped to construct social relationships and exchange information, which reduced free-riding behaviour, thus making human groups stronger. In support of this, research by Dunbar (1996) suggests that 75 per cent of human communication involves gossip about other people, which tends to spread and strengthen social norms. It is not certain when language emerged in humans. Estimates range from 2 million years ago when the human brain probably satisfied the conditions for language, to 100 000 years ago when necessary throat conditions were satisfied (Boyd and Silk, 2002, Chapter 15).

Animal Versus Human Languages

Comparison of humans, apes and monkeys by behavioural ecologists has shown the need to correct the belief in an enormous gap between animals and humans. This is supported by the fossil record on brain size, approximated by a clear sequence of increasing skull sizes of ape-like predecessors to modern humans. One surprising unique feature of the human brain, though, seems to be what has been called 'lopsidedness' or asymmetry of the brain, notably a small groove that runs deeper along the right side of the human brain than on the left side. Not only do other primates lack it, but it is associated with skills in which humans are more sophisticated, notably language and social interactions. In addition, it is connected with right/left handedness.

Many human traits are already present in primitive form in apes. Monkeys show much more intelligence than other mammals, and apes are intellectually more capable than monkeys. Rudimentary elements of abstract symbols are found in all apes. Especially chimpanzees and bonobos have mental features that were previously only considered applicable to humans. Examples of the latter are: having a sense of self-awareness, having complex emotions, understanding motivations behind others' behaviour (empathy), and fooling and manipulating others. In addition, they have the ability to learn simple language structures – although high early expectations regarding this have been reduced over time. This all explains why primates have a complex social structure and communication. The evolution of consciousness and brain on the one hand, and of social behaviour and systems are really two sides of the same coin. The reasons are clear: to survive in a complex society one needs a large brain; and such a society can only exist if its individuals are sufficiently intelligent.

Human language required, in addition, biological evolutionary changes in humans, not just of the brain but also of a range of anatomical and nervous system characteristics. As a result of such changes, humans possess a unique capacity to reproduce sounds they hear. This raises the question whether others in the human lineage possessed the ability to speak. Although Neanderthals did not have the exact same adaptations as humans (Cro-Magnons), what we know about their hunting and tool-making makes many experts suspect that they possessed some capability of communicating by simple speech.

Theoretical Aspects of Language Evolution

The capacity to exchange language with others may have started through a combination of meme and sexual selection that ultimately caused the brain to become large and the mind to become intelligent (see Sections 4.7 and 7.6). Human language is one among several major transitions brought about by evolution. According to Yun Ding et al. (2015) language reflects a unique 'unbounded combinatorial nature' possessed by humans, meaning that smaller elements are combined using an internal grammar to produce a hierarchy of words and phrases. Martin Nowak (2006, p. 249) underlines that human language is open, making 'infinite use of finite media'. In contrast, 'animal languages' are closed or even pre-programmed (i.e. instinctive) in the sense of involving a limited number of sounds, chemical or physical signals – as found in mammals, birds or social insects – connected to particular, concrete issues related to food, danger, sex and conflicts. An exception is formed by some birds that have a more extended capacity to 'culturally learn' songs from other birds and even show, like humans, song dialects within and across countries.

According to Nowak (2006, Chapter 13) the analysis of language can only be fruitful if it adequately combines formal language theory (describing the general, fundamental features of languages: alphabet, words, sentences, sets of sentences and grammar), learning theory (how people learn) and evolutionary dynamics (how variations in language come into existence, are copied, imitated and diffused). Language change involves alternations in vocabulary, grammar and spelling. Some believe that grammar has universal features. Generally, universal grammar is seen as a restricted set of grammatical rules to match the limited capacity of the brain. For example, Futrell et al. (2015) compared 37 languages and found they shared a tendency to place syntactically related words near each other in a sentence – i.e. 'dependency lengths' were shorter than what would be statistically expected, most likely to reduce pressure on memory capacity. For the same reason, humans need to categorise objects using associated words, analogous to how visible light is categorised in discrete colours by the brain to reduce the cost of information processing.

A universal grammar may be the outcome of evolution having selected, in accordance with Zipf's (1949) law, a distribution of word use such that communication requires least effort. Individuals communicating effectively and efficiently in comparison with others might have had a small selective advantage, causing this feature to spread through the human population. Universal grammar has likely evolved gradually along with the human brain. For example, brain regions in humans essential for language control like Broca's and Wernicke's areas correspond – in evolutionary jargon, they are homologous – to regions in monkeys with similar location, two-way connections and cellular composition. Moreover, monkeys use these regions to distinguish calls from other monkeys (Pinker, 1994, p. 348). Such manifold correspondence would unlikely be coincidental if it were true that human language had appeared out of the blue once the brain was sufficiently complex. A generalisation of this homology view is offered by Nowak (2006, p. 252), who argues that the brain possibly developed grammar or a classification of the world to communicate information between brain modules (hearing, smelling, seeing, feeling, walking, other actions), which then generated neural algorithms that served as a good basis for communicating with other brains.

Software–Hardware Coevolution

Language is like software jointly evolving with hardware – a kind of coevolution or perhaps more correctly co-adaptation. Note that the required hardware includes not only the brain but also the physiological features of mouth and throat, in particular tongue (to modulate sounds), lips, control of respiration and a descended larynx or 'voice box', which hosts the vocal folds. It has been suggested that the complete human vocal apparatus is sort of digital, as each element can adopt only discrete positions. This limits the number of combinations of options, which might explain the notion of a universal grammar. However, these ideas are controversial. Evolution transformed the existing designs for respiration and ingestion to also include the capacity to articulate speech sounds (and combine them well). A clear proof of language as an adaptation and not a by-product of human intelligence is that the capacity to speak goes along with the unique human incapacity to close the trachea (windpipe) for food, which provides a risk of suffocation, a clear evolutionary disadvantage which must be compensated by an adaptive advantage like language.

Linked to brain evolution is the evolution of language as a system of verbal–oral symbols. This is now widely agreed not to have been culturally formed but be the result of natural selection. Support comes from the observations that all human individuals are competent users of language, all societies possess languages with the same degree of complexity, and acquiring language is innate rather than the result of explicit teaching. The similarities of languages, which inspired the universal grammar theory of Noam Chomsky, point to a biological evolutionary origin. Recently, a study provided empirical evidence for this by connecting information on speech and cortical activity to track the time course of linguistic structures at different hierarchical levels, such as words, phrases and sentences (Ding et al., 2015). It shows that hierarchical linguistic structures in speech have boundaries that are defined not only by acoustic cues but also by internal grammatical rules. Such universal grammar does, of course, not deny that culture and its evolution have played a crucial role in the evolution of language and vice versa. Neither does it mean that languages are completely biologically programmed in the brain. Instead, they are plastic and leave freedom for individuals in using them. The sounds of speech alter over time and so do written forms. Part of this resembles a process of drift. Selection comes into play through conventions and formal rules about language.

The Evolution of Writing

The development of writing represents a recent phase in the evolution of human language. A short, accessible account is given by Diamond (1997a, Chapter 12). Almost all writing systems are related through ancestry, for two reasons: it is not easy to invent writing, and once invented it easily spreads because of its social benefits. Regions where writing possibly originated independently are ancient Egypt around 3400 BC, ancient Sumer (southern Mesopotamia) around 3000 BC, the Indus Valley around 2500 BC, China around 1300 BC and Mesoamerica around 300 BC. It must be added, though, that exact dates and potential interdependencies continue to be debated. One can divide writing systems into three types. The least common is a set of signs for each syllable, as found in ancient Linear B writing in Mycenaean Greece and in Japanese kana scripts for telegrams. A second system consists of logograms or ideograms, i.e. pictures and symbols for each word, as reflected in ancient Egyptian hieroglyphs or in modern Chinese characters and Japanese kanji. A third system is the most common nowadays, namely an alphabet with letters – often also using diacritics or combinations of letters – to represent units of sounds (phonemes), while constructing meaningful words with these letters. This writing system took off with the Phoenician (or Proto-Canaanite) alphabet which influenced later alphabets, such as the Aramaic one, from which modern Arabic and Hebrew scripts stem, and the Greek one, which gave rise to modern Cyrillic and Latin (or Roman) alphabets. Writing systems further include a particular organisation of symbols in rows or columns, and a requirement to read symbols in a given direction, such as from left to right and top to bottom.

The success of alphabet-based writing, generally involving between 20 and 30 letters, is due to its flexibility in terms of recombining letters, syllables and existing terms into meaningful words, and possibly also due to its moderate use of brain memory relative to logogramic systems. Among the 26 alphabets currently in use, Arabic, Cyrillic and Latin ones dominate. The term alphabet comes from the first two letters of the Greek alphabet, *alpha* and *beta*, which in turn derive from the first letters of the Phoenician alphabet, *aleph* and *bet*. In the Latin alphabet these have been simplified to *a* and *b*. Many other Latin letters can be traced back to Phoenician ones. The latter derive from either Sumerian logograms or Egyptian hieroglyphs. This involved a shift from concrete images to more abstract ones, based on playing around with images for words with similar sound or meaning. An intermediate stage combined various signs, namely logograms, phonetic signs and unpronounced 'determinatives' to resolve grammatical ambiguities (e.g. a verb versus an associated noun or adjective). This evolution of writing symbols ultimately culminated in completely abstract letters to denote a sound (phonograms). This represented a major transition in the evolution of languages. Because of the advantages of the alphabet, it was positively selected and diffused.

Language Diversity in the World

Worldwide, the geographical distribution of languages is strongly skewed. The countries with the most languages are Papua New Guinea and Indonesia, each with more than 10 per cent of all languages in the world, followed by India and Nigeria, each with more than 5 per cent. As new species often emerge in geographically isolated areas, languages tend to separate on the basis of natural boundaries – rivers and particular mountains. One can see this clearly in the geographical distribution of the languages in Switzerland: German (with three dialects), French, Italian (with three dialects) and Romansh (with five distinct dialects). Modern languages change even if they are bounded by strict grammar and spelling rules. So imagine how quickly non-written languages could change in the past. For example, Celtic is still spoken by several hundreds of thousands of people in remote parts of Ireland, Scotland, Wales, Cornwall and Brittany (France) as well as on certain UK islands, but is in decline. Lithuanian, Icelandic, Latin (in Vatican City) and Frisian are among the languages which have changed considerably less than others, for different reasons. The first can even be seen to reflect traces of the original Indo-European language, Sanskrit, old Hinduism that served as the *lingua franca* of South East Asia ('greater India'). According to Bill Bryson (1990, p. 13), the Basque language, called 'Euskara' by

Basque speakers, might well be the last survivor of the Neolithic languages that were spoken in Europe during the Stone Age.

The case of Papua New Guinea with about 830 languages[6] is surprising at first sight, especially given that its population is fairly small, namely about 7 million people, implying on average about 8400 people per language, with some languages being spoken by only dozens or fewer people. The explanation for so many languages is multifold. New Guinea is the world's second largest island (after Greenland), and it includes more than 600 small islands. More than 85 per cent of the population is rural, about half of the population is illiterate and population density is low, about 16 people per square kilometre. The island is characterised by sharp mountains, many peaking above 4000 m and several above 4500 m, creating deep valleys and dense, tropical vegetation (including, among others, rainforest, savannah and mangrove). This keeps local populations isolated, and calls to mind evolutionary isolation as described by biogeography (item v in Section 3.5). Isolation continues up to the present day due to a lack of paved roads and railways, the construction of which was complicated by the rugged terrain. To illustrate, the capital Port Moresby is not linked by road to the rest of the country.

In 2015, the world had more than 7472 languages and dialects in 228 language families, of which 7102 were living. Of these 1531 are in trouble, and 916 dying (https://www.ethnologue.com/world). The precise number of languages in the world, and in many large countries, is debatable as the distinction between a language and a dialect is fuzzy. For example, some sources claim more than 500 languages in Nigeria, while according to others there are fewer than 70 as many so-called languages are really dialects. A few languages dominate in the world: Arabic, Chinese, Spanish, Russian and English, and are now spoken in many countries and continents beyond those of their origin. Dominant languages tend to create pressure on small languages – witness the Catalan independence movement in Spain: it can be partly traced back to the fact that the Catalan language was forbidden in public use during the Franco dictatorship.

Change in languages has many causes: tribes going extinct or being assimilated into larger groups, with or without violence; slow changes in language habits by users over time; dominant influence of most common language in areas where people speak multiple languages; or integration of multiple languages. Modern English, the *lingua franca* of the world, is an example of the latter. It is the outcome of a combining Anglo-Saxon (with German-Frisian origins) and Norman-French languages, and as a result is a vocabulary-rich language. In the process, the original Celtic language was replaced by English, rather than assimilated.

Languages can evolve rather rapidly. This is illustrated by 'Cité-German' in Eisden-Tuinwijk, a former mining village in the Belgian province of Limburg. At its height, it comprised a few thousand inhabitants. The cité (French for 'city') was founded in 1911, offering work mainly to 'gastarbeiter' from central Europe. It enjoyed economic self-sufficiency, guaranteed by the mining company. Moreover, it was

---

[6] Not including the 257 languages of Papua Province (formerly known as Irian Jaya), i.e. the Indonesia western part of New Guinea.

geographically isolated, being bordered by a large forest on the one side and the river Meuse on the other, the latter also forming the border with the Netherlands. As a result, in just two generations the mineworkers and their families developed a new, common language. As many of the immigrants spoke or understood some German, it came to sound much like German, but it incorporated many elements of the regional language Limburgish (itself a mixture of Dutch and German), Polish and French. In the process, grammar was simplified, as happened with other hybrid or recombinant languages, such as Creole and Papiamento. In 1987, the mine was closed and the local community dissolved. The language is expected to die along with the few second-generation survivors, surely making it one of the more ephemeral languages in human history.

Irregular Verbs and Cursing

To illustrate the practical use of evolutionary insights about language, we consider two issues. First, the existence of irregular verbs has been explained using an evolutionary angle. In most if not all languages the ten most frequently used verbs are irregular and the two by far most used ones, 'to have' and 'to be', are among the most irregular of all. The reason is that irregular verbs have to be learnt by heart by young children, and parents and teachers have to repeatedly point out incorrect uses such as children's tendencies of applying rules for regular verbs. Now if verbs are frequently used, children can learn the irregular form easily, but if verbs are rarer in use, this will not be the case. It is hypothesised that verbs historically started as irregular, but when the number of verbs and variation in irregular forms grew while the use of some was infrequent, regular forms were needed to avoid problems of memory and confusing communication with others. To illustrate, in English several verbs have shifted from the irregular to the regular form in the last centuries (e.g. I help / I holp / I have holpen), or exist in both forms (e.g. dreamt and dreamed). Only a few verbs have changed in the other direction, namely due to a form similar to an existing irregular verb (e.g. 'to wear' following the rules of 'to bear'). This illustrates that irregularity often has a degree of regularity, following grammatical rules, to make language tasks acceptable to the brain.

As a second example, the evolution of cursing is interesting. Swear words change with cultures and times. What does not change much is the motivation for using such words, as well as the effect they have on others – indeed, every society has taboo words. Evolution of swear words is partly dependent on the creation of variation of words in subcultures, fascination of young children and adolescents with such words, and imitative behaviour causing diffusion among a larger population. Nowadays, television and movies may help in this process. When religion still dominated Western societies, many swear words originated around religious core terms. These were considered blasphemous, such as words or expressions including the term 'god' or 'Jezus'. Related evolved terms like 'gosh' and 'gee', however, both express a feeling of surprise, while they are generally not considered profane. Many European languages incorporated Latin words to replace taboo words that denoted sexual organs or activities, like 'penis', 'vagina' and 'copulate'. Some words are so strongly taboo that one would expect them to have completely disappeared from active language. An example is 'nigger' (the N-word), owing to its association with slavery. But it is still in use in a neutral or even positive way among certain groups of black people to express companionship (McWhorter, 2015).

Many aspects of language evolution are still unclear or debated. Its study combines insights from many disciplines, including human biology, medicine, cognitive science, anthropology and linguistics.[7] Since the origin of human language left no consistent traces in the fossil record, we likely will never uncover its complete history.

---

[7] See, e.g. the *Journal of Language Evolution*: jole.oxfordjournals.org.

6 Group Selection in Biology and the Social Sciences

6.1 Forgotten Groups

Dominant theories in the social sciences, particularly in economics, are based upon self-referential individual behaviour and neglect the role of groups. As a result, such theories are dictated by the idea of upward causation. Here, instead, it is argued that group-level phenomena are relevant to social science because the presence of groups can change the behaviour of, and interactions among, individuals. In turn, this may affect the social system and its dynamics. The combination of individuals and groups means that upward and downward causation operate simultaneously. This has been elaborated in group or multilevel selection theory. It addresses the emergence, growth and selection of groups, including unions, takeovers and conflicts. Arguably, it offers the best available framework for thinking about group dynamics and the interface between individuals and groups. Various behavioural and social sciences have drawn upon the large and growing literature on group selection, which has given rise to the distinction between genetic and cultural group selection. Combining individuals and groups – or within-group and between-group processes – in a single analysis results in a multilevel approach. It describes the interaction and net effect of V-S-I-R processes (Section 2.1) at multiple levels.

Such an approach may enhance the study of a number of topics relevant to social science, in four different ways. First, inclusion of groups can clarify the impact they have on cooperation, the structure of institutions and conflicts over distribution. Second, it can help to design adequate institutions or public regulations for dealing with collective action dilemmas. Indeed, standard public policy solutions to common dilemma-type problems are based on models assuming purely self-regarding preferences. Such policies may fail to address real-world situations characterised by norms influencing individual preferences and interactions (Bowles, 2008a). Third, combining group and individual levels of description allows for the analysis of countervailing forces of within- and between-group processes. Recent theoretical and experimental findings indeed suggest that groups, norms and social context are essential to explanations of individual choice (Akerlof, 2007; Fehr and Fischbacher, 2002). However, the dynamics of, and interaction between, multiple groups is rarely considered in theoretical explanations of these empirical studies. Finally, group evolution illuminates the complex organisational structure of human economies, involving nested structures, conflict between groups and the coevolution of different sets of groups and individuals (Hannan and Freeman, 1977, 1989; Potts, 2000).

The combination of evolution and groups means a focus on ultimate, as opposed to proximate, explanations. This is increasingly accepted as the most suitable way to understand the fundamental nature, history and dynamics of complex systems. Of course, one can simplify and assume away dynamics or just pose mechanistic dynamics (i.e. absence of populations, diversity, innovation and selection) in any particular analysis, but only an encompassing evolutionary framework is able to clarify the margin of error and the conditional range of explanation resulting from such a simplification.

An important basis for thinking about groups is the debate on group selection in biology and the behavioural and social sciences (Bergstrom, 2002; Boyd and Richerson, 1990; Henrich, 2004; Sober and Wilson, 1998; Wilson and Hölldobler, 2005; Wilson and Wilson, 2007). Despite continuing disputes, the extensive literature on genetic and cultural group selection is now an integral part of the large body of evolutionary thinking.

Group selection has received little attention in the social sciences. Rudiments of cultural group selection arguments were implicit in F.A. Hayek (1976 and later work). He argued that customs, morals, laws and other cultural artefacts are subject to group selection, generally surviving and being replicated if they benefit survival and expansion of the human groups carrying them. For a critical view on Hayek's ideas, see Steele (1987). Others have used the term group selection but with a loose interpretation that does not always clearly separate between kin selection and group selection applied to non-kin groups (e.g. Samuelson, 1993). They nevertheless seem to support the view that multilevel phenomena deserve more attention in the social sciences.

Parsimonious modelling, characteristic of both theoretical economics and theoretical biology, runs into problems when groups and resulting multilevel phenomena are added to the picture. Simple models are unable to adequately address group-related phenomena, such as synergistic interactions among individuals, relative welfare and status-seeking, clustering of individuals due to spatial isolation, multilevel selection, and the combination of upward and downward causation (van den Bergh and Gowdy, 2003). Not surprisingly, one can find many different approaches to modelling group selection. It is likely that some of the results obtained with formal models in the 1960s and 1970s are not as general as once thought, because of the limitations of these models. Indeed, at the time, elaborate numerical analysis with complex, multilevel and spatial models was challenging.

Besides evolutionary theories of group formation and selection, there are numerous, less well-defined, theories about groups (e.g. Forsyth, 2006). These involve concepts like networks and hierarchies, and employ proximate explanations based on psychological, sociological and economic reasoning. In addition, experimental techniques have been employed to examine the effects of groups (Section 6.4). In mainstream economics, fundamental change at the level of groups or institutions is usually framed as a rational and deliberate choice among options, rather than an endogenous phenomenon as in evolutionary theories (van den Bergh and Stagl, 2003). For example, North's

(1981) original view was that institutions are 'chosen' based on efficiency, although later he changed this view dramatically (North, 1997). Institutional economics and related work in sociology employ a rich palette of in-depth, historical case studies of group phenomena which tend to be more descriptive than analytical and predictive. Evolutionary theories based on within- and between-group selection can improve upon these approaches.

In this chapter, we show the value of genetic and cultural group selection theories as applied to social science. Both types of selection have influenced human nature and culture, while cultural group selection is arguably most relevant to current and future social-economic phenomena.

6.2   Defining Group and Multilevel Selection

For an initial understanding of the debate on group selection, it is useful to take a look at some definitions and interpretations of it (D.S. Wilson, 1975, 2002; Sober, 1981; Grafen, 1984; Wilson and Sober, 1994; Soltis et al., 1995; Sober and Wilson, 1998; Gintis, 2000; Bergstrom, 2002; Henrich, 2004):

- Differential survival and reproduction of groups within a population. Dysfunctional or non-adapted groups become extinct and are replaced by relatively successful or well-adapted groups. Here adaptation is to a given environment in which the population resides. Such a group evolution process requires a diversity of groups in terms of composition of individuals, strategies or institutions. This will serve as our working definition here.
- The frequency of genes, or more precisely alleles, is affected by the benefits and costs they bestow on groups.
- The fitness of every member of a group partly depends on a common characteristic not isolated in an individual. This may be a 'group meme' such as a social institution. Examples of this are more evident in a socioeconomic than in a purely biological realm, so that this interpretation perhaps better fits cultural rather than genetic group selection.
- The fitness of every member of a group depends on the behaviour of the other members. In genetic group selection, this is made explicit through non-additive genetic interaction occurring between individuals within a group, notably with regard to altruism or aggression.
- Evolution combined with functionalism, where group functionality means that 'the whole is greater than the sum of its parts'. A particular example in biology is groups adopting the characteristics of a superorganism, as has been recognised in the case of colonies of social insects (Wilson and Hölldobler, 2005). This has been referred to as group-level functionalism (D.S. Wilson, 2000). The combination of group evolution and functionalism goes beyond traditional thinking in terms of functionalism as found in sociology. The latter lacked a dynamic framework to describe and explain the mechanisms giving rise to functions.

Such group-level functions co-determine fitness and thus sensitivity to selection pressure, which in turn alter group structure, size or functions.

- Groups function as adaptive units. Group adaptation is realised or improved through evolution – i.e. a combination of variety creation, selection and inheritance – at the level of groups.

Figure 6.1 gives a schematic account of multilevel selection as the combination of individual or within-group and between-group selection. Concepts mentioned in the figure are clarified and elaborated in subsequent sections.

The many interpretations of group selection in the literature suggest that it is multifaceted and that researchers are not always addressing exactly the same phenomenon. This illustrates that group selection is not easy to capture, which may partly explain the persistent debate about it. Different authors stress a subset of the many core mechanisms of group selection, as discussed in Section 6.5. Despite a steady increase in the number of articles providing theoretical arguments, formal models, empirical regularities and experimental findings in support of group selection, several respected

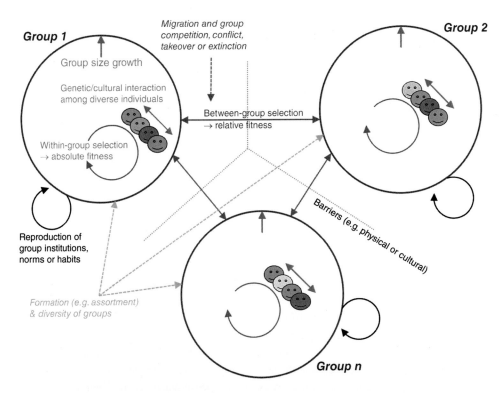

Figure 6.1 Multilevel selection.
Notes: The scheme depicts the evolutionary dynamics of *n* groups (with only 1, 2 and *n* shown) that are subject to between-group selection (red arrows) and within-group individual selection (green arrows). The meaning of coloured arrows or lines is explained by text in the same colour.
(A black and white version of this figure will appear in some formats. For the colour version, please see the plate section.)

biologists still see fundamental problems with it. Others consider it to be theoretically possible but an unimportant force of evolution. Yet, it is clear that group selection currently finds more broad support within biology than a few decades ago. For example, in an important turnaround, the eminent biologist E.O. Wilson now takes the position that genetic group selection was critical to eusocial evolution (Wilson and Hölldobler, 2005; Wilson and Wilson, 2007).

Group selection is often discussed and judged in terms of its application to altruism. However, group phenomena can be linked to a variety of social (other-regarding) preferences including reciprocity, reciprocal fairness, inequity aversion, pure altruism, altruistic punishment, spite, envy, comparison and status-seeking. Second, group selection can address a range of genetic or individual interactions, notably cannibalism (especially in insects), conflict (fighting) and complementary roles (labour division), such as in community selection (Goodnight and Stevens, 1997; Swenson et al., 2000). Any genetic or cultural trait might influence 'within- and between-group selection', while the directions of their effects may be opposite or not, depending on the issue considered. The two effects are opposite in the case of individual altruism that is group-beneficial, as is discussed below.

6.3   Lessons from the Debate on Group Selection

Natural selection, above the level of the individual, was a quite acceptable idea to Darwin and Wallace, who both believed in the differential survival of groups. Nevertheless, group selection was originally met with scepticism, if not outright hostility, partly owing to the publication of V.C. Wynne-Edwards (1962) book *Animal Dispersion in Relation to Social Behavior*. As discussed in Box 6.1, this was later evaluated as an incorrect statement of group selection and inconsistent with modern formulations of it (D.S. Wilson, 1983). Wynne-Edwards received strong adverse reactions from respected biologists, notably G.C. Williams (1966). They stressed that two widely accepted theories countered Wynne-Edward's version of group selection. One was kin selection, the idea that apparent altruism is genetically founded as altruists are actually protecting their own genes (inclusive fitness) by helping close relatives survive (Hamilton, 1964). The other was reciprocal altruism, the view that apparent altruism was based on the expectation that favours would be returned (Trivers, 1971). These two theories formed the basis of a new field, 'sociobiology', which initially was opposed to the idea of group selection, arguing that social behaviour is entirely the result of gene-level and individual-level selection.

The relationship between kin and group selection has been subject to much debate, especially in the context of altruism. The observed presence of high kin relatedness in a population should not be confused, as is often done, with kin selection acting as a crucial factor in the historical evolution towards the current system (Griffin and West, 2002). In the context of eusociality, it is now considered likely that group selection gave rise to kin selection which in turn produced a high degree of relatedness in colonies (Wilson and Hölldobler, 2005). Indeed, repeated group selection may increase

Box 6.1 The averaging fallacy and Simpson's paradox

While the critique of Wynne-Edwards's version of group selection was correct, it should not be misinterpreted as undercutting later, more sophisticated, conceptualisations of group selection. Wilson and Sober (1994) make a distinction between naïve and modern group selection theories. Wynne-Edwards (1962) is considered an example of the first as his theory does not account for effects of individual selection and essentially assumes that higher-level adaptation (even of a single, isolated group) is a fact. Modern group (or multilevel) selection theory instead recognises simultaneous processes of individual and group selection, as well as group selection requiring multiple groups. In line with this, it regards multilevel evolution as a combination of within- and between-group selection.

Williams' influential critique contained many misperceptions of group selection, one being the 'averaging fallacy'. This means that group selection effects are attributed to individual-level (i.e. between-individual) selection due to simply calculating the average fitness of individuals across groups, rather than decomposing it into average fitness per group and a between-group selection component dependent on the relative size of groups in the total population (D.S. Wilson, 1983). The averaging fallacy, in fact, merely defined group selection away, rather than showing it to be incorrect.

Williams assumed that when a trait is decreasing in frequency in every group, its population frequency must also decrease. Surprisingly, this is not necessarily true. The reason is differential growth at the group level, which causes the composition of the population in terms of relative group sizes to change. The net outcome may be that the trait will increase in frequency in the overall population. In statistics, this is known as Simpson's (1951) paradox. Suppose a population consists of two groups, then this paradox requires that the group with relatively many individuals with the trait grows relatively fast compared to the groups with few individuals with the trait; the proportion in the population of the first group will then increase (note that this is a necessary but not a sufficient condition). In mathematical terms, this can be formulated as follows. Suppose that the population size at time t is $p(t)$, consists of two groups with sizes $g_1(t)$ and $g_2(t)$, the number of individuals with the trait is $a(t)$, and is distributed as $a_i(t)$ in group $g_i(t)$ ($i = 1, 2$). Then $p(t) = g_1(t) + g_2(t)$. Now even if the changes over time in frequencies of the trait are characterised for group 1 by $a_1(t + 1)/g_1(t + 1) < a_1(t)/g_1(t)$ and for group 2 by $a_2(t + 1)/g_2(t + 1) < a_2(t)/g_2(t)$, then it is still possible that the trait's frequency in the population increases: that is, $[a_1(t + 1)+a_2(t + 1)]/[g_1(t + 1)+g_2(t + 1)] > [a_1(t) + a_2(t)]/[g_1(t) + g_2(t)]$, or $a(t + 1)/p(t + 1) > a(t)/p(t)$ holds. To illustrate this numerically: $1/4 < 2/7$ and $5/7 < 3/4$ while $(1 + 5)/(4 + 7) = 6/11 > 5/11 = (2 + 3)/(7 + 4)$. This example suggests that a rather special case is required, which is not surprising, otherwise it would not have been called a paradox. The paradox shows the power of group selection, namely that in principle it can increase a trait's frequency in a population even if its frequency falls in all groups. The net effects of group selection will be larger if the frequency of traits can also increase in certain groups.

relatedness in a group, through limited dispersal or inbreeding (Hamilton, 1975), providing a basis for more effective kin selection. Things get even more complex when group and kin selection function in tandem (Avilés et al., 2004). It is also often overlooked that the effects of kin selection, such as altruism benefits to kin and thus selfish benefits from the shared gene's perspective, are opposed by competition between kin (Frank, 1998). Of course, seemingly altruistic behaviour among kin-related individuals can simply be the result of direct, reciprocal or general group (i.e. not kin-group-specific) benefits (Griffin and West, 2002).

It has been suggested that kin selection is conceptually or mathematically equivalent to certain types of group selection. Kin selection is a subset of group selection when group members are close kin. Hence, the more general model seems to be group selection occurring through assortment (association or clustering of related traits or strategies), while kin relatedness can be seen as a special case of such assortment, namely of genetically similar individuals. This, however, does not clarify whether group selection is more or less useful as a formalised approach to study reality. Possibly, if kin selection is generalised as inclusive fitness theory where inclusiveness applies to assorted individuals (or genes), then group selection might be (seen as) a special case of it (Grafen, 2007). Others stress that kin selection does not require family relatedness caused by (recent) common ancestry, but only genetic correlation among individuals (Foster et al., 2005). This seems to come down to the same generalised inclusive fitness based on assortment. By stretching the meaning of kin selection in this way, the difference between group and kin selection becomes less pronounced.

It is commonly assumed that group selection involves assortment. Some think that such assortment is only kin based. However, group selection can operate through channels other than assortment and inclusive fitness. Van Veelen and Hopfensitz (2007) and van Veelen (2009) emphasise, as an important alternative mechanism, a shared interest or fate of group members. So the equivalence may hold, but only for a subset of group selection models, namely those in which assortative group formation is the key mechanism.

The model-based literature shows that formalisation of groups generally results in quite different mathematical formulations than those based on kin selection, and they are not necessarily reducible to one another. Moreover, they may employ different types of specific assumptions and, as a result, they can differ in accuracy of describing organisms and their evolution (Queller, 1992). Nevertheless, the similarity between Hamilton's rule and rules arising from group selection has been shown for certain types of group selection models (e.g. Frank, 1997; Ohtsuki et al., 2006; Grafen, 2007; Lehmann et al., 2007). The distinction in practice between group and kin selection (or inclusive fitness analysis) as a causal mechanism may ultimately depend on the type of data gathered, as has been argued by various authors (Colwell, 1981; Griffin and West, 2002). In this respect, the view by Martin Nowak et al. (2010, 2011) is relevant. They argue that inclusive fitness theory cannot serve as an alternative for studying social evolution (altruism, cooperation, eusociality, etc.). The reason is that it remains too aggregate about the nature of evolutionary processes, such as population structure (genetic and spatial distributions), non-additive fitness (e.g. due to synergy among different genes), social networks or spatial features of the selection environment. In contrast, modern group selection models explicitly describe these elements.

Whereas kin selection emphasises inclusive fitness and the impact of altruism on other individuals, group selection emphasises relative fitness in groups and impact of altruism on the (relative) productivity of the group (D.S. Wilson, 1983). They represent alternative ways of understanding processes that are similar when groups are defined as consisting of close kin. However, group selection has the advantage that it can also address the case of group members not being each other's kin. This is relevant, as we need to explain why individuals show kind or even altruistic behaviour towards strangers, as in human societies. The reasoning that humans possess an imperfect psychological capacity to distinguish kin from non-kin, or altruists from defectors, can count on little evidence. Henrich (2004) notes several empirical and experimental findings against this reasoning: individuals clearly distinguish close from distant relatives, have frequent encounters with strangers, do not cooperate to the same degree with all group members, and treat acquaintances and strangers differently.

Reciprocal altruism refers to behaviour based on the expectation that favours will be returned. This requires a certain minimum level of cognition. Reciprocal interaction has been invoked to explain altruistic acts towards strangers. The main problem raised by this is that reciprocity requires repeated interaction or trust that an altruistic act will be reciprocated. However, if individuals are strangers, neither condition is satisfied. The more credible explanation for generalised reciprocity as the basis of altruism in large groups is that it is an extension of evolved reciprocity in small groups consisting of individuals familiar with each other. Extended or indirect reciprocity, based on information of previous interactions with others, reflected in judgement of reputation and morality and involving trust, is apparently present in some animal as well as human societies. In humans, it has led to complex social interactions with correlated demands on individual cognitive capacities (Nowak and Sigmund, 2005). Indirect reciprocity, however, cannot serve as an ultimate explanation of human prosocial behaviour, as this requires the refuted 'big mistake' hypothesis, i.e. humans not treating acquaintances and strangers differently. This does not deny that both experimental and empirical studies indicate that individuals show altruism towards strangers in one-shot games (Fehr and Gächter, 2002). It should also be noted that indirect reciprocity cannot solve the n-person dilemma (note that most published models are two person models). The reason is that it gives rise to multiple stable equilibria (Panchanthan and Boyd, 2005). This means it requires an equilibrium selection mechanism, for which group selection serves as one possible candidate. In other words, group selection and indirect reciprocity are not alternatives (Henrich et al., 2006; Henrich and Henrich, 2007).

According to Alexander (1987), the evolution of moral awareness and behaviour in primates strengthening the structure of the own group received a major stimulus from the long history of violent interactions among ancestral primate groups. This is supported by asymmetric behaviour in conflicts among (living) apes and monkeys: conflict resolving inside the group and extreme brutality to outsiders. Similarly, humans apply ethics asymmetrically to insiders and outsiders of the group to which they belong. The most convincing examples of this are wars, notably when driven by religious and ethnic conflicts (de Waal, 1996, p. 29; D.S. Wilson, 2002). Group selection models have been invoked to show that group conflict between humans may be closely related to the evolution of parochial altruism (Choi and Bowles, 2007; Garcia and van den Bergh, 2011). Such parochialism has two faces, namely providing benefits to fellow group members and showing hostility towards outsiders, both at a personal cost. More generally, multilevel evolution means that 'virtues at one level become vices at higher levels: looking out for oneself becomes selfishness, looking out for one's family becomes nepotism, and looking out for one's village becomes corruption' (D.S. Wilson, 2007, p. 291).

Group competition and selection, together with reciprocal altruism, are considered by Frans de Waal (1996) as the core mechanisms that have driven the evolution of morality. On a psychological level, the stepping stone from non-moral animals to moral humans is the evolution of empathy – the ability to be affected by feelings of another organism or the situation in which it finds itself. Empathy goes along with self-awareness, i.e. distinguishing oneself clearly from the environment or the rest of the world, and with perception, i.e. positioning oneself in different roles or individuals to understand a situation or problem. Evolution of empathy ultimately gave rise to moral community, in which not only direct, one-to-one interactions occur among individuals but also indirect ones: care about good relationships between others, and mediation and arbitration in conflicts. De Waal (1996, p. 34) calls this 'community concern'. Humans add to these mechanisms of punishment and reward explicitly focused at maintaining or improving the social environment, sometimes referred to as meta-norms (Axelrod, 1986). Altruistic punishment leads to internalisation of norms and rules and guilt-like behaviour, common in humans and other primates, as well as in dogs.

D.S. Wilson (2005) regards the distinction between absolute and relative fitness as essential to understanding the impact of group selection. By increasing the absolute fitness of individuals within a single, isolated group to the same degree, their relative fitness does not alter, so that the fitness change will be without evolutionary consequences. But when adding other groups that interact (perhaps depending on the same scarce resources), the absolute change in fitness for the original individuals will mean an improved average fitness of the group relative to that of other groups. Then the group may grow more quickly than other groups and thus will increase its proportion in the total population. The effect of this is only interesting if fitness differences between groups relate to diversity in the structural, genetic or cultural, foundations of groups. Wilson notes that explanations based on individual-level selection tend to neglect the possibility of group selection effects because they focus on absolute instead of relative fitness improvements. Relative fitness of individuals may increase even if they reduce their absolute fitness, simply because the group as a whole benefits from their acts, such as those of an altruistic nature. The problem is that the identification of individual-level selection with absolute fitness improvement frustrates the search for decomposing evolutionary change into within- and between-group selection contributions.

So evolutionary dynamics may be seen as resulting from a combination of within- and between-group selection, irrespective of how weak each force is in a particular species or time period. This insight was formalised in the well-known Price (1970, 1972) equation which allows a decomposition of evolutionary changes into between-group and within-group effects. This should, however, not be considered as proof of the theoretical possibility or empirical effectiveness of group selection in real-world contexts. Henrich (2004) uses the Price equation to derive a set of necessary conditions for group selection to be effective. *Ceteris paribus*, the larger the variation between groups, the more opportunity for between-group selection to be effective and dominate within-group selection effects. Variation between groups is nevertheless hampered by migration, mixing and reformation of groups, which will receive attention later on.

The most common argument against group selection is free-rider behaviour. The idea is that free riders will profit from the benefits of being part of a group with genuine altruism and social institutions, without contributing to these or contributing less than average. As the relative proportion of free riders in the group increases, the benefits for the group will slowly disappear. Moreover, it is relatively easy to be altruistic when resource scarcity and competition are low, but selection pressure is then quite low as well, so that this altruistic behaviour will not quickly diffuse. On the other hand, altruism and altruistic punishment are less common when scarcity and competition are high, in which situation altruism implies a serious sacrifice. Because selective pressure is higher in the second case, individual-level selection will generally have a relatively larger impact. A major shortcoming of this reasoning is that it employs a single-level explanation that excludes group variation and selection. These make sure that groups with relatively many altruists grow faster than other groups, as a result of which the proportion of altruists in the population as a whole may increase. This is the earlier mentioned Simpson's paradox (Box 6.1). In addition, the reasoning about free-riding is restricted to altruism, whereas group selection has a broader significance.

Finally, to clarify the complexity of group selection, consider that to determine whether a particular trait evolves or has evolved by group selection, multiple groups must be identified. The relative fitness of individuals within groups also needs to be evaluated to assess the effect of individual selection; the relative fitness of groups in the population should be identified to assess the effect of group selection (usually measured by an aggregate group property, such as size), and the effects of group and individual selection have to be compared to determine the net impact of group selection on evolutionary change. These steps entail much more complexity than modelling individual evolution alone. To illustrate this, consider n different groups, each one consisting of many individuals employing one out of k strategies. Without group selection, evolution can be described by k variables and k equations (one for each strategy in the population), whereas with group or multilevel selection we require $n(k + 1)$ equations (and variables). Moreover, the number of interactions among endogenous variables without group selection is k^2, while it is $nk^2 + n^2(1 + k)$ with group selection, namely nk^2 interactions for within-group dynamics plus $n(n + nk)$ interactions for group dynamics. For example, in a simple case of three groups and two strategies, we obtain a system of already nine (nonlinear) differential equations, capturing 39 interactions among variables, while in the case of individual selection alone the system would merely comprise two equations and four variable interactions. This illustrates that complexity increases substantially when moving from individual to multilevel (individual + group) selection.

It suggests that one should be careful in deciding intuitively (i.e. without the help of a model) about the relevance and consequences of group selection.

An online essay by Steven Pinker (http://edge.org/conversation/steven_pinker-the-false-allure-of-group-selection), a highly respected evolutionary psychologist, should not be left unmentioned here. It is critical about the relevance of group selection. What struck me most, however, were the 23 supporting and dissenting reactions. These show immense variation in interpretations of group selection. Pinker expresses the opinion that 'the concept of Group Selection has no useful role to play in psychology or social science'. Throughout his essay, he deals, though, with a loose interpretation of group selection. For example, he never mentions core notions like assortment, between-group selection and within-group selection. But without these one can hardly get to the core of group selection thinking. Neither does Pinker provide convincing evidence that the social psychology of humans can be fully understood through inclusive fitness and reputation effects, central elements in his argument. The following responses to Pinker's essay clarify the different positions in the debate:

- H. Gintis: 'The first misconception here is the view that group selection is incompatible with kin selection. It is not. Kin selection says that the fitness of an individual depends on the genes of his kin and not just his own genes. Group selection says the fitness of an individual depends on the characteristics of the group he is in, not just his own genes. The second misconception is that group selection means that the group is a "unit of selection". This is not true. Group selection occurs when the fitness of individuals may be higher in one group rather than another, depending on the social structure of the group and its distribution of genomes.'
- J. Haidt: 'But if you examine the psychological traits that motivate and enable cohesion, trust and effective coordination, and if you do this during times of intergroup conflict, you will find many behaviours and mental mechanisms that are much harder to explain using only individual-level mechanisms. You will find yourself swimming among group-selected traits.'
- J. Henrich: 'Pinker wants banished . . . a modelling tool that has proved useful for breaking down and analyzing different components of a selective process. . . . When the situation under investigation involves something like group extinctions, due to war or environmental shocks, or biased migration driven by economic success, multilevel selection accounting can help isolate and analyse the impacts of different components of selection.'

We should get rid of the misconception that one has to choose between group and individual selection. They can be simultaneously at work. Unfortunately, the pro- and anti-camps in the debate on group selection are itself subject to a kind of group selection — you're for or against group selection depending on the evolutionary school you grew up in.

6.4   Experimental and Empirical Evidence

Evidence for group selection comes from various sources: artificial selection in experiments and observed selection in nonhuman nature (animals, plants, communities and ecosystems) and in humans. Goodnight and Stevens (1997, p. 62) review 11 experimental studies between 1977 and 1996, on beetles (various cases), domestic rats (group selection through male cooperation in mating), plants (leaf area) and chickens (interaction between egg production and aggression). They conclude that genetic group selection can be effective in influencing genetic and phenotypic change over time. Group selection is found to differ from individual-level selection in that it can affect interactions between individuals, unlike in experiments allowing only for artificial individual-level selection.

Michael J. Wade (1976) experimented with beetles (*Tribolium castaneum*). Selection in these experiments was artificial and focused on group size. After only a few generations, evident group selection effects could be observed. In this case, the concrete influence was altered rates of cannibalism. In evaluating these experiments, D.S. Wilson (1983) notes that group size regulation can occur by selfish behaviour in the form of cannibalism and altruistic behaviour in the form of voluntary birth control. He argues that if selfish and altruistic strategies are both present, then group selection will foster the selfish strategy. This is surprising and contrasts with most of the literature critical of group selection, which tends to incorrectly assume that selfish strategies are unexplained by group selection, while group selection is only needed to explain certain types of altruism.

A surprising finding is that group selection effects have been assessed in experiments of chicken egg production (Muir, 1996). Whereas artificial individual-level selection may maximise egg production for isolated chickens, group selection arranges a trade-off between productivity and aggression, which is relevant if chickens are to be housed in multiple-hen cages ('egg factories') characterised by aggressive interaction. Group selection realised a 160 per cent increase in group yield, i.e. group-average egg production, versus unselected controls, and likewise less aggression. The economic value of this result is evident: not only more egg production but also reduced hen mortality and less need for beak trimming.

The impact of group selection on plants is also surprising. Experimental results for groups covering low and high leaf area showed between-group selection to be more effective than between-individual selection (Goodnight and Stevens, 1997). Group selection theory has even been shown to apply to communities, i.e. involving interactions between individuals or populations of multiple species, resulting in community or ecosystem selection theory. This has also been confirmed by experiments (Goodnight, 1990a,b and 2000; Goodnight and Stevens, 1997; Swenson et al., 2000).

A necessary proviso of experimental findings supporting the potential effectiveness of group selection is that the artificial nature of selection likely means that effects will generally be stronger than when occurring under natural conditions. The more so since many experiments employ the so-called propagule model (explained in Section 6.5). Nevertheless, the experiments support the direction and potential effectiveness of group selection as an evolutionary force.

To obtain a more complete picture, additional evidence is relevant. Empirically oriented studies are reported by D.S. Wilson (1983). He mentions a commonly cited example of group selection, namely the evolution of 'avirulence' (i.e. a lack of spread through infection) in the myxoma virus that was introduced into Australia to control the

European rabbit. Each rabbit is a deme from the standpoint of the virus, so that the structured deme model of group selection can be applied. When a rabbit dies the associated virus group becomes extinct, as the virus cannot survive in a dead rabbit and mosquitoes necessary for diffusion do not bite dead rabbits. The virus groups that are alive are the least virulent. Therefore, even though avirulence has no selective advantage within a virus group, it arises through deme or group selection.

Field studies of group selection under natural conditions are summarised by Goodnight and Stevens (1997). They list five studies between 1989 and 1996, all of which supported the effectiveness of group selection. The studies deal with a number of traits and species, namely cannibalism in beetles, survivorship, flower production and fruit production in plants, as well as reproductive and worker allocation in ants. To assess selection effects, they use indicators like population density and percentage ground cover, and group fitness indicators such as mean leaf area, mean plant height and mean photosynthetic rate.

Another example of group selection under natural conditions often mentioned in the literature is biased sex ratios. Robert J. Williams (1966) showed that individual-level selection tends to give rise to an even sex ratio while group selection leads to a female-biased sex ratio as groups with more females can produce more offspring and thus grow faster. Hence, he regarded the absence of biased sex ratios in nature as evidence against group selection. However, later female-biased sex ratios were assessed in hundreds of species, which was then seen to reflect possibly an equilibrium between opposing forces of within- and between-group selection (Colwell, 1981; Frank, 1986).

With regard to social insects, Wilson and Hölldobler (2005) argue that genetic group selection needs to be invoked to offer a complete explanation of the evolution of eusociality. They regard close kinship likely to be a consequence of eusociality rather than a critical factor in its evolution. They think that only group selection is able to provide a consistent explanation for two central empirical facts: the rareness of eusociality and the ecological dominance of eusocial insects over solitary and pre-eusocial competitors. In particular, evolution of eusociality had to involve two phases of group selection: initially, competition between solitary individuals and cooperative pre-eusocial groups and, later on, competition between colonies, i.e. groups composed of individuals with strong cooperation and close genetic relatedness. Group selection created the conditions for kin selection to emerge in eusocial species. Comparison of the history of insects with and without eusociality yields some insights about preconditions for eusociality to evolve, namely individual pre-adaptations, such as building nests and feeding larvae, both of which foster cooperation within groups. Moreover, the key adaptation of eusocial species is defence against all sorts of enemies in which groups are better than individuals. It is long established that more advanced eusocial species are characterised by more labour division and subtle, chemical communication, which makes groups function more effectively in reproduction, conflicts and foraging.

Support for genetic and cultural group selection in humans evidently cannot make use of artificial experiments. Instead, evidence here takes a more indirect form, as discussed in Bowles et al. (2003) and Henrich (2004). A detailed study by D.S. Wilson (2002) of the evolution of religions offers one of the best arguments for the relevance of cultural group selection for the social sciences. He shows that there always has been a large diversity of religions, that religious groups are quite stable (existing for many human generations), that they compete and enter into violent conflicts (even within single 'meta-religions' like Christianity and Islam), that they bind groups strongly through fear and punishment and that they reproduce well. Propagation mechanisms include indoctrination of children, rules about partners and offspring, and active efforts to convert non-believers. Moreover, the suggestion that the most powerful and impressive god is a solitary god may have helped monotheist religions to become dominant in the world. Note that the evolution of religions is relevant to economics, as religions provide institutions and rules with important economic repercussions (Iannaccone, 1999). In a study of 'homogeneous middleman groups' supported by many historical examples of merchants/traders, J.T. Landa (2008) generalises the idea of selection of religious groups. For more details on four evolutionary explanations of religions, including by reasoning through group selection, see Section 7.7. More generally, Gürerk et al. (2006) show experimentally that a sanctioning institution comes out as the winner when competing with a sanction-free institution. In their experiments, despite initial aversion, the entire population ultimately chooses the sanctioning institution. This indicates the relevance of institutional selection for collective action and group dynamics.

The evidence coming from primate research is mixed. Some of it underpins the relevance of groups in behaviour and evolution. Silk et al. (2003) show that infants of more social female baboons have a greater chance of surviving to the first birthday. Melis et al. (2006) find that chimpanzees recognise when collaboration is useful or necessary, and know how to select among non-kin the best collaborative partners. These authors argue that since such skills are shared with humans, they may have characterised a common ancestor. However, in several experiments chimps act like 'homo economicus' (Silk et al., 2005). In addition, de Waal (2006, p. 16) notes that in primate species males or females often leave the group and join neighbouring groups. This suggests genetic isolation is imperfect and group selection effects are weakened. One should be careful, however, not to jump to the conclusion that such effects are entirely absent, as genetic differences between groups will not necessarily disappear completely.

Many reputable biologists have expressed the view that genetic and cultural group selection has likely played a role in the evolution of humans, but question the relevance for animals (plants are usually ignored). In addition, indirect experimental evidence is available. In psychology, group experiments have been performed and contrasted with similar experiments on the basis of isolated individuals (Bornstein and Ben-Yossef, 1994). Although not strictly about evolution, these studies make clear that interactions between individuals affect individual strategies. Competition among groups turns out to influence group outcomes. Even the awareness of the presence of another group makes a difference in terms of individual play.

In spite of the documented experimental and empirical evidence, one may wonder why attribution of experimental or empirical findings to group selection is still rare compared to individual-level selection. D.S. Wilson (2005) convincingly argues that this is largely due to ignoring the difference between absolute and relative fitness

(Section 6.3). Typically, researchers do not distinguish between group and individual-level selection effects on fitness but immediately calculate net effects and then automatically attribute these to individual-level selection.

6.5 Mechanisms of Group Selection

In this section we discuss V-S-I-R mechanisms (Section 2.1) that make up genetic group selection. Some of these are necessary for group selection to occur while others merely make it more likely or more effective. The following factors are considered: (1) migrant pool versus so-called propagule type of population dynamics (relating to group formation); (2) (non-random) assortment; (3) type of population structure (spatial, behavioural); (4) institutions; (5) splitting of groups; (6) group conflict; and (7) non-additive (genetic) interaction between individuals. Some of these factors have received more attention than others. No model captures all of them, so that there really is no complete model of group selection. This underpins the difficulty of the topic.

Whereas individuals are concrete, stable entities, groups are more vague and fluid, which raises the question of how groups come into existence. Groups can originate through random isolation, inbreeding, ecological specialisation or non-random assorting. Slatkin and Wade (1978) noted two alternative approaches to describe the colonisation of areas (habitats) referred to as migrant pool and propagule models. In the migrant pool model, all populations contribute migrants to a common pool from which colonists are drawn at random to occupy areas so that there is complete mixing of individuals from different populations. The haystack model by Maynard Smith (1964) is a well-known example. In the propagule model, a pool is made up of individuals derived from a single population so that there is no mixing of individuals from different populations. Through cumulative mutations in, and selection of, genetic or cultural characteristics, much more among-group diversity can be realised, and in turn group selection can be much more effective. Seen another way, in comparison with the propagule approach, the migrant pool model means that something which might be termed 'group heritability' or group reproduction is largely absent, suggesting that group evolution will be incomplete and thus ineffective or even non-existent. However, this may not be the case if group conflict is introduced (see below). Finally, repeated group selection in a propagule-type framework with isolated small groups sensitive to drift can produce increasing kin relatedness and ultimately give rise to speciation.[1]

---

[1] Traulsen and Nowak (2006) provide a model of group splitting that allows only for interaction between individuals within strictly separated groups. Groups split endogenously when (successful) groups reach a certain maximum size, while another group is eliminated so as to keep the total number of groups constant. This can be seen as a special type of propagule pool model. It is applied to prisoner's dilemma games between cooperators and defectors, and a fundamental condition for the evolution of cooperation is $b/c > 1 + n/m$, with b and c denoting the benefit and cost of an altruistic or cooperative act, respectively, and n and m the maximum group size (where probabilistically splitting occurs) and the number of groups. With smaller maximum group size or with more groups, the condition is more easily satisfied and cooperation is favoured

An important cause of group formation and selection is non-random, assortative interaction, which can be opposed to randomly remixing groups. Special cases are preferential assorting and common ancestry (kin selection). Assortative interactions lead to non-random variation among groups. This mechanism operates, for instance, if altruists are able to recognise other altruists. This works in any case within the context of extended families and small groups and for larger groups if individuals satisfy a minimal degree of cognitive abilities or (social) intelligence in combination with experience (learning). Random variation in small groups can produce outcomes that resemble non-random assorting, which is often referred to as genetic or cultural drift. Assorting can also occur through kin-recognition resulting in extended family groups, in which case group selection is equivalent to kin selection (Hamilton, 1964). In biological contexts, inbreeding is an important case of assortative grouping according to kin features. Other results can be obtained for different assorting rules. Bergstrom (2003) offers a generalised account based on an 'index of assortativity'. He explains the evolution of cooperation under assorting as the cost of cooperating being compensated by higher probabilities and associated higher benefits of meeting a cooperating partner. In an economic context, assortative grouping will depend on group-specific institutions that promote cooperation and altruism, such as education, religion, political voting systems and free press.

Assortative interaction depends on the effectiveness with which individuals recognise others with similar characteristics. This problem is often formalised through signalling games. In a biological sense this can depend on chemical signals (kin selection – e.g. ants). In a cultural setting, it can involve location of meeting (a public space like a bar or political party meeting), or expression of one's convictions (verbally or through physical appearance, such as clothing style). In order to be stable, this requires a constraint on, or complete lack of, mutations. The 'green beard effect' as a special case of signalling is relevant here (Dawkins, 1976; Henrich, 2004). Suppose altruists have green beards which allow them to recognise and cooperate with each other; but due to mutation selfish individuals also start to grow green beards. The result is a steady increase of the proportion of selfish individuals in the population. This is a special case of free-riding behaviour as discussed earlier, but based on reaction to signals. The effectiveness of assortative grouping in creating a sustainable 'altruistic group' depends on the relative forces of assortative interaction and mutations in defectors that allow them to send out fake signals. If the first dominates, altruism can survive; if the second dominates, altruistic behaviour is unsustainable and driven out by selfishness. In general, altruism can be sustained only if signalling and assorting is cheaper for altruists than is faking the signal for selfish individuals (Henrich, 2004).

Pepper and Smuts (2000, 2002) argue that the existence of spatial or patchy environments provides a sufficient basis for group selection and that there is no need for *ex ante*, stable discrete groups. They also present an explicitly spatial model in which non-random assortment occurs through individuals reacting to local environments by

by multilevel selection. A modified formula is derived for the case with migration, which shows the effect of group selection then to be weaker.

migrating in such a way that those with similar traits end up relatively more abundant in a particular locality. The result is that environments are positively correlated with individual traits. Environmental change, such as environmental degradation or increasing resource scarcity, may induce individual responses that stimulate assortment. Pepper and Smuts' work can be linked to the growing literature on evolutionary, agent-based models with local interactions. Here, agents perceive costs or benefits that depend on strategies by other agents in their immediate environment. For example, punishing or being punished can both be associated with a cost that depends on how many defectors or enforcers exist in the local environment of an agent (Bergstrom and Stark, 2003; Eshel et al., 1998, 1999; Nowak and Sigmund, 2000). These studies illustrate the general case that a change at a higher level of groups alters the cost, benefits or fitness at the lower level of individual agents. In such models, the survival of a certain agent or strategy depends on the local population environment, which can be interpreted as a case of the general intrademic or trait group selection model. Trait here refers to a feature of individuals affecting the fitness of other individuals in the group, a case of so-called 'genetic interaction'. Eshel et al. (1998) show that in a world with local interactions between altruists and egoists, altruism is a strictly dominated strategy, but altruists can survive as long as they are grouped together and push up each other's performance. Using a spatial model, Noailly et al. (2009) show that local equilibria include a group with a protection layer of enforcers. While the literature on local interactions with evolutionary agent-based modelling does not make reference to group selection, connecting these areas may represent a fruitful direction for research.

The traditional method to study these issues, systems of dynamic equations such as theoretical population models and evolutionary game theory, cannot address multilevel selection based on local interactions and spatial heterogeneity of environment and populations (Pepper and Smuts, 2000). The traditional theoretical approach makes simplifying assumptions, consistent with the principle of parsimony (Williams, 1966), which may be wholly unsuitable for the analysis of group selection. Possible oversimplifying assumptions include homogeneous or well-mixed populations, fixed splitting up of groups and infinite population sizes. Most fundamental is that the groups are exogenous and fixed whereas, in more realistic spatial, local interaction or agent-based evolutionary models, groups are endogenous while boundaries between groups change. The neglect of these features partly explains why many biology theoreticians have insufficiently recognised the relevance of group selection. This is all the more surprising given that the crucial role of space for speciation was clearly recognised by the founders of modern evolutionary biology, Darwin and Wallace, and was later elaborated in the theory of island biogeography (item v in Section 3.5). Tarnita et al. (2009) provides a general alternative approach to homogeneous, well-mixed populations, which they call 'evolutionary set theory'. It illustrates that evolutionary dynamics are strongly affected by population structure, such as defined by groups, space or networks. Hence, a well-mixed population evolves differently from a structured population. The approach entails that individuals interact with others in the same set, that they can be part of multiple sets, and that some sets are sparse while others are full, all of which affect evolutionary dynamics. Since set membership may alter with evolutionary dynamics, the population structure is not fixed but is itself evolving, which further complicates things and makes differences with outcomes under homogeneous populations even larger.

Samuel Bowles and Herbert Gintis (1998) regard the spatial dimension as important to address four mechanisms employed by communities to solve coordination problems: (1) reputation (low cost information about other agents needed); (2) retaliation (frequent or long-lasting interactions needed); (3) segmentation (non-random pairing of social agents); and (4) parochialism (limited migration among groups required). The latter two imply spatial disaggregation of the analysis as a special type of what the authors call 'structured populations'. At a general level, group selection can be seen as the outcome of a spatial game characterised by local interactions, in which groups emerge as to some extent spatially isolated units. This model favours the emergence of unique local equilibria causing group diversity at the global scale. In this context, the common typology of speciation – the emergence of a new species – is clarifying: allopatric speciation denotes the case in which one species splits into two species due to (spatial) separation; sympatric speciation occurs without spatial isolation (difficult because of interbreeding); peripatric speciation occurs in meta-populations at the edge of (large) populations living in vast areas; and parapatric speciation occurs in continuous populations living in vast areas where subpopulations are subject to different selection environments. All these types of speciation require group formation (starting with meta-populations), in different ways. This does not hold for sympatric speciation as here newly forming species are in the same geographic location.

A paper by van Veelen and Hopfensitz (2007) suggests that the most important distinction between group selection models of altruism is, in fact, between 'standard models' in which assortative group formation is the key to group selection and models in which the interest or fate of group members is aligned or shared. The first type of model is close to kin selection models, since here assortment is the outcome of kin relatedness. Assortment causes altruists to interact relatively much with other altruists, giving a selective advantage. Assortment requires a propagule pool structure. The main (or perhaps only) example of the second type of model is group conflict, as the fate of individuals in winning or losing groups are perfectly aligned. Assortment is not required here and group selection through group conflict might even leave an impact with a migrant pool structure, i.e. random remixing of groups in each period. The authors stress that it is not unlikely that groups are generally more competitive and prone to conflict than individuals, owing to a sharp in/out-group distinction. The latter is known as parochialism, which may stimulate behaviours such as xenophobia – also due to territoriality which is part of human-animal nature.

The distinction between group selection by assortment and shared fate bears a relation to Boyd and Richerson's (1990) discussion of multiple stable equilibria (MSE). The 'shared fate model' is just one way to get MSE. Boyd and Richerson show that in systems with evolution of social behaviour that have more than one evolutionarily stable strategy, such as coordination, reciprocity and sexual selection games, selection among groups can cause the spread of the strategy that is most likely to contribute to the formation of new groups. A condition for this is that processes increasing the frequency of successful strategies within groups, like behavioural variation through cultural acquiring, need to be strong relative to intergroup migration. Henrich and Boyd (2001) find that inclusion of competing, realistic strategies of cultural transmission (namely, copying the most successful individual and/or the most frequent behaviour) in a cooperative dilemma generates a MSE problem at the multi-group level. Cultural group selection acting upon this can generate a unique equilibrium for the population.

Avilés (2002) distinguishes between altruism and 'groupishness'. This allows her to trace the co-adaptation of the two characteristics of individuals. Groupishness is taken to mean that individuals are prone to join larger groups. In addition, she assumes that group size influences the fitness of individuals. A main finding is that freeloaders – groupish non-cooperators or free riders – increase in frequency in the population if they are rare, but are selected against beyond a threshold frequency because of a reduced productivity of the groups that host them. The result is periodic cycles in the population composition.

Finally, group selection is more effective in the presence of non-additive genetic interaction between individuals (in a group) or nonlinear cultural interaction between individuals (institutions) (Goodnight and Stevens, 1997). This may take various forms, such as genetic interaction within (epistasis) or between individuals. Cultural interactions may occur between individuals or within an individual when the synthesis of different cultural elements (e.g. 'culturgens' or memes) determines the cultural fitness of that individual. This aspect of interaction of genes or individuals has been generalised through the notion of 'social heterosis'. It denotes that synergistic effects or fitness benefits arise in groups as a result of interactions between genetically diverse individuals (Nonacs and Kapheim, 2007). In particular, groups can include phenotypic expressions through multiple individuals that cannot be combined in single individuals. This suggests a functionality of groups that is not simply the sum of the individuals. From an economic angle, one could interpret this as complementary specialisation or labour division, which may work to the benefit of all group members. Cultural interaction can result in an increase in group differences, a necessary basis for group selection to work upon. Unlike additive interaction (or independence of individuals), non-additive (or nonlinear) inter-action allows for a larger effect of group selection compared to individual selection. The reason is that the latter type of interaction means that there will be less similarity between individuals in different generations (parent and offspring). Individual-level selection then typically will be weaker. Moreover, group selection can influence or control the inter-actions between individuals, since it operates at the higher level which includes the interactions, so it can select among these. Individual selection cannot do this by defini-tion, as it works at the level below the group (i.e. the level of individual interactions). It is worth noting that many of the traditional writings on group selection – including those using simple models – neglect gene interaction and focus only on single genes or assume additive gene effects. Despite its empirical relevance, this issue seems not to have received sufficient attention in discussions of cultural group selection.

We have presented many processes and factors that may influence the feasibility and effectiveness of group selection. By way of summary, Table 6.1 contains some of the insights about group selection mechanisms.

Table 6.1 Factors determining the effectiveness of group selection

Factor	Negative if present	Positive if present
Population structure and dynamics	Migrant pool (temporary groups) Migration	Propagule pool (permanent, separate groups) (Non-random) assorting. Endogenous splitting of groups
Among-group diversity	Small Initially random	Historically large Initially non-random (e.g. assorting)
Spatial structure	No spatial barriers or separations	Spatial isolation Local interaction Spatial clustering/assorting
Selection pressure	No competition, conflict or further interaction with other groups	Direct interaction among groups, Group conflict Biased cultural transmission
Individual interaction	Genetic or behavioural independence of individuals	Non-additive (genetic) interaction between individuals (in a group) and cultural interaction between individuals (institutions) increase group differences
Group coherence	Weak internal group relations Migrant pool	Stable group, Propagule pool Strong internal relationships Group institutions (norms, punish, reward, sharing) Unique communication/signals (chemical, language, cultural habits)
Groupishness	Groupish free riders	Assortment of groupish altruists
Group size	Large group size negative as it hampers genetic or cultural drift, and may involve costs (anonymity, communication, coordination, monitoring, compliance)	Large size positive if synergy of (complementary) individual traits (e.g. labour division) Small group may allow for spatial isolation and random (genetic or cultural) drift

6.6   Cultural Group Selection

Group selection acting upon humans operates according to two broad categories of mechanisms, namely differential population growth of groups and cultural transmission. The first is known as 'genetic group selection' and the second as 'cultural group selection'. The latter was developed by Cavalli-Sforza and Feldman (1973a, b) and Boyd and Richerson (1985, Chapter 7). Genetic group selection or differential population growth is based on group differences in birth and death rates. It includes as specific processes intergroup competition (successful groups replace less successful ones through multiplication), intergroup conflict (possibly with extinction), and differential population growth through reproductive success or 'demographic swamping' (Henrich, 2004). Cultural transmission involves various mechanisms typically studied in social psychology, such as conformist and prestige-based transmission, following norms and punishing non-conformists or norm violators (Section 7.3). Also important is that people signal to others that they belong to a group using cultural clues such as clothes or behavioural habits. Relatedly, unlike other animals, humans form stereotypes of others. Both contribute to observable group differences which permit efficacy of cultural group selection.

Genetic and cultural types of group selection have both been active in human populations with culture. To make things more complex, they moreover influence each other, known as gene–culture coevolution (Section 7.4). For example, the historical diffusion of early agriculture may be attributed to both general mechanisms, where the relative contribution of each may have varied over time. When groups merge or one group takes over another, some rearrangement of behaviours and cultural habits will occur. In this context, a counterintuitive law attributed to Karl Marx is that conquest more often leads to adoption of the culture of the conquered than of the conqueror (Steele, 1987).

Cultural transmission based on the advanced cognitive and cooperative capabilities of humans makes group selection in humans possibly more pronounced than in other animals. This is reinforced by human intelligence enhancing group formation, by recognising like-minded individuals (assorting), coordinating actions (agreeing upon rules and standards) and organising complex labour division and institutions (language, planning, forward-looking behaviour). Cultural variation can more easily respond to group than individual selection because cultural differences can be maintained even if there is substantial migration among groups (Cordes et al., 2008). Humans have further been effective in organising war-like activities, causing them to be efficacious in cultural takeovers. Richerson and Boyd (2005) even suggest that human cooperation evolved through group selection processes. For all these reasons, it is believed that cultural group selection may have been more effective than its genetic counterpart (Cullen, 1995).

'Biased social transmission' is a dominant category of cultural acquisition and it may be important for cultural transmission between groups as well. Individuals are predisposed to adopt certain pre-existing cultural variants, and so these will increase in frequency. Boyd and Richerson (1985) distinguish three types (for more discussion see Section 7.3): (1) content (or direct) bias, where the adoption of cultural variants depends on the properties (attractiveness) of the variants (e.g. food characteristics); (2) prestige (or indirect) bias, the imitation of certain characteristics (e.g. style of dress) perceived to be associated with others regarded as attractive (e.g. fame, wealth, happiness); and (3) frequency-dependent bias, where imitation of the majority is dominant. The latter is also known as conformist transmission, a sort of rapid learning strategy which Henrich (2004) distinguishes from normative conformity, a strategy to avoid punishment and reap benefits of group membership. Content bias is more effective but involves more time and costs compared to the other two mechanisms.

To understand the difference between cultural and genetic group selection, one should realise that whereas differential population growth is characteristic of both genetic and cultural group selection, horizontal (from peers) or oblique (from non-parental adults, such as teachers) transmission is typical for cultural acquisition but rare in genetic evolution. The reason is that cultural habits can be changed through imitation, so that there is no need to replace their carriers, while a change in genes requires that their carriers (vehicles) be replaced.

Cultural group selection can build on the existence of institutions, leading to competition between, and thus selection of, institutions (Gürerk et al., 2006). This can be regarded as a proximate answer to many relevant research questions. An ultimate explanation of cultural group selection requires an explanation of the emergence of these institutions, which can be founded on genetic evolution, whether based in between-individual, between-group selection, or (most likely) a combination. To illustrate that institutions can support cultural group selection, note the role of organisation within groups, taking the form of hierarchical control, legislation and even representatives of groups interacting (negotiating) with each other. Institutions such as social norms homogenise groups, which in turn may lead to polarisation in a population consisting of multiple groups with distinct social norms. This can be magnified by dogmas and zealotry. We further know that languages of separated groups quickly become different (Section 5.5). Such processes contribute to the diversity of groups and possibly to group conflict, both increase the effectiveness of group selection. Incidentally, the human capacity for imitation plays an important role in creating diversity. Susan Blackmore (1999, p. 198) argues that meme selection (Section 7.6), driven by imitation, can contribute to increasing within-group homogeneity and between-group differences. Group selection is then strengthened by the presence of appropriate memes. This may mean that group selection has been, and is, more relevant to humans than to other animals as memes are especially associated with humans.

Frequency-dependent imitation or conformist transmission is a crucial mechanism in this process: individuals adopt a behaviour that is frequently observed. If cultural habits of dominant individuals differ among groups, groups will develop in different directions while remaining internally rather homogeneous (Henrich, 2004). Prestige-based transmission can have the same effect in terms of causing homogeneity within groups. It means that individuals copy the behaviour of an individual they regard as successful. In larger societies, network and information externalities may do the same job as conformist transmission. In the jargon of evolutionary economics, they cause a path-dependent development towards homogeneity within and diversity among groups (Arthur, 1989).

Henrich (2004, Section 6.2) discusses two other mechanisms: punishment of non-conformists (norm violators) and normative conformity. The difference between normative conformity and conformist transmission is a subtle one: the outcomes can be the same, but as opposed to the first the latter is based on the idea that the majority is indicative of the best choice. A relative gain (relative benefits or welfare) may be involved when the absolute cost that goes along with punishing felt by the enforcer is lower than the absolute cost of being punished felt by the victim. This is consistent with the widely documented human interest in relative payoffs and well-being (Bruni and Porta, 2005).

Various institutional mechanisms are responsible for creating group stability and replication. They can be seen as contributing to a kind of heritability at the level of groups, which enhances the effectiveness of group selection. For instance, the low direct fitness of individuals behaving altruistically is compensated by the social capacity to replicate pure altruism in subsequent generations, namely through social institutions and norms. This is reinforced by the existence of a meta-norm, the willingness to punish a person who did not enforce a particular norm (Axelrod, 1986). Norms are more stable under certain meta-norms. The fact that norms generate meta-norms can be regarded as an emergent property, or a new level in the multilevel evolutionary system of individuals, groups and group institutions and organisations. Other mechanisms supporting a norm system are dominance, internalisation, deterrence, social proof, membership, law and reputation. Persistent groups like religions are proof of the effectiveness of these (Section 7.7).

Next to selection, one can identify innovation at the level of groups. Innovation may be more rapid in groups than individuals, especially since many (meme) innovations spring from complementarity, recombination and cooperation. For example, conquest of one group by another can give rise to combinations of elements of two cultures that can lead to new group institutions. Apart from this, random cultural variation can occur, as cultural transmission involves 'errors' of various kinds. In fact, the rate of culturally transmitted errors is probably much higher than that in genetic mutation. 'Institutional drift', analogous to 'genetic drift', means that in small groups, cultural and institutional mutations may have a large impact, causing the respective culture to be less stable.

Once created, institutions can thus reinforce groups and support group selection effects along the line of group stability and inheritance, and between-group variation. Cultural variation can be created, maintained and enlarged through influencing offspring and immigrants from other groups. Certain groups adopt the more successful rules and habits of other groups. The intellectual capacity to learn, (re)search, understand, predict and communicate in a sophisticated way is required then, which, in turn, requires the more fundamental capacity of empathy and language. Herbert A. Simon (1990, 1993) argues that given bounded rationality the evolution of a trait like docility, being sensitive to suggestions, persuasion and information by others, was logical and inevitable, especially in children but also in adults. This contributes to the effectiveness of institutions fostering group selection.

The separation between genetic and cultural group selection does not mean that these are independent. Genetic diversity is likely to indirectly affect cultural diversity while cultural evolution – whether by between-group or between-individual selection – affects genetic diversity of humans. Indeed, the history of human evolution arguably shows a subtle interaction between cultural and genetic evolution. Such interaction between evolution through cultural transmission and differential population growth has been referred to as dual inheritance and gene–culture coevolution (Sections 7.3 and 7.4, respectively). Cultural traits have an impact on the survival and reproduction, or the genetic fitness of individuals, and, in turn, are influenced by these. For instance, certain food habits include tastes that are not easily learned but must have been selected, as they are related to toxicity. Boyd and Richerson refer to this as natural selection of cultural variants. Dual transmission can explain the enlargement of differences among groups. Cultural evolution is based on cultural acquisition or learning. Economic institutions can be regarded as creating ecological niches which affect the selection of individual traits, giving rise to a coevolution of individual behaviour and institutions (Bowles, 2000; van den Bergh and Stagl, 2003). Durham's (1991) typology of gene–culture coevolution illustrates the interaction dynamics that may involve group selection effects (see Section 7.4). One can have coevolution with one population being guided by individual-level selection and the other by group (or multilevel) selection, but it is also possible that both populations are subject to multilevel selection. The resulting combination of multilevel and coevolution leads to a system with a degree of complexity that is beyond intuitive comprehension.

Finally, groups can change their composition and structure not only through genetic and cultural group selection but also through goal-oriented planning and control, what Boyd and Richerson (1985) have called 'guided variation'. The latter is, of course, an institution that itself has evolved over time, but contributes to non-evolutionary change once having come into existence. As a result, over a sufficiently long time period, cultural and economic change, whether occurring at individual or group levels, is likely to be a combination of evolution and non-evolutionary forces.

6.7 Potential Applications in Social Science

Organisational Routines

The notion of a routine has been proposed as a property of an organisation, notably a firm. It consists of a complex set of skilled individuals interacting simultaneously and sequentially (Nelson and Winter, 1982; see also Section 8.4). The interactions depend on earlier contacts and the organisation-specific 'language'. If a routine breaks down it is not easy to restore, as it is the result of evolved cooperation, trust and mutual understanding. Organisational routines can be regarded as an emergent property of the group of relevant employees. New employees are influenced to fit into the existing group so as to contribute to its objectives, but may also contribute innovations to it. Routines can further transform due to interaction with other routines associated with competing groups of employees in the same or even other organisations. The composition of a group will affect the performance of the related routine and thus affect the between-routine selection. At a higher level, the competition between organisations can be seen as competition between group routines or sets of group routines in different organisations.

Transferring ideas from group selection theory might improve our understanding of how interactions of employees in a firm translate into firm performance, dynamics and interaction with its environment, including competition with other firms. This would imply a multilevel approach to the study of the firm, which would enable examining the net outcome of within-group selection of employees and between-group selection of routines or firms. This can address questions such as: Which type of selection will dominate under what conditions? And what is the relevance of firm type, market structure, public regulation, consumer behaviour and public opinion? Group or multilevel selection therefore allows opening up the black box of routines. It may clarify how routine (group) selection affects individuals and how individual selection affects routines. The resulting multilevel evolutionary perspective may enrich Nelson and Winter's basic framework of how firms operate and change over time.

Institutions and Public Decision-making

The main approaches to studying economic, social and legal institutions assume either purposeful or even optimal planning or individual-based evolution (Wohlgemuth, 2002; Hodgson, 2004). But the emergence and dynamics of institutions can also be understood as involving intergroup competition (Gürerk et al., 2006). Multilevel selection would add to this that not only groups interact, but within-group composition changes along with it it. For example, in political negotiations to strike coalitions, debates and opinions within a party are affected by interactions with other parties and vice versa.

This suggests a role for group selection in the analysis of public decision-making. Public choice theory recognises the role of multiple stakeholders, but regards them as individuals. So it overlooks the multilevel nature, which entails that individuals interact within groups, and the composition of opinions or individuals in these groups alters over time. A group selection approach combined with other-regarding preferences and behaviour, might better deal with the upward and downward causation that characterises political and public decision-making systems, and thus offer a better tool for understanding political dynamics.

If institutions are strongly connected to specific groups, then dominance, recombination or convergence of institutions may result from competition between such groups. Diffusion and disappearance of institutions due to group selection is an idea especially relevant to understand social-economic history. Indeed, the latter can be characterised by local economies and groups competing for scarce space, resources, market demand, and political power (Safarzynska and van den Bergh, 2010a), which in turn has affected all kinds of institutions. This has perhaps been elaborated in the finest detail for religious organisations (Sections 6.4 and 7.7), but is equally relevant to other kinds of institutions (van den Bergh and Stagl, 2003).

Such research might be embedded in a modern social science setting by creating links to topics such as interest groups and their interactions, the dynamics of lobbying power, the formation of coalitions in governance, and the emergence of international institutions, including the negotiation and stability of multilateral agreements.

Common-pool Dilemmas

Various communities around the world are involved in what has been called self-organisation in the context of common-pool dilemmas (Ostrom, 1990). Although the evolution of institutions has received ample attention in the literature studying such dilemmas (Noailly et al., 2006, 2009; Sethi and Somanathan, 1996), no group selection argument has been employed. A key question is whether resource conflicts and overuse should be addressed through policies set by higher-level governments, or instead by relying on the endogenous formation of use regimes. Case study research and evolutionary reasoning show that externally imposed rules and monitoring can destabilise cooperation (Ostrom, 2000). Under imperfect monitoring, external regulation may be undesirable as it will be ineffective and even harm the self-organisation process underlying the emergence of norms.

The relevance of a group selection approach lies in the fact that use regimes and local institutions are often connected to specific local groups. In the presence of multiple groups, self-organisation may become more effective, as multiple experiments are done in relative group separation, while the most successful ones can spread to the other groups – through a process of group selection. Boyd and Richerson's (1990) finding suggests that in the case of a problem characterised by MSE, group selection may act as an equilibrium selection mechanism. On the negative side, intergroup competition might be fiercer than competition among individuals, with possibly negative consequences for resource sustainability. These are interesting issues for research, which could involve studying the role of group size, within-group heterogeneity and hierarchy, the degree of conformism or imitation, social norms, punishment and the effect of external regulation on multilevel selection. Furthermore, the effect of the type of interactions between individuals and groups might be examined, such as the role of multi- versus unidirectional ('downstream effects') interactions, extended networks and distant versus proximate (neighbour) contacts.

Social scientists can learn much from the group selection literature. Genetic group selection is now regarded as theoretically feasible though still debated, even if supported by a large number of experimental and empirical studies. Cultural group selection is less debated and potentially more relevant to the dynamics of human societies. In applying group selection to socioeconomic phenomena, care has to be taken and specific problems have to be recognised. For example, it is not immediately clear what the simultaneous participation of individuals in modern societies, in multiple overlapping groups, means for the effectiveness of group selection. Knowledge from economics, sociology and social psychology may have to be combined to guarantee fruitful applications of group selection theories in the social sciences.

Part IV

Evolutionary Social Sciences

7 Evolutionary Theories of Human Culture

7.1 Nature and Nurture

The study of human cultures and associated economic, technological and political systems is dominated by theories of proximate causes and effects, and mechanistic types of development. These lack systematic attention for ultimate explanations and historical context. This is not to deny the relevance of such theories. Indeed, non-evolutionary social science is complementary to evolutionary thinking, as it allows linking a rich set of proximate to ultimate factors. Without considering these ultimate factors, however, social science comes down to what Durham (1991, p. 31) termed 'cultural creationism'. The alternative is to recognise that, like biological change, cultural change is characterised by heritable variation and descent with modification.

Along with marginalised attention for evolutionary thinking, the dominant conception of the social sciences is nearly devoid of information from the natural sciences. This can be interpreted as reflecting an extreme nurturist view. It denies the relevance of nature, claiming in essence that 'all culture comes from culture' (D.S. Wilson 1998, p. 130). But natural science teaches us that both genes and environment affect how human individuals develop. An impressive meta-analysis by Polderman et al. (2015) of almost all twin studies (2748) published in the past 50 years combined altogether more than 17 000 traits from more than 14 million twin pairs from 39 different countries. The results indicate that heritability across all traits is 49 per cent, suggesting the influence of genes and environment is approximately equal. Hence, there is no such thing as genetic, or environmental, determinism. All human traits turn out to have a heritable component. Moreover, relative influences of genes and environment are not randomly distributed across traits but clustered in functional domains, such as skeletal, reproductive, metabolic, ophthalmological, dermatological, respiratory, cardiovascular, endocrine, gastrointestinal, nutritional, neurological, cognitive, psychiatric, social values and social interactions. Yet the gene–environment dichotomy is overly simple. We have to add learning to the equation. It is partly instinctive, which means that ontogenetic development is influenced not only by genes and environment but also by instinctive learning, that is, individual learning which in turn is affected by genes. An example is human language, which children learn instinctively. By observing the interaction among learning capacities of millions of humans one can begin to understand the complexity of human cultures.

Many social scientists prefer to see culture as divorced from biology. Of course, this considerably limits the challenges faced by the social sciences. If one must gather and integrate knowledge from biology and other natural sciences, one's task becomes overwhelming (as at times it felt to me during the writing of this book). But this view of a divorce between the social and natural sciences is counterproductive and outdated. As we will see, culture can be better understood if it is embedded in a framework that gives attention to ultimate, including genetic and cultural evolutionary, explanations. This chapter will question the widespread belief that the evolutionary history of humans has been, like that of animals, shaped by genetic evolution. Evidence will be offered to show that once culture emerged in human groups, natural and cultural selection started to join forces in directing human evolution. The importance of culture for human past, and probably also current and future, evolution derives from the unique human ability for social learning and imitation, of which culture is the ultimate expression. Jointly with genetic evolution, cultural evolution has formed the brain–mind, behaviour and even physiology of humans. Therefore, the study of past and future human evolution requires, in addition to biology, insights from the social sciences about culture and cultural change.

Human culture is a relatively young phenomenon in evolutionary history. It can be seen as information that passes between humans due to a set of mechanisms that is variably called social behaviour, social information acquiring or social learning. Culture assures that the whole (society) is more than the sum of its parts (individuals). From an evolutionary angle, by being part of a larger cultural system, humans tend to have a better chance to survive than by living a solitary life. Thus, culture generates an adaptive advantage that may possibly enlarge with the size of a cultural group – owing to labour division, specialisation and increasing returns to adoption – which might explain the most recent cultural trend of globalisation. Of course, some aspects of human culture may be maladaptive, that is, feedback negatively to humans, as we will address in Chapters 13 and 14.

A core question in the context of culture and evolution is: why is there so much diversity in human behaviour at the level of individuals and cultures? The majority of writers on social evolution seem to agree that cultural differences do not have a genetic basis, for the simple reason that genes only shape culture at a general level. In line with this, genetic biological evolution is seen as useful mainly in offering an explanation of the commonality of cultures. Nevertheless, the debate on this is not over. For instance, similarity among distinct cultures that have been isolated, and thus have not influenced each other, may be interpreted as the result of a similar or overlapping genetic basis, similar non-genetic social dynamics, or a combination of these. Likewise, differences between cultures can be due to genetic differences, distinct social dynamics or a combination. Social-cultural evolution is therefore best seen not as the application of biological evolution to cultural phenomena but as a merger of evolutionary biology, social-cultural evolutionary thinking and social science theories focused on proximate explanations.

It is not uncommon in the social sciences to come across uses of evolution that reflect a broad, loose and unspecific meaning, drifting away from the strict, algorithmic interpretation adopted in this book. Here we do not use social evolution to mean just 'history' or 'development', terms that are common within all of the social sciences. Instead, cultural evolution denotes the social transmission of a diversity of patterns of knowledge, beliefs, behaviours and preferences along with the products they generate. This points to the existence of non-genetic diversity and changes in it caused by random processes, goal-oriented human creation and selection through markets, prevailing norms, governance, scarce material resources and the natural environment.

This chapter covers a broad array of ideas and related literatures, as social and cultural evolution have been addressed by many different disciplines using many types of theories and concepts. These are not all necessarily in agreement but neither entirely inconsistent either. Among others, we will pay attention here to dual inheritance, gene–culture coevolution, the meme notion, learning versus evolution, Lamarckian versus Darwinian views on social evolution and various more specific approaches that have arisen in sociology and anthropology. We will show that many of these explain specific aspects of human social evolution or generate non-trivial hypotheses for future research. In addition, we will devote attention to the evolution of religions, about which one can find several competing theories, and the evolution of aesthetics – notably music and humour, key elements of human cultures but which seem to have received fairly little systematic attention from an evolutionary angle.

The relevance of biological evolution for human culture can be most immediately seen by noting that our evolved psyches, with their preferences and fears, determine how we interact and form cultures. This surely leaves a lot of freedom in the sense that not one single, deterministic culture appears. Witness the many different cultures that can be observed around the world, or the many more that existed at one time during the long course of human history. One cannot exclude that some have disappeared precisely because they moved too far away from and then escaped the evolutionary survival logic – expressed by E.O. Wilson's 'genetic leash' (Section 7.4). For example, it has been suggested that modern societies defy genetic kin logic (Section 5.1), among others, by moving in the direction of more than half of all couples separating so that roughly half of all children live with a step-parent. We will examine to what extent cultural evolution, whether conceptualised in terms of selfish memes or other notions, can have its own dynamics that conflict with the 'interests of genes'. This raises various fundamental questions: which forces dominate, genetic or cultural, and under what conditions; which spectrum of cultural expressions can evolve given a genetic starting position and the coevolutionary nature of culture–gene interactions; how far can culture move away from the genetic leash; and can cultural evolution dominate or overcome genetic evolution? Note that these questions are not currently central to social sciences. In other words, an evolutionary angle can broaden the scope of social science.

One way to understand humans is by recognising a dual inheritance, involving transmission of genetic and cultural information pieces. This began with the evolution of social behaviours in larger groups and emergence of early social institutions. These likely had genetic fitness effects that promoted their diffusion. Only later on, cultural evolution became more independent or distant from biological evolution.

But the two never completely separated in humans, which is why the notion of culture–gene coevolution should be taken seriously, not just in historical or tribal (anthropological) settings but in any society (Section 7.3). While the notion of natural selection has been widely accepted since Darwin, dual-inheritance theory adds the role of cultural selection, a process by which certain socially learned beliefs and chunks of knowledge increase or decrease in frequency due to being culturally selected, that is, adopted or rejected by individuals. The combination of cultural and natural selection will be shown to complicate views on recent, current and future human evolution. Figure 7.1 offers a graphical representation of cultural evolution and how it, together with biological-genetic evolution, has shaped human behaviour. Its elements will be clarified and elaborated in subsequent sections.

"If Newton could not predict the behaviour of three balls, could Marx predict that of three people?" This question, by Poston and Stewart (1981), suggests that the challenges of the social sciences may surpass those of the natural ones. Three reasons are as follows: describing interactions between unpredictable humans implies a highly complex system; performing controlled experiments with humans, or human groups, remains a huge challenge, despite many efforts in this direction; and cultural–economic–technological reality is subject to rapid change, which defies the formulation of clear laws in social science. Hence, the question becomes whether Darwin – read: evolutionary thinking – could have better predicted the behaviour of three people than Marx. This chapter aims to provide some clues.

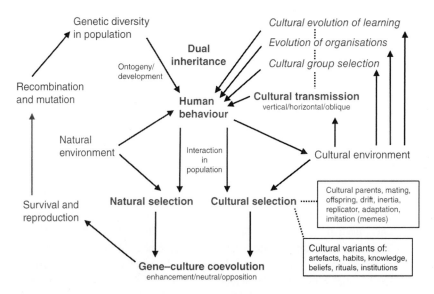

Figure 7.1 Combined cultural and biological evolution of human behaviour.
Notes: Green text and arrows denote biological evolution and blue ones cultural evolution, while blue-green is reserved for combined ideas.
(A black and white version of this figure will appear in some formats. For the colour version, please see the plate section.)

7.2   Evolutionary Thinking in Sociology and Anthropology

Sociology and Evolution: A Difficult Relationship

Evolution did not receive any attention from classic writers in sociology. Auguste Comte (1798–1857), who invented the term 'sociology' and even referred to 'social physics', wrote before Darwin published his *magnum opus*. Émile Durkheim (1858–1917), who emphasised the importance of empirical-statistical work, did not explicitly refer to evolution even though he clearly recognised the rapid societal changes at the time, including the progressive division of labour. Karl Marx (1818–1883), regarded as both economist and sociologist, presented an almost physical-biological viewpoint when formulating his materialist theory of history, according to which social change is mainly the result of economic class struggles. This gave rise to the first theory of capitalist development. However, rather than following an evolutionary approach, Marx adopted what one might call a 'revolutionary angle', suggesting the inevitability of a socialist revolution.

Max Weber (1864–1920), who, like Karl Marx, recurrently crossed the border between economics and sociology, attached much weight to ideas and values. In particular, he considered the influence of Christianity and Protestantism as crucial for the domination and success of the West European economy. In line with this, he stimulated comparative research of cultures and religions. In addition, he saw the rise of bureaucratic organisations in modern economies as both unavoidable and useful. Next, Talcott Parsons (1902–1979) was an important proponent of functionalism, according to which society consists of interacting parts, including institutions and organisations, which render continuity to social change. Functionalism can be linked to adaptation, thus adding an evolutionary dimension to it – which has especially influenced anthropology. Among later writers, Michel Foucault (1926–1984) was interested in the role of power in society. The main thesis of Jürgen Habermas (born in 1929) is that the focus on economic growth tends to destroy the norms on which it is based, so that it is unsustainable in the long run. Neither author explicitly invoked evolutionary approaches in their work. As can be noted from the previous selection of authors, sociology has its roots in France and Germany. Interestingly, evolution has its roots in England; this may, through the influence of the classical economists, be linked to the Industrial Revolution, which also started in England (Section 2.2). Finally, early sociologist Herbert Spencer cannot be left unmentioned. Box 7.1 is devoted to him.

Earlier writers such as Tylor (1871) and Morgan (1877) explored the potential of evolutionary thinking in sociology and anthropology. The issue was taken up by later authors, such as Steward (1955), White (1959), Sahlins and Elman (1960), Levy (1966), Parsons (1966) and Carneiro (1968). This is, however, not really about using core evolutionary concepts such as population, diversity and selection, but instead focuses on progressive stage-like change. Many studies can indeed be easily criticised for misusing the term evolution and, instead, representing an *ad hoc* approach to describe aspects or phases of human social-cultural and sometimes political-institutional history. During the 1970s silence surrounded social evolution, motivated by concerns about

Box 7.1 The 'evolutionism' of Herbert Spencer

Along with Thomas Henry Huxley (1863), known as 'Darwin's bulldog' because of his backing of Darwin's theory of natural selection, the early sociologist Herbert Spencer (1820–1903) was one of the influential initial supporters of Darwin. Spencer's education in physics and mathematics allowed him to have a deep understanding of the state of the natural and social sciences at the time. He tried to broaden the relevance of evolution and develop the notion of cultural evolution. While he had many original ideas, these have not well survived time. Nevertheless, Spencer did inspire later applications of evolutionary thinking in the social sciences. Hodgson (1993a, p. 81) notes especially his influence on the economists Veblen, Marshall and Hayek (Section 8.4).

Spencer proposed integrating the natural and social sciences for a better understanding of human societies. But perhaps it was too early to accomplish this, as evolutionary theory was still incomplete, notably lacking genetic foundations. Without any doubt, Spencer was much ahead of his time, daring to think along lines that were only picked up about a century later, through notions such as systems theory, coevolution and integrated assessment. Darwin supported many of Spencer's ideas on social evolution and developed some himself in his later writings (notably *The Descent of Man, and Selection in Relation to Sex* (1871)and *The Expression of the Emotions in Man and Animals*). In addition, he adopted some Spencerian notions, such as dynamic equilibrium, reflecting a sort of moving equilibrium as a balance between opposing forces (Ruse, 1999, pp. 79–80). Spencer played a critical role in making the term 'evolution' popular. He suggested to Darwin that it should be used along with 'survival of the fittest'. According to Hodgson (1993a, p. 82), however, it was Wallace who finally convinced Darwin to do so, based on the idea that selection could be easily misinterpreted as the act of an external, god-like entity.

Spencer strongly believed in evolution as continuous progress, also known as 'evolutionism'. This is because Lamarck and Malthus influenced his thinking more than Darwin. In fact, he developed his ideas on selection rather independently from Darwin (Ruse, 1999). One reason is that Darwin's *On the Origin of Species* was published after Spencer's earliest writings on evolution (namely *Progress: Its Law and Cause* in 1857). In line with his evolutionism, Spencer proposed an anarchistic, laissez-faire approach to social problems which later on became known as 'Social Darwinism'. Diverse groups, such as socialists and business people, were equally stimulated by his ideas, though for different, often opposite, reasons. Unfortunately, his ideas especially fascinated people with less academic or even malicious intentions. Connected to the negative sound of Social Darwinism, Spencer's ideas about laissez-faire arguably received too much attention at the cost of his other ideas, due in part to his popularity in the United States – where he outsold even Darwin. Besides, Spencer did not express a straightforward laissez-faire policy, as he believed in the dominance of Lamarckian over Darwinian aspects of human evolution. This means he saw – incorrectly – humans as capable of passing on their

---

Box 7.1 (*cont.*)

adaptive characters to their offspring. His focus was not on the 'out-selection' of the unfit, but the 'in-selection' of the fit. This gave rise to views that were similar to those of economists, going back to at least Adam Smith, who said that state regulation would weaken the competitive nature of the economy and ultimately reduce social welfare.

One of Spencer's main weaknesses was employing a form of biological reductionism in social analysis. He wanted to found social phenomena entirely on biological features of human individuals. Hodgson (1993a, p. 89) refers to this as 'natural causation' but perhaps a better characterisation is 'biological determinism'. In line with this, notions of cultural, non-genetic evolution remain outside of Spencer's thinking. One can hardly blame him for that, because no one before him had ever written about such ideas. Spencer can further be criticised for trusting more in Lamarck than Darwin, resulting in imprecise evolutionary mechanisms. Buskes (2006, Chapter 15) suggests that 'social Lamarckism' would be a more adequate characterisation of Spencer's work than Social Darwinism.

Social Darwinism and the sociobiology debate. But in the 1980s, evolutionary thinking was revived, involving consideration of anthropological, macro-sociological, historical, archaeological and biological aspects (Cavalli-Sforza and Feldman, 1981; Boyd and Richerson, 1985; Eibl-Eibesfeldt, 1989; Trigger, 1998; Runciman, 1998).

For a long time, evolution has not been well received by the social sciences. It has been confused with 'Social Darwinism' as ideology. Equating social evolution to Social Darwinism is, however, incorrect (see also Section 1.4). Social Darwinism is a collection term with a negative sound that is predominantly used in a rhetorical sense. At the most extreme, it expresses the worry that protection of the socially weak would give rise to a kind of physical and mental 'degeneration' of humans, where 'socially weak' is uncritically equalled to 'biologically weak'. Social Darwinism has a political overtone, being associated with idealistic, normative viewpoints such as laissez-faire, 'healthy society', racism, Hitler and neo-Nazism, fascism and eugenics. This suggests that it cannot be identified with a univocal, homogeneous movement. On the contrary, it is consistent with a broad range of conservative and progressive political movements. The flaws in thinking along lines of Social Darwinism and eugenics became more evident when the understanding of evolutionary biology and genetics improved over time.

The name eugenics – meaning 'favourable inheritance' – was coined by a half-cousin of Darwin, Francis Galton, in 1883, one year after Darwin's death. Although controversial since its inception, it gained some status as it became a discipline at some universities and was promoted early on by the American and British Eugenics Societies and, later, by the International Federation of Eugenics Organisations. One concrete policy suggestion arising from eugenics, sterilising mental patients, was implemented in various countries. Another, encouraging reproduction by people with desirable traits, a kind of selective breeding to improve the fitness of humans, turned out to be more difficult to apply. After 1930, eugenics lost credibility and support, partly because of its association with racial policies in Nazi Germany. Currently, eugenics is widely considered to be unethical. Some argue, though, that recent advances made in genetic screening and engineering open the door to new forms of eugenics. Indeed, in many countries unborn babies are already tested for various genetic deficiencies, and aborted, depending on the outcome. The ethical discussion is merely about which diseases to include in the testing. In addition, current motivations draw purely on genetics rather than ideas about evolutionary progress.

A central theme in sociology, which distinguishes it clearly from economics, is power. Goudsblom (2001) focuses the attention on the role of (changes in) power relationships and distribution due to social evolution. He regards power as the capability to influence the outcome of an interaction among humans. Power can be either unequal or balanced. The combination of muscular power of an individual, of a group of individuals, of (un)shared knowledge, technology and organisation gives rise to a certain distribution of wealth and power among humans and other living creatures, as well as among human individuals, groups, cultures or states. Power distribution between human groups created a selective advantage for the own group, notably in conflicts. The extinction of other hominines than *Homo sapiens*, i.e. *Homo (sapiens) neanderthalensis*, also fits in this scheme. Probably, technology played an important role in power distribution early on, notably through the use of fire, agriculture and energy sources. Unequal power distribution, and its counterpart solidarity, are intricately connected with the social stratification that became possible once agriculture created food surpluses. The distribution of power influences the organisation of a society and vice versa. This interaction leads to a co-dynamics of organisations and power structures. A topical example is the political trend of deregulation and market liberalisation which has created, among others, exorbitantly high pecuniary rewards for top managers and a reduction of the share of the public sector in the total economy.

An evolutionary view of society connects with political science perspectives that use concepts such as 'policy learning' and self-organisation, coevolution and 'second-order cybernetics'. Coevolution is a commonly used concept here, as it is coherent with the idea that governments cannot (perfectly) control social changes but rather change along with these. Such a viewpoint can explain the regular failing of governments and public organisations, as well as the call for dynamic governance. The work of Teubner (1993), based on the heritage of Habermas and Luhmann, forms the starting point of this approach. It assumes a diversity of interaction systems in a modern, complex society: politics, legislation, science and economy. 'Reflexive justice' is the result of 'external effects' of one subsystem on the others and is viewed as a 'third movement' next to, as well as a reaction to the failure of, the laissez-faire economy and the top-down organised welfare state. Of course, this failure or 'regulation crisis' is debated (Wilthagen, 1992). An analysis of reflexive justice can occur through the use of a model of 'social self-organisation' (Luhmann, 1984) or 'autopoiesis' (Maturana en Varela, 1980). Reflexive policy fits within the general idea of self-regulation, in which a main objective is to promote communication among subsystems and individuals (Leydesdorff, 2001). See further Section 15.4 on evolutionary policy and politics.

Perhaps the most interesting examples of evolutionary sociology can be found in *Evolution for Everyone* by evolutionary biologist D.S. Wilson (2007). He is known to believe in extending the scope of evolutionary thinking, while recognising that modern sociology is not very open to this. This book casts an evolutionary angle on group decision-making, art and music, variation in human social environments, the evolution of religions, vices and virtues, cooperation and violence, inequality and egalitarianism, oral cultures versus literate societies, democracy, and foreign policy and international conflicts. A nice illustration is the case of brainstorming and groupthink (Wilson, 2007, Chapter 26) – a theme tangentially allied to group selection theory, of which he has expertise. Using insights from many scientific studies, notably experimental ones, Wilson shows that groups are more intelligent than individuals, also known as the 'wisdom of crowds' (Surowecki, 2004), if individuals in the group can think freely and express themselves. There are various reasons for this. Individuals in the group will inspire or ask questions of each other, thus stimulating new thoughts, leading to lateral thinking. Moreover, the interaction among group members causes good ideas to be swiftly separated (selected in/out) from bad ones, and rough ideas to be refined or combined. An effective group decision process, i.e. collective intelligence, resembles a process of evolution according to the V-S-I-R algorithm: generate many ideas in an initial brainstorm (variety creation), followed by an evaluation phase (selection), which involves improving the best ideas (mutation, recombination and replication and accumulation). It delivers various interesting insights. It is wise to express minority positions early on, not when the group is about to reach a final conclusion. The quality of the group decision is negatively affected by a group leader prematurely stating his/her opinion, as this will inevitably constrain others in openly sharing their ideas. More generally, groupthink, the striving for harmony and consensus and avoiding conflict, can lead to bad group decisions or outcomes. Both leader dominance and groupthink increase the likelihood that creative, nonconformist opinions are suppressed. To illustrate practically the power of groups, Wilson refers to a game in which a list of job titles has to be created. Here, groups tended to produce considerably more job titles than isolated individuals. Moreover, they were better at the harder task of generating obscure job titles, such as 'bricklayer'. Finally, Wilson notes that evolution is relevant not only as an algorithm for group functioning but also in a historical context: whereas we have to learn mathematics, we automatically do well working in groups, as we evolved in small communities. Other studies offer nuances to these insights. For example, Kao and Couzin (2014) show that small groups have an advantage over larger ones as individuals can more easily escape the dominance of correlated and inaccurate information.

Evolutionary Thinking in Cultural Anthropology

Cultural anthropology studies historical and currently living primitive, pre-literate people who are close to nature and use simple technologies from before the times of industrialisation. It has a close interaction with biological/physical anthropology, which gave rise to the notion of gene–culture coevolution (Section 7.4). Many studies in cultural anthropology emphasise that cultural evolution underlies transitions from one stage to another. This often combines cultural and biological reasoning, as illustrated by the following stylised facts from Harris (1977):

- Humans shifted from a low-carbohydrate diet, largely based on hunter-gatherer sources, to a high-carbohydrate diet when intensive agriculture took off. Such a diet change resulted in more body fat, which accelerated the first menstruation (menarche) and shortened periods between pregnancies, thus increasing the birth rate.
- Pork became a taboo food in ancient societies of Israel, Egypt and other Islamic areas, because pigs are poor grazers in a desert environment, compete with humans for grain, produce no milk, unlike horses serve no transport purpose, and unlike cows/oxen do not provide labour.
- Cows became holy in Hindu culture, preventing their slaughter for meat, because they had a long-term capital value through providing farm labour. Maintaining large populations of cows for food purposes (i.e. animal husbandry) was unfeasible given the lack of stable grasslands due to regular droughts.

However, most studies in anthropology, as in sociology, that employ the term 'evolution' do not strictly adhere to the central concepts of evolutionary theory, such as population, diversity, selection, innovation and inheritance or replication. More than in biology, in the social sciences a distinction is made between evolution and revolution, based on a separation between gradual and radical change. This is somewhat confusing, as evolution does not preclude radical changes. Neither do most socioeconomic revolutions occur overnight. What, at a distance, in time or space, looks like a socioeconomic revolution often is a rather long process of small, cumulative changes preparing the way to a change that is *ex post* labelled a revolution. The Agricultural and Industrial Revolutions were more transitions than revolutions, similar in style to the major transitions in natural history – even though with unique time scales. For this reason, Goudsblom (2002, p. 35 and p. 42), among many others, prefers to avoid the terms 'Agricultural Revolution' and 'Industrial Revolution' and speaks instead of 'agrarianisation' and 'industrialisation', which signal better that they do not denote one-time events but ongoing processes over a long period of time. The Industrial Revolution was definitely preceded by a period of centuries during which significant cultural and economic developments took place (Section 12.1). In addition, it was pre-dated by a series of major inventions: the sea-going ship, the windmill, the clock, etc. On the other hand, seen from an evolutionary time frame, both agricultural and industrial systems have appeared on the scene quite rapidly, so that the term revolution still serves a purpose.

Naroll (1956), Freeman (1957) and Carneiro (1970) have proposed measurable indicators of cultures and their change, based on functions, defined as the production of goods, and organisational levels (or hierarchy). The minimum number of functions in any culture is two, since tasks differ between men and women in any society. In fact, two functions are characteristic of many hunter-gatherer societies. In tribal societies, the number of specialisms is in the range of six to ten. With agriculture, the number further increases and with industrialisation the number explodes. The quantity of specialisms and tasks also tends to increase with the number of individuals in a society (Carneiro, 1967). The reason is that demand must be sufficient to meet specialised supply. This explains why in larger cities more specialised shops are found than in small villages or cities. Another indicator is the number of organisational levels in a society. Also, a direct relationship exists with the size of the population, since a sufficiently small group can be overseen and ruled by one person, while the larger a group gets the more hierarchical and complex its organisation and communication structure needs to be to assure its smooth functioning. Hierarchical levels are found to differ as much as from 3 for populations around 100 individuals to 17 for populations of several hundreds of thousands. Freeman (1957) proposed additional indicators for tracing the level of cultural-economic development, such as secondary tools to produce primary tools, and the relative proportion of fulltime bureaucrats and fulltime priests. Finally, Leslie White (1943) suggested that cultural progress is positively related to the use of energy per person and the efficiency of techniques to access workable energy. Other quantifiable indicators relate to characteristics of language, such as the number of colours distinguished.

Cultural evolution has also been suggested to go through particular stadia. Morgan (1877) suggested a main distinction between savagery, barbarism and civilisation. Service (1963) proposed a categorisation based on social complexity and institutions: band, tribe, chiefdom, city state (with surrounding countryside), and regional or national states. Lomax and Arensberg (1977) distinguished gathering, hunting and fishing, arable farming, herding, farming with ploughs and industrial society. By studying six areas with early civilisations, namely Mesopotamia, Egypt, India, China, Peru and Mesoamerica, Steward (1955) arrived at the following classification: hunting, gathering, the beginning of agriculture, formation of small states out of villages, regional diffusion of a particular dominant culture, era of conquest and imperialism, and decay (medieval times in Europe). Stadia suggested by Marxian anthropology include (Claessen and Kloos, 1978, p. 39): the Primitive, Asian, Classical, Feudal, German, Bourgeois, Capitalist, Socialist and Communist production methods. Essential to this categorisation is the distinction between private and communal property.

An obvious question is whether all civilisations have gone through every stage, known as unilinear evolution. Although this seems to be true for most civilisations, exceptions are due to unique environmental conditions, leapfrogging and path dependence. As part of the latter, if a system is functioning satisfactorily, then it cannot be easily improved. One example is that the development of suboptimal pictorial scripture by the Egyptians hampered the development of a more flexible and efficient phonetic scripture. Another is England, being the first country in which industrialisation took place, which after some time meant historical lock-in into outdated technologies, creating a comparative advantage for latecomers such as Germany, the United States and Japan. Box 12.1 offers a similar example of a 'first-mover disadvantage', namely of why the Low Countries did not spawn an Industrial Revolution.

Related is the notion of independent innovation, known as polygenesis or multilinear evolution. This is an alternative to diffusion of innovations. An example of polygenesis is found in feudal systems in which vassals hold lands from rulers in return for military and other services. This developed independently in various regions in Europe and other continents, as it served a solution to the problem of a large area owned by one ruler without advanced means of quick and frequent communication. As opposed to polygenesis, diffusion implies that some stages can be skipped. Think of tribal people immediately shifting to city life. Another relevant notion is differential evolution, which denotes that some aspects of a society change at a different pace than others or, unlike others, do not change at all. Other terms sometimes used in the context of social evolution are devolution and involution. The first points at regressiveness, meaning that social evolution can be reversible. The second denotes persistent change that does not stabilise into a new, recognisable form. This can result from cultural drift, resembling genetic drift in genetic biological evolution (Section 4.2).

7.3   Dual Inheritance

Here we will consider the notion of cultural evolution as cumulative social learning, which is systematically studied in anthropology since the 1980s. This will be done in the context of approaches known as dual-inheritance theory, evolutionary anthropology or cultural Darwinism (Cavalli-Sforza and Feldman, 1981; Boyd and Richerson 1985; Sperber, 1996; Mesoudi, 2011; Paul, 2015; Brewer et al., 2017). It considers genetic and cultural evolution as complementary and both necessary to offer a full explanation of human evolutionary history. The first can explain human universals and cultural similarity, while the second variation of cultures as well as particular behaviours of individuals in them. Dual-inheritance models show that the mean and distribution of features of human individuals can shift away from the phenotype favoured by genetic selection towards that favoured by cultural selection, as long as cultural selection forces are relatively strong or operate on a shorter time scale. Indeed, cultural change is generally faster than genetic change, causing genes to be out of 'evolutionary equilibrium' as defined by non-cultural conditions.

Culture denotes a set of socially acquired and thus shared behaviours, beliefs and artefacts. It requires the existence of a social group, as, by definition, a solitary individual cannot possess any cultural features or habits. Boyd and Richerson (1985, p. 36) argue that the relationship between culture and behaviour is similar to that between genotype and phenotype as culture contains information that affects the behaviour of individuals. At a higher level, one might regard culture itself as a phenotypic response to environmental variation.

Here is a model of how genes might translate to culture (adapted from E.O. Wilson 1998, Chapter 7):

- Genes describe epigenetic rules in the brain and sensory organs that provide the 'channels' or 'constraining portals' through which communication with the outside world can occur and, in turn, allow for information processing and storage – understanding – in the brain. Epigenetic rules (Section 4.2) are hereditary regularities of mental development that guide, direct or bias the way humans approach the world and, hence, indirectly, they affect the way culture is formed and how genes are connected to culture.

- Through the communication allowed by epigenetic rules connected to the senses, humans acquire skills to perform tasks and communicate with others, and concepts that belong to the culture of which they are part.
- Elements of culture are reconstructed in the minds of individuals, through the interaction among individuals of overlapping generations, supported to a large extent by habits and artefacts containing information (writings, film, electronic files, art, buildings, tools and technology) and institutions (especially educational systems).
- Individuals belonging to a culture maintain, replicate and change the cultural artefacts and institutions. This leads to the accumulation of knowledge and technology.
- The existence of multiple cultural variants in one society allows for cultural selection and innovation at the individual and group levels. In addition, goal-oriented or planned cultural changes occur.
- Changed frequencies of cultural habits and artefacts, due to selection, influence the genetic fitness of individuals who belong to that culture.
- Genetic fitness leads through genetic selection to changes in the distribution – more or less common, or even disappearance – of certain epigenetic rules, which then can affect the rates of transmission of certain cultural habits.
- Some cultural change does not influence the fitness of individuals or groups and thus does not influence the genetic level. This is analogous to genetic drift and may be referred to as cultural drift.
- The most durable elements are the genes, epigenetic rules, artefacts and social institutions. Individual behaviour is more flexible or, in evolutionary jargon, has plasticity.

Transmission of behaviour modified by culture-specific V-S-I-R mechanisms (Section 2.1) shapes cultural evolution. Such mechanisms fall into five categories (Cavalli-Sforza and Feldman 1981; Boyd and Richerson 1985; Henrich and McElreath, 2003):

1. *Random cultural variation*: Cultural transmission involves 'errors' of various kinds. In fact, the rate of culturally transmitted errors seems much higher than that of genetic mutations.
2. *Institutional drift*: In small groups, cultural and institutional mutations may have a large impact, causing the respective culture to be less stable. Small, isolated human societies provide examples of this. Note the similarity with the biological notion of genetic or molecular drift (Section 4.2).
3. *Biased social transmission*: Individuals are predisposed to adopt certain pre-existing cultural variants, so that these will increase in frequency. Several types can be distinguished, such as: (a) direct bias, where the adoption of cultural variants depends on the properties (attractiveness) of the variants (e.g. food characteristics); (b) indirect bias, i.e. imitation of certain characteristics (e.g. style of dressing) that are perceived to be associated with others that are regarded as attractive (e.g. fame, wealth, happiness); special cases are prestige bias, i.e. imitating behaviours that are regarded to have high status, and success

bias, i.e. imitating ones considered to be successful; and (c) frequency-dependent bias, where imitation of the majority is dominant. Biases (b) and (c) invoke increasing returns to scale, path dependence and lock-in (Section 10.5).

4. *Guided (Lamarckian) variation*: Humans can consciously and purposefully change their behaviour, rules and norms through learning by doing (trial and error) and communication. This involves self-generation of alternatives, distinguishing it from 3a, although involving possibly the same or similar cognitive capabilities. It is often referred to as the Lamarckian aspect of cultural evolution, because it is a source of purposeful creation of variety. This label has been criticised, though, as reflecting a misunderstanding of the meaning of Lamarckism (see Sections 1.4 and 7.5).

5. *Genetic–cultural (Darwinian) coevolution*: Cultural traits have an impact on survival and reproduction, or the fitness of individuals, and in turn are influenced by these. For instance, certain food habits include tastes that are not easily learned but must have been selected, as they are related to toxicity. Another example is the evolution of cooperation. Boyd and Richerson refer to this as natural selection of cultural variants.

Mechanisms 3 and 4 are especially relevant to explaining the fast dynamics of social organisation, subcultures and institutions in our modern world. Social learning is critical to the operation of mechanism 3, whereas mechanism 4 is more dominated by individual learning, aided by social learning. Social learning is traditionally the domain of social psychology. It occurs through various mechanisms. Vertical cultural transmission from parent to child dominates at young ages. As children get older, horizontal transmission between peers and oblique transmission with non-parental adults, such as teachers, gain in importance. It should be noted that an empirical distinction between vertical cultural transmission and genetic transmission is difficult, as the majority of parents are biological parents. The combination and synergy of both is responsible for empirically assessed high parent–offspring correlations for many cultural traits, notably religion, political preferences and food habits. Various other empirical regularities, as shown in Table 7.1, indicate the cultural importance of horizontal and oblique cultural transmission.

Each of the five mechanisms above implies a particular cultural selection and diffusion pattern. Some authors have, however, focused more directly on the type of selection encountered in cultural contexts. For example, Fog (1997), makes – in analogy with the *r*- versus *K*-strategists in ecology (Section 4.6) – a distinction between cultural *r* and *K* selection. The first means that a group has the opportunity to greatly expand or decline, for example, because of external conflicts or even war. This suggest a group selection process is at stake (Chapter 6). So interactions with other groups dominate the course of the respective group. Fog suggests such cultural *r* selection is associated with central governance, monotheism, imperialism, strict rules, morals and sexual conventions. Cultural *K* selection, on the other hand, occurs when a group has no conflicts with other groups or is even isolated (e.g. on an island), so that group-internal processes and external selection through scarce resources determine the group's dynamics. Fog suggests that this tends to be associated with polytheism, tolerance, peace and creativity. Of course, this framework can easily be criticised as overly idealised and dichotomous.

Table 7.1 Empirical support for horizontal, oblique and vertical cultural transmission

Type of transmission (or lack of)	Examples
Vertical	- Boys with persistent delinquency lack competent male role models.
	- In the absence of a male role model, daughters lack skills in interaction with males.
	- Parental child-rearing styles explain much of the variation in children's moral development.
	- Sons often have the same occupational status as their fathers.
	- Children with high self-esteem have parents with high self-esteem.
	- Parents' and children's religious and political attitudes are related. It was found, for example, that politically active students of the 1960s tended strongly to have liberal to radical parents, relative to other students. In general, a high parent–offspring correlation can be observed for many cultural traits, including not just religion and political preferences, but also drinking, eating and smoking habits.
	- Sex role perceptions are affected by mothers' employment status.
	- Children's fears are significantly related to mothers' fears.
Horizontal	- Sex role behaviour is partly learned from older siblings.
	- Dialect variation spreads rapidly through interaction among peers (students).
	- Children's games and rituals are transmitted from older to younger children, not from adults to children.
	- There is a high correlation for many cultural traits among subcultures; an example is dress habits.
Oblique	- Self-reporting of individuals suggests that at a young age parents are important role models, while at an older age peers, teachers and employers are influential.
	- Children tend to adhere to a strict moral norm only when an adult model both preaches and practises the norm.
	- Experienced, reflective teachers caused a decline in children's impulsiveness after several months in the teacher's classroom.
	- Changing social values among college youth spread to non-college youth with a time lag of 4–5 years.

Based on discussion in Boyd and Richerson (1985, pp. 46–55).

Someone in favour of evolutionary thinking about culture, Dan Sperber (1996, 2006), is critical of the psychological content of it. He has argued that a 'deep understanding of cultural evolution' means going beyond 'shallow psychology'. One may wonder if this does justice to the fact that many evolutionary models include stylised facts from social psychology, as illustrated by the social transmission mechanisms mentioned above under 'Biased social transmission'. Henrich et al. (2008) offer an elaborate defence against this type of criticism. Of course, Sperber's comments can be positively judged as inspiration for future research.

Although cultural traits do not have the stability of genes, they generally do not change instantly in response to altered environmental circumstances. Instead, culture is subject to a degree of inertia and path dependence. The latter means that the historical order of events is crucial for the present state, as is represented by the phylogenetic tree of organic life. An analogous tree can be drawn for cultural development, incorporating the emergence of new or subcultures through cultural branching, as well as dead ends – i.e. being stuck (locked-in) in a certain branch of cultural history (see Figure 10.1). Most cultural traits continue to exist for longer than a human generation so that some have a long history. The inertia of culture can be seen, for instance, in the continuation of habits and rituals of people after having migrated to areas where different ones dominate. Archaeological, ethnographical and linguist studies show that certain cultures and culturally transmitted traits, artefacts and languages have persisted for a long time, even under changing environmental conditions. Owing to such cultural inertia, a diversity of cultures can be persistent in a single environment. This may relate to food habits, religion, work–study ethic, need for achievement, mastering of languages, tolerance to other cultures, etc.

Observing history, it is evident that as a result of dual inheritance and, arguably, group selection (Chapter 6), the role of non-kin relationships has become more important in the social organisation of humans. We started with families and bands, followed by tribes (one village), chiefdoms (multiple villages) and states (Diamond, 1997a). The current process of globalisation represents a next stage in which communication and cooperation involve millions to billions of people as well as many different ethnicities. This has given rise to the emergence of supranational and even global institutions for cooperation, trade, mutual aid, and conflict resolution, notably the European Union, United Nations, Organisation for Economic Co-operation and Development (OECD), the World Trade Organization (WTO), World Bank, International Monetary Fund (IMF), and many international agreements.

Genetic and cultural inheritance mechanisms have some fundamental differences (Boyd and Richerson 1985, pp. 7–8). 'Cultural mating' is different from biological or genetic mating. First of all, an individual ('cultural offspring') has more 'cultural parents' than genetic parents. In the 'cultural mating family' not only the nuclear family but also the extended family, leaders and prestigious individuals may influence the cultural development of an individual (Table 7.1). Nowadays with advanced communication and transport technologies (television, the internet, travel) influences have become more complex and less localised. Both horizontal and vertical transmission are part of cultural evolution, whereas biological evolution is dominated by vertical transmission (although there are exceptions, as noted in Section 4.2). The first of these is seen in fashions and diffusion of technologies. The second is determined by genetic ties, involving information flow from parent to child, from one generation to the next.

Over the course of human history, cultural evolution has involved mainly local adaptations, for three reasons. First, natural barriers like mountain ranges, rivers and seas, together with low population numbers, divided humans into fairly isolated groups. One has to imagine here the absence of modern communication and transportation technologies. Second, a very long history of small groups meant that social interactions were confined in space. Third, humans in different regions had to deal with distinct climate, soil and biotic conditions, causing some to adapt better than others in certain activities, such as using boats (depending on access to a lake, river or sea) or clothing and shelter (in cold climates). In fact, only through cultural adaptation have humans been able to occupy territories with harsh conditions like the Arabian Desert or the Arctic. Over time, as cultural groups increased in size, cultural diffusion and similarities extended over larger areas. Nevertheless, regional and even local cultural peculiarities and dissimilarities have endured.

The temporal characteristics of cultural evolution differ from those of genetic evolution in many respects. Cultural transmission of behaviour is faster and more flexible than genetic transmission. This implies that culture can never be subject to detailed genetic control, and that sociobiology cannot provide a complete underpinning of cultural and social theory, but can merely identify the boundaries of cultural evolution. Generations in cultural evolution can be longer than in biological evolution. For instance, parents are also grandparents and thus transmit culture to their grandchildren as well as children. Human children can also influence their parents and, in general, younger people can influence older ones. Moreover, unlike genetic transmission, cultural transmission can skip one or more generations because it can function through artefacts such as written text or art, nowadays even in electronic form. Seen at the scale of a human life, cultural transmission is a slow process that takes much time. In fact, it never stops during one's life. These features are opposed to genetic transmission, which occurs in a single moment, known as conception. As in genetic evolution, one can find temporal features, such as historical constraints or lock-in, in cultural evolution. They are due to individuals or local communities rarely being able to unilaterally step away from acquired cultural habits. Moreover, such habits are interwoven in a system of cultural habits and institutions, meaning that changing only a particular habit is difficult if not impossible.

Some strongly reject evolutionary thinking about culture. In a fundamental and critical discussion, Lewens (2015) reviews the main arguments. He concludes that many are flawed or not well informed. Other responses to critiques can be found in Henrich et al. (2008), Acerbi and Mesoudi (2015) and Richerson (2016). See further Section 1.4.

7.4   Gene–Culture Coevolution

Here we consider the coevolutionary connection between human genetic evolution and human cultural evolution. Although the terms 'dual inheritance' and 'gene–culture coevolution' are used by many authors as equivalent, the first can be seen as broader or more general, stressing that human behaviours are affected by both genes and culture, with explicit focus on the evolutionary dynamics of cultures. Gene–culture coevolutionary analysis is a special case of dual inheritance in that it explicitly devotes attention to two evolutionary processes, cultural and genetic, and how these influence each other over time. This complicates the picture considerably, as coevolution produces highly nonlinear dynamics (Section 4.6).

The notion of coevolution was originally proposed in ecology to refer to the joint and interactive evolution of species, reflecting an integration of elements from ecology and evolutionary biology. More generally, coevolution means that two separate populations exert selection pressure on each other and, thus, influence each other's evolution

(Section 4.6). Human invention of agriculture, involving domestication of animals and especially plants, and subsequent cultural-economic developments can be regarded as a special case of coevolution among humans, animals, plants, fungi and bacteria. For example, humans depend on the cultivated and selectively bred plants and the plants depend on human control, i.e. neither could survive without the other (a case of mutualism).

But gene–culture coevolution is also used to denote that both genetic and cultural evolution are at work in human populations, which is not strictly coevolution between two species. The term co-adaptation (Section 4.6) captures this better. On the other hand, if culture or its symbols and artefacts are regarded as species separate from humans – similar to language sometimes being regarded as a parasite thriving on human hosts – then human gene–culture interactions can well be regarded as coevolution in a strict sense. Various examples illustrate the relevance of these two interpretations when considering cultural-genetic interactions (Boyd and Richerson, 1985; Durham, 1991; Jablonka and Lamb, 2006):

- The frequency of sickle-cell anaemia among populations in West Africa depends on their means of subsistence. Slash-and-burn agriculture in Africa and the cultivation of yams created pools of standing water, in turn promoting colonisation by mosquitoes. The latter transmit malaria against which sickle-cell anaemia provides protection. Effectively, combined natural and cultural selection has resulted in a trade-off between two disadvantages.
- Among Ashkenazi Jews, Tay–Sachs disease, a type of genetic recessive disorder, is likely due to providing a selective advantage for heterozygous carriers of the associated gene, in the form of higher survival chances during a history of living in slums with a high propensity of tuberculosis.
- While most human infants can digest cows' milk without problems, adult humans show great variation in this ability due to physiology differences. The rise of agriculture, notably of dairy farming, has created a combined natural-cultural selection environment in which the proportion of individuals with genes allowing absorption of lactose, common in dairy products, could increase. Natural selection alone cannot reasonably explain this. Cultural selection is needed, which involved offspring of dairy users having a large probability of using dairy products too. As evidence, a strong correlation exists between the frequency of genes for lactose absorption and a history of dairy farming. This frequency ranges from less than 20 per cent in populations without dairy traditions to 90 per cent in those with such traditions.
- Living for many generations in densely populated societies with domesticated animals, such as cows, pigs and sheep, has resulted in contagious diseases being transferred from animals to humans (Wolfe et al., 2007). Owing to associated epidemics, which decimated human populations, surviving humans and their descendants possess resistance against these diseases.
- Current phenomena can be explained with gene–culture coevolution. For example, Jamaicans are disproportionately present among the best sprinters worldwide (Box 7.2).

Box 7.2 A gene–culture coevolutionary explanation of Jamaicans' excellence in sprinting

The cross-Atlantic slave trade from the sixteenth to nineteenth centuries involved 10 million people transported by ships under horrible circumstances, causing more than a million to die before they arrived in the Americas. As the last stop for the Caribbean slave ships was Jamaica, the people who made it there were relatively strong. Slave trade may thus have served as a selection factor in favour of a gene associated with powerful sprinting, namely the angiotensin-converting enzyme (*ACE*) gene. Individuals with the D-allele variant of this gene are more likely to have a larger than average heart, and hence the capacity to quickly pump highly oxygenated blood to their muscles. In people of West African origin, the frequency of the variant is slightly higher than in those of European origin, while in Jamaica it is somewhat higher than in West Africa. An additional effect is due to the *ACTN3* gene. This contributes to muscles being able to generate strong, repetitive contractions. The desirable variant for a sprinter is known as *577RR*, which is relatively common in all Jamaicans, not only athletes. A complementary factor is environmental, namely the bauxite-rich soil in Jamaica, causing food crops to contain relatively high concentrations of aluminium. The *ACTN3* gene makes the essential difference during the first months of pregnancy when relevant muscle fibres are determined. It is suspected that aluminium in the mother's diet promotes the gene's activity.

So slavery, in combination with geographical and environmental conditions, may have had impacts that last for many generations. Of course, we have to add to this the cultural dimension. The history of athletic successes by Jamaicans stimulated ever more enthusiastic participation in athletics among young people, as well as contributing to professionalisation of training programmes and early selection of talented youngsters. The British introduced athletics competitions to its colony of Jamaica in the late nineteenth century and it quickly spread to the non-white classes. In the 1950s, head of state N.W. Manley, himself an outstanding athlete during his studies at Oxford, strongly promoted track and field athletics. Jamaica is now the only country in the world where this type of athletics is the most popular sports event.

The combination of all these factors can explain the statistically unlikely successes of Jamaicans in athletics. To illustrate, at the Olympics in 2008 and 2016, Jamaicans won all the gold medals for 100 and 200 metres sprint for both men and women. In 2012, Jamaicans won three out of these four medals. While this involved different women in each Olympics, the same man, Usain Bolt, dominated all three. Incidentally, he is the tallest world-class sprinter in Olympic history. Furthermore, in ranking countries on medals won per capita or GDP in the 2016 Olympics in Brazil, Jamaica comes in third. Finally, 77 out of a total of 78 summer Olympics' medals for Jamaica, over time, were achieved by athletes, in large majority sprinters.

Sources: McClelland (2009), Scott et al. (2010), Brooks (2014), Kirk (2016), and Patterson (2016).

- Additional examples address issues like basic colour terms, skin depigmentation, plural marriage, incest taboos, excess female mortality, headhunting and cannibalism (Durham, 1991). Further cases of gene–culture coevolution are discussed below.

Once culture emerged in early human groups, it selected out unfit behavioural features, which caused individuals to become better adapted to the prevailing culture. If social norms sanctioned certain behaviours, then the latter – and the supporting genes – would disappear. Over time, cultures changed in response to intrinsic cultural dynamics as well as the changing distribution of individual behaviours and genes. The increased fitness of verbal communication skills invited for the selection of improved cognition and linguistic abilities. In other words, human individuals and cultures coevolved at a grand scale, involving interactions and co-adaptations between physiology, brain, behaviour, communication, social norms and institutions. Not only human behaviour (or psychology) but also the physiology of speech and facial communication were shaped by gene–culture evolution. As the fitness of humans in groups depended increasingly on communication, selection occurred in favour of genetic changes supporting relevant physiological features – in face, mouth, throat (larynx) and tongue – and associated brain processes.

To further clarify the relevance of cultural-genetic coevolution, note that the 'pure environment' view of culture is based on the idea that various types of learning, including guided variation and biased transmission, provide a sufficient explanation of cultures together with environmental and geographical conditions. According to this view, all diversity of cultures is due to diversity of the latter conditions, which include physical, biological and ecological features such as climate, resources and available flora and fauna, both in terms of their roles as providing food and serving as predators. On the other hand, a 'pure genes' view would reflect a kind of ultra-sociobiological view. Instead, according to a 'genes + culture view' cultural transmission dominates in the short run and genetic transmission in the long run, with the latter influencing the direction of cultural evolution.

The standard view of how genes affect phenotypes in terms of all possible expressions – physiological, morphological or behavioural – is relevant to behaviour and culture as well. E.O. Wilson (1998, p. 137) refers to the 'norm of reaction', which reflects that identical genotypes are not blueprints but can express different phenotypes in different environments, also known a 'phenotypic plasticity'. As an example in the realm of culture, Wilson notes a birth-order effect, referring to a large number of historical case studies of famous individuals undertaken by Sulloway (1996). This effect is that later-born children grow up in a different environment than the first born. As a result, they often identify less than their older siblings with the beliefs of the parents, thus allowing them more often than on average to become innovators, or even revolutionaries, in science and politics. This is despite the fact that the first born have, on average, higher intelligence, arguably owing to more parental attention in the initial phase of life and more care-taking responsibilities towards younger siblings later on.

At a social level, denying the genetic foundations and striving for control of environmental factors can be counterproductive. For instance, submitting every child to the same educational environment would lead to genetic variation completely determining the variation in social-economic performance of individuals and groups. In other words, any variation in the latter could be entirely attributed to genetic factors. The alternative is testing children at a young age so as to guide them to appropriate educational and professional environments. This would increase both environmental variation and heritability. If an egalitarian society is sought, in the sense of aiming at uniformity of outcomes (not merely opportunities), then as a third option talents and lack of them can be compensated by relatively much and little education, respectively. The result is that environmental variation increases and heritability is reduced (E.O. Wilson 1998, p. 141).

A well-known categorisation of gene–culture coevolution was proposed in a study of biological and environmental anthropology by Durham (1991):

1. *Genetic mediation*: Genetic changes affect cultural evolution. One can regard this as biological limits being critical or operative. The ability to talk, for instance, allowed for more subtle forms of communication and cooperation among humans, ultimately leading to advanced and cumulative social learning. The ability to walk on two legs allowed humans to keep hands free and develop capacity to do many things with hand-fingers (development of hand tools). The capacity to distinguish and recognise colours allowed for a classification and valuation of colours.

2. *Cultural mediation*: Cultural changes affect genetic evolution. This can be considered as culture changing the biological limits. For example, domestication and living close to animals led to contagious diseases, and ultimately to human genetic resistance against these diseases. Domestication of animals and dairying ultimately also led to a larger proportion of individuals capable of adult lactose absorption.

3. *Enhancement*: Cultural change reinforces natural evolution. This means that cultural and biological factors or effects add up. For example, a taboo on incest reduced the chance of unfavourable combinations of alleles. Another example is the spread of agriculture owing to the intrinsic growth of the population of farmers (natural selection: higher survival and offspring due to more food being available) and transformation of hunter-gatherers into farmers (cultural selection: imitation of successful strategy). Enhancement will result in a positive correlation between genetic and cultural influences – think of talented children being selected for specialised education in certain environments, such as music, arts and science. The respective genes will, in this case, find greater expression than otherwise. Testing for the relative influence of genes versus culture is then difficult.

4. *Opposition*: Cultural change goes against natural evolution, i.e. they are 'negatively additive'. For example, modern medicine and technology allow individuals who would have died under natural conditions to survive and reproduce. For instance, the author of this book has bad eyesight, which, if mainly due to genes rather than culture, would make it difficult to survive in an environment where spotting predators and prey is essential for survival. Fortunately, culture has evolved myopia-correcting eyeglasses.

5. *Neutrality*: Cultural change is independent of biological evolution or natural selection. One can regard this as biological limits being non-critical or non-effective. Perhaps a

major and increasing part of higher culture falls into this category. Nevertheless, it is difficult to prove that certain cultural change is neutral in this sense, as indirect genetic effects on a population level in the long run cannot be easily traced.

Galor and Moav (2002) propose a speculative illustration of the 'enhancement mode' (category 3) with far-reaching implications. Their thesis is that the struggle for survival, which characterised most of human existence, generated an evolutionary advantage to human traits that were complementary to the process of economic growth. This arguably triggered take-off from an epoch of stagnation to sustained economic growth (see Section 12.1).

Understanding of the neutral mode (category 5 above) can be enhanced by examining the role of positive feedback that leads to more complex systems. In this context, concepts such as meta-system transitions (Heylighen 1996) and autocatalytic cycles (Kauffman 1993) have been proposed. The beginning of the Agricultural and Industrial Revolutions show many such positive feedbacks. More recently, the advent of computers and information technology has created a growth cycle involving positive feedback through the incorporation of more and more individuals in a global network of communication and cooperation. This may explain why cultural evolution can follow its own dynamics. This can go as far as modifying the requirements for genetic fitness. For instance, imagine humans becoming infertile and completely dependent on biological and genetic engineering to generate offspring. Evidently, being infertile means zero fitness from a purely biological angle, which is, however, fully compensated through technology. In addition, if technology allowed the correction of imperfections in human DNA, the genetic connection between individuals and their biological parents would become weak, causing genetic evolution to become directed by technological evolution, an extreme example of cultural mediation (category 2).

Gene–culture coevolution has received criticisms. One claim is that neutrality is the dominant mode. E.O. Wilson (1978, p. 167) responded to this as follows: 'The genes hold culture on a leash. The leash is very long, but inevitably values will be constrained in accordance with their effect on the human gene pool.' Durham (1991, p. 35) poses three associated questions: Is the leash the same length at all times and places? The answer is probably no. Can the causal chain of the leash be reversed, so that culture leads genetic evolution? The answer is yes in the case of the cultural mediation mode. Is the leash ever so long as to be ineffective, providing no boundaries to cultural evolution? It seems that, in the long run, culture cannot escape the genetic leash. This suggests that at best the neutral mode only applies during a limited period of time.

7.5   Evolution and Learning

Interpreting Lamarckism in Social Science

Many commentators on the evolution of socioeconomic, cultural, institutional and technological systems express the belief that it is Lamarckian in nature. In biology, Lamarckian inheritance means that acquired properties at the phenotypic level are subsequently encoded in the associated genotype which then can be passed on to a next generation. In biology, since Weismann (Section 3.3), this view is rejected as a feasible or important evolutionary mechanism (but see Section 4.2). In fact, Lamarckism reverses the core of the evolutionary logic, which is that genotype influences phenotype, and not vice versa. In line with this, it is more precise to oppose Lamarck to Weismann rather than to Darwin.

Many authors have noted that a phenotype–genotype distinction is required to be able to distinguish Lamarckian from Darwinian evolution in settings studied by the social sciences. Even though analogies to the genotype have been identified at the social level (e.g. memes, routines), Hodgson and Knudsen (2006) conclude that the Lamarckian notion cannot be usefully applied here. Blackmore (1999, p. 62) suggests that one could regard Darwinism versus Lamarckism as copy-the-instructions versus copy-the-product. But she immediately admits that this is bound to lead to debate about what are the instructions and the product, and whether this is an adequate analogy of the genotype–phenotype dichotomy.

A defence of the label of 'Lamarckism', in a social evolutionary context, is that it reflects conscious decision-making, anticipation, planning and learning by humans. This view is mainly attributable to later 'Lamarckians', such as Herbert Spencer, Samuel Butler and Georges Cuvier. Of course, Darwinism is not inconsistent at all with such human behavioural features, and social evolution might be seen as being Darwinian as well as Lamarckian in nature: Darwinian in the sense of depending on the principles of variation, selection and inheritance; and Lamarckian in a figurative, not literal sense, dependent on learning. Lamarckism then reflects the capacity of humans and their organisations not only to learn, but also to transfer the results of – coincidental or deliberate – learning to cumulative knowledge, available to humans later in time. This means that the learning outcomes serve as an input to social evolution according to Darwinian selection. So, here, Lamarckism and Darwinism are then seen to interact or work together in social evolution. Lamarckism, as formulated in this way, adds a category of social development and progress, next to Darwinian selection of variety. One might criticise that learning is only loosely related to Lamarck's original ideas, and therefore, the use of the term 'Lamarckism' can be confusing rather than clarifying. On the other hand, the use of Lamarckism is so widespread, that trying to fight it is probably futile, while one will gain little from it. It should, nevertheless, be recognised that Darwinian evolution is not inconsistent with goal-seeking elements in human behaviour, for two reasons: these elements have themselves evolved through Darwinian evolution of systems lacking such goal-seeking elements; and Darwinian selection can work upon any heritable variation, regardless of whether or not this is the direct outcome of goal-seeking (Section 4.2).

The Baldwin Effect and Genetic Assimilation

Three specific, not entirely independent, processes seem to support a kind of Lamarckian evolution in the more strict sense: namely, the so-called Baldwin effect, genetic assimilation and epigenetic inheritance. While initially each of these met with strong resistance, currently they are more widely accepted as relevant mechanisms that affect evolution. This can be partly explained by biology traditionally focusing on natural selection operating on 'adult phenotypes', while later more attention was devoted to integrating insights from developmental biology and psychology about 'young phenotypes'.

The Baldwin (1896) effect[1] (labelled so by Simpson, 1953) was for some time considered – incorrectly –a kind of Lamarckian mechanism that undercut standard evolutionary biology. It denotes that learned behaviours can influence the direction or speed of genetic evolution (Weber and Depew, 2003). The Darwinian explanation of the effect is that selection will assure that genes which make individuals better at learning will become more frequent in the population as individuals possessing them have a higher fitness. Note that the Baldwin effect may be particularly appropriate for explaining certain aspects of social evolution and learning. In this respect, Papineau (2005) notes that if individuals are unlikely to acquire solutions to relevant problems by themselves, and can only learn these socially, this will create selection pressures for genes that make individuals better at socially learning. But these genes would not have any selective advantage without the prior culture of solutions, as then there is nothing to learn about or from. In view of this, the Baldwin effect is not just feasible but may have played an crucial role in the evolution of learning, languages, social interaction and human intelligence. A general argument in favour of the Baldwin effect is that it can explain fast, radical changes in evolution, through an evolved crane, to paraphrase Dennett (1995). More generally, one might regard social/cultural/technological evolution as a Baldwin effect on a grand scale that influences the course of human evolution. Learning might, though, also prevent genetic adaptation, which could be seen as the opposite of the Baldwin effect. For instance, human advanced knowledge, as in modern medicine, allows for artificial control of harmful pathogens, which reduces selection pressure that could create genetic immunity against them in the long run.

The preceding argument clarifies that it is not necessary or correct to interpret the Baldwin effect as a case of Lamarckian evolution. Indeed, it clearly involves Darwinian variety and selection. The Baldwin effect has been controversial, though. One reason is that learning has an energy cost, which tends to be higher for individual learning by trial and error than for social learning by imitation. Evolution has addressed this cost of learning by freeing learning capacity in a simple brain through evolution of certain instincts focused on features of the environment that show little variation. This allowed learning to focus on issues related to more variable environmental features.

A closely related mechanism is genetic assimilation. This denotes a process by which a phenotype appearing under, or triggered by, certain environmental conditions subsequently becomes genetically encoded via selection. Through repeated selection, genetic canalisation of developmental pathways is realised. In the process, development or

---

[1] Child psychologist J.M. Baldwin proposed it in 1895/1896. It was suggested by others around the same time, namely animal ethologist C. Lloyd Morgan in 1895 and palaeontologist H.F. Osborn in 1896 (see chapter 1 in Weber and Depew, 2003). Jablonka and Lamb (2006) mention biologist Spalding (1873) as the first to write about it. He connected the evolution of instincts to sexual selection. This shows not only early multi-disciplinarity in the research on evolution but also that when the conditions are right, a good idea is stumbled upon by independent thinkers. The most famous example is, of course, Darwin and Wallace discovering independently in the same period the mechanism of evolution by natural selection.

phenotypic plasticity is usually reduced. Evidence for genetic assimilation is stronger for artificial than natural selection. Various authors suggest that genetic assimilation is not really different from the Baldwin effect. Whereas the Baldwin effect was developed before genetics, genetic assimilation makes explicit use of genetics, and has been suggested to serve as a more precise way of formulating the original effect (Crispo, 2007).

Social Learning

Since Lamarckism is often associated with learning, in the remainder of this section we will consider different types of learning. In psychology, a common distinction is between non-associative learning like habituation, i.e. discontinuing after a while in response to a repeated stimulus, and associative learning, i.e. about a connection between two stimuli, or an action and a stimulus. The latter is arguably more important to evolutionary mechanisms. Here one has to distinguish between classical and operant conditioning. The first means that two, in principle, unrelated stimuli become associated by repeated experience. In particular, when a previously neutral stimulus is again and again presented jointly with a stimulus that elicits a reflex, then sooner or later the first stimulus will elicit a behavioural response on its own. An example is pets responding to sounds that they associate with food. But human children and adults show similar behaviour. In classical conditioning some aspect of the environment has been copied into a brain, but it cannot be passed on by imitation. In operant conditioning, a certain behaviour is either reinforced or weakened by altering the chance that the original behaviour occurs, either by punishment or reward. This can be done by an external agent (individual or institution), or through trial and error by the individual. Institutionalised punishment–reward systems have been improved during both biological and cultural evolution. This is not surprising, as learning of this kind carries many benefits for survival and reproduction.

A third general category of learning is imitation or observational learning, which has been particularly, almost uniquely, relevant to human (social) evolution. One can distinguish here between contagion of innate behaviours ranging from instinctive imitation to pure imitation. Imitation plays a core role in the meme theory of social evolution that is discussed in the next section. Another concept is social learning. This is similar to observational learning but stresses interaction and communication with others. It includes not only pure imitation but also learning about the environment and the opportunities it offers through observing others, and then combining this with already present knowledge or abilities, or with individual learning such as trial and error (Blackmore, 1999, p. 49).

One should avoid the misunderstanding that if an aspect of human brain–mind phenotype does not appear from birth on, it must be learned. As Cosmides and Toobey (1997) argue, this does not mean it is not part of our evolved architecture. They refer to the physiological example of teeth:

[N]ew-born babies do not yet have them but they are nevertheless part of humans' evolved architecture. Similarly, language and other aspects of the human mind are not simply socially learned but reflect the combination of evolved brain architecture and social-cultural factors. To call this learning would be inaccurate. The brain architecture can be seen as hosting innate learning mechanisms or learning instincts.

The authors add that more learning capacity means, in effect, that more nature allows for more nurture, suggesting the two do not necessarily have a zero-sum relationship.

Social Learning in Animals

Studies of various species, including birds, monkeys, apes and other mammals, have generated a strong body of evidence for the 'cultural intelligence hypothesis'. According to it, since social learning is more effective than independent trial and error by individual organisms, species with frequent opportunities for social learning will have, through selection, acquired more learned skills, notably vital ones such as associated with obtaining food. This is illustrated by primates lacking adult role models acquiring fewer learned skills (van Schaik and Burkart, 2011). In addition, better social learning capability tends to go along with enhanced asocial learning skills, as individuals are more intelligent or employ effective learning routines.

For animals, Jablonka and Lamb (2006, Chapter 5) distinguish between three types of behavioural transmission and learning: transfer of behaviour-influencing substances, socially mediated learning from experienced individuals, and imitation from others. They note that the separation between these is not sharp and that many concrete cases of learning involve several of these mechanisms. An illustration of the first mechanism is early maternally induced food information through the placenta, milk or faeces, which allows offspring to prefer nutritious, non-poisonous and easily available foods. Without these chemically learned food preferences, they would have to experiment, running the risk of being less well fed, weaker and thus having a lower fitness. Even insects seem to be able through transmission of substances to develop preferences for plants as food, mating partners and locations for egg laying. The second category of learning works through observation. A well-known case of this type of social learning is the diffusion of the capacity of a bird, the blue tit, to open milk bottles. This has been named 'cultural evolution in birds', as it involved a cultural rather than genetic invention (mutation). It not simply concerned imitation as birds merely learned about the opportunity rather than opened a bottle in an identical manner – indeed, birds went through their individual trial and error process of opening a bottle. An important difference with the first learning mode is that here information diffusion is not vertical (parent–offspring) but horizontal, from more to less experienced individuals. This learning even carried over to other bird species. An example of the third type of learning, through imitation, is imprinting of young by old animals. This happens in many species of birds, certain apes, rats and, possibly, in dolphins and whales. Human babies show also a great capacity to imitate sounds and movements. It is difficult, though, to prove that specific instances of capabilities are learned through perfect imitation, as an additional, complementary role of learning through trial and error can often not be excluded.

These examples of animal culture mean that not all hereditary variation is genetic in nature, but social learning through behavioural transmissions plays a role as well. Such cultural evolution, related to many different behavioural features, has now been shown to exist for various species, with primates offering the clearest examples. In particular, development and spread of food habits have been well documented, such as potato washing, wheat–sand separation by water, and catching and eating fish by macaques on the Japanese island of Koshima. Many such changes are cumulative, i.e. depend on earlier inventions, which is characteristic of cultural evolution. Often, new behaviour spreads from youngsters to mother to offspring, while adult males interacting less with youngsters tend to learn more slowly or not at all. Another feature of such cultural evolution is that it is faster than genetic evolution, simply because it can occur in a generation. Unlike in humans, though, cultural evolution in animals involves cumulative processes across generations that have a clear limit, explaining why animal cultures are much less complex and less prone to change.

To understand the power of the interaction of evolution and learning, Dennett's (1995) tower conception is instructive. It provides an analogy of how evolution generates new levels of intelligence and learning capacity: on the ground floor are Darwinian creatures who do not learn; on the next floor are Skinnerian creatures who have evolved trial and error learning; next, there is a floor with Popperian creatures that can solve problems by thinking about them; and on the top floor are Gregorian creatures (humans) who can imitate ideas, solutions and tools from others. The Baldwin effect – i.e. evolutionary selection favouring better learning capacity – acts as a mechanism to move creatures to higher floors. At the first three levels, nothing fundamentally changes in evolutionary terms. That is, Skinnerian or Popperian learning merely affects the selection environment for thoughts in the human or animal brain. But, on the top floor, a new replicator, the meme, appears, which according to some, alters the basic rules of evolution and fundamentally changes the selection environment for humans. This is what we turn to now.

7.6   Imitation and the Selfish Meme

An inevitable question in the context of cultural-social evolution is: which exact information unit is culturally evolving? Various terms have been proposed. E.O. Wilson (1998, p. 136) mentions 'mnemotype', 'idea', 'idene', 'sociogene', 'concept', 'culturgen' and 'culture type'. Nelson and Winter (1982) suggested 'routines' for the particular case of organisations, notably firms. Aunger (2000, p. 6) refers to 'memotype' and 'phemotype'[2]. Probably the best-known term now, and one which has generated various responses in the literature, is 'meme'. It was suggested by Dawkins (1976, Chapter 11) to show that not everything in evolution is about genes and that there may be some

---

[2] Similar is 'femotype' suggested by Mikael Sandberg (personal communication).

value to 'universal Darwinism'. According to Laland and Brown (2002, p. 200), 'As scientific concepts go, the meme had the best possible start – it was launched in one of the most popular scientific books of the twentieth century.' In the remainder of this section, I will try to convince the reader that the notion 'meme' helps to churn up old issues in the social sciences.

Dawkins' Basic Idea

Dawkins proposed the notion of 'memes' as 'cultural replicators', analogous to genes as biological replicators. Memes can be seen as elements or recognisable small pieces of culture, but also as instructions for human behaviour. The name 'meme' was a clever choice as it combines a sound similar to 'gene' with a link to words that refer to 'memory', sameness (French 'même') and imitation (μίμημα / mimema in Greek). A meme is part of a pool of cultural memes that are supported by the brains of people belonging to the respective culture. It is socially transmitted between these brains (or people). Dawkins argues that memes share a number of characteristics with genes: they are durable, fertile, replicable with high accuracy and selfish. Incidentally, Henrich et al. (2008) argue that replicators of this kind are a sufficient but not necessary condition for cultural evolution.

According to Dawkins, for more than 3 billion years, the gene was the dominant, if not only, replicator. That is, until the origin of human culture, which involved the emergence of a second important replicator, namely the meme. Given that humans have memes and genes, they can leave two things behind, genes and memes. The latter, if innovative and powerful, may have a long-lasting influence – witness the notion of natural selection proposed by Darwin.

Dawkins (1993) has suggested that memes have an information-epidemiological or virus-like character, that is to say, their diffusion is driven by a virus that jumps from one mind to another. In line with this, he has characterised religion as a 'virus of the mind', a disease which needs to be approached from a medical angle. In particular, he suggests curing 'faith-sufferers' from their 'infected mind'. Vulnerable brains of children are infected by adults and after these children grow up they will infect the brains of children in the next generation. Dawkins thus provides an analogy with how we approach biological and computer viruses. Children are more vulnerable to memetic contagion as their 'meme-immune' system (a term from Blackmore, 1999) has not been developed.

So, should all knowledge be considered a meme? Below, we will see that the answer to this is negative. Some knowledge is meme- or imitation-like in character while other knowledge diffuses according to distinct mechanisms. This further underpins the relevance of evolutionary versus learning versus epidemiological models in the social sciences. Probably all of these are needed to arrive at a full explanation of information flows and accumulation in human cultures. For instance, if there are multiple memes competing for attention in a population, their diffusion can be understood as an evolutionary process with a diversity of memes in individuals and selection pressures through social norms, judgement capacity of humans and distinct social transmission routes.

An early form of memetic type of thinking is found in evolutionary epistemology (Campbell, 1974; Plotkin, 1982; Hull, 1988, 2001). This considers scientific ideas as 'replicators' and scientists as 'vehicles' or 'interactors' that cause development along species-like lineages. This involves evolutionary mechanisms, such as diversity, innovation and selection through citation and peer review – implying selection-by-rejection, as well as pressure for conformism or imitation. In addition, there is hereditary accumulation of information through teaching, lectures and storage on paper (books and journals) and in a variety of digital forms. Cloak (1975) proposed a meme-like notion, suggesting that cultural instructions are a second replicator, without using this exact terminology. He considered them as parasites that control some of their hosts' behavioural features to get themselves spread. He also recognised that cultural instructions do not necessarily benefit individuals.

It is fair to say that, according to Dawkins' (2006, p. 196), his original intention never was to promote a theory of memes but only 'to counter the impression that the gene was the only Darwinian game in town – an impression that *The Selfish Gene* was otherwise at risk of conveying'. Elsewhere, he noted, though, 'But I was always open to the possibility that the meme might one day be developed into a proper hypothesis of the human mind [. . .] Any theory deserves to be given its best shot' (foreword to Blackmore, 1999, p. xvi).

Elaboration by Others

Various authors have expressed support for the meme as a basis for thinking about cultural evolution (Dennett, 1991, 1995, 2006; Brodie, 1996; E.O. Wilson, 1998; and several scientists in Aunger, 2000). Additional opinions can be found in *The Journal of Memetics – Evolutionary Models of Information Transmission*, which was active from 1997 to 2005.[3] Blackmore's (1999) book *The Meme Machine* represents the most elaborate — and in my view most convincing — effort to construct a theory of cultural-social evolution based on the meme notion. She shows that meme thinking has a wide reach and generates a number of non-trivial hypotheses for empirical testing.

The basis of her theory is that humans differ from other mammals and animals in an unprecedented capacity to imitate behaviour, language and sounds of others. True, some birds can imitate songs from others, but apart from this there are not many examples of perfect imitation in other than human species. Reader and Laland (1999), though, take issue with the idea that pure imitation is rare in animals, or that this should be the basis of memes. They suggest the notion of 'reconstructed memes' as a sort of mutant memes, arguing that humans rarely imitate perfectly and that memes therefore will always slightly vary between individuals. This is one reason that jokes improve over time, when spreading through a population (Section 7.9). Reader and Laland think, unlike Blackmore, that evolution-of-memes theory does not fall apart by allowing for such

---

[3] See http://cfpm.org/jom-emit and http://www.lycaeum.org//~sputnik/Memetics/index.html.

impure imitation. They also suggest that learning can create experience or the capacity to imitate memes better or improve them, as in cooking recipes, jokes or music.

From a meme selection angle the emergence of language was critical to the explosion of memes in human groups. This is not to say that all memes are associated with language – certainly not, as they can also pertain to visual information. But language magnified the human capacity for imitation, and thus for meme diffusion. From a meme perspective it is not strange, therefore, that silence is rare among humans. In fact, silence has to be enforced by rules (in schools, libraries, cinemas). From a selfish meme angle this makes sense as silence hampers meme spreading.

Meme thinking follows the V-S-I-R algorithm (Section 2.1), where replication takes the form of imitation. For this reason, Blackmore spends much time explaining different types of imitation. She makes a clear distinction between learning, teaching, classical conditioning (rewarding and punishing) or trial and error, and effortless, spontaneous and immediate imitation. Classical conditioning is common in animals (mainly through parental upbringing), while many animals are capable of trial and error (operant conditioning). But learning through pure imitation is rare in animals and common in humans. She argues that this is the reason why only humans are capable of extensive meme transmission, and connects this to the unique evolutionary pattern of humans, both in biological and social (including economic and technological) terms. Modes of social learning other than pure imitation do not allow for replication with a degree of heredity as characterises transmission of memes through imitation.

According to Blackmore, Dawkins (1976, p. 192) formulated the basic idea of universal Darwinism as 'differential survival of replicating entities', which he suggested to be still in the stage of the 'primeval soup of culture'. In her Chapter 6 on 'Big brain', Blackmore argues that the memes did not await the finishing of human brain evolution, or the creation of modern man. Instead, in line with Dennett (1991) in his book *Consciousness Explained*, she thinks that a process of brain–memes coevolution caused brain and memes to become adapted to each other, for the purpose of spreading memes more easily. High intelligence and big brains are needed to imitate accurately, as it requires various skills: deciding what to imitate, adopting distinct perspectives on an action to be imitated, and coordinating one's hands or other body parts well.

Hence, essential for meme evolution is the easy and immediate imitation that is natural to humans. It means that memes can jump rapidly from brain to brain. This stimulates Blackmore to develop in various chapters something of a theory of human brain and language evolution as the result of an interaction of genetic and memetic factors, or gene–meme coevolution, a special case of gene–culture coevolution (Section 7.4): genes in support of brains that can copy others well stand a better chance to replicate and multiply under memetic selection pressure. Genes for imitation will be favoured by genetic selection, which is essentially the Baldwin effect (Section 7.5). Blackmore argues that meme evolution influences gene evolution as much as vice versa, while recognising that since memes can change more quickly, genes are not always able to track them well. This does not deny, of course, that genetic evolution had to arrive at some stage to assure that memetic evolution could emerge, or that if, along the way, memetic evolution would have destroyed genetic conditions it would have come to an end.

Within these boundary conditions, she proposes a form of genetic–memetic coevolution that affects human evolution in all its facets.

In line with this, Blackmore (p. 33) is critical of the already discussed thesis by Wilson (1978) that genes hold culture on a leash (Section 7.4). She interprets the statement to mean that genes are the 'final arbiters', which she is unwilling to accept. The traditional, dominant view is that memes need a genetic basis, otherwise they cannot function. In other words, there is asymmetry in the relation between genes and memes. To paraphrase Wilson and Lumsden, the leash can be extremely long, but the memes will never be free. This asymmetry denies the possibility that the memes become the dominant force in mutual control of genes and memes. One could also say that after the rise of the memes, the genes are no longer free either. This suggests, then, a two-way dependence or symmetry rather than a hierarchy.

Cultural evolution does not happen to improve the performance of anything, not the human species, human culture, human individuals or human genes. Dennett (1995) adopts a clear selfish gene/meme angle when he says that 'the first rule of memes, as it is for genes, is that replication is not necessarily for the good of anything: replicators flourish that are good at ... replicating!' Adopting this angle, social science cannot expect an invisible evolutionary hand that assures desirable social outcomes (see further Section 15.2). Wilkins (2005) suggests that there are really three interpretations of memes: as gene-like replicators, as germ-like pathogens (both suggested by Dawkins, 1976), and as prion-type replicators (Aunger, 2002). The memes-as-germs metaphor is behind the mentioned 'virus of the mind' idea. According to Wilkins (p. 586) this means that 'instead of us having ideas, the ideas have us' and, similarly to a virus, are just selfishly trying to maximise their own reproduction by using our brain.

Memes spread through social interactions between individuals. Easy imitation means few barriers, in which case good, neutral and bad memes can all be propagated – the memes do not care as they are sort of selfish, aiming at maximum diffusion. In fact, many harmful memes are found in reality, such as chain letters, pyramid-selling scam methods, dubious doctrines and fake medical treatments. Mobile communication and internet technologies mean a new way for quick dissemination of such memes. According to Blackmore, as biology had difficulty making the step from studying organisms to recognising genetic evolution, the social sciences will have a hard time making the intellectual transition from addressing ideas as creations by humans to recognising the presence of autonomous, selfish memes that only try to get themselves copied and multiplied. Such a paradigm shift will require new assumptions, vocabulary and hypotheses, and overcoming what Blackmore calls 'meme fear'. It arguably will improve explanations of a number of features of human culture, notably the evolution of the enormous human brain, the origins of language, our tendency to talk and think so much, the changing role of sex in human relationships and society, the scope of human altruism, the emergence of institutions as diverse as religions and the internet, and why birth control and smaller families have become popular even when they do not clearly contribute to fitness – such as in rich countries.

Not all thoughts are memes. Perceptions and emotions that cannot be easily or directly passed on to others are not part of the meme collection. The brain and sensory systems of humans act as a selection environment: memes can only spread if they fit with our senses, memory capacity, attention mechanisms, processing capacity and limits to imitation. Memes further need to be part of groups of cooperating or mutually reinforcing memes, i.e. memeplexes, to survive and spread – similar to how genes cooperate and join in DNA or chromosomes and in cells. Together, the diversity of memes, their heredity, their changes and selection imply an evolutionary process as a result of which memes emerge, persist and disappear. Blackmore proposes considering technologies and practices as memes or memeplexes, while Dawkins uses the term 'co-adapted complexes of memes' to describe the body of thought of political parties and organised religion. The meme's power derives from the fact that the mere idea that an innovation or a religious act would improve the conditions of humans – i.e. not necessarily their actual beneficial effects – is sufficient reason to adopt them. In addition, memes often 'fool' humans just to assure that they are better diffused by them. Not surprisingly then, religion has been seriously addressed from a meme(plex) perspective (Section 7.6).

Immature meme theory already offers clear insights. Meme pressure is higher in cities than in the countryside, among others, because of higher population density. Even within cities and organisations meme diffusion will have its dominant channels. This means there is a spatial structure and network, call it a geography, of meme diffusion defined by a combination of physical infrastructure and communications technology. On the internet, memes show features that relate to those which have predominantly affected our genes – consider the most googled term, 'sex', and the most played computer games that involve violent conflict, killing and even war. Although people think they make voluntary and deliberate choices about many important things in life – such as a job, a house, a partner or clothes – such choices are often strongly influenced by memes associated with lifestyle and fashion. Blackmore (Chapters 17 and 18) goes even so far as to suggest that there is no conscious, deliberating 'I' that has free will and makes decisions but a 'selfplex' of memes that helps memes to spread – hence the opinion that humans are meme-propagating machines. This is in line with Dennett's (1991) view that human consciousness is a product of memes with the aim of making the brain more suitable as a habitat for, and diffuser of, memes.

From Concept to Empirical Study of Memes

'Meme' can play the same central role as 'gene' and 'genotype' did in biology. Nevertheless, the meme has not been enthusiastically received by the majority of social scientists and biologists. The main reasons are that it is a not well-demarcated concept and that it cannot be unambiguously identified in reality. It should be recognised, though, that the notion of 'gene' was usefully employed before the discovery of the physical structure of the DNA, so against this background the use of a non-operationalised notion of meme is not necessarily problematic, as long as it sharpens our thinking about cultural evolution. Moreover, there are several proposals to operationalise the meme (Box 7.3). Memes in scientific language have also been called 'shorthand abstractions' (SHAs), a term attributed to James Flynn. Examples are

Box 7.3 Proposals to empirically operationalise memes

In analogy with the gene, a meme needs to satisfy features such as durability and variability, in order to serve as the basis of cultural evolution. Three main candidates for serving as the empirical counterpart of the theoretical meme concept are (E.O. Wilson, 1998):

1. *Molecular (chromosomes, DNA or proteins)*: This may be the ultimate level with which to start, given that these molecules provide the code for human individual development, including the brain and sensory organs, and, therefore, for communication with the outside world that makes exchange and development of memes possible. However, this is probably a step too far.

2. *Patterns of brain or neural (nerve cell firing) activity*: Clusters or regions of activity have been identified that can be associated with certain movements or thought processes. Perhaps smaller units of activity – cycles or networks of nerve cells – that form the building blocks for such clusters or regions can be identified. For instance, automatic behaviour seems to be consistent with predictable or rather fixed patterns of neural activity in the brain, whereas new experiences and conscious thoughts involve adding connections throughout the brain. Important issues to initially consider are: differences between sleeping and being awake (consciousness); differences between brain activity in a newborn and an adult; general differences between mentally healthy and sick people; general differences between men and women; differences due to culture versus genes, by comparing identical and fraternal twins; increasing differences arising in identical twins over time; differences in how similar concepts in distinct languages translate to the level of brain activity (e.g. by studying bilingual individuals).

3. *Basic or general elements of human cultures, such as assessed by anthropologists*: The 67 universals of culture proposed by Murdock (1945) may serve as a starting point. These include, among others, cooking, courtship, etiquette, calendar, sports, games, dancing, decorative art, joking, tool-making, community organisation, government, education, division of labour, trade, status differentiation, hygiene, medicine, ethics, law, religious rituals, marriage, funeral rites, language, greetings, inheritance rules, personal names, property rights, sexual restrictions and various taboos. But if one takes the meme as something amenable for direct, effortless imitation then more appropriate examples are perhaps musical tunes, simple storylines, catch-phrases, games, rules or skills and habits picked up from others. Dennett (1995) and Blackmore (1999) both argue that candidates for memes should neither be too small nor too large. Too few notes do not sufficiently characterise a melody, while too many words cannot be remembered. To shine more light on this, one might compare similarities and differences among existing cultures or with cultures in the past.

environmental externalities, global warming, carbon pricing, prisoner's dilemma, group selection, naturalistic fallacy, sexual selection, survival of the fittest and wicked problems. Many creative suggestions for potential memes or SHAs that might improve science, society and political decision-making are offered by the over 150 essays by leading thinkers in the collection by Brockman (2012).

The search for an empirical equivalent of the meme may be motivated by history showing that fundamental foundations have been critical to scientific progress. Evolutionary theory was developed by Darwin without any knowledge of genetics, let alone the molecular basis of genes. The development of Mendelian and later population genetics, and the discovery of the genetic code, has tremendously increased our understanding of the specific mechanisms underlying evolutionary change, leading to an explosion of insights and useful applications in medicine, agriculture and industry. In chemistry, understanding the physical processes underlying molecular ones has helped to better understand and predict chemical processes under different circumstances. The increasingly important role of psychology in the new paradigm of economics, driven by bounded rationality and other-regarding preferences (Fehr and Fischbacher, 2002), is another notable case. Similarly, one should expect that a better grip on memes will contribute to progress in the social sciences.

A distinct view is that cultural evolutionary thinking can do without an information carrier like a gene or meme. This viewpoint is defended by Henrich et al. (2008). They argue that cultural evolutionary analysis as described in Sections 7.3 and 7.4 represents a better approach. This raises the question how the notion of 'cultural variant', common in this approach, relates to that of 'meme'. It seems that the first notion is broader and includes the second. Edmonds (2005) distinguishes between the broad and narrow approaches to memetics. The first he defines as 'modelling communication or other social phenomena using approaches which are evolutionary in structure'. He thinks this is done already using other frameworks, and so it does not require the label 'meme (tics)'. The narrow approach based on perfect imitation, as addressed by Blackmore, had a short life and lost momentum after 2000. Edmonds attributes this to a lack of additional explanatory power. An additional reason is a lack of empirical measures of memes, so that memetics had a problem of moving beyond 'just so stories'. Yet another reason may have been that memetics was not well connected to other approaches in the broader area of evolutionary thinking about culture (see Sections 7.2 and 7.3 and Chapters 8–10).

Not all operationalisations of the meme as suggested in Box 7.3 imply a sharp separation between meme, meme-phenotype (phemotype) and meme vehicle.[4] Many authors see memes as information prescribing behaviour which is carried by meme vehicles, such as books, computers, memory sticks, tools, machinery, buildings, drawings, photos, slide shows and videos. Some would reserve 'meme' to denote information in the brain and regard the associated behaviours as 'phemotypes'.

---

[4] A related older distinction is 'i-culture' versus 'm-culture', denoting instructions in the head and behaviours, respectively (Cloak, 1975).

7.7   Evolutionary Explanations of Religions

Religions are pervasive in human societies and have, like few other institutions, influenced our history up to this moment. Moreover, they set us far apart from other animals. Human cultural features such as language, tool-making, music and other arts can arguably be found in primitive forms in some animals, notably primates and birds.[5] But this does not hold true for religious belief and institutions. Religion can be differentiated from other human traits and organisations in that it enforces and replicates a set of norms, rules and beliefs – which constrain and guide human conduct and relations – that are believed to be defined by, or derive in some way from, a supernatural authority.

Proximate explanations of religion range from indoctrination of children to psychological need for answers to questions about origins and meaning of life. Here we will focus the attention on ultimate explanations of religion, using a V-S-I-R framework (Section 2.1). Human history and current times show a great diversity of religions. Possibly some 100 000 religions have ever existed. This is partly due to the fragmented nature of humanity characterised by many small isolated groups during the greatest part of human history. This prevented the emergence of large religions. With the rise of agriculture this changed, as human settlements grew in size and with them the dominance of certain religions. This picture of large numbers and changing diversity of religions hints at the relevance of evolutionary thinking.

Social scientists have developed various non-evolutionary theories of religions. This area of research is sometimes referred to as 'religious studies'. It combines elements from anthropology, economics, neurology, philosophy, psychology, sociology, political sciences and history of religion. The difference with theology is that it represents secular analyses and thus proceeds without adopting any dogmatic religious stance or without assuming that particular religious books, like the Bible and Koran, are truthful about history. In line with this, researchers do not need to be religious believers themselves. In fact, it may be seen as an advantage to be non-religious so as to be better able to develop a neutral, non-prejudiced outlook on religious facts and texts. The relevance of social science perspectives on religions is clear: religions have conquered the world, dominate norms and politics in many countries, affect access to resources, contribute to gender inequality and are a significant factor in many persistent conflicts around the world. As Roman philosopher and statesman Seneca stated that 'Religion is regarded by common people as true, by the wise as false, and by the leaders as useful.'

Famous early voices expressing scientifically motivated statements on religion are Karl Marx in *Theses on Feuerbach* (1845), Max Müller in *Introduction to the Science of Religion* (1873), Cornelis Tiele in *Outlines of the History of Religion to the Spread of the Universal Religions* (1877), William James in *The Varieties of Religious Experience* (1902), Max Weber in *The Protestant Ethic and the Spirit of Capitalism* (1905), Émile

---

[5] Some would say that cooking, language, art/aesthetics and humour are uniquely human too. See Sections 11.4, 5.5, 7.8 and 7.9.

Durkheim in *The Elementary Forms of the Religious Life* (1912), Sigmund Freud in *Totem and Taboo* (1913), and Carl Jung in *Psychology and Religion: West and East* (1938). These tend to offer proximate explanations of why many humans are religious believers. These mostly vary between functionalism and rational choice theories. Durkheim's idea that something as pervasive as religion needs to be functional goes a long way in the direction of adaptive, evolutionary explanations.

Although Darwin mentioned religion as a culturally evolved human feature in his 1871 book *The Descent of Man, and Selection in Relation to Sex*, elaborate evolutionary theories to explain the origin, existence and dynamics of religions are of a more recent nature. Evolutionary explanations of religion focus on ultimate rather than proximate explanations. Most such approaches come down to recognising the adaptive value of religion, in terms of enhancing individual or group fitness. They connect to functionalist perspectives in sociological theories, such as by Durkheim, in the sense that an evolutionary process expounds how functions of religions came about and alter over time. Others argue that religion is a by-product of the evolved human brain or of evolved culture. Particular views come from evolutionary psychology (Section 5.4), memetic theory (Section 7.6) and group selection theory (Chapter 6). It is likely that several mechanisms stressed by such distinct theories (elaborated below) have jointly, as complements, contributed to the rise of religions in human groups and societies.

Adaptive Values of Religions

An important reason for an adaptive, functionalist explanation of religions is that they generally represent very costly activities, in terms of time, money and energy, up to the point of extravagance – think of the many cathedrals that the Catholic Church has brought forth. This suggests clear benefits of religions that outweigh their costs. Another reason to seriously consider an evolutionary, adaptive explanation is that religions are found in all historical and current human societies.

Many observers have identified specific functions of religions, such as supporting morals, reducing group-internal conflict and contributing to group members' happiness. These need to be explained by a process generating such functions. I will argue here that adaptation through evolutionary selection is among the best candidates. Van Straalen and Stein (2003) mention the following adaptive values of religion or religiosity:

- *Acceptance of authority*: The belief in supernatural power is a correlate of the propensity to accept authority, which has the evolutionary advantage that it controls internal group conflicts and enhances social-cultural transmission. In this respect, it is helpful that humans can be easily indoctrinated. This allows religion to become a system that effectively imposes norms and values on a group, which contributes to a good performance of the group and successful competition with other groups. This provides the right conditions for a process of evolution by group selection (Chapter 6).
- *Explaining natural phenomena*: Darwin noted that humans have always tended to project human characteristics onto the environment. Human-like images of gods

that are found in virtually all religions fit this pattern. This arguably contributed to the mental stability of humans.

- *Sublimation of passions*: Ethologist Lorenz has stated that rituals have emerged to sublimate strong passions related to hunger, fear and aggression because otherwise these would hurt humans. Religion can be seen as a set of rituals that help to avoid conflicts and aggression among individualistic, egoistic and opportunistic individuals.

Such explanations can be linked to the reality of Pleistocene humans, which were characterised by close groups, possibly slowly increasing in size. Here the emergence of authority, the sublimation of passions, competition among groups and control of the environment were important for survival of individuals.

A more down-to-earth, adaptive value has also been suggested, namely that religious people are generally less depressive, live longer and have more children (Buskes, 2006). Another suggested adaptive value is that religion contributes to economic success – witness how trust in honest behaviour of orthodox religious Jews helped them to gain a comparative advantage in, and thus dominate, the diamond trade during a large part of history. In turn, they could provide a good livelihood for their families and offspring (Richman, 2006). As part of a broader area of study of cultural evolution, religious practices, institutions and rituals are often depicted as contributing to economic prosperity and social stability (Rappaport, 1999). Rituals coordinate individuals and reduce the cost of decision-making, or align individuals psychologically, thus promoting in-group pro-sociality (Whitehouse et al., 2015). Ethnic markers such as rituals, religious practices and clothing, demarcate religious groups visibly, reduce free-riding and facilitate in-group cooperation. Anti-syncretic rules resisting the blending of belief systems add to this (Wilson, 2002). Studies have further found that ecological information can become encoded in certain rituals. A classic example is ritually structured distribution of irrigation water in Balinese water temples that existed for over 1000 years, and which functioned better than systems based on modern scientific knowledge (Lansing and Kremer, 1993).

An evolutionary explanation of religion means that religions had adaptive functionality for individuals or groups in the past. Since this applied to other social and environmental contexts, as well as to small kin groups or tribes, one cannot assume that the same adaptive value extends to current societies made up of very large numbers of unrelated individuals. Of course, the fact that religion is on its way down in most countries around the world might be interpreted as it being maladapted to current economic and social environments. To this should be added that other institutions — such as democracy, public media, science, education and legal bodies — have taken over the role of religion as a stabilising factor of human societies.

Four Evolutionary Theories of Religion

Theories of individual selection offer a first explanation of the emergence, dynamics and diffusion of religions. The idea here is that religions enhance individual adaptation to societal conditions, leading to an evolutionary advantage for the respective individual. Such theories emphasise one or more of the adaptive functions mentioned above. Religiosity can alleviate stress associated with fear of death, which appeared in intelligent humans with self-reflection and planning for the future. Indeed, religions propose answers to fundamental human questions along with sacred restrictions as well as threat and punishment against questioning them. According to Johnson (2015) angst for supernatural punishment is ingrained in our brains through evolution, and can be seen to trigger conformism, docility and submission. In addition, religions offer protection to individuals against daily stresses, by allowing their believers to more easily accept misfortunes in their lives than non-believers. On the other hand, the feeling of guilt is common in believers, which likely enhances stress. Studies indicate that religious people on average, nevertheless, are slightly happier than non-religions individuals (Lim and Putnam, 2010). This type of research has been criticised, though, for confounding community with religion, both of which involve like-minded people in a group offering support to co-members. When statistically controlling for social relationships, Diener and Seligman (2002) find that religiosity does not influence well-being. A more nuanced insight comes from Okulicz-Kozaryn (2010), who assesses that for relatively non-religious countries religion does not significantly affect life satisfaction, while for countries where religion is the social norm, which includes the United States, it does. Apparently, it is harder for non-religious people to be part of an extended social network in the latter type of countries. Another individual selection view on religion is that by fostering belief in a 'moral god', who is judgemental about human behaviour, one encourages such behaviour to be in line with the god's morality. This coordinates individuals, keeps them in line, and assures that those who do not conform will be socially handicapped, if not become outlaws. It implies a replication mechanism that causes the religion to be stable or even growing and dispersing. Several arguments, though, go against individual selection explanations of religion. One is that religions contain many anomalies, more so than in other human groups or institutions, that go against the individual fitness of its members: celibacy, sacrificing oneself for one's religious groups in conflicts, not eating certain types of animals (Hindu cows, Jews and Muslim pigs) even though they provide an excellent source of nutritious food, contributing years of toils or taxes to support ostentatious buildings to glorify the religion, sacrificing food or animals for the gods and spending much time on praying or other rituals. This suggests that group or meme selection, or both, may have played an important role. These can also explain why religion goes beyond kinship groups that are dominated by individual selection.

Group selection as an evolutionary explanation of religion shifts the question from 'Are religious individuals fitter and better adapted to their environment than non-religious ones?' to 'Are religious groups fitter and better adapted to their environment than non-religious ones?' The various mechanisms underlying group selection were discussed in detail in Chapter 6. Particularly relevant is 'cultural group selection' (Section 6.6) driven by social transmission of information and institutions contributing to social order. Intuitively, a group selection account makes sense as religions are able to regulate group norms and hence individual behaviours. Moreover, there have been, and still are, a great number of religions or religious groups on Earth, implying potential competition or even conflict. Indeed, many religions are expansive, and show distinct behaviour to insiders and outsiders. Where there are multiple religions there is often, in the words of the theologian Mitri (1999), 'face-to-face war or living back-to-back in peace'. Biologist Richard Alexander (1987) goes as far as saying that humanity is unique among animals in that other groups of their own species, often with another religion, represent the most life-threatening of all factors.[6] Altogether, these factors create ideal conditions for a process of 'religious group selection'. It means that religious communities are adaptive units, like organisms and social insect colonies, suggesting that their evolution creates group-level functions which assured humans in the respective groups to become better adapted to their natural and social environment. This idea is elaborated and empirically illustrated for many religions in D.S. Wilson (2002). Group selection does not mean individual selection is inactive but rather that selection operates at multiple levels, which can reinforce or counter each other in terms of adaptive direction. For example, individual selection works against a high degree of altruism while group selection can stimulate it, resulting in behaviours such as sacrificing oneself for the good of the group, reinforced by hero- or holiness-status. Wilson claims that group selection explains the mix of blessings and horrors that characterise religions much better than any other hypothesis (p. 3). Indeed, most religions show some form of preferential treatment for co-religionists, and are even characterised by in-group morality and out-group hostility. More generally, human minds show parochial altruism, that is, an asymmetric response to in-group and out-group individuals. Whereas from a moral angle, this asymmetry is hard to understand and is easily judged as inconsistent or insincere, it makes perfect sense from a group selection perspective. The reason is that it coordinates individuals, creates group coherence, reduces group-internal conflicts and strengthens them in competing with other groups, all contributing to evolutionary success of the group. Group selection further connects well with the psychology of social control as groups are effective in indoctrinating their members to believe and act in ways that supersedes voluntarism. In fact, all religions ask individuals to sacrifice individual for group interests. As opposed to individual selection explanations, group selection ones are much better able to clarify why there have been, and still are, so many different religious groups, and what evolutionary impact resulted from a long history of religious conflicts among such groups. Individual selection explanations have nothing to say about the latter.

According to a third theory, *evolution of memes* (Section 7.6), a religion works to the benefit of religious memes. Examples of such memes are 'your mind cannot die but is eternal', 'non-believers or members of other religions should not be respected but conveyed to our religion or killed', 'faith is a virtue and deserves the highest respect',

---

[6] Many current major conflicts in the world are due to, or reinforced by, religious differences. We know of various ancient religious conflicts, and by extrapolation have to assume they were common in prehistory as well. The Muslim conquest starting in the seventh century and Christian Crusades in the eleventh century come quickly to mind, but only because they are relatively recent and well documented, and their consequences are still visible.

'supposed miracles should not be understood or criticised but treated as a mystery' and 'self-sacrifice for your religion (martyrdom) will be rewarded in afterlife' (for a longer list, see Dawkins, 2006, pp. 199–200). According to Blackmore (1999) genetic adaptation – whether based on individual or group selection – and by-product theories (see below) may all contribute to an explanation of religion, but have likely interacted with meme mechanisms to become more powerful. For example, religious memes might improve selection in favour of 'religious genes' that program for the capacity of having religious experiences. An important feature of memes is that they reduce within-group differences and, at the same time, increase between-group differences, in terms of beliefs, norms, food habits or sexual practices. Thus, memes can reinforce the efficacy of group selection. We find such polarisation in the form of conformity within religious groups, and distrust and conflicts between them. From a memetic angle religions are here to stay, not because they hold the truth, but as our minds are adapted to accept their 'meme tricks' which ensure their survival and reproduction. More generally, memes, particularly religious ones, distort our capacity to rationally study and manage reality as they include many misleading and false pieces of information. Religions are especially good at preventing their 'theories' from being tested – witness the discussion about creationism in Section 2.4.

A fair number of authors support the idea that religion is a *by-product* of the evolved brain rather than being adaptive itself (e.g. Hinde, 1999; Boyer, 2001; Atran, 2002). This takes the route of offering psychological explanations for why children or adults tend to believe. A central argument here is that the evolution of the brain, and associated social evolution, has gone along with the capacity of children to learn fast and be docile. An evolutionary explanation is that it contributes to their survival. As a consequence, children are prone to accept information provided by parents and other adults even when it is incorrect. When they are old, they offer the same information to their children, guaranteeing its dissemination through subsequent generations. When a whole group is caught in the untruth, it becomes locked-in, hence difficult to escape from. In this context, Dawkins (1993) has characterised religion as a 'virus of the mind', spreading selfishly like normal infective, chemical viruses without evident benefits or even with damage to its hosts. At this point, the by-product and infective meme explanations join forces. Relevant associated psychological considerations to support the by-product explanation are that children grasp intuitively more easily purpose (teleology) than undirected evolution (complex causality). In a review of evidence from developmental psychology on childhood origins of adult resistance to science, Bloom and Weisberg (2007, p. 996) conclude that a degree of resistance to scientific ideas is a human universal. The reviewed studies indicate that when one asks children about the origin of animals and people, their intuition steers them in the direction of creationist type of explanations.

Another suggestion in the context of by-product-of-the-brain explanations is that religious faith resembles the irrationality of falling in love (Dennett, 2006, p. 256). Arguably, some irrationality mechanism in the brain is responsible for falling in love. The term 'irrational' denotes that despite counter-evidence one believes that one's religion, god or partner is the best. We do not act so irrationally and monogamously

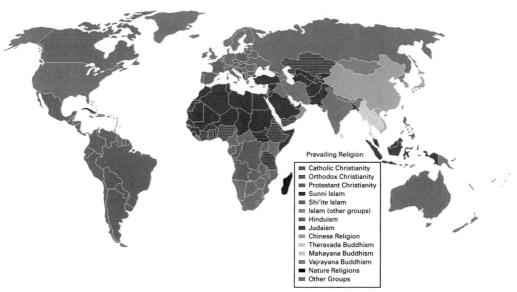

Figure 7.2 Geographical distribution of main religions in the world.
Source: https://en.wikipedia.org/wiki/File:Prevailing_world_religions_map.png; part of
Wikimedia Commons, a freely licensed media file repository.
(*A black and white version of this figure will appear in some formats. For the colour version, please see
the plate section.*)

with regard to other choices in life, such as music, food, wine or friends. Dawkins
(2006, p. 185) states that love, and sexuality associated with it, is such a potent force in
the brain that it should not come as a surprise it was opportunistically used by 'religious
viruses' to benefit from. This is supported by documented religious meditation and
other experiences by deep believers showing characteristics that resemble those of
sexual love. Some speak in this context of a 'God spot' in the brain, a location with
high activity during meditation or prayer, as confirmed by neuroscience. But perhaps
more significantly, religious conviction appears as less activity of the anterior cingulate
cortex, which can be interpreted as the absence of a 'cortical alarm bell' that normally
rings when an individual has made a mistake or experiences uncertainty. The Dutch
brain researcher Swaab (2014, Chapter 16) notes that political conservatism goes along
with similar reduced activity in the same location of the brain. He further draws
attention to the fact that the borderline between religious-spiritual experiences and
psychopathology is fuzzy, and that patients with psychoses are in larger part religious
compared to the general population. On the other hand, there is also research suggesting
that religious beliefs protect against depression (Muramoto, 2004).

It is worth noting that spatial and temporal patterns documented for religions provide
further support for an underlying evolutionary process (Park, 2004). One can observe
clear geographical structure in historical patterns of most religions, pointing at an
underlying spatial diffusion or expansion process. Figure 7.2 shows that all major
religions have spread rather continuously over space and as a result are confined to

particular (sub-)continents. This holds also for 'sub-religions' of Christianity, like Catholicism (dominant in South Europe and Latin America), Protestantism (North Europe and North America) and Orthodox Christianity (Eastern Europe and North Asia). Christianity has been able to diffuse across a wider scale globally than other religions, with the exception perhaps of Islam, due to past colonial activities of countries dominated by Christian religion.

Monotheism Versus Polytheism

Many other questions related to religion can be tackled with evolutionary thinking. For example, why are monotheistic religions currently dominating the world? In the pre-agricultural world humans lived in many isolated tribes, each with its own religion and unique gods. In most cases these took the form of polytheisms worshipping multiple deities – often gods as well as goddesses. After the rise of agriculture, this continued for a while, as can be seen in the well-known cultures of the Egyptians, Greeks and Romans – the latter two sharing the same gods although using different names for them. Slowly, however, monotheistic religions became more widespread. The first of the Ten Commandments, 'Thou shalt have no other gods before me' is telling in this respect as it clearly suggests a historical context of competition or even conflict between monotheistic and polytheistic religions, providing an ideal basis for a group selection type of evolutionary process as discussed above.

Several explanations have been given for the change to monotheism (Moor et al., 2009). One is that a single god represents a more powerful image than multiple gods who have to share the power – hence are not almighty. One can see the monotheistic concept as representing the highest form of patriarchal power, also as the monotheistic god is commonly portrayed as being male. A related explanation is that the complex societies formed after the rise of agriculture required more hierarchy and control which, through a so-called sociomorphic translation, promoted the image of a god with full control. Another explanation is that complex societies beyond the family-based tribe required mitigation of conflict through moral norms. This arguably was easier with the image of single, good and moral god issuing rules for conduct, such as the Ten Commandments in Christianity. In other words, religions and god images evolved in close interaction with social and economic changes, suggesting a sort of coevolution of religion, society and economy. From an evolutionary perspective, the spread of mono-theistic religions can further be understood from their being less tolerant to other religions than polytheistic religions. Hence, they had a strong propensity to grow and diffuse, resulting in their current dominance worldwide.

Trinity or the 'Trinitarian God' in Christianity – Father, Son and Holy Ghost (or Holy Spirit) – can be seen as providing a bridge between poly- and monotheisms. Two possible explanations of Trinity clarify this. First, Trinity may be the result of merging ideas such as: there is one almighty god, Jesus was the son of god, Jesus was god, Jesus was created by god, Jesus existed eternally with the Father, etc. The integration of the three elements in Trinity enhanced support from, and avoided conflict among, the different factions behind each particular idea, and made sure the early Catholic Church did not fall apart into competing religions. For this reason, Trinity became part of the official doctrine of the church after the Council of Constantinople in the year 381. Another explanation is that Trinity reflects the adoption in early Christianity of the common idea found in many Egyptian, Sumerian, Assyrian, Babylonian and even Etruscan and Roman polytheistic pagan religions, namely that gods formed a triad of similar rank (Hislop, 1959). Note in this respect that the representation of a triad of gods by the equilateral triangle is still common in Christianity. Incidentally, the meaning of the Holy Spirit remains unclear up to the present day. Armstrong (1993, Chapter 4) suggests it should be understood to reflect the mystical or spiritual experiences of religious believers. According to McGiffert (1932), early Christians regarded the Holy Spirit not as an entity or experience but as the divine power working in the church and world at large.

7.8   Evolving Musical Sense, Styles and Technologies

Aesthetics is not unique to humans. It plays a role in the courtship behaviour of bowerbird males. They build a structure and decorate it with sticks and brightly coloured objects to attract a mating partner. Some even create optical illusions by arranging objects from smallest to largest, which holds the attention of the female longer, leading to higher mating success for the respective males. More generally, many if not all animals tend to recognise aesthetic features of potential mates as these reflect health and suitability to become a parent. This indicates that sexual selection is behind the capacity to recognise and produce aesthetics. Perhaps even sexual attraction is the ultimate aesthetic experience.

What is uniquely human is that aesthetics goes beyond sexual partnerships, being a pervasive element of human cultures, reflected by clothing, furniture, technologies and art. One might claim, though, that all of this is merely a sublimation of sexual feelings, expressed through making oneself attractive, such as through cars or garments. Add to this that early painting may have been a form of early pornography. Witness also the effect of female lovers on the work of painters throughout history – the case of Pablo Picasso being most notorious (Hudson, 2016).

In this and the next section, we will deal with two examples of human aesthetics and language that strongly depend on interactions between people in social groups or societies, namely popular music and humour. Both illustrate the power of non-genetic evolutionary processes while also drawing attention to genetic roots. Music may be seen as part of a wider set of art or aesthetic expressions by humans, including painting, sculpture, theatre and film. All of these can be argued to have, not quite unlike scientific knowledge, the features of complexity that results from a long, historical process of cultural evolution. So, what is presented below can be fairly easily generalised or transferred to the other arts (e.g. Pinker, 1997, Chapter 8). This does not deny that each art form has its 'evolutionary peculiarities', owing to unique technological and psychological drivers and barriers. Music is perhaps the most interesting case of all art forms to consider here as it awakens some of the strongest emotions in us.

Opposing views on the evolution of art in a broader sense are expressed by Boyd (2009) and Verpooten (2011). Boyd argues that art is a biological adaptation, which arguably could happen as art hones cognitive skills, arouses creativity, fosters harmony and cooperation in groups, and contributes to status. My view is closer to that of Verpooten, who instead regards the arts as 'culturally evolved practices building on pre-existing biological traits' (p. 176), though potentially exempting music. He notes that art in general having been selected for utility is inconsistent with distinct art forms being associated with separate cognitive mechanisms and brain regions. Recognising the tight fit between art and cognition he proposes that art was culturally selected to fit human cognition.

Research supports the idea that animals, including primates, show no or little interest in music, that is, rhythmic melodies with repetition and surprise, harmonies and textures of sound, and varying levels of volume to create tension, created by instruments or voices and characterised by distinct combinations of tones and overtones. Animals seem to ignore music, not being negatively or positively affected by it. This is not to say that animals are unaffected by sounds associated with what we humans call music – it is trivial that these enter through their ears in their brains and trigger neural responses. There is a widespread confusion, though, that the fact that certain types of music create excitement, nervousness, calmness or sleepiness in animals indicates they are able to enjoy music. On the internet, for example, one can find many claims that dogs are calmed by classical music, reggae or even soft rock while they are agitated by heavy metal and grunge. This may be true, but not being annoyed is not the same as understanding and enjoying music. Sleeping in response to music can certainly not be taken as evidence for enjoyment – it might as well signal boredom or disinterest. While some humans are equally agitated by heavy metal, others delve deep into its specific musical features and enjoy it immensely – something which no dog has been reported to do. Apart from the fact that the complexity of music – in terms of rhythm, melody and harmony – likely goes beyond the mental capacity of dogs, they are simply unable to perceive changes in pitch as we do. While humans can distinguish one twelfth of an octave dogs cannot do better than half to a third of an octave, so even if they could identify some melodic resolution they are just unable to enjoy music in the way humans can (Fritz, 2008). Dogs are obviously sensitive to sounds, including ones that humans cannot hear because of low volume or too high pitch (frequency), which causes their perception of human music to be even further removed from ours. Dogs' sensitivity to sounds explains why they often bark when we do not hear anything that warrants a response. Many people suggest their dogs respond to music, but one cannot exclude that they are just reacting to their owner's responses to music, in terms of serenity, happiness or excitement, whether in a role of listener or instrument player. In addition, like wolves, dogs howl at times in response to certain sounds, notably of wind instruments, but again this cannot be taken as evidence for musical sense or enjoyment. For dogs, howling simply serves a means of communication. It is safe to say that music that creates emotions in humans will not create the same emotions in other animals. Of course, one cannot exclude the reverse is true as well. Indeed, some suggest the existence of 'species-specific music'. One study found that cats seem to be responsive to cat-specific music, which to our ears sounds as something in the direction of ambient music (Snowdon et al., 2015). One can, though, not be certain that cats perceive this really as music, i.e. something clearly different from general sound.

The two logical places in the animal world to search for musical taste are birds and primates. Some research suggests that similar parts of the brain of birds and humans react positively to 'music'. In particular, it has been found for white-tailed sparrows that females respond positively when they hear males sing, while males respond negatively when they hear competitive males sing – pointing at sexual selection as the ultimate causal mechanism (Earp and Maney, 2012). The researchers conclude that the same neural reward system is active as in humans. With regard to primates, a recent experiment found that chimpanzees stayed longer in areas where African or Indian music sounded and avoided those with Japanese music or silence (Mingle et al., 2014). The authors argue that this points at possible homologies in acoustic preferences between nonhuman and human primates. But this does not clarify why humans generally get so excited and happy from certain music. All in all, more research is needed to settle the issue of whether animals can actually enjoy music.

A genetic biological evolutionary explanation for interest in, and enjoyment of, music by humans is that they are an adaptation as music serves a clear function. One social cohesion hypothesis is that music, notably rhythmic drums together with dancing, evolved to create a feeling of closeness and solidarity in ancient tribes. Another is that musical taste derives from the sexual interaction of male and female, involving coordination, rhythm, shuffling (dance) and use of seducing language characterised by soft, sweet and melodic sounds. This mate-choice explanation arguably makes sense to both the Pleistocene and modern times. Indeed, there is considerable archaeological evidence for prehistoric instruments, notably for flutes made of bones. Illustrating current relevance, Miller (2000a) points to rock stars attracting groupies for sexual intercourse, which at times gives rise to offspring, causing genes that contribute to musical talent to be replicated. The strong emotional power exerted by music on human brains is consistent with the arms race, which is triggered by sexual selection (see Sections 4.7 and 11.3). Incidentally, Darwin hinted at sexual selection in relation to human sensitivity for music in his book *The Descent of Man, and Selection in Relation to Sex* (1871). He noted that love is the most common lyrical theme found in music, and this is still true today. In further support of sexual selection, Snowdon et al. (2015) collected various studies showing that women around the time of ovulation feel more attracted to men showing a capacity for handling complex music. Nevertheless, a sexual selection explanation for human music is still debated among biologists. Huron (2001) and Scherder (2017) review various other proximate explanations and their linkage to evolutionary processes. They suggest that evolution and musicality can also be understood from the angles of bonding between parent and child, music as a strategy to fight psychological distress and music as a game to stimulate cognitive development. But these authors do not really offer a convincing alternative evolutionary mechanism to sexual selection that can explain our strong interest and emotional responses to music.

As argued elsewhere in this book (Sections 5.4, 5.5 and 11.3), major features of the human mind, notably problem-solving, social interaction and language capability, show a clear adaptive design by natural selection. But this does not mean anything the mind can do is an adaptation. It is possible that our taste for music, like our capacity for mathematics, is a non-adaptive by-product of the complex brain and mind. This may explain why for most of us learning to perform calculus or to play an instrument represents a challenge. The by-product explanation leaves some things unexplained, though. For instance, how can it then be that we can remember so easily melodic phrases as opposed to merely verbal phrases? More so, few people know whole poems by heart, but many can replicate song texts, apparently without great effort. It would seem that all of this requires a lot of – unconscious – information processing by the brain.

As an alternative explanation, art has been proposed to be a cultural or socio-economic, not genetic, adaptation driven by status-seeking. The idea here is that rich people with enough leisure time can afford actively making, or passively listening to, music. Buskes (2006) mentions that, consistent with human sexual selection, until recently influential creative artists have been in the majority male, as in the prevailing culture they derived much social status from being a successful musician, painter or scientist, which guarantees interest of women in them. So the idea here is that art is a way to attract women. While a speculative theory, it is not easy to put it aside as complete nonsense. On this theme, see also Section 4.7 on sexual selection and Box 4.1 on its cultural implications.

Our capacity to recognise different rhythms, frequencies and volumes of sounds in spoken language may mean that we can also use this capacity to appreciate what we call music. All of us can sing to some extent, which is not surprising as the separation between speaking and singing is fuzzy. Indeed, some languages sound more like music than others. More generally, rhythm and intonation are as important to speaking as to singing, allowing the expression of relevant emotions that make language maximally clear and convincing to others. Most of us will agree that a human voice generally adds a unique and agreeable character to music, which is not surprising if our capacity to interact through spoken language serves as a basis for our love for music. The evolutionary changes in physiological features like face, mouth (including smaller teeth creating more internal hollow space), throat and tongue that allowed for speaking equally made singing easier. Also bipedalism may have helped as the human body achieved a position that improved vocal sound as well as permitted tapping rhythmically with a foot.

Pinker (1997, pp. 534–538) mentions potential reasons why humans are affected by music: similarity to language, creation of an auditory scene, reflecting acoustic signatures of emotional calls, resemblance to environmental sounds such as thunder, wind, rushing water, bird songs, animal growls and footsteps, rhythm being associated with motor control and heartbeat, and music triggering resonance of neurones in the brain firing in synchrony which contributes to emotions. Another explanation is that music binds humans in groups, which could have evolved through a process of combined genetic-cultural evolution (Section 7.4). Rhythms contribute to order, purpose and discipline, used in an extreme manner in military marches, while singing creates feelings of belonging and cohesion. Both can have made groups fitter in competition with others, allowing for a cultural group selection type of explanation (Chapter 6). That music is not a purely cultural phenomenon in humans is suggested by the fact that new-born babies already show sensitivity to variation in rhythm and melody in music offered to them (Honing and Ploeger, 2012).

Music is capable of triggering deep emotions in us, more so than any other art. This often works by music building up tension through difficult melodies and harmonies or unstable notes, which is then followed by resolving this tension through a tender melody, a succession of harmonies or evident end notes. Jazz has experimented with this in the most extreme form, by letting improvisations walk far away, and for an extended time, from the original melody and even rhythm, creating sometimes almost unbearable tension, which is resolved then by returning to the original melody. This explains why some find it hard to enjoy particular types of jazz, which is something that needs to be learned. This holds more generally for music, which ultimately is a set of related but distinct languages, such as classical, folk, jazz, blues, rock and pop, with many dialects. If one has heard a lot of jazz at a young age, one knows its idiom and can more easily enjoy it than when one hears it for the first time at an adult age. If one is not accustomed to a certain musical language it may sound difficult or even irritate. This suggests cultural evolution: we like the music in which we have been brought up, but we do not necessarily feel the same pleasure and emotions when hearing music from other cultures, particularly when it uses tonalities to which the listener is not accustomed.

Listening to music, and even singing, may be easy for most of us. But playing an instrument is a completely different matter. It requires a long process of education and practice, which reflects an underlying cultural process of diversity, evolution and accumulation of information about musical practices, theory and instruments. It is, of course, no surprise that humans did not develop a biological aptitude for handling musical instruments as the invention of such instruments is of a rather recent date. The historical expansion of music in human societies has indeed all the characteristics of an non-genetic evolutionary process, with 'populations of musical styles', characterised by internal diversity, creativity that increases existing diversity, selection and diffusion through human tastes, record contracts, reviews, radio and television broadcasting and record/CD markets, and path dependence due to fashions, imitation and coevolution with musical instruments, recording, composition, mixing, storage and playback technologies. This is schematically illustrated in Figure 7.3. The evolutionary mechanism of competition through innovation is central to the music industry. This explains both the high rate of new bands appearing, as well as existing bands trying to innovate their style. Miles Davis is a good example in jazz music, somewhat similar to Picasso in painting. Davis contributed to the emergence of new forms of music, such as cool jazz, hard bop and modal jazz in the 1950s and jazz fusion at the end of the 1960s. This often involved combining elements from previous musical styles, explaining the use of the genetic term 'crossover'.

Humans have probably always made some type of music, involving rhythms, singing and dance. Musical styles during the last millennium include medieval (with Gregorian chant being best known as it survived through the church), Renaissance (1400–1600),

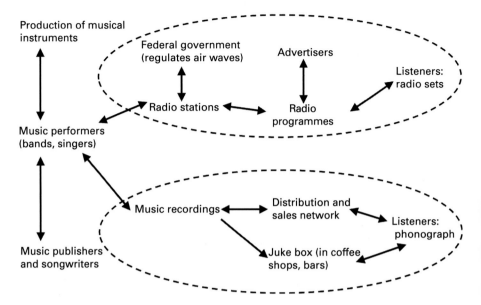

Figure 7.3 Factors involved in the evolution of modern music.
Source: Geels (2007, Figure 1, p. 1413).

Baroque (1600–1750; J.S. Bach, Handel, Telemann, Vivaldi), Classical (1750–1830; Haydn, Mozart); Early Romantic (1830–1860; Beethoven, Berlioz, Chopin, Liszt, Mendelssohn, Schubert, Schumann, Verdi); Late Romantic (1860–1920; Brahms, Bruckner, Dvořák, Mahler, Ravel, Strauss, Tchaikovsky, Wagner); and Modern Classical (1920–; Britten, Gerschwin, Prokofiev, Stravinsky). Each period is characterised by use or dominance of particular instruments, combinations of instruments with particular sounds (harmonics), voice and choral styles, use of musical scales, and style of musical pieces (structure and length). There was clear path dependence (Section 10.5), because of strong rules for music-making, lineages of master–pupil relationships, regional tendencies, and the connection of music to worldly and religious powers.

The cultural evolution of modern popular music is of recent date. One can study it using cultural-phylogenic approaches (Tëmkin and Eldredge, 2007; Mesoudi, 2011). It started with African-American slaves singing spirituals and, from there, many species of music developed in partial semi-isolation and subcultures, such as gospel, country, blues, jazz, rock and punk. Subsequently, these interacted, creating fusion styles: e.g. folk rock (Bob Dylan), country rock (The Byrds), blues rock (Jimi Hendrix or Fleetwood Mac), rhythm and blues (The Rolling Stones), jazz rock (Return to Forever), soul rock (Otis Redding) and music that is more difficult to classify (Pink Floyd and more recently Radiohead and Wilco). Many 'subspecies' can be found, such as progressive death metal (Opeth), folky symphonic rock (Jethro Tull), funk metal (Living Colour) or punk blues (Jon Spencer) – the list is endless. Or as Baba Brinkman sings in 'Natural Selection', a song from his 2009 album *The Rap Guide to Evolution*, rap has evolved from a common root into a wide diversity of styles. The cultural evolution of musical styles thus appears as a tree combining descent and recombination resulting in new genres, sub-genres, sub-sub-genres, etc. This very much resembles the evolution of technology as is depicted in Figure 10.1. Hence, one can easily imagine a tree of music depicting its genealogy. In fact, one can find many instances of it on the internet (e.g. https://musicmap.info) and in the form of posters of classical music, country or jazz. In line with this, culture can be seen to have species, such as classical, jazz and rock/pop in the area of music, although these are not so absolutely separated as biological species. To illustrate, there are various crossovers: Gershwin combined jazz and classical music, Miles Davis started jazz rock ('fusion'), there is also the distinctive 'rock jazz' of Steely Dan (arguably the apex of human evolution hitherto), recently Kendrick Lamar virtuously blended rap, soul and jazz, and we should certainly not forget 1970s symphonic rock which integrated elements of classical into rock/pop music.

The role of talented individuals – many self-taught and so less constrained by rules, or youngsters deliberately trying to bend the rules, both guaranteeing innovation – has been critical to modern music. In addition, heritage and learning from great masters have been important forces – witness subsequent Miles Davis' bands offering a school for young musicians who later became successful in their own right: pianists Chick Corea and Keith Jarrett; saxophone players John Coltrane and Gerry Mulligan; guitarists John McLaughlin and John Scofield; bassists Dave Holland and Marcus Miller; and drummers Tony Williams and Billy Cobham. One could call this a form of replication with mutation.

The Beatles are widely regarded to form a major node in the transition from the rock'n'roll of Chuck Berry and Elvis Presley to pop-rock. An unlikely combination of talents, complemented by ambition and hard-working spirit, and almost encyclopaedic knowledge – especially by George Harrison – of guitar licks and melodies before them, and not to forget the guidance by a trained musician like George Martin ('the fifth Beatle'), translated into an entirely new musical style. It changed the course of modern music fundamentally and irreversibly. The Beatles' lineage is reflected in virtually every pop song since.

One can regard music technology, culture and styles to be subject to a coevolutionary process: each of them consists of a population of diverse elements while they exert mutual selection pressure on one another. Modern blues, rock and pop could not have developed without the electric keyboard/piano, electronic synthesiser, electric guitar and bass guitar. Electrification not only created new sounds but also, as microphones had done before, louder and more sustained sounds creating strong emotions in listeners. Jazz fusion and jazz rock were the result of incorporating electric rock instruments into jazz. But even earlier, classical music developed in interaction with technological inventions such as the harpsichord (around 1400), which combined elements of the organ (notably its keyboard) — some form of which exists since millennia — and the harp (or more generally the psaltery), common since the twelfth century but of much older origin. In turn, the organ itself can be traced back as far as the panpipe, used first by the ancient Greeks. The piano followed much later, around 1700, but clearly built upon these earlier inventions. Music technology thus can be seen to have a clear evolutionary, cumulative character, as holds for technology in general (Chapter 10), while it interacted with the evolution of music-making.

Like biological evolution, musical evolution has 'cultural-geographical', similar to biogeographical (item v in Section 3.5), features. In each (sub)continent, one can find types of music that are unique and quite distinct from those in other (sub)continents. Even within smaller areas one can find spatial varieties. Illustrative are the names given to various types of blues: African blues, British blues, Chicago blues, (Mississippi) Delta blues, Detroit blues, Kansas City blues, Memphis blues, New Orleans blues, Texas blues, West coast blues, and so on.

A relevant question in 'culture-biogeography' is why England has been dominant not only in the early period of industrialisation (Section 12.7), but also in rock/pop music since its inception. An island biogeography angle suggests it might be related to England being part of a large island, able to foster innovations owing to its cultural isolation, and with enough space and people to support a viable musical industry. Other factors are a strong cultural connection and overlapping language with the United States, where modern music styles like gospel, blues, jazz and rock'n'roll arose.

To end, it is worthwhile noting that evolutionary algorithms are being used to generate musical melodies and rhythms in what is known as 'evolutionary music' (Miranda and Biles, 2007). It is part of a broader area of evolutionary art, in which genetic algorithms assist in generating and improving designs of all sorts (Romero and Machado, 2007). One could say that cultural evolution has generated evolutionary software that in turn contributes to further cultural evolution through musical and other innovations.

7.9   Evolution of Humour, Jokes and Laughter

To illustrate the potentially wide reach of evolutionary thinking, we consider here human humour from an evolutionary angle. Humour triggers laughing responses in humans through elements such as surprises, unlikely events, logical paradoxes, absurdity, surrealism, irony, sarcasm, deadpan delivery, figure of speech, metaphors, hyperboles and double meanings of words. As cultures change, humour adjusts, since it is fuelled by context, general education, history and tolerance. Humour has always enjoyed economic value — from Ancient Greek comedy to troubadours in medieval times. During the twentieth century, humour-related money flows have swollen immensely due to a booming entertainment industry, catalysed by television and movie technologies. This has contributed to a huge diversity of humour, spanning a spectrum from political satire as in the television series *Yes Minister* to sexual explicitness as climaxed in the series *Californication*. Public humour in the form of sketches, movies and sitcoms is the result of a process of competition, interaction and accumulation of ideas, with historic contributions by such diverse actors and directors as Charlie Chaplin, Buster Keaton, The Marx Brothers, Billy Wilder, Woody Allen and Monty Python. A recent trend is dark comedy — witness the Coen Brothers, Quentin Tarantino and Vince Gilligan (of the TV series *Breaking Bad*).

Proximate explanations of humour are psychological in nature, regarding it as a mechanism to reduce stress or anxiety in individuals or as a defence mechanism to avoid difficult or sensitive conversations, personal critiques or sexual themes. Jokes are a way to trigger laughter in a controlled way, which creates desirable social situations and psychological states in others. More generally, humour and laughter help in creating a close bond with others. They facilitate social interactions among strangers by creating a relaxed feeling and trust, while removing tensions in demanding social interactions such as political negotiations. These social functions of humour are suggested by the fact that it triggers strong facial and bodily responses as well as loud laughing noises.

A fascinating finding of 'laughter research' is that laughing removes the dominance of the left or right hemisphere of the cerebral cortex. This may foster creativity, i.e. the emergence of new insights, as rivalling concepts can be present simultaneously or quickly alternate. As in patients with manic depression and schizophrenia, the frequency of this alternating process considerably deviates from the average person, therapies have been proposed that stimulate patients to partake more in laughing (Gibbs, 2001).

The evolution of humour is rooted in the evolution of laughing, which has received attention from researchers for several decades already, being part of comparative primate studies of facial expressions (van Hooff, 1972). Human laughing can be triggered by humour, but also by other surprises, shifts in meaning, tickling or a nice social setting. Primitives of human laughter are found in apes and arguably also in rats and dogs, suggesting that laughter evolved before humans appeared. Chimpanzees show two types of laughter: bare teeth associated with fear to show submission and non-aggression, and open mouth when relaxed and having fun. While they definitely laugh when playing and being tickled, it has been reported they can also laugh when understanding a joke or some other surprise. Although humans show a greater diversity of laughing and smiling types, and can laugh about themselves, there are clearly many similarities and common roots (Davila Ross et al., 2010). For example, chimpanzees use laughter to signal that aggression in play fights is not for real, which prevents escalation. This functional value suggests it is an adaptive trait. Likewise, human children who fight with each other, or their father, frequently engage in laughter. It has also been suggested that humour and laughter increase the chance of being found attractive by members of the opposite sex, and may therefore have been affected by sexual selection. Not surprisingly perhaps, human sex is a frequent theme of jokes. For instance, how can you determine the sex of a chromosome? Answer: Pull down its genes. Incidentally, most laughing in humans is not provoked by humour. Empirical research has found that speakers laugh much more than their listeners, while women laugh more than men when in mixed company (Provine, 2000). An evolutionary explanation for the latter might be that men finding laughing women more attractive, or that laughter, like smiling, can function as a smoothing strategy in new, potentially threatening situations.

Various cultural evolutionary angles on humour have been proposed as well. From a memetic angle (Section 7.6), humour can be seen as a set of parasites thriving on the human mind that easily spread. From a cultural group selection angle (Chapter 6), humour contributes to assortment, i.e. putting minds in the same direction. A brief rhetorical joke by a talented speaker can sometimes help to convince more people than a long elaborate argument. In addition, laughing about the misfortunes of others might be explained as a mechanism to discourage inequity in human groups. One may wonder if distinct types of humour are a response to different biological or cultural selection pressures. Humour as a by-product of brain–mind evolution could be motivated by arguing that our capacity to reason sharply, needed for survival and partnership, means we have a derived capacity of easily recognising paradoxical, unlikely, surprising, absurd and other humorous situations. Human intelligence means that, unlike other primates, humans have the capacity to laugh at themselves. It also explains why humans enjoy so many distinct types of humour: slapstick, sexual humour, or humour about other people's imperfections, other groups, countries, religions, cultures, political views, etc. These explore the friction between different dimensions of human nature, such as nature versus culture, male versus female, us versus them, or group versus individual. The latter opposition was treated in a paradoxical manner by the Monty Python team in their movie *Life of Brian*. Its main character Brian, born on the same day as Jesus of Nazareth and in an adjacent stable, is followed by a crowd that mistakes him for the Messiah. One morning Brian wakes up and finds the crowd below his window. Irritated, he calls out to them: 'You don't need to follow me, you don't need to follow anybody! You've got to think for yourselves! You're all individuals!' The crowd answers with one voice: 'Yes, we're all individuals!' Brian goes on: 'You're all different!' and the crowd responds: 'Yes, we are all different!' with the exception of one man who utters in a low voice: 'I'm not.'

Cultural evolution in the context of humour works even on a short time span. Indeed, one can regard most verbal jokes as being the outcome of an evolution process, involving the accumulation of contributing ideas by many individuals, rather than as the invention of a single individual. To see this, one has to understand that at each time there is a variety of similar jokes around, since all persons who recount a joke to others will add their small innovations to it, simply as it is virtually impossible to perfectly replicate a joke one has heard. Among the set of similar but slight variations of a particular joke, the ones most appreciated will be more replicated and diffused. As a result, their share in the 'population' of jokes increases. Hence, the surviving jokes tend to be funnier.

To illustrate that there is always a diversity of similar jokes around, consider this one, which I have told to others since I was young. Coming from the south of the Netherlands close to the border of Belgium, it was common in my region to make jokes about Belgians, such as this one:

A Dutchman and Belgian are sitting in a train. The first is eating apples all the time. The second asks why he does this. The Dutchman answers that apples contain many pits, and if you eat these you become cleverer. The Belgian suddenly shows a lot of interest. 'Well,' says the Dutchman, 'in the next station the train will stop for a while – you'll have time to buy some apples in a shop near the station.'

When, somewhat later, the Belgian sits down and grabs an apple from his bag, he hesitates: 'But I don't really like apples.'

The Dutchman proposes a solution: 'I can eat them and spare the pits for you.' So it happens.

After a while, the Belgian says, 'Wait, I paid for the apples, and now you are eating them, while I only get to bite the pits.'

The Dutchman responds, 'See, it is working already!'

There are many variations on this joke, such as a Jew selling intelligence-improving fish-heads to a Russian for a value above the market price of the associated fish.

One can also recognise 'species' of jokes with a similar structure or theme. Think of lightbulb jokes expressing stereotypes about certain groups. This even includes various subspecies for defined groups. For example, many lightbulb jokes make fun of equilibrium thinking in mainstream economics. Take this one: How many economists are needed to change a broken lightbulb? Answer: None, the market will arrange it. Or more appropriately, how many evolutionary economists does it take to replace a light bulb? Answer: None, given sufficient time a new, functioning light bulb will evolve.

If humour is in our genes, then it is not surprising it is ubiquitous, found in all people and cultures. It seems, though, that some appreciate it more than others. One may wonder, in fact, if humour includes distinct national or regional species. In a funny little book called *English Humour for Beginners* George Mikes (1980) tries to disprove the existence of an allegedly distinctive English brand of humour, but then, nonetheless, characterises it in rich detail. He claims (p. 5) that 'In other countries, if they find you are inadequate or they hate you, they will call you stupid, ill-mannered, a horse-thief or a hyena. In England, they will say that you have no sense of humour.'

7.10   A Comparison of Four Approaches to Cultural Evolution

An early basis for thinking about human culture was created by sociobiology, and somewhat later by its descendant evolutionary psychology. These approaches were discussed in Chapter 5. Two other approaches received attention in this chapter, namely dual inheritance (including gene–culture coevolution) and memetic thinking. All use concepts like population, variation, selection, information transmission, and adaptation, so that they fit in the evolutionary basket of methods. This final section aims to discuss differences and similarities between them, as summarised in Table 7.2. All theories emphasise that human behaviour is not the simple consequence of culture (social learning), individual learning (e.g. trial and error) or environmental conditions. All three determinants are generally recognised as relevant, but the specific weight assigned to each differs among the theories. In addition, some theories propose that culture itself is subject to a V-S-I-R type of evolutionary process (Section 2.1), or that even something like gene–culture coevolution is going on.

Sociobiology and evolutionary psychology express the idea that human social behaviour can to a great extent be explained by our genes as these reflect historical selection and adaptation conditions. According to cultural evolutionary theories, cultures have a significant effect on individual behaviour, explaining diverse habits and practices among societies. Within this set of theories, dual inheritance and culture–gene coevolutionary theories stress that genetic and cultural information, and thus natural and cultural selection, are both required to offer a good explanation of human behaviours and cultures. Another theory in this set, memetics, considers cultural memes as subject to their own evolution that is rather independent and free from genetic constraints, and which may even direct genetic evolution to facilitate the diffusion of memes. Of all approaches, dual inheritance is perhaps the broader one, and might be seen to encapsulate sociobiology, evolutionary psychology and memetics as special if not extreme cases.

Table 7.2 Evolutionary approaches to human behaviour and cultural change compared

Feature	Sociobiology	Evolutionary psychology	Dual inheritance (including gene–culture coevolution)	Memetic thinking
Main assumptions, concepts and mechanisms	Gene's eye view: kin selection, direct and indirect reciprocity, reputation	Brain modules and associated behaviour genetically adapted through natural selection during Pleistocene	Genetic and cultural influences during recent/ ongoing evolutionary histories (Holocene); social transmission biased by evolved learning rules; cultural group selection; mutual selection of genes and cultural variants	Social transmission through unique human capacity to imitate certain cultural traits or behaviours – 'memes' – from others; memes carry cultural information, but not any cultural information is a meme
Core insights	Humans are social animals driven by selfish genes; they possess cultural universals due to evolution under similar ecological conditions	Humans are psychologically Pleistocene hominids with cultural universals; they show evolved inclinations and are maladapted to modern life	Genes and cultural variants both matter; human adaptive capability increased by cultural evolution; the theory explains uniqueness and diversity of cultures among regions	Memes are parasitic; culture is an autonomous evolutionary system; memes can dominate and direct genetic evolution
Methods	Gene accounting; comparison with animals, notably primate lineage	Comparison with primates; questionnaire surveys; lab experiments; link to psychological insights about proximate factors	Mathematical modelling; field and laboratory experiments; ethnographic data; integrating insights from social psychology; testing genetic data	Argumentation and 'meme accounting'
Fair criticism	Too dominant role for genes; ignores behavioural effects of culture and group selection	Underplaying cultural factors and dynamics	Solid evidence for certain aspects of cultural evolution and gene–culture coevolution; imbalance theory and empirics	Mainly 'just so' stories without strong empirical or experimental evidence
Unfair criticism	Genetic determinism (environmental factors are considered)	Only 'just so' stories	Too complex, inaccessible	No useful insights beyond the other three approaches

Table 7.2 (*cont.*)

Feature	Sociobiology	Evolutionary psychology	Dual inheritance (including gene–culture coevolution)	Memetic thinking
Strengths	Learning across animal species; exploring the explanatory power of the selfish gene approach (i.e. genetic accounting)	Integrates evolutionary insights about human evolution into large body of psychological research; aids in assessing genetic constraints on altering human conduct	Giving balanced attention to genetic and cultural factors: integrating social psychology in population models of social transmission; understanding current/ future human behaviours cultures	Exploring how far one can go in explaining behaviour and culture through human ability to imitate; assess cultural effects of meme-spreading technologies (e.g. Internet)

Because of an integrated approach to gene–culture evolution, it delivers potentially strong explanatory models. On the other hand, it considerably complicates things causing its insights to be rather theoretical and hypothetical. But this holds to distinct degrees for all four theories, hence does not represent a very useful criterion to favour one over another.

What makes dual inheritance unique and powerful is that it does not just say that environment, including the social environment or culture, affects human behaviour – all areas of study agree on this. It goes beyond this by explicitly describing cultural change as cumulative, involving never-ending social learning that depends on a variety of individual behaviours and attitudes, thus involving transmission and inheritance of cultural variants. This process is guided by selection of these variants as well as of any associated genes, through a combination of cultural and natural environmental selection factors. As a result, dual inheritance allows distinguishing among, as well as interlinking of, the dynamics of environmental, genetic and cultural influences on humans. Thus, it can explain why the human capacity for culture is a unique adaptation, why distinct behaviours and cultures exist in similar environments, why genetic adaptation can be superseded by cultural adaptation, or why seemingly maladaptive behaviours continue. As opposed, evolutionary psychology puts culture in the same basket of environmental factors, which usually is considered to be rather static. In this sense, it remains closer to evolutionary biology while dual inheritance has a foot in both biology and social sciences.

Memetic thinking is more situated in the social sciences. This does, however, not mean it completely omits genetic evolution. Witness Blackmore's (1999) conjecture that memes catalysed the genetic evolution of the human brain–mind, which then facilitated meme spreading. But memetics is definitely closer to the common approach in the social sciences to consider culture, economy or technology as subject to changes that are fully determined within the wider social system. If dual inheritance is the correct general model, then such conventional approaches to the social sciences are prone to make errors, due to disregarding the role of genes and the environment, or the evolutionary dynamics underlying culture, or more basically, as in fields like economics or innovation studies, the influence of culture at all. Mesoudi (2011) suggests that by overcoming these various shortcomings, cultural evolutionary theories can stimulate an 'evolutionary synthesis' in the social sciences, reducing fragmentation and increasing coherence among them.

Table 7.2 shows that the four evolutionary approaches are to some extent complementary, but also somewhat overlapping and inconsistent. For example, memetic thinking can be seen as a special case of dual inheritance with a focus on imitation as the dominant way of social learning. Evolutionary psychology can further be seen as the human version of sociobiology, where genes affect behaviour, along with development factors influenced by the natural environment. Evolutionary psychology has given considerable attention to the conditional role of culture and environment in causing genes to come to expression. This suggests there is a grey zone of studies in between evolutionary psychology and dual inheritance. Dual inheritance can be seen as a broadened version of evolutionary psychology, which extends the environment to include the cultural context, and moreover makes culture an endogenous variable that in turn affects the genetic factor, thus recognising a process of gene–culture coevolution. So moving from evolutionary psychology to dual inheritance means changing from one-directional to two-directional effects. Finally, less complementariness and more incongruity can be seen among memetics, on the one hand, and sociobiology and evolutionary psychology, on the other. Altogether, one might consider dual inheritance and gene–culture coevolution as the most flexible approaches that occupy a balanced middle ground between theories focused purely on genetic or memetic processes. In addition, they arguably best explain human evolutionary history since the Pleistocene, when cultural factors came into play.

Further comparisons of these approaches are offered by Blackmore (1999) and Laland and Brown (2002). The latter includes human behavioural ecology as a fifth approach. Much of the literature, however, does not identify this category very clearly. Laland and Brown present it as a variation of sociobiology using also ethnographic data. It is indeed very close in spirit to traditional sociobiology, in that it considers humans as animals, focuses more on behaviour than on culture, and regards adaptive behaviour as strongly influenced by ecological circumstances. One might further regard evolutionary economics, organisational theory and technology as additional or sub-approaches of cultural evolution. To identify one difference, they tend to focus more on current and future conditions than long-term history. In fact, one might argue that the early evolutionary history of human sociality can be reasonably well understood with sociobiology and evolutionary psychology, the next cultural phase by dual inheritance, and the recent economic phase by evolutionary economic, organisational and technology studies. But it does not stop here. In addition, evolutionary psychology and memetics might identify genetic and memetic maladaptation by individuals to modern life, as in relation to overconsumption or addiction to digital gadgets. This may provide useful information for understanding present-day societal phenomena and solving any associated problems with appropriate policies. Subsequent chapters elaborate these ideas.

8 Evolutionary Economics

8.1 A Typology of Evolutionary Economic Thought

The world economy has, since the end of the eighteenth century, undergone extremely rapid developments, which still have not come to a halt. These developments are characterised by changes in diversity at various levels, steered by innovations and various selection processes, leading to cumulative causation in terms of knowledge, technologies, organisations and artefacts. This suggests a process of economic evolution. It can be observed everywhere in the horizontal organisational structure of the economy, including in science, firms, markets, the legal system, consumer preferences and institutions. While economic evolution is mainly about non-genetic changes, it may involve or interact with genetic evolution – whether in humans or other species. This fits within the broad pattern of gene–culture coevolution of humans discussed in Section 7.4, and is elaborated in Chapter 13 under the label of economic-environmental coevolution.

Despite a fairly long history, specialised journals, several academic organisations and frequent conferences,[1] it can be debated whether evolutionary economics is a mature field. For example, it lacks a good textbook treatment, let alone competition between textbooks that offer distinct viewpoints and foci. On the other hand, it has produced various collections of articles that serve as a kind of introduction – an early one being Dosi et al. (1988). Hodgson (1993a) is a comprehensive overview of the field, rich in historical context. There are also various compilations of classic articles on evolutionary economics: Hodgson (1995), Witt (1993) and Dopfer and Potts (2014). This chapter provides a general introduction to, and brief historical overview of, evolutionary economics. It devotes attention to the most important contributions, as well as a number of core themes, including similarities and differences with biological evolution, types of selection processes in the economy, evolutionary foundations of economic growth theory, spatial and network aspects of economic evolution and the difference with equilibrium economics.

---

[1] Several societies (The European Association for Evolutionary Political Economy; International Joseph Schumpeter Association) and the *Journal of Evolutionary Economics* are specifically devoted to evolutionary economics. In addition, the *Journal of Economic Behaviour and Organization* and *Research Policy* frequently publish studies adopting an evolutionary economics approach.

Several types of evolutionary economics can be identified (more details are in the historical review of approaches in Section 8.4):

- Evolutionary economics with an institutional bend, following Veblen (1898).
- Neo-Schumpeterian theories of technical change, stimulated by the work of Schumpeter (1942) and more recently of Nelson and Winter (1982), addressed in Chapter 10.
- Evolution according to the (Neo)Austrian school, with Menger and Hayek as the main exponents. The work by Faber and Proops (1990) is in this spirit.
- Evolutionary game theory, with its origin in evolutionary biology (Section 8.5).
- Complex adaptive systems, operationalised through (multi-)agent-based models (Section 8.5).
- Evolutionary dynamics of economic organisations and institutions, addressed in Chapter 9.
- Coevolution of economic and environmental systems, like agricultural pests and pesticide control techniques, discussed in Chapter 13.

Economic evolution must be distinguished from development, change, history and growth, notions which are frequently used in the social sciences. Sometimes a loose use of 'evolution' substitutes for these notions. Economic evolution, as elaborated here, will be seen to have a more specific meaning. Indeed, it will be argued that economic evolution is not a metaphor but is real – because it is V-S-I-R algorithmic in nature (Section 2.1) – just like biological evolution.

8.2   Building Blocks

As discussed already in Section 2.2, Darwin and other early evolutionists were influenced by the economic thinking at the time – notably Malthus' theory of human population growth – as well as by industrialists in England during the nineteenth century. Nonetheless, economic evolutionary theorising missed its chance and currently lags far behind biology. Although, in 1898, Veblen suggested the option of economics becoming an evolutionary science, a coherent development of this idea did not take off until the work by Nelson and Winter in the 1960s and 1970s. It blended with the earlier work of Joseph Schumpeter, to give an impetus to the neo-Schumpeterian school of evolutionary economics. Economic evolution involves a number of complementary V-S-I-R elements and processes as shown in Table 8.1, while in Figure 8.1 these are connected to provide a systems view on economic evolution.

Selection in economic evolution concerns multiple selection factors. Selection by markets already leads to a multidimensional selection environment as it involves final goods, intermediate product and factor, financial and labour markets. Selection by financial markets means, among others, the judgement by banks and investors of future expected cash flows and profitability. Market selection is more effective in changing the distribution of product characteristics than that of processes, as process costs often have an indirect relation with product cost and output prices. This holds especially true

Table 8.1 Key concepts in evolutionary economics

---

1. Diversity

- Population with heterogeneous elements
- Elements may pertain to organisations, techniques, products, agents, behaviours/strategies, institutions, rules, etc.
- Diversity components: variety (number of different types), balance (distribution) and disparity (distance/difference)

2. Selection

- Physical (e.g. thermodynamic boundaries)
- Technological (technical feasibility, costs)
- Geographical features (including soil, water, wind and sun)
- Business features (organisational structure)
- Market (relative prices, market power distribution)
- Institutions and public policy
- Cultural and preferential change (e.g. fashions)

3. Innovation

- Public and private research
- Combination/cross-fertilisation of existing ideas or artefacts
- Creativity, curiosity and serendipity
- Education
- Isolation (spatial, economic)
- Cooperation
- Risk-seeking behaviour (venture capital)
- Niche markets
- Future visions
- Public policy (e.g. subsidising R&D or technological diffusion)

4. Inheritance

- Reproduction
- Imitation
- Memory
- Artefacts (art, technologies, buildings, infrastructure, books, recordings, digitised information)
- Education
- Rules, norms and laws

5. Bounded rationality and social interaction

- Time preference and time horizon (myopia, discounting)
- Habits and routines
- Social interaction networks
- Other-regarding preferences
- Imitation (diffusion) and conformism to group
- Social comparison and status-seeking
- Perception of, and attitude to, uncertainty
- Simplifying complexity and using emotions to arrive at decisions

6. Learning

- Individual, organisational and social learning
- Cheap (imitation) and expensive (search, investment) learning
- Horizontal, vertical and oblique transmission (peers, parents and teachers)

Table 8.1 (*cont.*)

- Learning about facts, concepts, procedures, rules
- Learning to reason and learning to solve problems
- Information exchange and learning in networks

7. Adaptation

- Repeated interaction of selection and innovation
- Change in population distribution
- Individual and social learning
- Adaptation to multiple selection factors: trade-off

8. Path dependence and lock-in

- Change in diversity of elements in a population
- Irreversible change
- Increasing returns to adoption (economies of scale, imitation, learning and positive network externalities)
- Lock-in
- Extended level playing field

9. Coevolution

- Interacting subsystems with heterogeneous populations
- Negative or positive feedback
- Demand–supply interactions
- Joint evolution of complementary technologies
- Economy and environment/resources

Source: Revised and extended from van den Bergh et al. (2007, Table 2.1, p. 20).

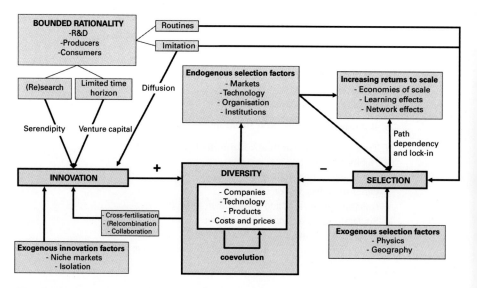

Figure 8.1 Interactions among core elements in an evolving economy.
Source: van den Bergh et al. (2007, Figure 2.1, p. 35).

for multi-product firms, in which overhead costs are assigned in a rather arbitrary way to products. Coevolution of demand and supply in markets means that selection pressure works in two directions. For more discussion, see Box 8.1.

Public policies contribute to the overall economic selection environment. Particularly important are competition (antitrust) legislation, policies outlining what is patentable and what is not, and safety, labour and environmental regulations. As an illustration of evolution of public policy, consider the case of air quality regulation in the United States (Nelson and Winter 1982, pp. 372–379). This reflected an interaction between public awareness, scientific understanding of environmental issues and public policy. Broader institutions can act as selection forces. This involves educational systems, unions, scientific research institutions, non-governmental organisations, social norms and public opinion on environment.

An evolutionary economic system functioning according to a V-S-I-R algorithm, as shown in Figure 8.1, undergoes continuous change in structure and complexity along a non-equilibrium dynamic path. The creation of diversity, through various mechanisms of knowledge, organisational and technical innovation, functions as a disequilibrating force. On the other hand, selection generally reduces diversity and can be considered as an equilibrating and directive force. This is one reason why understanding of economic diversity is relevant for understanding if not predicting economic dynamics.

8.3   Evolutionary Economics and Biology Compared

To further clarify evolutionary economics, it may be useful to consider basic similarities and differences between economic and biological evolution (Hodgson, 1993a; Eldredge, 1997; Gowdy, 1997; and van den Bergh and Gowdy, 2000). Arguably, a chief difference is that in biology the genotype–phenotype distinction is clear-cut, while in economics no similar sharp division exists. The reason is that there is no singular equivalent in economics to the most basic unit of biological selection – the gene. Nevertheless, transferable tacit knowledge, through education and apprenticeship, and transferable codifiable knowledge through traditional and digital publications can serve as the analogy of genotype information. Moreover, routines, science, art, rituals, institutions and politics all represent durable information affecting artefacts and behaviours in the economy. Some propose 'economic memes' (Section 7.6) but the requirements for these may not match the way information is conceptualised in evolutionary economics. According to Norgaard (1994, p. 87), 'One type of gene is no more real than the other.' This might be taken to mean that one can freely choose the appropriate 'economic gene' depending on the context or question. From all these angles, the lack of a unique equivalent to the gene in economic evolution does not appear to be a major shortcoming (Winter, 1964; Hodgson, 1995). It means, though, that economic evolution is fundamentally different from biological evolution. This is not to say that a strict genotype–phenotype analogy has not been explored in economics (Boulding, 1981; Faber and Proops, 1990), but it has had limited success and influence.

Innovation and learning are central to both economic and biological evolution. The term 'mutation', common in biology, is often used by evolutionary economists, along

---

Box 8.1 Demand–supply coevolution, price diversity and innovation

The dominant approach in economics to describe markets is representing consumers by a single, downward sloping demand curve and producers by an upward sloping supply curve, as in Figure B8.1.1. The intersection of these curves defines the market equilibrium characterised by an equilibrium price and quantity traded. Such an equilibrium approach treats both consumers and producers as aggregates. An alternative offered by evolutionary economics follows a more disaggregate approach, namely describing interacting populations of consumers and producers. This gives rise to a coevolutionary, dynamic approach to demand–supply dynamics. This, arguably, is more realistic and accurate.

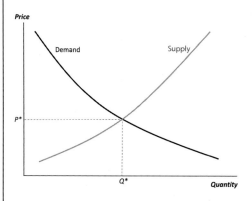

Figure B8.1.1 Standard partial equilibrium diagram of a market with equilibrium (P^*, Q^*).

Coevolutionary demand–supply models have a number of unique advantages, allowing for relevant applications in economics and innovation studies. They can describe a rich set of interactions within each population of consumers and producers. For example, producers can be modelled as competing, by pricing or product innovation strategies, and consumers as imitating or comparing themselves with others. A basic model was developed by Windrum and Birchenhall (2005).

Coevolution can even be used to describe transaction prices. Here the price will typically be in a range with the minimum willingness-to-accept value by the seller as the low end and the maximum willingness-to-pay value by the purchaser as the upper end. Whether the actual transaction price will be closer to the lower or upper bound will depend on the relative negotiation power of seller and buyer: If the first (latter) is the better negotiator, the price will be closer to the upper (lower) bound. Multiple transactions thus lead to unique prices, which over time can converge, diverge or remain in a stable range. The outcome depends on diffusion of information about prices in the populations of consumers and producers. Figure B8.1.2 provides an illustration obtained from exercises with the well-known Sugarscape model.

Coevolution is also relevant to better understand innovation. Main approaches in studies on technical change (Chapter 10) can be categorised into demand pull and technology push. Each simplifies the complexity of consumer–producer interactions. The first approach is sometimes criticised for overly relying on the rather mechanical

Box 8.1 (*cont.*)

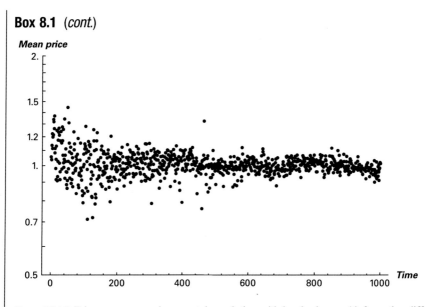

Figure B8.1.2 Price convergence in economic evolution with local prices and information diffusion. *Source*: Epstein and Axtell (1996, Figure IV-3, p. 109).

reactiveness of technology to market conditions, while the second stipulates the one-way causal determination from science to technology and production, hence ignoring the role of demand factors (Dosi, 1982). A coevolutionary demand–supply approach can address the impact of producer–consumer interactions on innovation processes. For instance, knowledge generated through learning by using can only be transformed into new products if producers have direct contact with consumers (Lundvall, 1988). Producers may monitor consumers to assess their competences, i.e. the learning potential of the market to adopt new products. Consumers may be willing to pay higher prices for similar but innovative products, which stimulates Schumpeterian competition rather than price competition by firms. All these mechanisms can be well integrated in a coevolutionary framework.

Safarzynska and van den Bergh (2010b) use such a framework to study major economic-technological transitions in the presence of increasing returns to adoption on demand and supply sides (Section 10.5). In order to design transition policies, e.g. for enhancing a low-carbon economy (Chapter 14), one needs to understand how combinations of increasing returns at demand and supply sides affect the probability of market lock-in. Witt (1997) suggests that the capacity to pass a critical-mass threshold in terms of the number of potential adopters of a market alternative is key to the success of escaping a situation of technological lock-in. A demand–supply coevolutionary framework allows analysis of policies that can bring about such critical mass, in a context of multiple lock-in factors and under various types of consumer and producer behaviour. Various studies now offer coevolutionary market models along such lines (e.g. Janssen and Jager, 2002; Windrum and Birchenhall, 2005; Schwoon, 2006).

Table 8.2 Economics faces more challenges than physics or biology

Feature	Physics (& Chemistry)	Biology	Economics
Constants and laws	Many	Few	Rare
Evolution	No	Slow	Quick
Novelty	No	Regularly	Often
Emergence	Limited	Important	Important
Diversity	No	Yes	Yes
Controlled experiments	Yes	Limited & partial	Very limited and partial

with such terms as 'invention' and 'innovation'. The traditional distinction is between invention, as a novel technological breakthrough, and innovation, as its application and spread – diffusion through imitation. It has been recognised that their distinction is fuzzy as innovation processes feedback to inventions, also known as a nonlinear model (Nelson and Winter, 1982, p. 263). The biological mechanism of sexual recombination to generate innovations has an economic analogue. Indeed, the idea of combining existing modules is crucial in explaining economic and technological evolution (Weitzman, 1998a). The idea that learning capacity in organisms can evolve over time (the Baldwin effect; see Section 7.5) is found in economic-technological systems as well: in the variety, competition and development of both educational and R&D systems, both in industrial and public settings.

Evolutionary thinking in economics is consistent with the view – central to behavioural economics – that human behaviour is boundedly rational and other-regarding. An obvious question then is whether humans are less bounded in their rationality than animals, and how this affects evolutionary outcomes. It is clear that humans, because of their intelligence, can anticipate problems and opportunities. As a result, they possess unusual problem-solving and learning capacities and can show goal-oriented and planning behaviour. These features are, though, a far cry from the 'homo economicus' model of mainstream (neoclassical) economics, which suggests that we have well-defined preferences and unlimited intellectual powers, allowing us to arrive at individually optimal decisions. Such a purist neoclassical economics view does not go well together with either an evolutionary biology or evolutionary economics approach. Burnham et al. (2016) plead in favour of an evolutionary approach to behavioural economics. They argue it should address Niko Tinbergen's 'four why questions' (Section 2.3) about functional value, development history, and proximate and ultimate causation.

Finally, evolution in economics is generally occurring at a fast rate, which complicates the study of social-economic systems. Inspired by Section 3.4 of Faber and Proops (1990), Table 8.2 offers a comparison between non-evolutionary physical/chemical systems, on the one hand, and evolutionary biological and economic systems on the other. The presence of quick evolution, novelty and emergence in, as well as the impossibility of controlled experiments with, complex economic and biological systems makes their study considerably more challenging than that of purely physical-chemical systems (i.e. without life processes). Based on the differences in the table, one might further infer that economics is more difficult than biology.

8.4   A Brief History of Core Contributions

Evolutionary Economics Until the Early Twentieth Century

Adam Smith (1723–1790), generally considered as the first complete economist, was not familiar with biological metaphors. In his main work, *An Inquiry into the Nature and Causes of the Wealth of Nations* (1776), he emphasised the importance for social well-being of free markets, notably the combination of economic competition and rational self-interested behaviour – notwithstanding his *The Theory of Moral Sentiments*, published 17 years earlier. Smith developed the famous metaphor of the invisible hand of the market. According to Ofek (2001, Chapter 3), both natural selection and the invisible hand reflect a design through spontaneous order. Various studies in evolutionary economics have linked evolutionary perspectives on markets to the notion of self-organisation (Foster, 1997; Witt, 1997).

Karl Marx (1818–1883) seems to have been impressed by Darwin, but thought that conceptualising of capitalist development as a series of class struggles was more appropriate and richer – faintly reminiscent of cultural group selection. His theory allowed for non-gradual patterns (Hegelian sudden changes), which show some resemblance to the notion of punctuated equilibria. Marx regarded gradual change relevant in the context of technological development, but he saw radical change as inevitable in relation to social-economic developments. Unlike Darwin, Marx was idealistic and an activist. In his writings, he focused on the end of capitalism and a transition to a socialist society, a theme later studied along more evolutionary lines by Joseph Schumpeter (see below).

After the classical economists, Marshall (1842–1924) emerged as the most influential economic scholar, notably because of his textbook *Principles of Economics* (1890). His famous aphorism 'the Mecca of economics lies in economic biology rather than economic dynamics' was motivated by his conviction that economics and biology share a reality that is constantly changing internally and externally. He recognised that selection due to competition among firms carried elements of Darwinian gradualism.[2] As discussed in Section 2.2, this did not translate into an evolutionary economics at the time for various reasons. Social evolution was interpreted as Spencerian in spirit, which had a strongly negative connotation. Moreover, Marshall realised that biological framings would imply more serious analytical challenges than mechanical ones. He thought that mechanistic analogies would be useful to describe early stages of economic development, while biological ones could serve a useful purpose in understanding later stages. His focus on mechanistic analogies may be motivated by addressing the initial stage first, or by striving for mathematical rigour and developing economics into an exact science, which required keeping formalisations as simple as possible. The latter culminated in Walras' general equilibrium theory (1874), which was mathematically much more ambitious than Marshall's partial equilibrium approach. Even though

---

[2] According to Groenewegen (2001), Marshall read Darwin's *Origin of Species* as well as early evolutionary strides in the social sciences, namely Herbert Spencer's *First Principles* and an economic textbook with Darwinian elements, *Plutology*, by the Australian scholar W.E. Hearn (published in 1864).

Marshall wrote after Darwin, evolutionary biology was only partially developed, lacking the modern synthesis, linking it up with genetic theory and molecular biology. Evolutionary thought did not have a good reputation, and this holds even more so for its application to social sciences. This might explain the little interest economists showed in it at the time. In fact, for the same reason, Joseph Schumpeter, who developed a kind of evolutionary theory of economics several decades later, still had serious reservations against using biological analogies. Interestingly, Nelson and Winter (1982, p. 45) note that their evolutionary theory, in many respects, is closer to the ideas of Marshall than is mainstream economics. As an example, they mention that Marshall tried to incorporate certain dynamic aspects relating to technological change in his static formalism, which later was erroneously considered by economists to be an incorrect approach to static analysis.

Carl Menger, the founder of the Austrian School (of economics), proposed in his book *Money*, in 1892, a kind of evolutionary theory of the emergence of money, which he compared with the evolution of language and law. For him, evolution means not an arrangement of an institution by a government or ruler, but the unplanned outcome of interactions among many individuals. In particular, Menger saw money as a specific case of a good that can be exchanged for another. Money has the attractive characteristics of being light and small (easily transported), durable (use can be delayed, allowing for storage and accumulation) and divisible hence flexible. Initially multiple currencies coexisted next to barter trade. This can be explained by spatial isolation and path dependence. Menger's evolutionary thinking is considered to belong to the 'Austrian School', along with that by Hayek (see below).

Thorstein Veblen (1857–1929) is often regarded as the first evolutionary economist, because he explicitly used the term 'evolution' and was inspired by Darwinian evolutionary theory. He was equally critical of neoclassical economics, the capitalist economy and Marxism, even though he was impressed by Marx's work. Veblen's approach is sociological in nature, focusing on institutions, the interaction between culture and the economy, behavioural strategies of businessmen and conspicuous consumption driven by status-seeking – which he considered wasteful. He devoted much attention to technical change, which at the turn of the nineteenth century was recognised as an important driver of society, but about which a coherent theory was still lacking. In 1898, he wrote the article 'Why is economics not an evolutionary science?' in *The Quarterly Journal of Economics*. His view of evolution was that of a causal process characterised by an unfolding sequence and cumulative causation.

It is not common to mention John Maynard Keynes (1883–1946) in relation to the birth of evolutionary economics, but he certainly showed a serious interest in Darwinian thought. He was a socialist who saw the economic problem as solving the human struggle for employment and existence in general. Keynes respected Malthus and recognised the influence of population pressure on economic problems as well as on conflicts and wars. According to Laurent (2001, p. 65), in his book *The Economic Consequences of the Peace*, Keynes (1919, pp. 8–13) proposed that population pressure in Germany had not been a major factor contributing to the outbreak of World War I, drawing support from arguments by Malthus. Like many others socialist thinkers at the time, Keynes was not afraid to render support to evolution-inspired eugenics (see Section 7.2). Incidentally, Keynes was in personal contact with some of Darwin's children and grandchildren – his brother was married to a granddaughter of Darwin. This illustrates how highly connected intellectual England was at the time.

The Legacy of Joseph Schumpeter

Joseph Schumpeter (1883–1950) was without any doubt the most influential of all early evolutionary economists, both owing to his general standing in economics, in Europe as well as the United States, and because of the many inspiring concepts and ideas that sprang from his mind. Schumpeter questioned the static approach of standard economics, showing a great interest in economic dynamics, in particular of the capitalist system, in all of his major works (Schumpeter 1934, 1939, 1942). He considered qualitative economic and technological change in a wider context of social change, and devoted attention to the psychological features of the innovative entrepreneur (Schumpeter 1934; first published in German in 1911). Schumpeter regarded capitalistic economic change to be the result of revolutionary forces within the economy that destroy old processes and create new ones. He called this 'creative destruction' and characterised it using language from evolutionary biology:

The opening up of new markets, foreign or domestic, and the organisational development from the craft shop and factory to such concerns as U. S. Steel illustrate the same process of industrial mutation – if I may use that biological term – that incessantly revolutionises the economic structure from within, incessantly destroying the old one, incessantly creating a new one. This process of creative destruction is the essential fact about capitalism. (Schumpeter, 1942, p. 83)

Schumpeter further argued that major innovations involve a kind of recombination, what he called 'neue Kombinationen' in German. He elaborated these themes in his later studies of business cycles.

The foregoing illustrates that, contrary to what is sometimes claimed, Schumpeter did not completely avoid the use of evolutionary notions. This is underpinned by a statement in Schumpeter (1954, p. 82): '[T]he essential point to grasp is that in dealing with capitalism we are dealing with an evolutionary process.' At other times, however, he explicitly rejected the use of biological concepts in economics. According to Hodgson (1997) this may have been due to the fact that Schumpeter misunderstood Darwinism. He refers to a passage from Schumpeter (1934, pp. 57–58): 'With all the hasty generalisations in which the word "evolution" plays a part, many of us have lost patience. We must get away from such things.' This opinion might have been motivated by the fact that evolution was used by many writers in the social sciences to denote general concepts like development, change and history. Another explanation may be that Schumpeter's main ideas were formed before the evolutionary synthesis of genetics and Darwinian theory – which ultimately led to broad acceptance of evolutionary theory within biology.

Inspired by Schumpeter's work, it is now common to distinguish between Schumpeterian and equilibrium or price competition. The first denotes that a competitive advantage is realised through innovation or early adoption of a new product or process, rather than through price competition as in the second case. This distinction has consequence for deriving entry/exit conditions for markets to function well and for the design of antitrust laws (Futia, 1980). It also links up with the theory of market contestability, which says that few market players may not lead to market outcomes far removed from competitive equilibrium as potential entrants put pressure on incumbents which constrains market power (Baumol, 1982). Schumpeterian competition can be seen as similar to a Red Queen effect in sexual selection (Ridley, 1995). Namely, when innovation by market competitors becomes more intense, a firm has to increase its R&D efforts to stay in the same place, that is, maintain its market share (Cullis, 2007, Section 6.4).

With Marx, Mill and Ricardo, Schumpeter shared the idea that the economy would ultimately reach a steady state. In Schumpeter's view, it is characterised by technological progress being the result of carefully planned team research under a socialist organisation of society. Like Marx, Schumpeter gave much thought to the process of change from the capitalist to the socialistic economy. Although he realised that discontinuities play a role, he did not assign them the critical role as Marx did. Instead, he believed that political responses would lead to a gradual transition. The convergence of many Western countries to a social welfare state characterised by institutions promoting equity renders some support to Schumpeter. On the other hand, income disparity looms still large at the international level as well as in many developed countries. In other words, Schumpeter's hypothesis is not confirmed by all the facts.

It has been said that there are 'multiple Schumpeters' (Freeman, 1990). The young Schumpeter worked with notions such as entrepreneur and small innovative firms, whereas the old Schumpeter was interested in big monopolistic firms. In addition, seemingly at odds with his ideas on evolutionary economic change, Schumpeter maintained a belief in the value of static equilibrium analysis, in particular the Walrasian approach of connecting all markets within a single model. Possibly, like Marshall, Schumpeter must have suffered from internal conflicts, being attracted by the mathematics and aesthetics of equilibrium economics, but realising that there was much to say in favour of a dynamic evolutionary approach.

Hayek's Spontaneous Order and Evolution of Institutions

The work of (von) Hayek (1944, 1948) is best known for the idea of 'spontaneous order'. Unlike Adam Smith's 'invisible hand' this entails an explicit dynamic view on the self-organisation of economic systems through markets. Hayek regards change as persistent in economic systems, and a liberal society as more effective than central planning in realising the economic changes required for social goals. He sees markets as communication networks to share and process information. Since he regards the state as not being better informed than all private agents together, it cannot be as efficient as the market process to coordinate decisions among consumers, producers, investors and innovators for the good of the society. Although Hayek does not offer an explicit evolutionary approach, many of his ideas reflect notions similar to diversity, selection and path dependence. Hayek clearly recognised that habits, rules and institutions are being selected and evolve. He depicted institutions – such as legislation, norms, markets and money – as culturally fixed after having been experimented and tinkered with through successive generations. He even invoked a group selection type of argument (Chapter 6), claiming that institutions and practices are preserved as they enable a group in which they arose to prevail over other groups. But he did not elaborate this idea through explicit group selection mechanisms (Hodgson, 1993a, p. 171).

Hayek's (1967) grand cultural-institutional development theory has been elaborated, among others, in the neo-Austrian economics school (e.g. Faber and Proops, 1990). This stresses the roundabout, multistage nature of production, temporal aspects of production, production processes as compositions of discrete techniques and market processes in time instead of equilibrium. A shared feature with evolutionary economics is a Schumpeterian type of competition between old and new production techniques, where the first are threatened and slowly replaced by the latter – which are more advanced and usually more roundabout, i.e. involving more intermediate production connections.

Owing to his repeated argumentation, in earlier and later works, for a liberal society and respect for tradition and against the feasibility of central planning, Hayek is seen by many as the preeminent advocate of laissez-faire and minimal government. Hodgson (1993a, p. 183) notes that Hayek's political ideas arose from his fear of Nazism and collectivism, as had emerged in Germany and the Soviet Union, and which had made him intolerant to socialism. Hayek was, though, not completely against regulation of markets as he perfectly understood the benefits of guaranteeing labour, health and safety conditions. He recognised the need for governments to assure the basic conditions for a stable market economy in terms of property rights or removing market imperfections, and accepted that this involved a certain degree of centralisation and power concentration at the level of national states. According to Bowles et al. (2017), Hayek's views on the social benefits of markets were ahead of their time and complementary to arguments for markets based on general equilibrium thinking, and suggest evaluating them as independent from his political-philosophical ideas. They note that Hayek's insights about markets as information-processing institutions are only now being rigorously tested using evolutionary agent-based models.

Alchian and Friedman on Rationality Through Market Selection

Since the 1950s, there has been a slow increase in publications on economic evolution. This can be partly explained by the success of evolutionary biology, the limits of the dominant school of neoclassical economics, and the search for evolutionary underpinnings of economic behaviour. An early effort to realise the latter was the work by Alchian (1950). He argued that it is unnecessary to assume rationality of firms in the form of profit maximising, or even profit seeking, as a fundamental and universal characteristic of firms because those that are successful in realising profits will be selected (Alchian uses the term 'adopted') by the mechanism of market competition. Milton Friedman (1953) went further than Alchian and argued that profit 'maximisation', rather than 'seeking', is selected by the market, in this way trying to find support for the idea that profit maximising, a cornerstone of neoclassical economics, is the outcome of selection by markets. The Alchian–Friedman thesis can be seen as an early, non-technical version of evolutionary game theory (Section 8.5). Indeed, both emphasise equilibrating selection working upon a given pool of diversity, which allows an evolutionary process to result in a particular equilibrium. More generally, both approaches attempted to found equilibrium theory on evolutionary principles.

In a persuasive critique, Winter (1964) argued that Alchian and Friedman employed the selection analogy from biology without complementing it with a clear sustaining or replication mechanism, such as reproduction or inheritance. This is required to assure that profit maximising behaviour of firms is replicated. Winter argued that firms' success depends to a large extent on random circumstances, regardless of whether they are seeking profits or not. The more uncertain the world, the more winning will depend on pure luck. Hence, winning in one period will then be weakly related to winning in another period. In addition, successful profit-making is often unconscious or at least something that cannot easily be codified, and thus passing it on to others – in the same firm or in other firms – is far from straightforward (Hodgson, 1988, p. 78). As a result, winning remains a largely haphazard process, as shown by the profits of many firms fluctuating erratically over time. This is reinforced market selection being weak, for various reasons: competition between firms may be lax in general, or additional selective forces like policy constraints counteract market selection (Foss, 1993).

Altogether, the defence of profit maximisation as the behavioural model for firms using a market selection argument is not convincing. By implication, profit maximisation cannot serve as a good approximation of reality. Instead, one would do better to model the actual bounded rationality and heterogeneity of firms. This will provide a more realistic picture of how markets function. By adding market and other selection forces, one can then arrive at a genuinely evolutionary and more accurate approach to describe markets.

The market selection argument neither rescues the other core assumption of neoclassical economics, namely that consumers or households maximise their utility. Friedman did not devote any attention to utility maximisation, while in footnote 3, Alchian (1950) says that 'In the following we shall discuss only profit maximisation, although everything said is applicable equally to utility maximisation by consumers.' However, it is not clear how the mechanism of competing firms can be transferred to interaction between consumers. A consumer not maximising utility would not have a clear disadvantage in relation to other consumers that do maximise their utility. Survival of consumers is likely to depend on satisfying basic needs rather than seeking fulfilment or maximisation of desires beyond those. One might, moreover, argue that due to technology and the social welfare state – i.e. selection factors beyond the market – selection pressure against non-maximising utility behaviour has weakened over time.

Nelson and Winter's Routines, Search and Selection

The most cited and influential work in evolutionary economics since the 1950s has been that of Richard Nelson and Sydney Winter, which culminated in their celebrated book of 1982, *An Evolutionary Theory of Economic Change*. It proposes a formal, axiomatic approach to evolutionary economics that allows for empirical operationalisation, and is suggested to be able to do most of what neoclassical theory can do and much more. In a nutshell, Nelson and Winter interpret economic evolution as gradual microlevel changes of routine-like behaviour through search processes. Firms in their framework are motivated by profit but are not profit maximising over well-defined and exogenously given choice sets as is the standard assumption in neoclassical economics. Nelson and Winter's theory of microevolution consists of three building blocks (Box 8.2 provides a formal representation).

(i)   *Organisational routines*: This is about the way in which businesses make their decisions. Routines can apply to production, sales (e.g. mark-up pricing), inventory control, purchasing, marketing, human resource management, investment and even R&D. Routines reflect a bounded rationality according to which firms do not alter their behaviour in the short run. This is because of barriers to change, in an abstract sense transaction and switching costs, owing to firm politics, vested interests, avoidance of conflicts, financial costs of change and management control systems. Also contributing to constancy of routines is their reproduction through hierarchy, promotion, teamwork and training. In view of its stability, a business routine can be considered as the equivalent of the gene in biological evolution. Although a routine of an organisation has been equated to the habit of an individual, it is defined by Nelson and Winter as being more complex, namely combining multiple people with particular skills. Here a skill reflects tacit knowledge that does not involve deliberation and conscious choice but operates rather automatically. A routine can then be seen as an adapted complex set of skilled individuals that interact simultaneously and sequentially. The interactions are crucial and depend on learning from earlier contacts and organisation-specific 'language'. They involve formal and informal contacts, guided by control mechanisms and trust, respectively. In line with the foregoing, routines are often unique which contributes to the diversity of businesses, motivating the relevance of an evolutionary framework. Seen this way, organisational routines work according to a decentralised or self-organisation mode, even if the organisation has a hierarchical structure.

(ii) *Search behaviour*: It is clear that routines create stability and continuity in the firm's behaviour and sometimes undesirable inertia. They assure that employees do not free ride and enter into conflicts too much but work towards a common goal. However, routines can and do change, both unintentionally and intentionally. With regard to the first, many changes in routines are non-directed and accidental, being merely the result of solving problems in the organisation or the organisation's performance; or of old employees leaving and new ones entering, bringing along new skills and creating new patterns of interaction. With regard to intentional change, firms may evaluate their current routines and decide to adapt or replace them or to imitate routines, or elements thereof, from other, more successful firms. Intentional change may involve innovation driven by R&D or discovery (e.g. routines of other firms) through organised search. To be effective, R&D and search can be based on separate organisational units which themselves follow routines. This is a major activity in many firms, aimed at changing existing routines in production, marketing, internal organisation, etc. To make things more complex, routinised innovation of this type is something that is subject to innovation itself – as is clear from the evolution of a small entrepreneur-owner firm to a large company with a R&D department. An important source of organisational changes is new combinations of existing routines (Schumpeter, 1934), analogous to the notion of recombination in biological evolution. This may take the form of novel patterns of information and material flows among existing (sub)routines. Routine change through replication of products, processes, strategies or organisational structures of other firms can be called copying or imitation. Since replication is imperfect, inevitably small innovations (mutations) are introduced along with it (Nelson and Winter, 1982, pp. 117–118). For instance, through reverse engineering a new competitor's product, one may learn all about its ingredients or components, but little about the exact engineering processes that underlie them. Given the complexity of most routines – involving technologies, people, skills and communications – it will be impossible to perfectly copy them from another organisation.

(iii)   *Selection environment*: The selection forces to which a firm is subject influence its performance as well as changes in its organisation and activities. Selection results mainly from demand and supply on markets, from institutional and policy constraints, and from the (strategic) behaviour of other firms. Nelson and Winter (1982, pp. 142–143) distinguish selection from imitation (which is debatable), and argue that the first dominates in biological systems, whereas both are relevant to economic systems. Selection functions as a kind of filter that makes well-adapted firms – i.e. with good routines – survive and ones with inadequate routines disappear (exit or being taken over). Due to industries consisting of many heterogeneous firms being subject to continuous selection, which in essence is a process of negative feedback, diversity is reduced and efficient equilibria may be approached. This contrasts with the undesirable, locked-in states that result from positive feedback driven by increasing returns to adoption (Section 10.5). Of course, both neglect continuous innovation that keeps economic systems away from a perfect equilibrium state.

Schumpeter's influence in Nelson and Winter's work is most clearly present in their elaboration and formalisation of Schumpeterian competition (Nelson and Winter, 1982, part V). It comes down to an interaction among market structure, R&D spending, technical change and industry performance (e.g. profit-making). This approach allows questions to be addressed such as: what is the best market structure for innovation, what is the impact of the initial industry structure on long-term outcomes, and what are the conflicts between market concentration for R&D and attaining social welfare?

Nelson and Winter (1982, part IV) have further developed an evolutionary (simulation) model of economic growth. Its purpose is to generate and explain patterns of aggregate outputs, inputs and factor prices. This will be discussed in Section 8.6. Finally, the neo-Schumpeterian school of technological change and evolutionary economics continues the work of Schumpeter and Nelson and Winter. Chapter 10 is devoted to it.

Box 8.2 Models in evolutionary economics illustrated

Nelson and Winter's General Theory

The three basic concepts of Nelson and Winter (1982) are organisational routine, search and selection environment: they affect the performance of the firm and its changes. A decision rule on output and input is

$$X_i/K_i = D(P, d_i)$$

with X_i a vector of outputs, K_i a capital input, P a vector of output and input prices, d_i a vector of decision rule parameters, and i denoting firms. Aggregation over firms gives:

$$X/K = \Sigma_i D(P, d_i) K_i / K$$

with variables without index representing equivalent aggregates. One can decompose the change in the output-capital ratio between times 0 to T as follows (with interpretations of the terms to the right):

$$(X/K)^T - (X/K)^0 =$$
$$\Sigma_i [D(P^T, d_i^t) - D(P^0, d_i^t)](K_i/K)^0 \quad \textit{Firms moving along the decision rule}$$
$$+\Sigma_i [D(P^0, d_i^T) - D(P^0, d_i^0)](K_i/K)^0 \quad \textit{Decision rule evolving through search}$$
$$+\Sigma_i D(P^0, d_i^T)[(K_i/K)^T - (K_i/K)^0] \quad \textit{Selection of firms changing the}$$
$$\textit{distribution}$$
$$+ \textit{rest term} \qquad\qquad\qquad\qquad\qquad \textit{No disjunct decomposition}$$

This model is elaborated in Nelson and Winter (1982, Chapter 9) in the context of their evolutionary theory of economic growth.

A Model of Innovation and Imitation

Iwai (1984) presented a general model of innovation and imitation. The model describes the change in the frequency of unit costs $f_t(c_i)$ of production methods co-existing in a particular industry, where $c_n < c_i \leq c_1$, i.e. with c_n the unit cost of the best-practice method. The cumulative frequency function is then $F_t(c) = \Sigma_{i=k,\ldots,n}$ $f_t(c_i)$ for $c_k < c \leq c_{k-1}$. Dynamics follow Schumpeterian competition, meaning that each firm continuously tries to reduce the unit production cost by employing two strategies: innovate or imitate. These jointly determine the dynamics of technology and industry.

The probability of imitation, adoption or copying a particular production method – obviously with lower unit cost than its present one – is assumed to be proportional to the frequency of firms which employ that method at the time. The proportionality parameter μ can change over time owing to R&D and investment activities by the firm. This can be shown to generate logistic (S-shaped) growth, through a set of differential equations:

$$dF_t(c_i)/dt = \mu F_t(c_i)[1 - F_t(c_i)] \, for \, i = 1,\ldots,n$$

Box 8.2 (cont.)

Here, $[1 - F_t(c_i)]$ is the relative frequency of firms amenable for imitation, i.e. with unit costs higher than c_i, where $F_t(c_i)$ reflects the relative frequency of firms that can be imitated, i.e. with unit costs lower than c_i. The relative frequency of a certain inefficient method ($i < n$), i.e. the difference between two adjacent logistic curves, initially expands due to absorbing firms with even less efficient methods, but ultimately approaches zero, when all firms have adopted the most efficient method (n).

The latter equilibrium will not be attained since innovation acts as a disturbing force. The innovation (i.e. first implementation) by a certain firm of a cheaper method with cost c_{n+1} ($< c_n$) and probability ζ is the beginning of a process of creative destruction. Imitation of the new technology will create another logistic growth pattern. If innovation occurs by the 'most efficient firm' then another logistic curve is added, starting at initial frequency $1/M$ (M is total number of firms). If a less efficient firm with cost c_i 'jumps' immediately to the new method then logistic curves 'in between' ($I + 1, \ldots, n$) experience a discrete jump. If the unit cost of the best practice resulting from innovation is as follows:

$$c_{min}(t) = exp(-\lambda T),$$

with λ the rate of decline of minimum unit costs, then innovation time T can be written as a function of unit cost:

$$T(c) = -ln(C)/\lambda$$

Now the impact of the parameters on the distribution of methods or cost levels in the industry (population of firms) can be examined: imitation/adoption rate μ, innovation probability ζ, and rate of decline of minimum unit costs λ. Combining all the previous results into a long-run distribution (density) function leads to alternating phases of widening and narrowing distributions of characteristics, i.e. interaction between equilibrating and disequilibrating forces (innovation: λ, ζ) and equilibrating (adoption: μ) forces.

The *NK* Model

Kauffman (1993) proposed the well-known *NK* model. It describes complex decisions and evolutionary change as combinatorial problems, which can be regarded as one way to operationalise bounded rationality. The term *NK* denotes the total number of dimensions (N) of the problem and the number of dimensions that on average determine the function of the system in each dimension (K). The value of K can thus be seen as a measure of the interdependence of dimensions, i.e. the complexity of the problem. $K = 0$ means minimal complexity, and $K = N - 1$ maximum complexity. If in each dimension at least two discrete options are available, then the total number of decisions or systems is at least 2^N. The dimensions can be thought of as components in products, sub-processes in production, individuals in

Box 8.2 (*cont.*)

groups or organisations in market networks. The model further includes the evolutionary notion of fitness. Two levels are defined, namely fitness per dimensional choice and average fitness of combinations of choices over all dimensions. This allows, for instance, the examination of whether a continuously ascending path in terms of overall fitness exists that results from single mutations, i.e. changes in choices within only one dimension. The *NK* model can further be used to illustrate path dependence and evolution as going through or ending in local optima. Kauffman initially focused on applications to biology and chemistry, such as molecules in networks of chemical processes, and genes and proteins in living organisms. It has also been applied to economics, particularly innovation trajectories (Frenken, 2001). Section 12.4 discusses its application to the early development of the steam engine.

Evolutionary Finance

The study of financial markets has been pursued with an evolutionary approach, which can be seen as a cousin of behavioural finance (Shiller, 2003). The latter looks at how different types of bounded rationality explain investor's behaviour. For example, as men show more overconfidence than women they tend to invest more aggressively and achieve higher profits. Or theories about behavioural biases under uncertainty explain equity puzzles that are unsolved by traditional rational-agent theories (Benartzi and Thaler, 1995). Evolutionary finance instead focuses on social interactions between agents to explain a variety of financial market phenomena. An evolutionary approach is logical here as it typically accounts for networks of interaction between multiple, boundedly rational agents, resulting in imitation ('bandwagon effects'), behavioural diffusion and information externalities. As part of this, investors make decisions that must be regarded as deviating from rational, well-informed behaviour (Sahi et al., 2013). Overall, such phenomena add up to positive feedback mechanisms that easily destabilise economic-financial systems – accounting for both boom (high growth) and bust (recession) cycles.

As opposed, conventional financial (and macro) economics pay little to no attention to such networks of influence and consequently have been unable to explain the most recent financial crisis (starting in 2008). The classical analytical approach of finance is dominated by von Neumann and Morgenstern's rationality under uncertainty, namely expected utility behaviour. This has given rise to standard financial models, such as the capital asset pricing model (CAPM), the Black and Scholes option pricing model, the Modigliani and Miller model of capital structure and arbitrage pricing theory. These models should be regarded, though, as normative rather than positive, in the sense that investors would like to possess the ambitious information and calculation capabilities that the expected utility maximisation model takes for granted. But things are more complicated, as the traditional financial models have influenced reality, simply because they have been available on calculators and investors have assumed that competing investors act in accordance with them. This has partly contributed to self-fulfilling prophecies that are typical of financial market phenomena. To thoroughly understand these, notably their underlying imitation and dissemination processes, actually requires a population-based approach.

A typical evolutionary model thus starts with defining a population of investors who show boundedly rational behaviour, from which one can derive properties of market processes at a higher level. It builds on insights from behavioural economics and economic psychology. An main source of inspiration is Prospect theory by Kahneman and Tversky (1979), which offers a basic model for understanding behaviour under uncertainty, including in financial markets. Among others, this theory accounts for overweighing probabilities of uncertain outcomes that are considered more certain than others, a preference for a sure gain over a probable gain with a larger expected value, less risk-aversion and more risk-taking in the domain of losses than gains, and generally a tendency to avoid losses. These psychological features of decision-making under uncertainty can explain why many people hold losing stocks for too long and sell winning stocks too soon. Another interesting decision anomaly, from the viewpoint of expected utility theory, is that inconsistent preferences result from individuals basing their choices on differentiating characteristics between alternatives, giving less attention to shared characteristics. This is, arguably, stimulated by a strategy to simplify comparison of alternatives in a complex decision context.

Since it is difficult for analytical work to go systematically beyond the usual rational behavioural assumptions, numerical simulation is required. It can, moreover, examine a wider range of behavioural assumptions, and especially the interaction of many agents with heterogeneous behaviour features in line with Prospect theory. This motivates the use of an agent-based modelling approach, also referred to as 'microscopic simulation' (Levy et al., 2000). This allows a more precise examination of the volume of trade and the pattern of prices, over time, in stock markets. Particular phenomena that can be better explained, or even should not occur at all according to traditional finance models, are excess price volatility, heavy trading volumes and endogenous market crashes. In addition, in contrast with option pricing models that unrealistically assume investors have complete information about the value of assets underlying the options, heterogeneity of investors' belief leads to more realism regarding option trading volumes and volatility of option prices.

Evolutionary models show coevolution of market indicators and investor strategies. They can elucidate specific details of strategies by investors, such as the number of assets held in a portfolio. Whereas CAPM suggests that all assets are included in the portfolio – for any investor, regardless of risk attitude – empirical research indicates that investors on average hold just about three or four different stocks in their portfolio. Evolutionary models are more flexible in that they allow for heterogeneity of investors. This can explain, for example, why more wealthy investors can hold more assets in their portfolio. Evolutionary models are also ideally suited to examining the relationship between the number of investors and price fluctuations. Most studies so far have found a negative relationship, some even that fluctuation disappears, for a large number of investors. Nevertheless, other studies suggest that periodic crashes appear for sufficiently large numbers of individuals. Evolutionary modelling has shown that under certain conditions adding agents changes the dynamics of the system. Indeed, an evolutionary approach characterised by explicit consideration of population structure represents an effective tool to deal with this topic.

Evolutionary finance has been extended in more complex models to explain cascades of failures in the banking sector and even the macroeconomic implications of such financial collapse (Neveu, 2013). This overcomes the shortcomings of traditional aggregate or rational-agent approaches to macroeconomics that have been criticised for being unable to explain and foresee financial-economic crises. In one model of this type, Safarzynska and van den Bergh (2016) describe the coevolution of four populations, namely of heterogeneous consumers, industrial firms, power plants and banks, which interact through interconnected networks. This allows the assessment of the macroeconomic impacts of energy policies in four dimensions: environmental impact, financial stability, employment and income distribution. Finally, because of its ability to address fundamental uncertainty and bounded rationality, evolutionary finance is well capable of studying investment in innovations and new markets (Speidell, 2009).

8.5 Evolutionary Games and Agent-based Models

Any evolutionary theory begins with defining a population of similar but distinct items or individuals. This immediately clarifies an essential difference with traditional micro-economics, which is built around representative or average agents. Contrary to common belief, such a microeconomics is not really as micro as is possible. In fact, evolutionary theories can be said to be 'more micro', because they describe populations with behavioural or technical diversity among individuals or firms rather than aggregating these into representative agents or average behaviour. The latter would mean losing descriptive detail that may be essential to describe the dynamics of the system.

A population or V-S-I-R approach (Section 2.1) can be operationalised in two important ways (for a richer typology, see Safarzynska and van den Bergh, 2010c). One is evolutionary game theory. It defines a limited number of distinct sub-groups, each of which is internally homogeneous – representing a rather aggregate approach. Evolution then takes the form of a change in the population composition in terms of the share of sub-groups in it. A common choice is two sub-groups, as these can be described by one dynamic (difference or differential) equation, allowing for explicit analytical solutions. As a variation, one can introduce stochastics and define a probability distribution of characteristics in the population, which then changes over time under the influence of innovations (mutations) and selection forces. An example of such an approach is the model by Iwai (1984) in Box 8.2.

A second approach is more disaggregate, representing in fact the most thorough micro-approach conceivable. It takes the form of multi-agent systems (or agent-based modelling), in which each individual is explicitly described and can be assigned unique features. This microscopic approach allows agents to be defined in a setting of entirely random interactions (gaseous cloud) or systematic interactions through a network structure or a spatial grid (lattice). Note that the term 'multi-agent' is easily misinterpreted: one should not confuse multi-agent evolutionary models with traditional multi-agent (general equilibrium) and multisector (e.g. dynamic macroeconomic or input–output) models in economics. The latter two types of models are based on complementary and representative agents (e.g. associated with different production sectors, or with financial, labour and goods markets). These are fundamentally different from multi-agent evolutionary models which instead include populations with heterogeneous agents. Below we briefly discuss core features of both approaches.

Evolutionary Game Theory

Evolutionary game theory is the preferred method of evolutionary analysis in mainstream economics, as it provides for a connection between equilibrium and evolutionary concepts. It is also known as equilibrium selection theory: it solves the problem of multiple Nash equilibria common in nonlinear economic equilibrium models, by examining evolutionary paths to identify feasible equilibria (Samuelson, 1997).

While traditional game theory studies conflict and cooperation between intelligent agents, as initially developed by John von Neumann and Oskar Morgenstern in their book *Theory of Games and Economic Behavior* (1944), evolutionary games address populations of organisms, individuals or strategies. The method of evolutionary game analysis came to maturity in biology (Maynard Smith, 1964, 1982; Maynard Smith and Price, 1973). Subsequently, it was also applied to economic problems, by authors like Reinhard Selten (Economic Nobel laureate) and Kenneth Binmore (an early survey is Friedman, 1991).

Three assumptions characterise evolutionary games. In the first place, behaviour is characterised by rules, norms, imitation, trial and error, mistakes or analogy with similar situations. Second, the context is non-cooperative and there is no influence on others' actions. And third, situations should be frequently occurring, so that learning and selection work. This means one-shot games are excluded. Nor are repeated games among a given set of players relevant. Instead, individuals from large population are randomly paired.

The central concept in evolutionary game theory is an evolutionarily stable state or strategy (ESS). Its definition is as follows: given fitness function $f(r,s)$ state s is an ESS if $f(s,s) > f(x,s)$ or $f(s,s) = f(x,s)$ and $f(s,x) > f(x,x)$. An ESS is able to resist all intrusions of 'mutations' either because they are less fit in general or when they are widespread in the population. The conditions can be seen as combinations of Nash equilibrium[3] and stability-type requirements. Stability relates to disturbance by a small change in the population distribution, caused by mutations, recombination or immigration.

The dominant type of game studied in the literature is symmetric bi-matrix games. Strategically there are three types: a prisoner's dilemma,[4] which has a unique

[3] In a Nash equilibrium economic agents, given the choices of others, cannot improve their profit or utility by changing their decision.

[4] Two suspects of a crime who are separately interrogated by the police face the dilemma that, whatever strategy the other suspect follows, each is better off confessing. As a result, they do not reach the most desirable outcome for both, which is denying having committed the crime – this would require some form of cooperation, coordination or mutual trust. A similar game problem appears in many other settings, such as

equilibrium in pure strategies which is ESS, namely defection; a hawk–dove game, which has three equilibria, of which one in mixed strategies is ESS, and two in pure strategies is not; and a coordination game, which has three equilibria, of which two in pure strategies are ESS and the mixed strategy is not ESS. Many other games build upon these results and can, in fact, be seen as combinations of them. Gintis (2000) presents a rather complete array of examples of games. For applications in biology, see Nowak (2006).

According to Maynard Smith and Szathmáry (1995, p. 261) the evolutionary importance of the prisoner's dilemma has been overestimated. They make a distinction between sculling and rowing games. The first is an example of the familiar prisoner's dilemma, whereas the second one is perhaps a more general description of interactions among humans. Examples of related payoff matrices are ((3,3) (0,5); (5,0), (1,1)) and ((3,3) (0,1); (1,0), (1,1)), respectively. These show that whereas in the prisoner's dilemma defection, i.e. non-cooperation, is rational since it is the best strategy whatever the other player plays, in the rowing game both cooperating and defecting are ESS, with cooperating having higher payoffs.

The power of evolutionary game theory has to do with its analytical approach. This is due to the fact that it adopts an aggregate approach to describing evolutionary economic phenomena. Groups are considered to consist of identical individuals; in this sense, evolutionary game theory really is a compromise between representative agent and fully fledged evolutionary models. A related feature is that interactions among individuals and with the environment are only described implicitly and jointly. In evolutionary games, the replicator equation is the most common representation of selection dynamics (Taylor and Jonker, 1978). The basic replicator equation formalises the idea that individuals with above-average (below-average) fitness will increase (decrease) their proportion in the population, or: $dx_i/dt = x_i[f_i(x) - \bar{f}(x)]$ with $\bar{f}(x) = \sum_{j=1}^{n} x_j f_j(x)$, where x_i is the share of individual i (=1,..., n) in the population, $f_i(x)$ represents the fitness of individual i, and $\bar{f}(x)$ denotes the weighted average fitness of all individuals in the population. Individual fitness depends on the population distribution of characteristics, or the diversity of individuals, because of their interaction (e.g. competition). The aggregate character of a replicator equation allows for multiple interpretations of selection, namely as exit of the system, individual learning or imitation. Other than replicator dynamics have been studied as well, such as best-response, imitation or adaptive dynamics, and various types of stochastic dynamics (Nowak, 2006, Chapters 2–4; Safarzynska and van den Bergh, 2010c).

Evolutionary game theory focuses on the existence of asymptotic equilibria. These are possible because no attention is given to a structural process of diversity generation, as a result of which selection completely dominates system dynamics. In other words, the interaction between innovation and selection, typical of a complete evolutionary process, is missing. Innovations are treated as shocks to the system dominated by selection dynamics. A more suitable name for evolutionary game theory, therefore,

price competition in markets, voluntary contributions to a public good or solving environmental problems such as global warming (Chapter 14).

would be 'selection game theory'. Biological models exist, though, that combine innovation and selection in an abstract manner as found in evolutionary games. These include the quasi-species equation, replicator–mutator dynamics and recombination dynamics (Jacobi and Nordahl, 2006; Nowak, Chapter 3). These can be transferred to economics if appropriate changes are made (Safarzynska and van den Bergh, 2011a). The resulting models can address mutations or recombinant innovation, which allow the consideration of continuous innovation processes or the introduction of mistakes due to the bounded rationality of agents. The realism of such models is that they explain under which conditions diversity in systems can remain high over time. They have been applied to address questions of optimal diversity of investment strategies for a transition to renewable energy (Safarzynska and van den Bergh, 2013).

Discrete Mathematics and Agent-based Models

To contrast the approach of evolutionary game theory, it is worthwhile to consider Potts' (2000) methodological proposal for evolutionary economics. In his view, economic systems are complex 'hyper-structures', i.e. nested sets of connections among components. Economic change and growth of knowledge are thus in essence a process of changes in connections. In line with this, Potts calls for a new microeconomics based on the technique of discrete, combinatorial mathematics. As opposed, traditional microeconomic equilibrium theory in essence assumes a continuous reality, which is convenient as it allows the application of techniques like integral and differential calculus. But these cannot well address heterogeneity, networks, modularity and decomposability which are typical of evolutionary systems (Birchenhall, 1995). Potts makes the ambitious claim that discrete mathematics operationalised through agent-based models can address the core of all main heterodox economic schools, which have, in distinct ways, tried to provide an alternative for neoclassical economics. Arguably, the lack of a consistent formal approach has hampered broad acceptance of heterodoxy, because it gave rise to an unpractical heterogeneity of concepts, theories and methods. With agent-based modelling a unified heterodox microeconomics may be in reach, which would come down to an evolutionary approach.

Agent-based models – sometimes referred to as 'artificial life' or 'artificial world' simulation models – include microscopic modelling (Levy et al., 2000), social simulation (Troitzsch et al., 1996), modelling of complex adaptive systems and genetic algorithms (Holland, 1975; Janssen, 1998, 2002). Potts argues that such models allow addressing complexity, defined as a dynamic interpretation of balance between order and chaos and a sort of a counterpart of equilibrium in a static setting. Here he closely follows the work of Kauffman (1993), who proposed that order relates to the presence of few connections among system components, and chaos to the presence of numerous connections. Complexity is then associated with a certain range (of the number of connections) in between. 'Underconnection', caused by e.g. path dependence, means an inflexible, non-adaptive system; 'overconnection', like in some unregulated financial markets, means continuous and unpredictable change (chaos); complexity, in between, means a system with a relatively stable structure that has the capacity, within boundaries, to adapt to external changes.

With the advent of the personal computer, agent-based modelling – involving a numerical V-S-I-R approach (Section 2.1) to the study of interactions among large numbers of adaptive agents – has become easier. Epstein and Axtell (1996, p. 4) refer to the multi-agent models as 'laboratories', because they can 'grow' social structures in the computer and create an artificial history. Some basic questions when designing a multi-agent model are:

- Who are the agents or what are the microscopic elements? Which elementary behavioural characteristics do agents have? And how do they interact?
- Do agents move, and if so, what makes them move, and how quickly and far can they move?
- Do agents appear and disappear? In other words, can they exit or enter, and be born or die?

Schelling (1969, 1978) is regarded to have offered the first application of multi-agent modelling to social-economic phenomena. He developed a spatial model of local interactions among agents who prefer some fraction of their neighbours to have the same 'colour' as themselves. Despite computational power at the time being limited, he was able to show that the result is segregated neighbourhoods. The well-known Sugarscape agent-based model developed by Epstein and Axtell (1996) covers many issues within the broad scope of social science, such as economic trade, cultural change, conflict, exploitation of renewable resources and the impact of pollution. It is a showcase for what can be achieved with evolutionary multi-agent models. Sugarscape refers to sugar as the renewable resource and scape as the landscape in which activities take place. Individuals are characterised by sex, metabolic rate and vision (myopia), which are fixed for life, and vary among individuals. On the other hand, their preferences, wealth, cultural identity and health can change. Agents are born and can die. In certain versions of the model, birth can result from mating and involves genetic inheritance. Behaviour is assumed to be boundedly rational, expressed by myopia in space and time. The natural environment provides the sugar which the agents need to survive, which is unevenly distributed over space. In some model versions, a second resource, namely spice, is available which leads to local trade and prices. The landscape is the medium in which interaction and moving occurs. A spatial grid allows tracing the development of networks of communication, relationships (friends, family) and trade. As a result, evolution in Sugarscape is multilevel and occurs in three ways. First, selection changes the distribution of fixed characteristics – one might say that society learns. Second, behavioural rules are adapted due to experience and learning, or imitation of other individuals (culture). Third, transmission of genetic traits occurs via sex. Some results of Sugarscape are as follows: carrying capacity becomes evident, i.e. the environment can only support a limited number of people; migration occurs, in response to seasonal patterns or exhaustion of the sugar resource; selection changes the population distribution of fixed agent characteristics, increasing the frequency of low metabolism and less myopic individuals; the distribution of wealth tends to be highly skewed, owing to differences in capabilities and initial positions in the Sugarscape; spatial concentration occurs, or subpopulations emerge; the resulting congestion causes carrying capacity to go down; finally, the inclusion of pollution as an externality associated with sugar use reduces concentration (density) of people.

A version of Sugarscape with two commodities is particularly interesting as it introduces real economics, namely trade and negotiated local prices. Agents trade at non-equilibrium prices. Internal valuation is based on marginal rate of substitution (MRS) of spice for sugar (the marginal value of sugar in spice units). The MRS is based on a combination of metabolism (biology) and preferences (culture). Local trade leads to an approximation of a local Pareto optimum. As a result, trade increases the carrying capacity of the system, but it also can contribute to more social inequality. Furthermore, continuous preference change stimulates trade since possessions may no longer satisfy wants, in turn destabilising prices. Local trading prices cause total trade to be smaller than demand and supply under perfect coordination and a (single) equilibrium price. Finally, local trading prices may statistically converge, or not, depending on how information diffuses. Anyhow, no *ex ante* equilibrium price is assumed.

Because of its numerical approach, agent-based modelling is capable of addressing a wider variety of assumptions than is common in economics, including those that do not allow analytical treatment. While insights obtained with numerical analysis are less general, this can be compensated by sensitivity analysis using parameter ranges informed by empirical data. This, however, often runs into limits as empirical information on individual-level data – needed to know the actual diversity of behaviours, wealth and so on – is limited or classified. Agent-based models are now applied to deal with original research questions in a variety of areas: finance, labour, industry dynamics and energy (Heath et al., 2009).

8.6　Evolutionary Growth Theory

The analysis of economic growth is dominated by aggregate models of exogenous and endogenous growth in equilibrium. Although these have generated many clear insights, they suffer from two problems. First, they leave out relevant issues because of their bare-bone description of reality. As a result, certain policy-relevant aspects disappear from the radar. Second, they make many assumptions that are convenient but erroneous, causing its results to be questionable. Examples of debatable assumptions are representative agents, rational behaviour, perfect information, an aggregate production function, growth in equilibrium and reversible growth.

The attractive feature of evolutionary growth theory is that it integrates micro and macro aspects of technology and its change over time. Changes in the state of a sector follow probability rules, modelled as a stochastic process with time-dependent probabilities that depend on search behaviour, imitation, investments, entry and selection. If firms make sufficient profits, then they do not search or imitate others, otherwise they do. Search is local, implying small improvements and staying close to the present production technique. Imitation can focus on either the average or the best practice.

Fundamental to an evolutionary approach to growth analysis is the recognition of the diversity of firms, continuous technical change, and the absence of any notion of economic equilibrium.

The first formal evolutionary model of economic growth was developed by Nelson and Winter (1982: part IV, esp. Chapter 9), combining their theory of the firm, which consisted of routines, search and selection (Section 8.4; see also Box 8.2). Their model was built on a population of boundedly rational firms, assigned a central role to investment of a part of firm profits in technological innovation, and modelled market competition as selection working upon variation of firms. Bounded rationality of firms expressed itself in rules of thumb to allocate investments between innovation, imitation and expansion of production capacity. Nelson and Winter compared the results of their model with those from Robert Solow's aggregate growth model from 1957, for which he later received the economics' Nobel Prize. This model was intended to explain empirically observed patterns of aggregate output, inputs and factor prices. Nelson and Winter criticised the Solow type of growth accounting for explaining only about 20 per cent of productivity growth, based on movements along an aggregate production function due to factor input changes, leaving 80 per cent as residual, referred to as 'technological change', but without really offering any explanation. Nelson and Winter's evolutionary model output qualitatively resembles the aggregate data used by Solow. But the results are also consistent with both firm decision-making (routines, search), with empirical observations on factor use and efficiency features across sectors and empirical patterns of technological innovation and diffusion.

Other formal evolutionary models of growth have been proposed. For example, Conlisk (1989) works with a probability distribution of productivity of firms. The growth rate can be analytically derived as being dependent on the rate of diffusion of innovations and the size of innovations as indicated by the standard error of the productivity probability distribution. Silverberg and Verspagen (1994) have proposed an evolutionary growth model that starts from the well-known Goodwin (1967) model. The latter revolves around a formalisation of the illustrious Philips curve that depicts the relation between unemployment level and wage change: the idea being that the higher employment, the higher wage increases due to inflation. In all these models, the modelling of a population of a large number of firms and their behaviour in terms of fixed rules generates industry dynamics. An important behaviour rule is that new capital follows from profit accumulation, so that relatively profitable types of capital accumulate relatively fast. This can be regarded as reflecting a process of selection in which a technique with a relatively high fitness spreads quickly. This is often modelled through replicator dynamics. In addition, the introduction of new firms and technologies in the economy follows from firms undertaking R&D to improve labour productivity, the outcome of which is stochastic (e.g. a Poisson process). The stochastic character is a way to reflect extreme uncertainty associated with innovation – in the form of surprises and ignorance. Spill-overs are taken into account, through which a firm can profit from other firms' R&D. The capacity of the respective firm to assimilate spill-overs is measured by its own R&D. Firms can employ two strategies for innovation: mutation or imitation. The probability of imitation depends on the gap between the firms' own profit rate and the maximum profit rate in the population. This follows the general evolutionary model by Iwai (1984) of innovation and imitation (see Box 8.2).

To understand better the evolutionary approach to growth theory, we consider the main similarities and differences with neoclassical endogenous growth theories (Mulder et al., 2001). Both explicitly address the fact that growth is fuelled by technical change (Aghion and Howitt 1998; Barro and Sala-i-Martin 1995), and endogenise technical change (R&D) by using outlays on R&D as a core variable. But whereas neoclassical economic growth theory defines R&D at the aggregate level, evolutionary growth theory derives production as well as technical change from the population of firms thus explicitly modelling R&D as being undertaken inside productive firms. Evolutionary models generally address behavioural heterogeneity, whereas in neoclassical theories representative or identical agents are assumed. The latter is implicit in the argument that microlevel production functions can be replicated, resulting in constant returns to scale at an aggregate level (in the absence of positive R&D externalities). While evolutionary models further assume bounded rationality, usually in the form of routines and learning through imitation, neoclassical models assume, by definition, individual rationality (marginal decision rules) and social (intertemporal) optimality. This explains the neo-classical growth theory focus on equilibrium growth paths, as opposed to the non-equilibrium features of evolutionary growth. The hybrid growth theory of Aghion and Howitt (1992) incorporates some elements of heterogeneity and destructive or vertical innovation – creative destruction à la Schumpeter – in a neoclassical type of model, but maintains the assumption of the rational agent. Mulder et al. (2001) refer to this as a 'neoclassical Schumpeterian approach'.

One may consider the idea of directed technical change by Acemoglu (2002), elaborating on Hicks' (1932) notion of induced technical change, as another hybrid theory in this vein. It suggests that innovation will focus on old technologies through a price effect in response to scarce factors or resources – e.g. energy or skilled labour – if one cannot easily substitute these by more amply available new factors. In this case, one effectively deals with complementary production factors, which moves away from standard neoclassical assumptions. On the other hand, if replacement is feasible innov-ations will be directed at technologies that use the new factor as then a positive market size effect can be enjoyed. An example is the historical transition from firewood and charcoal to coal that went along with innovations in the steam engine and textile technology during industrialisation. Note that the first type of innovation, associated with the old scarce factor, improves the efficiency with which it is used and thus its productivity, while the second increases factor and technological diversity.

A crucial concept in conventional growth theory is the aggregate production function. The Cambridge capital debate had already assessed that it was a theoretical construct which gave rise to the internal inconsistency of growth theory (Harcourt, 1972). Neither single firms nor the aggregate of all firms can move along an aggregate and continuous production function, because they possess only information or knowledge about a limited and discreet number of production techniques. This idea is also recognised by the neo-Austrian approach, which is formalised using an activity analysis type of model (Faber and Proops, 1990). Evolutionary growth theory avoids an aggregate production function by describing changing diversity of production relationships in the pool of individual firms. Consequently, aggregation is avoided at the level of assumptions (i.e. functional specifications) but achieved at the level of results (i.e. generated numbers).

It is fair to say that several of the promises of evolutionary growth theory still need to be delivered. So far, two important bottlenecks seem to have prevented progress. First, many data are needed to formulate models with technological or firm heterogeneity. Second, evolutionary growth models offer a wide degree of freedom and hence still suffer from a lack of agreement on a common approach, resulting in the inclusion of many ad hoc elements. An additional shortcoming is that most evolutionary growth models focus on diversity and innovation in a single sector context. In other words, a single population of heterogeneous firms is described. To address changes in sector structure, a multi-sectoral and thus multi-population approach would be required. The recent wave of complicated macro-evolutionary models goes in this direction (e.g. Delli Gatti et al., 2011; Dosi et al., 2013; Lengnick, 2013; Chen et al., 2014; Safarzynska and van den Bergh, 2016). These are intended to serve as an alternative to both general equilibrium and mainstream macroeconomic models. Several of them include a detailed banking sector (central and private banks), being motivated by recent macroeconomic debates on the financial crisis. As a result they are, unlike mainstream economic growth models, able to incorporate domino effects due to diffusion of information and expectation effects that characterise periods of economic downturn. This makes such models ideally suitable for studying the interaction of technological innovation, long waves and economic business cycles (Section 10.4).

8.7   The Geography of Economic Evolution

Economic structure is closely associated with spatial organisation. Assuming economic space away, or to be homogeneous, leads to a significant loss of information about diversity and dynamics in the economy. In fact, economic dynamics pertains not only to time but also to space. Moreover, spatial heterogeneity affects evolutionary temporal dynamics of economies in essential ways. The reason is that space generates both opportunities and barriers for evolutionary change. Consideration of space is much more common in evolutionary biology than in evolutionary economics. Nevertheless, regional economics and economic geography have developed their own branch of evolutionary analysis (Boschma and Lambooy, 1999). An early contribution in this vein is Pred (1966), who stressed the role of coincidence, selection and sub-optimality, presaging notions of path dependence. Spatial evolutionary economics can be seen as a merger of regional and urban economics, international trade theory, and evolutionary economics. One can identify several spatial regularities of evolutionary economic processes which illustrate the relevance and insights of this synthesis.

Regional specialisation and domination is driven by a combination of local fixed conditions (climate, geography, soil and other natural resources), comparative advantages and path dependence (including historical accidents). Domination can be with regard to products or innovation activity. To illustrate, both in the United States and Europe few regions dominate in generating patents: for example, with regard to electronics and ICT, in Europe the leading regions are Bayern, south-east England (London), Ile-de-France (Paris), and the south of the Netherlands (Eindhoven) (Boschma et al., 2002, Section 5.2.1). Halfway through the nineteenth century, particular regions dominated in traditional industries, such as mining, steel and textiles, but subsequently lost their weight: notably, the Ruhr area (Germany), Twente (the Netherlands), Wallonia (Belgium) and Birmingham/Manchester (UK). Regional specialisation involves location choice by firms. The product life-cycle theory of Vernon (1966) suggests that this depends on the phase of this cycle. When products are not standardised but still are subject to frequent changes, location near local suppliers and customers has the advantage that quick and intensive exchange of information needed for deciding about such changes is feasible. When products and production processes are more standardised and stable, this proximity is less important. Moreover, the associated larger firms and demand make it more difficult for firms to be proximate to all their customers and suppliers. As price and cost competition will, in this phase, be driving the sector dynamics, firms will give much weight to the availability of low-wage labour in their location decisions. The latter explains why multinationals often opt for locating factories in countries with emerging economies. Regional specialisation further involves the development of young, starting companies. If these are under the umbrella of a larger, international company, they can adopt its routines (DNA), but over time these will adapt to the local conditions and challenges (for more details, see Boschma et al. 2002, Section 5.1.3).

Spatial interaction and diffusion form another central theme. Diffusion of knowledge, innovations or technology always has a spatial connotation and meaning to it. Diffusion occurs through communication of agents that are generally located at different, possibly distant, points in space. Diffusion follows a spatial pattern limited by the spatial structure or network of connections among agents. Here, local interactions often matter; proximity often means more similarity; and since imitation or copying is bound to involve errors, larger distances imply an accumulation of errors of local imitation and thus larger differences (in economic or cultural features). However, space plays a different role in economics than in biology as a range of modern communication technologies and human intelligence allow for direct non-local, and even global, interactions.

Spatial isolation is an important factor underlying 'economic speciation', giving rise to distinct cultures and economies, characterised by unique market or trade regimes. Note that both in biology and economics, factors other than space can isolate populations, such as behaviour, time (being active during night time instead of during the day) and subcultures (different religions, languages, dress codes). However, often these factors have a distinct spatial or network component: individuals of different groups, religions or cultures can live in the same area but hardly meet and interact because they avoid being at the same place at the same time.

Many spatial structures result from evolution implying multiple potential outcomes, which may take the form of spatial uniformity, spatial polarisation and spatial complementarity. Spatial uniformity can result from similar but independent developments or from spatial diffusion. Late adopters in more peripheral regions can enjoy learning benefits and use more efficient technologies, not being bothered by large capital invest-ments in the past (sunk costs). Major innovations that cause creative destruction often go along with a geographical relocation of activities and shift in dominant regions. This is well illustrated by the rise of high-tech, notably micro-electronics and computer-related hardware and software firms in Silicon Valley (an unofficial name for the southern part of the San Francisco Bay Area in California).

Certain patterns get stuck in a spatial structure that has all the features of a lock-in, which can be explained in terms of increasing returns to adoption, related to agglomer-ation or urbanisation economies (proximity of other activities), rigid relations between firms and their suppliers, and regional politics based on consensus and vested interests (Olson, 1982). Existing infrastructure and cities create positive externalities, due to existing services and cheap connections, such that new firms and households may find it attractive to be near them. The product life-cycle approach, applied to a regional scale, implies that old industrial regions (e.g. the Ruhr area) may be hampered in their innovative potential by unfavourable cost structures, oligopolistic markets preventing entry of 'innovators' and labour markets (and education) biased towards the old industry. More generally, given the typical historical process (path dependence) of industrial regions, more favourable regional features cannot be simply copied and implemented but need a long gestation period. For instance, a so-called eco-industrial park with unusual recycling and industrial symbiosis patterns, such as Kalundborg in Denmark has a coincidental and unplanned history, and cannot be simply imitated by other regions striving for better environmental performance (Jacobsen and Anderberg, 2006).

Lock-in further relates to historical spatial contingencies and 'accidents': natural infrastructure such as ports, rivers, climate, natural resources; and barriers like mountain ranges, rivers, seas and lakes, etc. cause some economic outcomes to be more likely than others. The importance of such factors was, of course, recognised long ago in regional and international economics, reflected by the notions of comparative and absolute advantages. An evolutionary perspective can be seen to add a historical dimension to this. In addition, spatial accidents involve a critical mass of successful firms breeding a derived industry, as happened with the car industry and its suppliers in Detroit, after the 'Big Three' (General Motors, Ford and Chrysler) started to dominate the car industry and influenced each other in terms of technology and management. This, combined with Detroit being close to relevant activities (coal, iron and copper mining), and easily accessible by water and land, created the right conditions for Detroit's dominance in the auto-making industry within the United States.

Boschma and Lambooy (1999) suggest a temporal phase, called a 'window of locational opportunity model', to explain escape from lock-in. This can involve 'spatial leapfrogging' driven by local positive externalities due to the proximity of many different economic activities/industries. More generally, coincidences, creativity and agglomeration effects can stimulate new sectoral developments in regions that alter spatial dominance patterns, though with a small probability – witness the long duration of industrial dominance of many regions. Beyond the early development of a new industry, the window of opportunity will be closed and spatial dominance will be fixed.

Long-established centres of production are important factors in communication networks among firms, whether competitors or suppliers. This gives rise to localised 'collective learning', also known as the Silicon Valley effect. Local interactions matter as firms located near to each other have a high chance of hearing sooner about relevant information than others. Inventions and innovation are therefore often concentrated in regions. Terms such as 'learning regions' and 'technology districts' reflect this (Castells, 1996). These are characterised by a high rate of inventions and innovations, and local spin-off of small risk-taking firms from larger ones. The internet hosts many such innovators, previously working and building useful experience and networks in either product- or (computer) technology-oriented companies. Ironically, at the same time the internet seems to reduce the need for local contacts and proximity of firms in a communication network.

Finally, connecting to the previous section on growth and the above notions of spatial evolution, we consider briefly how evolutionary economics looks at the issue of convergence of rich and poor countries, or catch-up of poor countries, in terms of incomes, which is at the core of growth and development studies. The arguments about regional dominance and spatial lock-in point in the direction of persistent divergence, whereas those about spatial interaction and diffusion suggest convergence is possible. Specific features of technologies and market barriers, associated with sectors and products, play an important role in determining what is more likely to happen (Fagerberg, 1988). Convergence in terms of average income will mean that lagging countries will have to develop activities in industries that are at the beginning of a life cycle, likely with considerable involvement by national governments in the form of industrial policy (infant industries), technology subsidies and appropriate education strategies. Examples of successful cases in this respect are South Korea's chaebol (business conglomerates), such as Samsung, Hyundai and LG, which have, over time, received intense support from the South Korean government.

Here we conclude the account of evolutionary economics, which has illustrated that it is characterised by a diversity of specific approaches. Their common denominator is a real micro-level approach, in the sense that diversity among individuals is not hidden behind representative agents. Moreover, they all share a focus on multiple agents, which integrates evolutionary and behavioural approaches. Evolutionary economic change is not identical with progress, because of bounded rationality, coevolution, local adaptation and path dependence, among other things. Adequate policies, as discussed in the final part of the book, can though help economic evolution go in the direction of what one would consider progress.

9 Evolution of Organisations and Institutions

9.1 Theories of Organisations

Modern organisations deliver products and services by combining complementary skills of individuals. They are equipped to realise useful adaptations and innovations in the face of known and unforeseen changes in their economic environment. Organisations are a logical response to the goal of reducing uncertainty, as well as transaction and operational costs associated with production and sales. This is recognised by many classic authors on firm organisation theory, such as Ronald Coase, Armen Alchian, Harold Demsetz and Oliver Williamson. Transaction costs arise from informational imperfections or asymmetries, and involve defining, as well as maintaining, rights. Transactions cover search, information acquisition, negotiation, transfer, monitoring, and legal protection and enforcement of rights. Organisations limit transaction costs by coordinating activities by many individuals in space and over time.

Harold S. Geneen, former chief executive officer of International Telephone and Telegraph Corporation (ITT), once said: 'Every company has two organisational structures: The formal one is written on the charts; the other is the everyday relationship of the men and women in the organisation.' Indeed, organisations often are black boxes to outsiders, which makes them a challenge to those trying to understand their structure, logic, causality and dynamics. It explains why human organisations have been studied from many different angles, including evolutionary ones. We review and compare these here. In addition, attention is devoted to a related topic – self-organisation. This denotes a set of theories explaining complex systems and their dynamics, whether pertaining to abiotic systems, living organisms or human organisations. In this setting, some speculative thoughts will be offered on the connection and differences between self-organisation and evolution.

Organisations come in a number of types. One can distinguish between firms (business corporations), unions, political parties, local, regional, national and international public governance, educational and research organisations (schools, universities), advisory boards, sports and leisure activity clubs, non-governmental organisations, prisons, hospitals, etc. Some of these are more bureaucratic (very large firms, government agencies), while others are subject to strict surveillance (prisons), and still others depend strongly on individual freedom and informal relationships (university research). The commercial firm is the organisational form most frequently encountered in current economies. Many theories exist about it within sociology, economics and business management. Firms are characterised by a large diversity of organisational structures. Their organisations are not planned but evolve, learn and adapt to changes in their environment. They are part of a population in which they struggle against external factors, including competition from similar organisations, to survive. This suggests that evolutionary thinking can offer a useful perspective on organisational structure and dynamics.

Organisations are often associated with bureaucracy. Indeed, large organisations tend to be bureaucratic, being guided by formal rules and procedures. Although bureaucracy has a negative connotation, implying inefficiency or even ineffectiveness, the famous sociologist and early writer on organisations Max Weber (1864–1920) compared the ideal type of bureaucracy to a sophisticated machine. Bureaucratic organisations have existed at least since Imperial China but have become a common phenomenon during modern times. Giddens (1997) characterises large organisations as having many of the following features: hierarchical, i.e. authority and a chain of command from top to bottom; having formal rules that are written down; subject to controls, checks and surveillance of goals and output as well as performance of individuals; having a separate administrative organisation with individuals employed fulltime as officials and managers; showing a clear separation between work in the organisation and private life outside it, which allows individuals to be a member of many organisations; and having a separation of ownership and control. To understand the great variety of organisation forms, business management has developed various organisational theories (Keuning and Eppink, 1996; Pugh, 1997):

1. *Scientific management*: Old school, starting with Frederick Taylor (1856–1915), expressing the idea of healthy management characterised by scientific identification of objective norms, selection of the right employee for the right job, manufacturing supported by labour division and specialisation, and efficiency and motivation through incentive wages.

2. *Management processes*: Based on Henri Fayol (1841–1925), this identifies six core functions in each company – technical, commercial, financial, safety, accounting and management; management links the other functions through the tasks of planning, prediction and analysis, design of organisation, guiding and motivating personnel (staffing and directing), coordination of activities and controlling.

3. *Strategic management*: Motivated by authors such as Igor Ansoff, Henry Mintzberg and Michael Porter, the external organisation and the role of environmental change receive much attention; organisations are seen as flexible and adaptive, but also goal-oriented in their decision-making and planning. This can provide linkages with an evolutionary approach.

4. *Administrative behaviour and decision theory*: Based on Herbert Simon's bounded rationality as well as the work of Richard Cyert and James March (1963); it has been said to match evolution as a blind process in which lucky organisations, coincidentally adapted to their environment, survive, while others do not (see also Box 1.1).

5. *Environmental theories*: A variety of approaches, including evolutionary ones based on population ecology (Hannan and Freeman, 1977), organisational fit

(Miles and Snow, 1984) and cultural environment (Hofstede, 2001). These are discussed in Section 9.2.

6. *People and personnel*: Thinking in terms of human relations, addressing issues such as health, safety, joy in working and interaction among people; this suggests approaches to select personnel, manage, direct, give incentives, etc.

7. *Systems theory*: Using tools of operations research and cybernetics, systems models describe information, physical and monetary flows and stocks, their relationships and feedback mechanisms; makes use of communication theories to explain information flows.

8. *Contingency theory*: This argues that no general model for analysis or design of organisations exists, since context and situation are essential. This provides some linkages with evolutionary theory, notably through a historical case study orientation.

Organisations delay and push economic and social changes and hence should receive serious attention in evolutionary social science. Opportunities for organisational change depend on the form of organisations –hierarchical, line-staff, matrix or flat – and on the geographical shape. New elements can appear or old ones can disappear, associated with staff, technical support, R&D, production and sales.

Evolutionary, V-S-I-R type of theories (Section 2.1) of organisations have been around for some time. Work by Schumpeter and Nelson and Winter in economics has already been discussed (Section 8.4). In addition, there is a large literature in sociology. Romanelli (1991) provided a broad overview of the early literature. He distinguished three approaches to variation in forms and other characteristic traits of organisations: (i) 'organisational genetics' stressing random variation; (ii) 'environmental conditioning' where variation is contextually constrained or even driven; and (iii) a social systems view, according to which variation is due to social-organisational interactions. I am not sure I can agree with this classification, though, as it somehow artificially separates the three core aspects of evolution: variety creation as stressed by (i), selection as stressed by (ii), and interaction of units in a population as stressed by (iii). I therefore find this distinction or characterisation of approaches rather unproductive. Instead, I see elements of (i)–(iii) jointly appear in many theories, which is logical as otherwise one cannot arrive at a genuine and complete evolutionary approach.

9.2 Population Ecology of Organisations

Population ecology of organisations is a field of research that uses models from ecology and evolutionary theory to understand the sociology of organisations (Hannan and Freeman 1977, 1989; Hannan and Carroll, 1989). It has both a theoretical and an applied dimension. The main motivation for this approach is that a wide diversity of organisational structures can be observed in reality, not only between but also within sectors. For firms of the same type, i.e. in the same sector, this is explained by these having faced and adapted to distinct environmental conditions. However, this does not

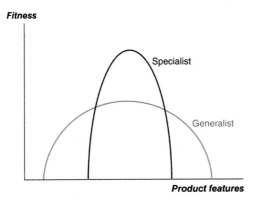

Figure 9.1 Resource niches of generalists and specialists.

explain that some firms are found to differ much even when their environments are rather similar. Additional explanatory factors are history (path dependence) and idiosyncratic behavioural or other constraints. Against this background, organisations are regarded as complex systems that have a limited capacity to respond and adapt to changes in their environment.

Population ecology tries to get a grip on these issues by combining evolutionary theory, population ecology, economics, sociology, and behavioural and organisational theories. Central are population and evolutionary dimensions, developed around the notions of variation, selection, competition, cooperation, fitness and adaptation. Much attention is given to the life cycle of organisations, including birth (founding) and death (failure). This involves linking of organisation theory and social history, by regarding diversity of organisations as the result of an accumulation of innovation and selection processes over a long period of time. Typical ecological concepts are used, such as niche and *r*- versus *K*-strategists. The latter are aimed to capture generalist versus specialists with regard to organisational strategies (Figure 9.1).

Some authors in economics have considered population issues in relation to industry. George Stigler (1958) introduced the survivor technique, which states that the size of surviving firms can be taken as an indicator of the minimum efficient scale. According to this idea, changes in population distribution would imply changes in scale economies. Baumol et al. (1982) have considered populations of firms in their theory of contestable markets. However, these approaches do not include explicit evolutionary mechanisms causing increases and reductions in diversity of organisational forms.

Hannan and Carroll (1989) argue that the use of population ecology models in organisational studies should not be considered as focusing on metaphors. Neither in nature nor in the economy do populations exactly follow growth curves as predicted by evolutionary models. In other words, the only metaphor is that of a mathematical model. Key notions in population ecology are as follows:

- *Levels of organisation*: Individual persons, plants, firms or populations of organisations with overlapping realities (e.g. markets).

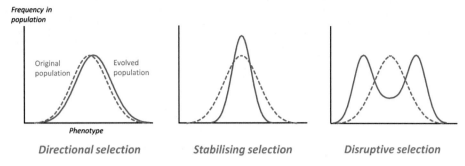

Figure 2.1 Change in phenotype distribution under three selection modes.
(*A black and white version of this figure will appear in some formats.*)

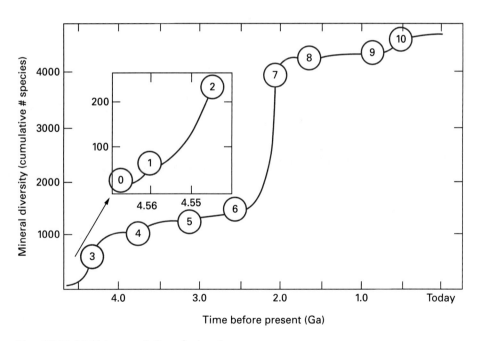

Figure B2.1.1 Multistage evolution of minerals.
Note: Biological processes appear in Stage 6.
Source: http://hazen.carnegiescience.edu/research/mineral-evolution. With permission of
R.M. Hazen.
(*A black and white version of this figure will appear in some formats.*)

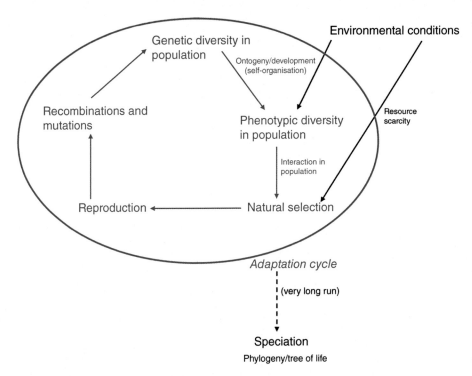

Figure 3.2 Neo-Darwinian evolution as a dynamic system.
(*A black and white version of this figure will appear in some formats.*)

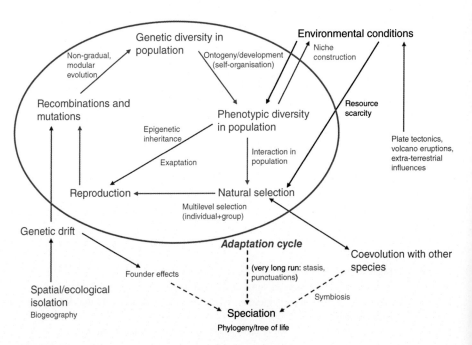

Figure 4.1 An extended dynamic systems view of evolution.
Notes: Extension of Figure 3.2 (blue and black objects) shown in purple. Broken arrows indicate effects over longer time periods.
(*A black and white version of this figure will appear in some formats.*)

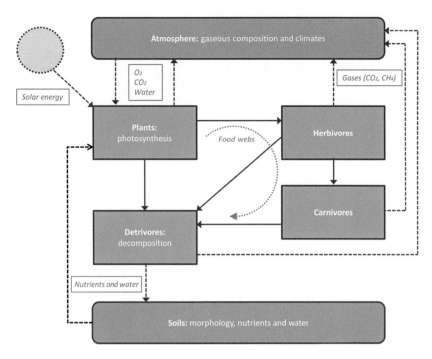

Figure 4.3 Main relationships between biotic and abiotic factors in a terrestrial ecosystem. (*A black and white version of this figure will appear in some formats.*)

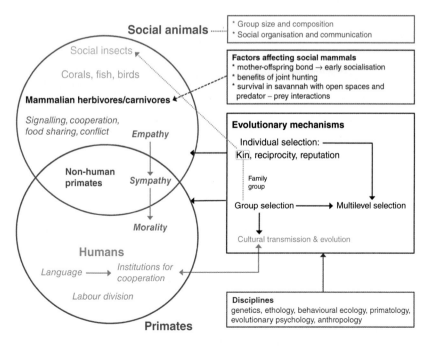

Figure 5.1 Evolution of social behaviour. (*A black and white version of this figure will appear in some formats.*)

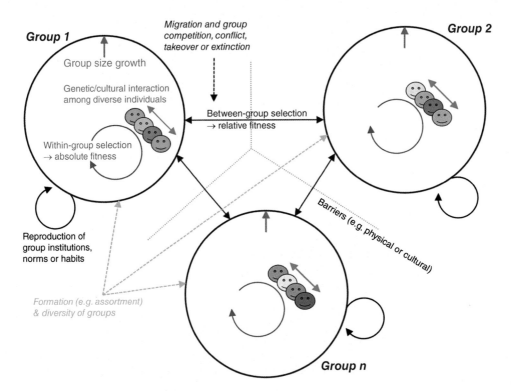

Figure 6.1 Multilevel selection.
Notes: The scheme depicts the evolutionary dynamics of *n* groups (with only 1, 2 and *n* shown) that are subject to between-group selection (red arrows) and within-group individual selection (green arrows). The meaning of coloured arrows or lines is explained by text in the same colour.
(*A black and white version of this figure will appear in some formats.*)

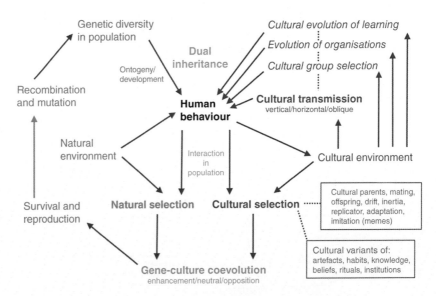

Figure 7.1 Combined cultural and biological evolution of human behaviour.
Notes: Green text and arrows denote biological evolution and blue ones cultural evolution, while blue-green is reserved for combined ideas.
(*A black and white version of this figure will appear in some formats.*)

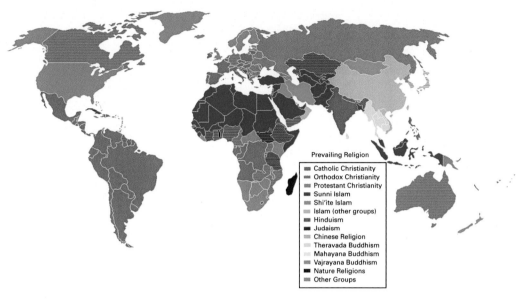

Figure 7.2 Geographical distribution of main religions in the world.
Source: https://en.wikipedia.org/wiki/File:Prevailing_world_religions_map.png; part of Wikimedia Commons, a freely licensed media file repository.
(*A black and white version of this figure will appear in some formats.*)

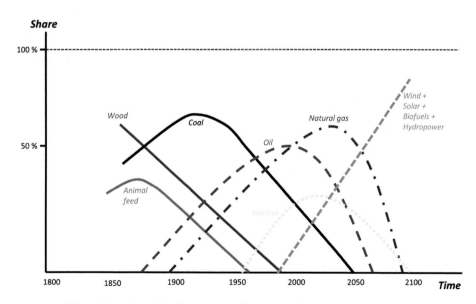

Figure 10.3 Historical and ongoing long waves of energy carriers.
(*A black and white version of this figure will appear in some formats.*)

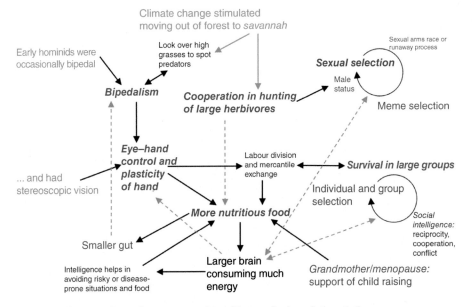

Figure 11.4 Systems dynamics representation of human brain–mind evolution.
Note: Bold, italics terms denote factors emphasised in the literature. Purple items are more specific factors. Some arrows are broken to avoid confusion about the direction of intersecting arrows.
(*A black and white version of this figure will appear in some formats.*)

Figure 11.5 Distinct attitudes to humans of dogs and cats.
Source: http://bizarrocomics.com. © Dan Piraro.
(*A black and white version of this figure will appear in some formats.*)

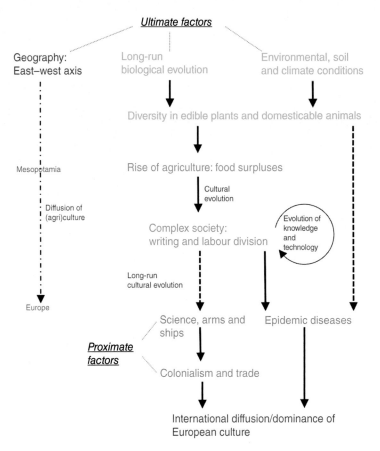

Figure 11.6 Proximate to ultimate factors behind cultural dominance of Europe.
Source: Elaboration of Diamond (1997a, Figure 4.1, p. 87).
(*A black and white version of this figure will appear in some formats.*)

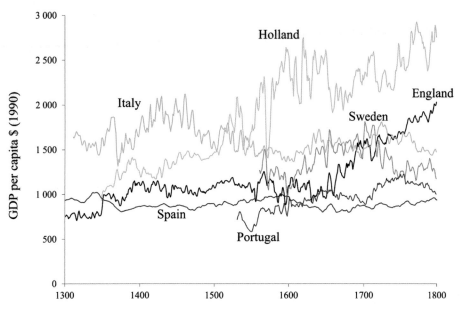

Figure 12.1 Gross domestic product (GDP) per capita in selected European economies,
1300–1800.
Note: Curves represent a 3-year average and for Spain an 11-year average.
Source: Fouquet and Broadberry (2015, Figure 1, p. 230). © 2015 American Economic
Association.
(*A black and white version of this figure will appear in some formats.*)

Figure 13.1 Devolution – evolution of environmental problems.
Source: http://bizarrocomics.com. © Dan Piraro.
(*A black and white version of this figure will appear in some formats.*)

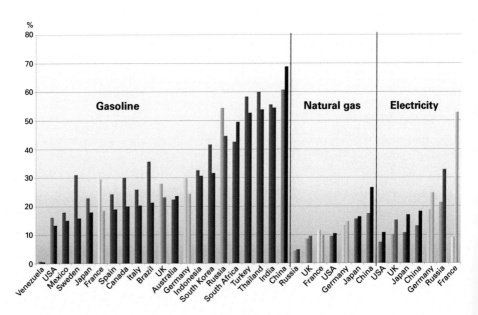

Figure 14.1 Re-spending rebound, for three energy carriers.
Notes: National averages, 2009; rebound for energy in first columns, and for carbon dioxide emissions in second columns; colours denote countries (standard blue means data was only available for gasoline).
Source: Antal and van den Bergh (2014, Figure 1, p. 586).
(*A black and white version of this figure will appear in some formats.*)

- *Organisational diversity*: Relevant for responding to uncertain future events and changes. Diversity means alternative options or solutions to certain problems and can be considered as a hedge against future risk and uncertainty. The diversity of organisations can be partly explained by the variety of individuals in it with different interests, tastes and capabilities. Diversity will generally be less pronounced in stable environments than changing ones.

- *Niche*: This is a protected sub-environment or competitive activity range. A niche is created by segregation, in particular resource partition, whereby firms divide the market by specialising in distinct, possibly somewhat overlapping, subranges of the total range of product features. One can distinguish between specialists with a narrow niche and generalists with a broad niche (Figure 9.1).

- *Selection*: This comprises internal factors, such as hierarchy, information, communication, culture and politics, as well as external factors, such as competition with other organisation types and regulation.

- *Organisational genetics*: This term has been used in analogy to the genetic structure in living organisms, as able to produce organisational forms and structure in these organisms. Blueprints might serve as the analogy since they influence or restrict the type of organisational structure that can form, and do not have a deterministic relationship with organisational phenotypes due to environmental influences. Operationalising this approach, as in biology, is hampered by the fact that complete blueprints for organisations do not exist or are unobservable.

- *Isomorphism*: It expresses the idea that under similar environmental conditions, organisations may evolve into similar forms. This notion is similar to convergent evolution in biology.

According to Hannan and Freeman (1989), there are two types of evolutionary theories of organisational change. The first, referred to as 'selection or population adaptation theories', is based on the idea that diversity arises mainly from new organisations. Existing organisations adopt a certain structure early in their lives and rarely change it, since they are more rigid than flexible. Indeed, major innovations in organisational structure and strategy occur early in the life cycle of firms and sectors. Given this setting, adaptation occurs at the population level, which involves the founding of new organisations and the demise of old, non-adapted organisations. On the other hand, in 'individual adaptation theories' individual organisations are assumed to respond and adapt to environmental changes, threats and opportunities by adapting their strategy and structure. The largest and oldest organisations, therefore, have superior capacity for adaptation. Specific theories in this respect are contingency theories (particular environments stimulate particular organisational adaptations), resource dependence theories (adaptation is aimed at neutralising environmental uncertainty), institutional theories (adaptation is guided by prevailing norms) and Marxist theories of organisations (organisations adapt so as to maintain control over labourers).

A study by Hannan and Freeman (1989, Chapter 9) illustrates the population ecology of founding of labour unions. It considers national labour unions in the United States over the period 1836 to 1985, during which 479 unions were founded, with the most dynamic period being 1883–1906. A method of arrival processes, involving queuing theory, is taken as the starting point to specify a model for statistical regression analysis. The main findings are that the founding rate of labour unions depends on the number of foundings in the previous period and increases at a decreasing rate with increasing density. Furthermore, competition among different organisational forms is found to affect the founding process. In particular, the population of unions is split into two subpopulations, craft and industrial unions, which are shown to have significantly different founding processes. The analysis shows that there is competition: initially, increasing density of industrial unions raised the founding rate of craft unions; beyond a certain point the impact is reversed, i.e. more industrial unions depresses the founding rate of craft unions.

An interesting question is whether initial founding conditions are crucial and which factors make for successful founding firms: pre-defined organisational form, employment of excellent scientists, incentive wage system, non-bureaucratic management, etc.? Arguably, this has to do with the relationship between these factors and derived features, notably adaptive flexibility and profitability. Inflexibility can arise through rigid (employment or other) contracts or autocratic and bureaucratic management. Other factors underlying flexibility are learning through networks with other companies and costs of organisational change. The evolution of organisational flexibility, analogous to phenotypic plasticity in living organisms, has adaptive value. Flexible organisation can better perform and survive under distinct environmental – micro-, meso- and macro-level – circumstances.

Organisational evolution means a change in forms, elements and connections. In addition, it can involve re-allocating resources or even splitting off parts. This can be a useful strategy if group size becomes too large. Similar phenomena are found in social animal groups such as primates. There is a large variation in organisational structures. This applies to various characteristics:

- *Forms*: hierarchical versus line-staff, fixed versus matrix; various geographical forms.
- *Core elements or units*: strategic top, staff, technical support, R&D and core operations (production, sales).
- *Connections*: formal versus informal; individual versus team; and fixed versus variable (matrix).

Table 9.1 illustrates the change in the form of business organisations over a long period of time, starting shortly after the Industrial Revolution at the end of the eighteenth century.

Organisations consist of formal and informal relationships. In the short run, the formal ones are fixed and create stability, while the informal ones allow flexibility. Over the long run, organisational change involves new formal relationships and, along with these, new informal relationships. The formal relationships can be regarded as partly the result of evolution of a population of organisations, while the informal relationships are subject to evolution at the level of individuals and groups within the organisation. The latter suggests that insights from group selection (Chapter 6), so far

Table 9.1 Evolution of organisation forms

Period	Product-market strategy	Organisation structure	Inventor or early user	Core activating and control mechanisms
1800–	Single product or service Local/regional markets	Agency	Numerous small owner-managed firms	Personal direction and control
1850–	Limited, standardised product or service line Regional/national markets	Functional	Carnegie steel	Central plan and budgets
1900–	Diversified, changing product or service line National/ international markets	Divisional	General Motors, Sears, Roebuck, Hewlett Packard	Corporate policies and division profit centres
1950–	Standard and innovative products or services Stable and changing markets	Matrix	Several aerospace and electronics firms (e.g. NASA, TRW, IBM, Texas Instruments)	Temporary teams and lateral resource allocation devices such as internal markets, joint planning systems, etc.
2000–	Product or service design Global, changing markets	Dynamic network	International/ construction firms Global consumer goods companies. Selected electronics and computer firms (e.g. IBM)	Broker-assembled temporary structures with shared information systems as basis for trust and coordination

Source: Miles and Snow (1984, Table 1, p. 19). © 1984, The Regents of the University of California. Reprinted by Permission of SAGE Publications, Inc.

neglected, may be relevant to the study of organisations. It emphasises that selection operates not only on the level of individuals but also of groups, giving rise to a process of multilevel selection. This might clarify the emergence of social mechanisms – think of social norms, education and punishment – to guide or control groups. Organisations or their units can be seen as groups subject to interaction with other groups, in terms of information exchange, people exchange, competition and even conflict. This does not immediately mean that group selection is at play, but its basic ingredients are definitely available, suggesting it is worthwhile to examine its potential application to enlighten the diversity and dynamics of organisations (see Section 6.7).

In the area of organisational studies one can find many hybrid approaches, which combine evolutionary concepts like search and the routines of Nelson and Winter (Section 8.4), evolutionary models of technological innovation (Chapter 10), classical organisational theories such as mentioned in Section 9.1, and even ideas from biology and complexity theory, such as adaptive landscapes, the role of gossip, multilevel selection and modular evolution (Levinthal, 1997; Ethiraj and Levinthal, 2004; Kniffin and Wilson, 2010). Others have invoked evolutionary psychology to examine if the organisation of current corporations resembles, or can learn from, prehistoric small-scale societies. This considers a wide variety of issues, such as mismatch of organisational structure to the business environment, effective leadership and followership, and workplace design to moderate work stress (van Vugt and Ahuja, 2011).

9.3   Demography of Firms

While population ecology has a theoretical focus, the study of organisational demographics can be considered as its empirical counterpart. Its roots are in the sociology of organisations, the economics of business statistics, the geography of economic activity and – obviously – demography. Although a combination of organisations and individuals make up the structure of modern society, the demography of human populations, in terms of individuals, has received much more attention than that in terms of organisations. Conventional demography has an empirical orientation, focusing on spatial and temporal regularities of populations, their size and structure, notably age distribution, and underlying processes of birth, death and migration. The demography of firms involves related but somewhat unique concepts, such as foundation rates, merger rates, relocation, disbanding and failure rates (Carroll and Hannan, 2000). In terms of distribution of characteristics, it employs indicators such as size, age, management structure, profits and specialisation. Geographical features include location factors and spatial heterogeneity of environmental conditions or growth factors. Such heterogeneity in turn allows for comparative research.

One may wonder about the relevance of an evolutionary, or population and demographic, orientation in organisational studies. It is evident that an average or representative firm does not exist. Even more so, firms show great diversity along a number of dimensions. Trying to aggregate or average out this diversity will inevitably lead to a loss of information and mean that one is less able to understand or predict the structure, dynamics and the aggregate behaviour of sectors composed of many heterogeneous organisations. This, in turn, could contribute to inappropriate public policies in the areas of innovation, merger, unemployment and regional specialisation. Silicon Valley and Hollywood, two Californian success stories, provide interesting examples of the importance of observing industry and firm demography, as illustrated by Carroll and Hannan (2000, p. xxi). Silicon Valley's demography includes large and small firms, rapid and slow growers, old and young firms, bureaucratic and flat organisations, and innovators and refiners. Its demography is moreover subject to swift dynamics, as many firms do not survive fierce competition. The structure of Hollywood's film industry has been and still is driven by nonstop reorganisations, mergers and takeovers. The latter involve elements of evolutionary recombinant innovation.

What form do demographic analysis and explanations take? Carroll and Hannan (2000) ask to identify the diversity of organisations and their interaction. This creates the basis for organisational population dynamics as affected by environmental dynamics. A main problem in operationalising such a model is that data are always imperfect. One needs sufficient temporal coverage, a representative sample and organisation-specific environmental conditions. The latter involve exogenous ones such as social context, public regulation, resource availability and technology, as well as endogenous ones such as the population of similar organisations. Endogenous processes are often modelled through density-dependent processes, such as the logistic growth curve. This predicts monotonic evolution to a steady state or carrying capacity. More advanced models try to address non-monotonic patterns, which notably decline after an initial period of growth. This can involve delayed density-dependent effects (the so-called Leslie model), which causes dampened cycles.

Basic demographic data include counts of organisation types, the number of firms and size in terms of monetary value, revenues or personnel. The dynamics of these, over time, allows for evolutionary type of analyses. Time-series of certain variables can be studied, such as average number of employees per firm or number of firms in certain industries. These often show a bell-shaped form. This has been documented, among others, for the automobile and brewing sectors (Carroll and Hannan, 2000). Such a bell-shape can be explained by shake-outs, i.e. many founding companies fail to survive. One can regard this as a necessary experimentation at the population level, where market selection filters out those firms that are inefficient, produce goods or services for which there is less demand, or that use inferior technologies. Alternatively, changes in the population might also reflect mergers and takeovers motivated by scale economies.

Can organisations grow in size without limit? One tends to find a broad distribution of firm sizes in any industry, which includes some huge companies. Some nowadays are even larger in terms of monetary value of total output than most countries in the world. According to one article, in 2014, 37 of the world's 100 largest 'economies' were corporations, most of which were oil companies and banks, but also companies from the automotive, electronics and food industries, such as Wal-Mart, Volkswagen, Samsung, Toyota, Apple, IBM and Nestlé.[1] Next to indicators of company size, structure and performance, comparative research on variety of organisations stresses the role of organisational culture, employing indicators of communication efficacy. Country culture exerts selection pressure on organisation forms and cultures. Understanding how this precisely works is of relevance to multinationals, because they have plants in distinct cultural environments. In this context, Hofstede (2001) has identified four factors to explain cultural differences that affect organisations: power distance, uncertainty avoidance, individualism and masculinity.

An interesting phenomenon in social systems is globalisation of organisations. In the economy this occurs through the emergence of international trade, activities by multinationals, international agreements, international organisations (OECD, World Bank, IMF, UN, etc.), aided by advanced transport and communication technologies. Two organisational structures are found among multinationals: polycentric and geocentric. The first acts as largely separate firms in each country; the second as internationally managed firms, with extremely mobile and communicative managers at higher levels.

---

[1] https://makewealthhistory.org/2014/02/03/the-corporations-bigger-than-nations.

The latter is best illustrated by Japanese firms, which are supported by their government through strategies to stimulate the spread of firms across the globe. This model has been copied with great success by other Asian countries, notably South Korea, resulting in gigantic corporations, such as Samsung, Hyundai, LG and Kia. In opposition to this, firms and government tend to be more separated in Europe and the United States, and to a lesser degree in France. Large multinationals may be compared to the dinosaurs of the past, in terms of size and power but perhaps also sluggishness and limited adaptiveness in the face of rapid environmental change. It would be unwise, though, to take this analogy too far.

9.4   Evolution of Institutions

Institutional changes have made up an integral part of the evolutionary history of social-economic systems. New institutions arise through different processes, such as techno-logical innovations, regrouping of individuals, or integration of distinct levels of existing human organisation. Often this happens in response to economic or environmental challenges, which gives rise to the notion of coevolution of institutions, economies and technologies. Institutions often involve enforcement of norms through rewarding individuals who follow the norms or punishing those who do not. Meta-norms may be at stake here, meaning that enforcers of a norm are rewarded or non-enforcers are punished. In Section 8.4 we discussed Hayek's evolutionary perspective on institutions. Here we will add some other relevant insights from an evolutionary angle.

Institutions include structures with human-organisational and physical characteris-tics, such as state departments or ministries implementing and monitoring economic, technological or environmental policies. Another type of institution is less tangible, namely norms and ideology, that is, a set of moral and ethical codes for members of a society. A dominant ideology functions as the cement of social stability which makes the combination of state and economy viable (North, 1981). Ideology may be religious, as it often has been in the past and still is in large parts of the world, but can equally be secular, as in the case of Liberalism or Marxism. Ideologies can encourage individuals, among others, to show altruistic, non-selfish behaviour. Several genetic and cultural evolutionary mechanisms have been offered as an explanation of their emergence and stability (Sections 6.6, 7.3, 7.6 and 7.7).

A third type of institution is law and associated legal arrangements which specify duties, rights and sanctions, meant to assure the smooth functioning of society. A basic legal institution is property rights, which many historians regard as having been critical for major transitions in human societies to come about (Hunt, 2003). The role of property is discussed for the rise of agriculture and the Industrial Revolution in Sections 11.6 and 12.1, respectively. Important historical innovations associated with property rights are:

- Communal land, excluding use by non-group members, during the rise of agriculture.
- State ownership in ancient civilisations with hierarchical organisation.

- Contractual rights during Roman times.
- Private land at the end of the feudal area.
- Patent law during the period of industrialisation.

Such property rights acted as incentives for individuals to assure economic decisions are systematically and structurally aimed at efficiency and calculated risk-taking, which, through investment and innovation, contributed to economic growth at the macro level. In addition, property rights allowed control over labour conditions – with extreme cases of peonage and slavery – and the lending of money, grain, other food or goods – with the extreme case of usury, i.e. asking excessive interest rates. Diversity in such practices gave rise to cultural and market selection, resulting in distinct diffusion patterns. The practice of demanding interest on loans slowly spread among human societies, giving rise to externally funded entrepreneurship. As part of cultural selection, rules for lending were set by leaders and religions, including the Roman Empire, the Catholic Church and Islam. In this respect, interest-free Islamic banking is noteworthy as a surviving exemption.

Nobel laureate Douglass North (1981) has proposed the following evolutionary theory for the emergence of property rights. Rulers have rivals, namely competing states or potential rulers within the state. Changes in institutions are, under normal conditions, gradual adaptations by the state to changes in the economy, in line with the ruler's objective. Succession rules and democratic institutions stabilise the state and minimise sudden change. Individuals do not have an incentive to influence the institutions in a significant way because of the free-rider problem: the personal costs of affecting institutions are relatively high for an individual whereas its benefits are diffuse, i.e. they fall mainly upon others. Only under extreme conditions will individuals organise themselves to replace a ruler, and cause a revolution that replaces or drastically alters existing institutions. In this sense, views on economic institutional change based in evolutionary gradualism and Marxist radicalism somewhat blend.

Property rights start with possession, a theme studied in biology as many animals tend to respect possession rights of other animals in their group, other groups or even other species (Stake, 2004). Indeed, such rights have an evolutionary advantage, as trying to obtain resources from, or defend them against, others goes along with a cost – of effort, conflict and possible harm. This explains why an evolutionary equilibrium can include a certain degree of respect for another's possession or property: it limits the overall costs of staying alive. Property is thus not a strictly human affair, even though humans have formalised and diversified it, through a combination of conscious design and cultural evolution (Krier, 2009). Transfer of existing, and establishment of new, property rights is usually a response to benefit–cost opportunities arising due to technological progress or threshold effects triggered by congestion or externalities (Demsetz, 1967). Transfer of property titles to offspring upon one's death, common practice around the world, makes sense from a genetic kin altruism perspective. It contributes to reproductive success as offspring are thus endowed with wealth, which improves their quality of life and longevity, in turn increasing the chance of offspring one generation further down the line.

Using a game-theoretic analysis describing costs and benefits of resource access, transfer and resistance, Hartley (2017) analyses the evolutionary dynamics of possessive instincts among animals, notably primates. Thus he can clarify the emergence of different property ownership institutions in historical human societies. He suggests that communal ownership evolved to govern large unpredictable resources, command ownership to manage resources that can be monopolised, and titled property to deal with resources that are expanding, such as in market economies subject to economic growth. DeScioli and Wilson (2011) perform experiments with students and find that ownership conventions are more likely to arise in patchy than uniform environments. The reason is that the first environment more easily gives rise to a natural separation of discrete land units.

In discussing the evolution of institutions, one cannot omit to mention the works of two seventeenth century writers, *Leviathan* by Thomas Hobbes (1651) and *Two Treatises of Government* by John Locke (1689). Both started from a concept of a state of nature to unravel the features of human existence in the wild, before and outside any culturally evolved society. Subsequently they try to set up an argument of how civil society came into being. While Locke tried to explain the origins of property, Hobbes focused on the consequences of civil society. Hobbes can be seen to actually combine elements of evolutionary and neoclassical economic thinking. He regarded the state of nature as characterised by scarcity, in which all individuals were continuously competing, or even at war, for control over resources. Locke, instead, thought resources, notably lands, were non-limiting for productive labour activity as long as individuals used them non-excessively. He suggested that property rights emerged to arrange a fair distribution of resources and foster productive capacity.

A main economic institution, which is both an outcome of long-term cultural evolution, while it has also driven more recent cultural-economic evolution, is the market. Its emergence contributed to the evolution of a diversity of goods and services. Properly functioning markets depend on well-defined and enforced property rights, as well as on accurate measurement of the dimensions of a good or service so as to realise fair price competition. The market can be seen as the highest level of economic organisation which allows firms and households to overcome the limits of their internal organisation and production capacity. The separating line between a market and a firm is an old theme in economics. According to Coase (1937), the existence of firms reduces transaction costs, associated with search, negotiation and contracts, compared to bilateral exchange. Firms are subject to decreasing returns to scale, though, due to the organisational bottlenecks and limited supply of production factors. The latter may alter over time under the influence of technological change, allowing already large firms to further expand. Incidentally, markets also reduce search and negotiation costs, notably when they offer standardised products and services with uniform prices.

The earliest markets allowed people to specialise in food production and enjoy economies of scale, while still being able to obtain a varied, healthy diet. The transition from barter exchange to price markets went along with the emergence of money and early banks. The Athenian agora in the sixth century BC may have been the earliest market with uniform prices for all market participants. Markets increased in importance during the period of industrialisation. Until then, mercantilist regulations included bad legislation and privileges, such as through guilds, which had acted as a constraint on fair market competition. Free markets, along with well-defined property rights, meant fewer constraints and better incentives for making efficient economic decisions. As a result, markets increased in size, up to international scales. As this increased potential profits, it enhanced the emergence of large firms and the funding of technological innovation. During the late nineteenth century product markets became immense and impersonal, and the same applied to production factor – capital and labour – markets.

Closely linked to markets is money, which may be regarded an institution in itself. Much has been written about money, but little from an evolutionary angle. It is not surprising that money evolved, as it serves many functions, including a measure and standard of value, a medium of exchange, a tool for durable storage and general flexibility – money might be said to create economic plasticity. According to Lea (2006), money is too recent an invention to have allowed for genetic human adaptation to it, but surprisingly most, if not all, humans show strong incentivised responses to money, or for that matter, monetary information. Moreover, indigenous people without experience of money generally seem to quickly adapt when confronted with it. Guided by Tinbergen's 'four why questions' (Section 2.3), Burnham et al. (2016) clarify the presence of money: it triggers brain responses in a Skinnerian sense, and its use is picked up from adults during childhood. Phylogenetically, it can be associated with reward systems in other species. Humans tend to show maladapted behaviour towards money, what Lea calls 'money as drug'. This might originate from humans being sensitive to money status or appreciating malleable reserves for uncertain futures.

The notion of social and political power has received much attention in sociology. It is too broad a theme to cover here. Power can take various forms, such as affecting the distribution of access to resources and information, lobbying influence on issues appearing in the political realm, influencing public opinion through public media, or restraining choice sets and decisions by individuals. Institutional centralisation and power tend to generate rents that can be used to maintain or reinforce those in power, contributing to increasing returns in support of lock-in. Safarzynska and van den Bergh (2010a) suggest that the co-dynamics of power relations and institutions can be well addressed within an evolutionary multilevel selection framework. The reason is that it allows for institutions to emerge from interactions of individuals in competing groups as well as for merger and spill-overs, i.e. mutual learning, among distinct institutions in interacting groups. This surely cannot address all aspects of institutional dynamics, but offers a potent framework to deepen its study.

Various proposals from an evolutionary angle have been made to create more robust institutions (Bednar, 2016). One is guaranteeing internal diversity and even redundancy – overlapping functionality but with low association of vulnerabilities – of agents, components and ideas. This arguably fuels useful institutional mutations, ultimately fostering adaptation to altered circumstances (e.g. Hong and Page, 2004). Another recommendation is to achieve modularity of institutional structure and independence of institutional branches. An example of this is the system, common in most countries now, of separate judicial, executive and legislative powers. This allows for mutual monitoring and correction of basic powers, and their adaptation over appropriate time scales. Modularity further allows for experimentation, possibly by independent teams, with particular components, without creating devastating pressure on the institution as a whole. Too much modularity may, though, generate a hard and costly coordination problem. This suggests some optimal degree of institutional modularity (see also Sections 4.3 and 10.7).

Finally, when elaborating an evolutionary perspective on institutions, Currie et al. (2016) propose that Tinbergen's 'four why questions' or levels of analysis (Section 2.3) are addressed, as illustrated already above for the effect of money at distinct time scales. More generally, these questions come down to relating the functions of institutions to their mechanistic day-to-day performance, their long-term 'descent with modification', and the psychological features, capabilities and learning processes of individual human agents who make up the institution. The authors apply this suggestion to a variety of cases, including the formation of the United States constitution and the international spread of democracy.

9.5   Self-organisation and Emergence

When reading about organisations, one occasionally stumbles upon the notion of self-organisation. Although this concept is hard to pin down, it has generated a considerable literature. Central to it is the idea that a particular organisational structure was not arranged with a preconceived goal or through hierarchical or central planning and control, but is the outcome of a set of unintended and unplanned processes. The term self-organisation is used in various disciplines and research areas, notably physics (especially thermodynamics), chemistry, biology, cybernetics, computer science and the social sciences (Ulanowicz, 1981; Luhmann 1984; Prigogine and Stengers, 1984; Silverberg 1988; Foster, 1997; Sawyer, 2005). As regards living systems, the idea of self-organisation goes back to at least Erwin Schrödinger 1944 book *What is Life? The Physical Aspect of the Living Cell*, which inspired Watson and Crick's search for DNA. In social systems self-organisation has been referred to by some authors as 'autopoiesis' (Maturana en Varela, 1980; Mingers, 1996). In mathematics, self-organisation has been studied with nonlinear models, characterised by concepts as attractor, positive feedback, bifurcation and chaos (Haken, 1977). This section briefly discusses the most common interpretations and building blocks, while the next section will concentrate on the relation and difference between evolution and self-organisation.

Self-organisation denotes that some higher-level structure arises out of a disorganised system of lower-level physical components due to spontaneous processes involving interactions between these components. Hence the popular phrase 'order out of chaos' (Prigogine and Stengers, 1984). The organisation of such a higher-level structure is unplanned and decentralised, also referred to as 'distributed control' or 'self-regulation'. The macrostructure can comprise physical, chemical, biological and economic arrangements or coordination of behaviours and activities of agents in a larger system. Even markets have been proposed as being self-organised, namely through the spontaneous organising force of the 'invisible hand'. An entirely different example of self-organisation is cellular automata that generate visual patterns or emergent structures like networks or groups. Work on them illustrates that simple rules at the micro level can produce surprising, emergent patterns at the macro level.

To characterise in general terms the mechanisms underlying self-organisation, one should note that it is driven by some type of gradient of energy, matter or information which causes associated currents, flows or streams. For example, convection cells involve a temperature current, biological cells a nutrient flow and social systems an information stream. A more abstract definition of self-organisation is as a wide variety of complex processes at the 'edge-of-chaos', characterised by a trade-off between stability and flexibility (Kauffman, 1995). The self-organised state is in between two extreme states: (1) a rigidly structured, ordered system with few connections and extreme stability but no flexibility; and (2) a random or chaotic system with many connections and extreme flexibility but no stability.

Another way to understand self-organisation is through the notion of 'emergence', the spontaneous appearance of a globally coherent pattern out of local interactions (Holland, 1998). This is reflected in the expression 'the whole is more than the sum of its parts'. Although the latter is also known as 'synergy', emergy denotes the appearance of qualitative novelty (Bunge, 2003), whereas synergy is merely pointing at a quantitative or positive scale effect from combining multiple inputs. Emergence has also been called a meta-system transition (Heylighen, 1996) and major transition (Maynard Smith and Szathmáry, 1995) – examples being molecules joining in cells, cells in organisms, and organisms in structured groups and societies (Sawyer, 2004). Emergent properties do not pertain to the lower-level units. For example, atoms and molecules do not have the colour that appears in materials made out of them. And a wheel can roll in a way that cannot be said to be the case for the molecules or atoms that together form the wheel. In economics one can think of labour division as an emergent property of interactions among individuals. De Haan (2006) suggests that the element versus system (i.e. lower/higher) level connection means emergence always appears pairwise. In other words, two associated novelties appear simultaneously. For example, the advent of formal markets went along with the emergence of the notion of exchange value of goods and services, which previously only had use value.

The debate on emergence is old – witness the British 'emergentists' (believers in emergence) from J.S. Mill (mid-nineteenth century) to C.D. Broad (early twentieth century). It addresses big questions: are all chemical laws fully explainable by considering the underlying physical processes; are all biological regularities understandable as chemical processes; and are all social processes clarified by the underlying biological phenomena? The answers are no, no and no, because each new level involves emergent properties. The nature of emergence means that lower levels cause higher levels. In addition, the higher levels can influence or constrain the lower levels. This is sometimes referred to as 'downward causation'. A complete description of a complex, multilevel system includes both upward and downward causation processes (van den Bergh and Gowdy, 2003). Emergence and subsequent downward causation both contribute to unpredictability and surprise, thus complicating the traditional view that microfoundations or micro-level descriptions are sufficient to understand the system at any higher level.

While evolutionary theory has its origins in life and social processes, ideas on self-organisation originate from the physics of energy, i.e. thermodynamics and chemistry. A formal, somewhat technical definition of self-organisation in terms of energy-related concepts is: systems open to a flux of energy may spontaneously generate a dissipative structure in a way that cannot be explained by conventional equilibrium thermodynamics as the resulting structure is far from equilibrium. The work by Ilya Prigogine and associates is usually regarded as an influential starting point for this kind of thinking. Prigogine received the 1977 Nobel Prize for his research on self-organisation in chemistry. Self-organisation is a process in which high entropy energy and matter are dissipated – especially degraded and possibly exported – while low-entropy energy and low-entropy matter are imported. For this reason self-organising systems are also referred to as dissipating (exporting) energy and matter. The boundary between the self-organising system and its environment may be regarded as a sort of a barrier, which prevents entropy from moving in the thermodynamically expected direction. As a result, there can be concentration of low entropy, structure and order in a confined space – be it a cell, organism, ecosystem or economy. Self-organisation in such systems goes along with a decrease of entropy, or an increase of order, and an increase of entropy in the environment. The basic example is crystallisation, where randomly moving molecules become fixed in a crystalline structure by passing on their kinetic energy to the surrounding liquid, leading to an entropy increase of the total system (crystal and liquid). Another example is that economic development and growth create environmental pressures and damages.

It is useful to distinguish between self-organisation that leads to static or dynamic structures. Crystals and magnets are examples of the first type, while weather patterns, life forms and economies are examples of the second type. Static systems reach an equilibrium situation where energy dissipation comes to a halt. Dynamic systems continuously generate and dissipate entropy. Such systems have the capacity, within limits, to resist change and absorb perturbations. This has given rise to the notion of resilience as an extended notion of system stability (Section 13.3). It denotes that a system is characterised by redundancy and distributed organisation, allowing it to absorb perturbations without changing its structure, namely by dissipating any changes in entropy. An example is sweating on a warm day. Resilience does not apply to static self-organised systems (e.g. crystals). These lack redundant features and fall apart when subject to a certain degree of external disturbance. Of course, many dynamic systems are not perfectly stable and respond to external influences by continuously moving from one attractor to another (weather patterns), or by adaptation to perturbations through slowly changing their subsystems or their interactions (ecosystems, economies). Such adaptation fits a framework of biological or economic self-organisation, enriched by notions such as evolution and resilience.

9.6   Adaptive Self-organisation

Evolution and self-organisation are the only processes we know that are capable of generating complex physical structures. Whereas evolution is clearly delineated by the core components selection, innovation (variety creation) and replication, self-organisation is more an umbrella term which points at a diverse collection of mechanisms. This is not to say that evolution is always the same: indeed, its components take distinct forms in particular settings.

There is no general agreement about whether evolution should be considered as a special case of self-organisation or vice versa, or whether it should be regarded as complementary or independent. It has been suggested that variation, central to evolution, can promote self-organisation. This is also known as the principle of 'order from noise'. The explanation is that through perturbations a system will move through its state space, thus allowing it to reach an attractor or equilibrium. This view suggests a similarity with the Darwinian mechanism of variety creation which also comes down to searching a space. This might suggest that self-organisation is a special case of evolution. In both, environmental factors play a crucial role: self-organisation can result from changes in energy boundary conditions, while evolutionary adaptation is a response to an environmental stressor. One might also consider self-organisation to involve, like evolution, interactions between elements of a population consisting of atoms or molecules.

According to Schneider and Kay (1994), both self-organisation and evolution create order, but whereas self-organisation creates order out of chaos or disorder, evolution creates order from order. The latter allows evolution, unlike self-organisation, to result in sustained accumulation of adaptations resulting in greater complexity and order. This view is consistent with evolution as a cumulative process taking place over a longer period and self-organisation as functioning over a relatively short period. It explains why evolutionary thinking can make statements about genotype changes and unique irreversible phylogenic history, whereas self-organisational thinking addresses relatively short-term and repeatable processes such as the ontogenetic 'unfolding' of a phenotype.

In line with this view – notably that evolution creates order out of order – is the fact that self-organisation can be regarded as a precondition for evolution. This is illustrated by the fact that both selection and recombinant innovation work on pre-existing, self-organised structures. On the other hand, evolution creates the genetic conditions that constrain and direct processes of self-organisation in terms of ontogenetic development. Hence, evolution and self-organisation complement each other. Without evolution, self-organisation may produce a variety of random macro-level phenomena, but these will not represent good adaptations without the repeated force of environmental selection, and they will not persist and accumulate without the mechanism of hereditary replication. Adaptive self-organisation requires continuous interaction with evolution, according to the following sequence: genes (+ energy, substances and environment) => self-organisation => ontogeny / phenotypes => replication => phylogeny => new genes, and the sequence is repeated, etc. To phrase it differently, evolution both builds

Table 9.2 Differences between evolution and self-organisation

Evolution	Self-organisation
Interaction between individuals and environment	Interaction between components
Competition or imitation	Exchange of energy
Focus on phylogeny	Focus on ontogeny/development
Relatively long term	Relatively short term
Unique history and unpredictable/stochastic	Repeatable and deterministic
Irreversible	Often reversible
Diversity essential	Diversity not necessary
Specific sub-processes (selection, innovation)	General processes (unspecified local interactions)
Order out of order	Order out of chaos
Roots in biology	Roots in thermodynamics and chemistry

upon and gives rise to structures that self-organise. Both processes are needed to realise observed organic complexity on Earth. Kauffman (1995) goes as far to argue that the interaction between selection and self-organisation might turn out to be an important additional evolutionary mechanism. He suggests that we need to invoke self-organisation to explain the puzzle that so many major evolutionary innovations (phyla) occurred during the relatively short period of the 'Cambrian explosion' (Section 3.6).

Note that much of what has been said here is subject to a degree of speculation, meant to stimulate the reader's awareness of the issues and questions. Table 9.2 adds to this by offering a comparison of main features of evolution and self-organisation. More insight into the relationship between evolution and self-organisation can help getting to grips with any complex system and its dynamics. Nevertheless, self-organisation still seems to be more frequently used when speaking about processes of physics and chemistry, while evolution is more common in biological and social contexts. It is fair to say that whereas evolutionary thinking has developed in a consistent framework with clear core elements and mechanisms, self-organisation remains rather ambiguous. This may explain its lesser popularity in biology and the social sciences, as well as why the distinction between self-organisation and evolution remains fuzzy.

It is worth mentioning here the notion of a 'complex adaptive system' (CAS), which is popular in both natural and social sciences. It denotes an arrangement characterised by multiple interactions over various networks among many heterogeneous components or individuals. These individuals can themselves adapt and re-organise, leading to a combination of self-organisation and evolution. CASs are operationalised through agent-based models describing interactions between microscopic agents. Such models have also seen application to organisations and their dynamics (Dooley, 1997). They combine elements of systems theory, evolutionary thinking and population ecology.

D. S. Wilson (2016) proposes to distinguish between two types of CAS. The first, CAS1, is adaptive as a system, due to past selection at the level of the entire system.

Examples are the brain, the immune system and social insect colonies. The second, CAS2, is a complex system of agents that are capable of undertaking individually adaptive strategies. Wilson, who is an expert on group selection theory, uses multilevel selection – i.e. combining individual and group/systems level selection – to understand the differences between the two CAS types. CAS2 tends to lack selection at the systems level, causing it not to be well adapted to its environment. This is of utmost relevance to social science, as it studies complex cultural and economic systems that are not, or ineffectively, selected by their environment, instead being dominated by internal processes based on human agents and organisations that adapt to their local environments. As a result, the cultural or economic system as a whole tends to be maladapted to its environment or, in Wilson's words, lacks adequate functional organisation. To overcome this shortcoming, public policies and governance institutions are needed, as a kind of substitute for (global) environmental selection. For example, world peace requires an active role of the United Nations, as individually adaptive strategies of countries sometimes involve conflicting behaviour. Or laissez-faire of the market economy will not serve the common good in the presence of market failures such as externalities, public goods and market power – which are no exceptions, and require corrective policies. We will further illustrate maladaptiveness of CAS2 in relation to environmental issues and climate change in Chapters 13 and 14.

Note that policies to guide CAS2 towards an adaptive-like outcome are not really an 'evolutionary solution', as this would require the global selection of, and competition among, multiple similar units, as happens in a population consisting of many humans with different brains and immune systems, or in population of multiple, slightly distinct, insect colonies. At the global level, however, there is no similar population of heterogeneous economies or biospheres to allow for this. Admittedly, one can observe numerous regional economies and ecosystems, but these cannot really be regarded as making up a population of units subject to a single selection environment. Instead, each economy is positioned in a distinct cultural and biophysical environment, and thus subject to unique selection forces.

When thinking of economies, if the CAS2 system is subject to selection at both lower and systems levels – i.e. multilevel selection – the first will undercut the effects of the second. So even in this case, which may best depict cultural and economic realities, the system will not be well adapted to its environment, making correction in the form of public governance necessary. To illustrate, despite global environmental feedback to the world economy, it is not well adapted and moves along an unsustainable path – unless we are able to strike effective climate and other environmental agreements that guarantee appropriate, harmonised regulations in all countries.

10 Technological Evolution

10.1 The Tree of Technology

While some question the tree of life, doubting whether all species in the natural history of planet Earth are related to one another (Section 2.4), few deny the affinity of technologies. There is strong evidence that technological change is a cumulative process characterised by building upon, and benefitting from, previous technologies and knowledge. One could speak of a 'tree of technology' (Basalla, 1989) to represent technological history. As captured by the right image in Figure 10.1, it is an odd tree. Whereas the tree of life on the left is 'normal', with independent branches and sub-branches, the tree of technological history has connections between branches and sub-branches to reflect that knowledge in distinct branches is at times combined to form new, hybrid information or technologies. This means that the biological notions of species and phylogeny do not have direct counterparts in technological evolution. Still, as we will see, many relevant aspects of technological dynamics, or its guidance by public policy, can be clarified by adopting a V-S-I-R algorithmic, evolutionary angle.

This chapter continues the discussion of technology that started in Chapter 8. As noted there, the historical basis for evolutionary thinking about technological change is provided by the classical work of Schumpeter and its extension by Nelson and Winter (Section 8.4). The dominant approach is hence known as neo-Schumpeterian economics. The motivation to apply an evolutionary approach in studying technological change is that, as can be easily observed, at any time there is a large diversity of technologies, whether associated with processes, products or services. This diversity changes through innovation and selection. One technology may come to dominate through hyper-selection or through an agreement to use a single technological standard. Joel Mokyr (1990, p. 9) notes that technological change is the study of exceptionalism that cannot be explained with the standard model of rational 'homo economicus' because 'technological change involves an attack by an individual of a constraint that everyone else takes as given'.

The term 'technology' will be used to include both knowledge and artefacts reflecting knowledge. Since the evolution of any industry, i.e. the change in the distribution of characteristics of firms in it, is closely related to the evolution of technologies used in that industry, certain aspects of industrial evolution will receive attention here as well. This, then, provides a connection with the previous chapter on organisations in industries. An evolutionary angle on industries has been elaborated in Andersen (1994).

Figure 10.1 Tree of organic life (left) versus tree of culture and technology (right).
Source: Kroeber (1948, Figure 18, p. 260).

He proposes a 'tree of industrial life' to express the (arguable) idea that industries descend from earlier industries.

The main sources of technological change were already identified by Schumpeter, namely entrepreneurship in small firms, R&D by large firms and science. Investments in basic research, notably in scientific institutions, provide the insights for major technological breakthroughs. Entrepreneurs and R&D departments tend to focus on applied research, while certain large firms invest in basic research as well. Next, firms invest considerably in the commercial implementation of applied research findings. This is captured by the notion of innovation, generally distinguished from the preceding stage of invention. However, as discussed in Section 10.2, this simple distinction does not do justice to the feedback from later to earlier stages in technological change. In order to examine the sources, characteristics and impacts of technological change, we consider four levels of analysis hereafter. Technological innovation is studied at the firm level. The interaction between innovation and imitation is dealt with at the level of markets and industries. At the macro level, growth and long waves are identified. In addition, path dependence and lock-in receive attention as they are critical for a complete understanding of historical patterns of technology. We end with deliberations on technological policy and optimal technological diversity.

10.2   Inventions and Innovations at Firm Level

Sources and Characteristics

The core concept in evolutionary theorising about technological change is 'innovation'. It denotes an invention that is implemented in, or leads to, a new operational technology. An invention can be seen as a surprising mutation. It differs from an innovation most clearly in the case of a new technology that is functioning but unused. However, the distinction is usually blurred, as the invention and innovation phases are not clearly separated, because inventions tend to get refined in the implementation phase. Mokyr (1990, p. 11) regards an invention as an individualistic activity, the success of which largely depends on the qualities of the individual inventor struggling with a physical-natural reality. To this, one has to add the roles of inspiration and luck. As opposed, an innovation is more a social and economic activity, the success of which is determined by markets, institutions and generally interactions with other individuals.

Roger Cullis (2007) offers a nice treatment of inventions, adorned by many examples of electricity, artificial lighting, electronics, computers and communications technologies. He suggests that the majority of important inventions responded to recognised problems. The personality of the inventor, which increases the chance of serendipity, is a main explanation for inventions. Serendipity denotes that an invention results from the combination of chance, insight and expertise (Fine and Deegan 1996). According to Merton and Barber (2004), favourable personality features of an inventor include: a genius or at least a logical thinker with a prepared, open mind and a sharp intuition, showing unconventional, lateral thinking and knowledge of different fields, and often in possession of polytechnic qualities. Being young, with few distractions, is also helpful. And we should not overlook the importance of courage, perseverance and independence (Darwin comes to mind), especially in view of established seniors often inhibiting young researchers from taking risks and going in new directions. Another factor is the environment, composed of the organisational, socioeconomic and scientific context. It involves not just the research setting (department or laboratory) but also the financial and market opportunities offered, the appearance of problems-requiring-solutions, and the previous knowledge and technology to build upon. Many good inventions do not come about owing to a lack of time and financial or other resources. The ones that do appear are successful because they are mature, complete and timely. In the case of unsuccessful inventions, at least one of these conditions is not satisfied. Untimeliness results from the disconnection of scientific knowledge and recognition of problems. Once the connection is made, the opportunities may have disappeared. This suggests a limited time window for many potentially successful inventions. Comparing the personality and contextual factors, the latter has undergone more change than the former during the past centuries. Moreover, socioeconomic and technological evolution bites its own tail through having drastically altered research contexts, in terms of laboratory equipment, access to knowledge (think of computers and the internet), diversity of research funding and the sheer number of researchers cooperating and interacting worldwide – as well as competing for scarce resources and attention space. Given this factorial background, Cullis (2007) proposes an eight-part typology of inventions as in Table 10.1. This covers various processes such as serendipity (i.e. stumbling on a valuable thing without looking for it because one is intellectually prepared), purely theoretical inquiry, extrapolation of knowledge, analogous thinking and even emulation (i.e. trying to equal or outdo one's peers).

Table 10.1 Sources, characteristics and timing of inventions

Type	Inventor	Rent magnitude	Risk	Stimulus	Frequency	Predictability	Timing	Example
Serendipitous	any	any level	high	revelation of problem, chance observation	sporadic	statistical	at any time, may follow directed effort	point-contact transistor
Theoretical	usually higher intellect	any magnitude, potentially large	low	inventor dependent	inversely dependent on height of inventive step	inventor dependent	at any time after identifying a problem	laser
Extrapolation	familiar with technology	any magnitude, usually small	low	deficiency in precursor	dependent on precursors	relatively high probability	increases with knowledge of invention	Pentium microprocessor
Product extension	familiar with technology	incremental	low	identification of application	invention dependent	depends on nature of invention	follows seminal invention	transistors used for space programme
Topological manipulation (geometry of technological components)	lateral thinker	any magnitude	low	fresh problem solvable by manipulation of existing technology	invention dependent	apparent from nature of invention	follows seminal invention	low-energy light bulbs
Analogue	familiar with technology, lateral thinker	any level	low	need for improvement of precursor	dependent on existence alternative technologies	dependent on alternative technologies	follows emergence of suitable technology	silicon transistor (replacing thermionic valve)
Emulation	lateral thinker	any level	low	established market	relatively rare	technology/ inventor dependent	depends on market need	incandescent filament lamp (replacing gas burner and kerosene lamp)
Non-working, 'half-baked'	acquisitive of patents	nil or latent	high	desire to increase tally	sporadic	inventor dependent	at any time	Edison's fluorescent lamp

Based on information from Cullis (2007, Tables 12.3 and 12.4 and Section 5.4).

Table 10.2 Characteristics of innovations

1.	Products versus processes
2.	Factor saving versus quality improving
3.	Incremental versus radical
4.	Single sector, multiple sectors or whole economy
5.	Codifiable versus tacit, i.e. well documented versus embodied
6.	Public versus private
7.	Unintentional versus organised
8.	Stimulated by institutions (markets, regulations, norms) or production bottlenecks
9.	Autonomous versus systemic
10.	Hard versus soft
11.	Demand pull, technology push or user–producer interactions

The relevance of evolutionary thinking about technology begins with the fact that more inventions are conceived than can be successful. Notice the analogy with Darwin's idea of the 'struggle for survival': more organisms of a species are born than can possibly survive. Therefore, some economic selection process is needed to translate the inventions into a limited number of innovations. Innovations are associated with imperfect and asymmetric information, as well as pervasive uncertainty. Unlike the technical part of production processes, they do not function by blueprints but by search and heuristics (Nelson and Winter, 1982).

Innovations come in many types, as summarised in Table 10.2. They may relate to: a product or process; to reductions in factory use, notably of energy and labour; to new services or functions provided for production or consumption; and to an increase in economic efficiency, or a reduction in costs. Technological change through innovation involves both incremental and radical changes. Incremental changes occur regularly, often taking the form of efficiency or cost improvements that arise from experience, namely through learning by doing or learning by using. Radical innovations relate to discontinuous events which often go along with so-called technology system changes that influence practices in multiple sectors. When such changes affect the entire economy, the terms 'new techno-economic paradigm' or 'technological revolutions' are employed. Innovations may further be codifiable, i.e. well documented, or tacit, i.e. learned or embodied in organisations (routines) or people (skills). This distinction is relevant because codifiable knowledge is more easily imitated and diffused, and hence more widely applicable, than tacit knowledge. Next, public innovation, such as scientific publications, can be contrasted with private ones, which are characterised by tacit, secret or patented information. In addition, one can distinguish between unintentional innovations arising from learning and serendipity, and intentional innovations through strategic planning in private firms or public research institutes (including universities). Next, innovations may address, and be stimulated by, institutional requirements (markets, regulatory, norms) and bottlenecks or imperfections in production. Finally, some innovations are autonomous or stand-alone, while others are systemic, requiring significant complementary adjustment in other parts of the production system to be able to function or from which to benefit. To illustrate the latter, electric cars require an appropriate infrastructure consisting of battery recharging stations.

To many, the term innovation suggests changes in a physical, machine-like artefact. But opposed to such 'hard innovations' there are also 'soft innovations', a subcategory of product innovations. These include changes of an aesthetic or intellectual nature, such as those associated with publications of intellectual ideas, movies, theatre plays, paintings, music, design (furniture) and architecture (buildings). Paul Stoneman (2011) argues that these have a number of peculiarities in terms of supply and diffusion that merit separate study. For example, their imitation is relatively easy (e.g. visual information about designs) or cheap (e.g. copying of electronic information), which affects the speed of diffusion. Moreover, such innovations are widespread, suggesting that the relevance of innovation for our culture and society is easily underestimated.

Another common distinction made in the context of innovation is between demand pull and technology push factors. The latter generally dominates in basic research. However, the pure forms have come under criticism as simplifying the complex drivers and interactions underlying many innovations in reality, giving rise to the idea of user–producer interactions (Bogers et al., 2010). These may involve positive effects through useful mutual exchange of information about needs, desires and capabilities. But they can also function as a brake on important innovations as the interactions can enforce existing technological patterns and even contribute to technological lock-in (Laursen, 2011).

The list in Table 10.2 is not exhaustive. More can be said about types of innovations. For example, Dosi (1988) categorises innovations by sector and type (product/process) as follows:

- Supplier dominated, mainly process innovation. Examples are textiles and clothing, printing and publishing, as well as wood products. Innovation here is dominated by diffusion of best-practice capital goods and intermediates and appropriability is low.
- Scale-intensive, both product and process innovation. Examples are transport equipment, electric consumer durables, food products, metal, glass and cement. Large firms assembly products through complex processes. R&D investment tends to be large here. Moreover, firms produce a relatively high proportion of their own process technology. Appropriability occurs via lead times, product complexity and scale economies.
- Specialised suppliers, notably product innovations. Examples are mechanical and instrument engineering. They produce mainly capital goods for other firms and other sectors. Appropriability is based on embodied expertise in individuals and organisation. Innovation opportunities are high.
- Science based, operating through new technological paradigms. Illustrative sectors are electronics and chemical processes. Innovations are based on R&D laboratories in large firms. Products are capital or intermediates for other firms and other sectors. Appropriability includes patenting, lead times and learning curves. Innovation opportunities abound.

Cultural, economic and geographical contexts are essential to innovations, leading to the much discussed notion of a national or regional system of innovation (Freeman, 1987; Lundvall, 1992; Nelson, 1993). This reflects the idea that innovation depends on networks of public and private organisations, science-based opportunities, institutions and economic conditions such as incomes and market regulation. Innovation systems affect both conception and diffusion of new technologies. Often they have a clear spatial dimension and sometimes they operate at a sector scale. Core elements are private firms, financial institutions, research institutes and universities. Much variation can be observed among innovation systems throughout the world. For example, the average OECD country invests much more in R&D than countries in the South; and certain regions have acquired a core innovation position even at an international scale (e.g. Silicon Valley). The German innovation system is dominated by regional governments, industrial conglomerates with strong links to investment banks, and a subtle multi-layered system of publicly funded research from basic to more applied research: Max Planck, Helmholtz, Leibniz and Fraunhofer organisations. The United States is characterised by relatively high public outlays on defence and aerospace, which stimulates particular industries, such as aircrafts and micro-electronics. A typical element of the Dutch innovation system is the so-called 'polder model', i.e. the network of consensus-oriented deliberation between unions, employer organisations and government, which may have hampered innovation and overprotected existing industries and vested interests (Box 8.3 in Boschma et al., 2002). Finally, in Japan and South Korea national governents have strongly supported industrial conglomerates and stimulated education to be well adapted to the particular needs of industries.

Organisation of Private R&D

The organisation of R&D in firms, involving both strategy and structure, can take various forms. R&D can be positioned at a primary level in the firm's hierarchy, next to personnel, finance and manufacturing (U-form). Alternatively, it can be defined at a second level, separated by product division (M-form). Owing to the pervasive uncertainty and secrecy surrounding the outcome of R&D its finance is usually dominated by firm-internal funds rather than external capital market support. Nevertheless, capital markets respond favourably to firms' R&D investments or signals about these, as it contributes to their long-run viability.

For similar reasons, R&D is usually undertaken by firms themselves and not outsourced through contract research. Contracts cannot specify the output due to the uncertainty inherent to R&D success, so that appropriability is not assured. Another reason is that there is a continuous interaction between all production stages on the one hand and R&D on the other hand, through learning (by doing) and embodied and tacit knowledge. A third reason is the presence of transaction costs, which are likely to be larger in the case of external than internal communication. Only when minor, not too fundamental, innovations are at stake, can outsourcing occur. Consulting aimed at improving organisational efficiency is an example.

Nowadays there is so much specialised knowledge that research collaboration is almost inevitable. This is reflected by a diversity of indicators, such as increasing research team size in science, joint R&D ventures and cooperation between government institutes, universities and firms. This is especially relevant to basic research since the output often is non-product specific and of value to many research participants. Collaboration sometimes means integration of activities. Horizontal and especially vertical integration is expected to enhance systemic or integrated changes to be recognised and managed or coordinated. Otherwise certain systemic innovations could not be realised. R&D by larger firms has this advantage because of more internal diversity, knowledge and room for experimentation compared with small firms.

Finally, market concentration is often suggested to contribute to innovation. It is widely believed that oligopolies combine the best of all, i.e. appropriability of benefits, diversity and competition. Monopolies strategically use R&D since there is no competition, although in line with the theory of contestable markets (Baumol, 1982), possible entrants may influence both price-setting strategies and R&D choices. Nevertheless, correlation between firm size and market concentration on the one hand and research intensity on the other is not confirmed in general by empirical research. Variation among industries is considerable.

Strategies of R&D

As opposed to investments in productive capital, such as machines and buildings, investments in R&D are characterised by uncertain time delays between investment and output, which affect appropriability (next item). Uncertainty is usually pervasive, immeasurable and non-insurable. Such uncertainty is essential for innovation success, because it means that there are fruits to be reaped by creating unique, first-mover advantages. Uncertainty involves a variety of issues (Nelson and Winter, 1982, pp. 128–129):

- Technical uncertainty: whether a motivating technical or physical problem can be resolved; and what the characteristics are of the resulting innovation.
- Organisational uncertainty: what the consequences are of the innovation for the functioning of the organisation as a whole.
- Market uncertainty: what responses the innovation will have from consumers (demand) and competitors.
- Financial uncertainty: whether the search or investment costs can be compensated by a flow of future benefits.
- Systemic relevance: whether suitable, complementary technologies will become available in time; and what the economy-wide effects of the innovation are.

These uncertainties have different degrees, ranging from mere probabilistic risk to ignorance. Each requires a specific evaluation and response. Hence, there is no single innovation strategy. This is also true for other reasons. Firms can focus on different objectives, among others, creating a new function, creating a new design, devising a more efficient or accurate process, perfecting an existing product, increasing market share, increasing product variety, leading the industry or following the leader. Of course, medium- to long-term profit seeking will often serve as a guide for private firms. Non-profit organisations may instead be guided by factors such as a cost-effective approach, a benchmark approach (national or international evaluation, comparison and target-setting), strong ideals to improve the world, and freedom and decentralised organisation (universities, including external or peer-review evaluation).

An innovation strategy involves many decisions, notably budget allocation to R&D, in order to hire experts with appropriate scientific backgrounds, to develop a network of contacts or formal cooperation with other firms, universities and government, and to set up an interaction among R&D, production and sales units for experimental implementation of new routines. A long-term viewpoint is relevant as R&D does not pay off quickly, but requires cumulative investments, so as to create a knowledge base. This means patience, foresight and adaptive flexibility required from research managers overseeing the entire process.

R&D is often thought to be focused on inventions. But it is also about innovations. The notion of technological or market niches is relevant here. It refers to an area in which interdependent new technologies are applied and can be experimented with. The initial stage is often a protected technology that is not yet cost-effective, and the final stage is a market niche. Niches facilitate processes of learning and economic integration, such as capital accumulation and overcoming barriers of entry such as economies of scale.

The Role of Patents in Appropriating R&D Benefits

A topic that traditionally has received much attention from economists writing about technology is the appropriability of the benefits of R&D. The combination of extreme uncertainty, huge investments and notable time delay between investment and results that is typical of basic, fundamental research is an important reason to organise it as a public activity. On the other hand, there are various factors that cause more applied R&D investments to pay themselves back, so that there is a motivation for firms to undertake them:

- *Lead times*: Competitors will have to undertake long preparations before an actual innovation occurs that ultimately leads to sales of related products or services.
- *Scale economies*: The first innovator can, through a large market share, benefit from scale economies. In fact, scale economies and lead times together create what is known as a 'first-mover advantage'.
- *Learning by doing*: The first innovator increases his head start by efficiency increases, through learning by doing. This includes both technical and organisational improvements, some of which are difficult to know or otherwise to copy by other firms.
- *Secrecy*: The details of many inventions and innovations are secret so that imitation will be ineffective or at best imperfect, resulting in a more costly

process, product or service. Reverse engineering can only uncover some aspects of a product, not the full set of problem-solving and production processes that have generated it.

- *Duplication being costly, imperfect or impossible*: Whether duplication (perfect imitation) is possible, and how costly it is, depends on some of the characteristics listed in Table 10.2, such as whether relevant knowledge is codifiable or tacit.

- *Patents, copyrights, trademarks*, and other forms of intellectual property, such as trade dress, trade secrets, industrial design rights and *sui generis* database rights: A copyright assigns exclusive rights to a creator of original intellectual or artistic work, for a limited period of time. A trademark is a unique, visually recognisable sign or design feature of products or services that allows them to be distinguished from similar ones of other producers. Industrial design rights protect the creation of two- or three-dimensional shapes and compositions of pattern or colours, which serve an aesthetic rather than utilitarian value of the products to which they contribute. More information on patents is provided in Box 10.1.

Box 10.1 History, design and effectiveness of patents

Patents have a long history, starting in the late Middle Ages in Venice, and developing more fully into a legal system in England during the early seventeenth century. National patent systems have shown great variety and were only slowly integrated through international agreements, a process which is ongoing. For example, the formal administration of patents varies between countries (e.g. in the United States it is the responsibility of the Department of Commerce, while in Germany it belongs to the Federal Ministry of Justice). During some periods, certain countries did not have or even discarded patent law. The Netherlands is a known case: in 1869 it was generally felt that patents were not in the national interest and the existing patent system was abolished. This allowed the young Philips Company to sell electric lamps more cheaply than competitors in other countries, contributing to its fast growth (Cullis, 2007).

A patent provides a legal barrier to costless duplication. It represents an exclusive right to a novel, potentially useful and non-trivial invention of a product or process, and is valid during a limited period of time. Hence, the patentee can, ideally, prevent others from commercially producing, using or selling a patented invention without his or her permission. This creates a sort of monopoly position, or raises competitors' cost of production through royalty payments. Ironically, since a patent requires the public disclosure of the invention details, others can learn about the invention and intelligently make changes so as to circumvent the patent's control, and thus indirectly benefit from it. This is why secrecy is sometimes preferred as a more effective way to protect an invention during a certain (early) period. Since many products and processes are subject to multiple patents controlled by different individuals or firms, cross-licensing agreements are not uncommon. Through it, parties grant a patent licence to one other, allowing each to commercially produce,

---

Box 10.1 (*cont.*)

use or sell the associated product or process. The alternative options for them would be to pay a (considerably higher) royalty or enter into a legal dispute. Nevertheless, patent litigation is still common practice.

According to Cullis (2007, p. 29), the impact of patent systems has changed over time: 'During the first stage of the Industrial Revolution (1780–1880) capability power protected markets. From 1880 to around 1914, patents were the main vehicle. After 1920, marketing power was dominant.' Even though one may debate this portrayal, it is likely that the patent system has not had a uniform effect over time, also given that it is just one among many appropriability factors. In fact, the effectiveness of patents to stimulate innovation and diffusion continues to be debated. Patents pose a trade-off between the social benefits from stronger incentives for invention and losses in consumer welfare as a result of higher product prices as patents create some degree of market power for the patent holders. Opponents of patents claim that they limit future opportunities for innovation as productive patent holders are likely to block innovations that compete with their own products while unproductive holders of broad patents can delay the emergence of new technologies and industries. In addition, patents raise innovation costs due to litigation over supposed patent infringements. Patent data are of limited relevance to test these ideas, as many innovations occur outside the patent system. Considering broad historical evidence, Moser (2013) finds that the majority of innovations in countries with patent laws indeed occur outside of the patent system and that secrecy has been the most important mechanism to protect inventions, although the degree to which this happened varies between industries. Historical evidence further suggests lessons for improving patent system design. For example, patent pools, created to avoid litigation risks in the context of overlapping patent grants between firms for the same technology, tend to discourage R&D by outside firms as the pools' licensing schemes favour their members. This means that the design of such pools needs close attention from regulators.

If duplication were perfect, costless and immediate, no incentive would exist to innovate, since appropriability of benefits would be zero. Owing to not having to cover R&D costs (fixed investment, threshold or entry costs) competitors could then generate products or services at lower prices than the original innovator. On the other hand, extreme appropriability hampers diffusion of innovations and is thus also detrimental to social welfare. The original innovator could dominate the market in an early stage of product development, which, in turn, would lead to stagnation of subsequent innovation efforts, and possibly monopolistic behaviour by the innovator. In a traditional welfare economics context, imperfect appropriability is conceptualised through the notion of positive knowledge or innovation externalities or knowledge spill-overs without any monetary compensation. Externalities can never be entirely avoided and it is widely accepted that optimally regulating them through subsidies or patents is an ideal that will never become reality.

Table 10.3 Industries within technological regimes

Regimes	Industrial sectors	Focus	Technological opportunity and entry barrier
Life-science based	Drugs, bioengineering	Product	High–high (knowledge)
Physical-science based	Electrical Telecommunications Instruments (photography and photocopying)	Product	High–high (knowledge)
Fundamental processes	Chemicals Mining and petroleum	Process	Medium–high (scale)
Complex systems	Motor vehicles Aircraft Computers	Product	Medium–medium/high
Product engineering	Non-electrical industry Instruments (machine controls, electrical and mechanical) Fabricated metal products Rubber and plastic products Other manufacturing	Product	Medium/high–low
Continuous processes	Metallurgical process (basic metals, building materials) Chemical processes (textiles, paper and wood) Food, drinks and tobacco	Process	Low–low

Based on combining information in Tables 5.1 and 5.2 from Marsili (2001).

Diminishing Returns to Scale of Innovation

Are innovation efforts subject to diminishing returns to scale? This question is not easy to answer. Innovations related to one product or process technology might be characterised by diminishing returns. This is reflected in the notion of product or industry life cycle, characterised by the notions of conception, growth, saturation, decay and senescence. An effective strategy to counter diminishing returns is to complement technological improvements by learning and enlarging the scale of activity, or by seeking new markets or even applications. In addition, market mechanisms offer a way out, namely when marginal returns from additional innovations start to fall rapidly, firms will be stimulated to shift to a new line of R&D or 'technological regime', or be selected against – i.e. exit or be taken over. Table 10.3 lists main technological regimes along with relevant industries. Diminishing returns may be enhanced by lock-in of technologies, as it allows marginal improvements that are likely to be subject to diminishing returns, while obstructing fundamental and radical innovations. This will be further discussed in Section 10.5.

10.3   Innovation Impacts and Diffusion at Market Level

Economic Impacts of Innovations

Innovations have many economic impacts. First, they give rise to diversification, i.e. more technologies and products, and an increasing division of labour due to new specialised tasks associated with producing, improving, repairing and using the new technology. Second, as inventions and innovations often build upon pre-existing technologies, more complexity of technological design results. This is reflected by growing interdependence of activities – among and between sectors – that produce specific components of a technology, which has been referred to as increased roundaboutness of production (Faber and Proops, 1990) or the number of connections and hierarchical levels among basic components (Potts, 2000). Third, innovations often contribute to increasing productivity – think of how personal computers make people more efficient at certain tasks. Fourth, they also contribute to economies of scope, as diversity at the firm level is augmented and flexible manufacturing systems are made possible. Fifth, innovations allow firms to broaden their range of activities and products so as to be resilient in the face of market and competitive selection, resulting in Schumpeterian competition (Section 8.4). Sixth, innovations affect the structure of markets and may give rise to temporal monopolies or oligopolies. In turn, such markets structures influence the rate and direction of innovation. Seventh, paraphrasing Schumpeter, innovations 'creatively destruct' old technologies, hence contributing to structural change at sectoral or economy-wide levels. Eighth, innovations cause asymmetry in technologies among firms, sectors and countries to become larger, altering comparative advantages and stimulating exchange and trade. On the other hand, trade contributes to diffusion of knowledge, which diminishes the asymmetry. Finally, innovations do not just change the spectrum of technological options, but also lower the prices of certain goods and services, thus expanding the opportunity space for producers and consumers.

Imitation and Diffusion

New technologies stimulate adoption decisions by firms and consumers that involve replication, copying and duplication – or more generally, imitation. This can take the form of using patented information, spying and reverse engineering. While the latter two are the cheapest, they tend to result in imperfect imitation. Diffusion, the spatial and temporal spread of earlier innovations, follows from imitation. Unlike innovations, diffusion follows a qualitatively predictable pattern, namely a logistic curve pattern, as shown in Figure 10.2. This reflects an epidemiological type of process, which is consistent with product life-cycle (trade) theory (Vernon, 1966) and life-cycle theory (Klepper, 1997). It covers the stages of conception or introduction, early and late growth and maturity or saturation. Stages go along with particular innovations in products and processes as well as specific industry and market structures. At an industry level, the cycle follows a pattern of many firms with much variation in the initial phase, followed by competition ending in one or a few dominant designs, production-scale increases,

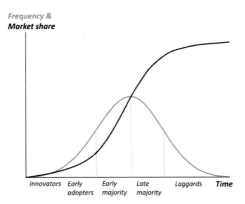

Figure 10.2 Distribution of adopters (light curve) generates logistic diffusion pattern.

more price competition and the exit of many firms. The latter is known as 'shake-out' and can involve the disappearance of 90 per cent of all firms within a few decades, as happened in the automobile industry after 1910, and later in many other sectors, such as television, radio, car tyres and internet-related industries. Shake-out can contribute to technological lock-in as technological diversity and competition are simultaneously reduced (Klepper, 2002). In addition, one can see a boom in product innovations followed by a flourishing of process innovations once few designs have survived and are to be improved. The reason is that process innovations require the product to be quite standardised, in order to be worth the effort and cost. Empirical research has shown that firms with more experience tend to have a higher survival rate in a shake-out process. The reason for this is that they are better in both technological and market strategies. This also holds for spin-offs of experienced companies, which suggests that routine-like DNA reflecting built-up experience can be transmitted to some degree.

According to Boschma et al. (2002, Section 4.4.5), product life-cycle theory is not suitable to describe the dynamics of services as the distinction between product and process innovations is less sharp here. They also suggest that the theory focuses on market concentration (horizontal) and is worse at predicting the degree of vertical integration in industries. The previous mechanical view on expected technological diffusion has been more fundamentally criticised for neglecting several subtle issues: discrete jumps due to income–consumption relationships; thresholds and entry costs in firms; interactions between supply and demand factors; new technologies stimulating improvements in the old technology that slow down their diffusion; and uneven and erratic diffusion of techniques. Hence, the pattern in Figure 10.2 is merely indicative and one should expect to find many variations of it.

International Diffusion and Trade

Diffusion is not limited by national frontiers. It occurs also at an international scale of economic activity. For at least a century most innovations were generated in a small number of OECD countries. This is because of a strong country specificity in technological expertise, sectors and trade, which derives from cumulative and historical, path-dependent features of technical change. Developing countries have a relatively low degree of innovation, due to a combination of lack of tradition of entrepreneurship, protectionism, lack of basic capital and technology endowments, as well as domination of Western-owned multinationals. As a result, technological diffusion has occurred mainly from OECD to developed countries. Krugman (1979) models North–South trade as a process of product innovation in the North which, through technology diffusion with some delay, leads to low-wage competition from the South. Continuous innovations are thus needed to maintain industry and welfare in the North.

Evolutionary trade theories argue that trade derives from technological differences between countries. Diversity of technologies among firms and countries contributes to international comparative advantages. Two types of technological differences, and their trade consequences, can be distinguished (Dosi et al., 1990): intersectoral, intranational differences give rise to sectoral specialisation patterns consistent with traditional comparative advantage trade theories; intrasectoral, international absolute differences lead to adjustments in world market shares, which is neglected in traditional trade theories (Daly, 1993).

Foreign trade patterns resulting from the foregoing cases, in turn, contribute to technology diffusion at an international scale. Two additional factors can enhance international diffusion of technological innovations. First, wide-scale use and refinement of technologies can be based on worldwide patents and standardisation. Second, multinationals with plants in many countries, including North and South, affect trade and diffusion. Often they are part of oligopolistic markets, work in consortia and joint ventures, can generate large investments in R&D, and are flexible in globally rearranging funding and, thus, responding to changing national opportunities, including those arising from new public regulations.

10.4   Long Waves at a Macro Level

Besides growth, long-run impacts of technological innovation may include so-called 'long waves'. These can be defined as cycles of prices, wages, outputs, foreign trade, interest rates and various other economic variables. The notion of waves, or cycles, suggests up- and downswing, also referred to as 'rise and decline' or 'boom and depression'. Many opinions have been expressed regarding the existence and origins of waves as well as the reasons for their existence (Freeman, 1990). Four types have been identified:

- The Kitchin cycle: about 40 months. It is suggested to be related to maintaining inventories. Nowadays similar short cycles may be due to political (election) cycles.
- The Juglar or business cycle: 7 to 11 years. Suggested to be related to adjustment of capital investments responding, with delay, to price changes driven by an imbalance of demand and supply.

- The Kuznets cycle: 15 to 30 years. It has been explained by waves of migration and weather conditions, notably 'lunisolar tides' affecting rainfall and in turn crop production (Tylecote, 1992).
- The Kondratiev (also spelled 'Kondratieff') long wave: 40 to 60 years. Mainly attributed to major technological innovations and scientific breakthroughs.

The latter long wave has been much studied and is most relevant in the current context of technological evolution. Its name derives from Nikolai Kondratiev[1], founder-director of the Institute of Conjuncture in Moscow, who, during the 1920s, identified three wave periods: for England and the United States running from approximately 1790 to 1814, for the same countries and France for the period 1844 to 1875, and for Germany and the world for 1896 to 1920. Schumpeter later accounted for these in terms of four phases – namely prosperity, recession, depression and revival (Kuznets, 1940). Table 10.4 summarises five long waves, and the beginning of a sixth one, during the past 250 years.

Table 10.4 A chronology of Kondratiev waves

Label	Features	Approximate start
Industrial Revolution (Britain)	Mechanisation, textiles, iron, canals, the steam engine and entrepreneurs	1770
Victorian boom	Coal, steamships, railways, small firm competition, shareholders and international free trade	1830
Belle Époque (Europe) / Progressive era (United States)	Electricity, electrical and heavy engineering, emergence of giant firms, monopolies and banks, R&D departments within firms, imperialism and colonisation	1880
Fordist mass production	Road and air traffic, urban growth, oil, synthetic materials, assembly line, multinationals, supermarkets, international tourism,	1930
Post-war golden age	Telecommunications, computers, software, internet, robotics, satellites, mobile technologies, digital services	1970
Low-carbon economy	Renewable energy, circular economy, electric vehicles	2010

Long waves can be partly understood through the theoretical notion of technological paradigms in combination with temporal clustering of innovations. Technological or techno-economic paradigms, analogous to scientific paradigms, denote major shifts in methodology. They reflect the role of demand–supply interactions and related price changes which, along with existing knowledge, determine the boundaries and opportunities for innovations. Mokyr (1990, p. 292) suggests the term

---

[1] Although others, notably Parvus and Van Gelderen, wrote about a similar long wave before Kondratiev, in his 'Business cycles' Schumpeter (1939) coined it 'Kondratieff wave'.

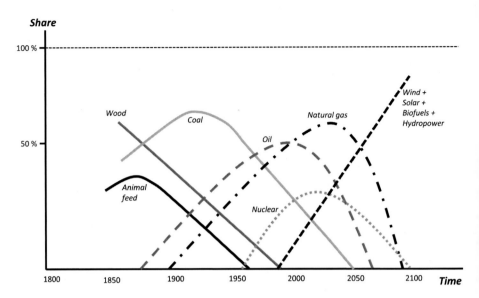

Figure 10.3 Historical and ongoing long waves of energy carriers.
(*A black and white version of this figure will appear in some formats. For the colour version, please see the plate section.*)

'macro-invention', as opposed to micro-inventions, to denote a radical new idea with huge technological and possibly economic implications. It is analogous to Goldschmidt's (1940) 'macromutation', associated with saltationist, non-gradualist views in evolutionary biology (see Section 4.3). Macro-inventions generally need micro-inventions to be turned into operational and profitable innovations. Both are essential to the success story of technology because '[w]ithout new big ideas, the drift of cumulative small inventions will start to run into diminishing returns' (Mokyr, 1990, p. 297). Hence technological change is a cumulative, path-dependent, directive process involving macro- and micro-inventions. It can take off in two ways, through a macro-invention or following a sequence of micro-inventions. Of course, one can debate the precise dividing line between the macromutations and micro-innovations. Moreover, rapid gradualism through micro-innovations may, from a distance – i.e. looking back at history without seeing all the detailed dynamics behind it – seem like a macro-invention. This suggests one has to be careful with the notion of macro-invention. As illustrated in a sketchy manner in Figure 10.3, energy carriers have followed a clear long wave pattern, driven by discoveries and macro-inventions giving rise to steel, electricity, the automobile and nuclear fission technology. Such a long wave is not surprising, as energy is the most essential primary input, and ultimate scarce factor, to the economy.

Explanations for Kondratiev long wave abound. The ultimate source of new paradigms is fundamental advances in science. These trigger radical commercial innovations in particular sectors and firms, which represent early stages of new technological paradigms. An innovative key factor or technology generates many related innovations, causing a clustering of innovations over time. Owing to a phenomenon resembling the product life cycle – notably a final phase with saturation, senescence, marginal improvements and diminishing returns to further investments in the dominant technology – a pattern with up- and downswings results that can be interpreted as a long wave. In addition, downswings or depressions stimulate entrepreneurs to be more adventurous, undertaking experimentation with radical organisational and technical innovations. This leads old technological regimes to be closed down and new technological routes to be opened up. In turn, this can launch a new wave (Tylecote, 1992). Various other explanations for long waves have been suggested:

1. Patterns of investment in long-lived capital assets such as buildings, fixed capital and infrastructure. Short-lived capital of eight years would be responsible for the Juglar cycle, while long-lived capital such as rolling stock of railways, with a life expectancy of 40 years, would be responsible for the Kondratiev cycle.

2. Sluggishness in changing economic structure, partly driven by lock-in factors (next section).

3. Slow diffusion, meaning that it takes a long time to replace an old technology with a new one.

4. International diffusion and inequalities, which causes a distinction between leaders and followers among countries.

5. International and civil wars, which in the past often occurred with some regular time intervals.

6. Demographic dynamics due to prosperity and depression having distinct influences on fertility and, possibly also, diseases.

7. Monetary factors, such as discoveries of gold induced by its relative price, or lending conditions by banks, which cause debt inflation and the credit cycle.

8. Optimistic borrowing and lending of land and property without real underlying value, which gives rise to speculative bubbles.

9. Fluctuations in trust and business confidence that affects the general investment climate for both capital and R&D.

10. Emergence of new key resources and sectors, such as coal, oil, plastics, automobile, information and communication technology.

An aggregate classification of ultimate factors behind waves is: natural environmental, social, technological and economic. Examples of natural factors are Jevons' sunspots and climate–weather patterns that arguably influenced the productivity of, and thus prices of, agricultural crops. Examples of social factors are wars, revolutions and opening up trade with new countries. Examples of technological factors are inventions and scientific discoveries, while examples of economic factors are infrastructure investments and monetary-financial processes. Combinations are also possible: for example, resource (e.g. gold and oil) discoveries combine natural and technological factors, while diseases can be seen as a combination of natural and social factors.

Through the downturn phase, long waves are tightly linked to economic crises. Perez (2013) casts a historical angle on the relationship between finance and long waves to answer the question of to 'unleash a golden age after the financial collapse'. She puts us with our feet on the ground by clarifying that crises are a recurrent historical event typically encountered in between successive technological revolutions. She provides evidence for the idea that crises following major technology bubbles contain much innovation potential that can be unleashed if the right public policies are implemented. Indeed, in the past there have been several times when financial recessions were followed by Golden Ages. Perez proposes that the current era offers the option of entering a 'green golden age'. In this respect, one may wonder whether the conditions for governments to become proactive are as favourable as in previous times, such as the period immediately following World War II. Economic revival in the form of a sustainability transition arguably requires new institutions, equivalent to Bretton Woods and the New Deal, which are able to unleash the right types of innovation and investment. An example is carbon taxation of markets (Section 14.5). A barrier to change is that the people who carry a chief important responsibility for the crisis are rich and powerful as well as in control of decisive levers almost everywhere in the economy –in business, finance and politics. Moreover, even though many observers have expressed the need for change, understanding of the fundamental nature of the changes required is generally poor. Perez suggests that a crisis may need to last longer and become deeper before people realise the magnitude of the institutional changes required.

10.5   Increasing Returns, Path Dependence and Lock-in

The more a technology is adopted the more attractive it often becomes to potential future adopters. This is known as self-reinforcement, positive feedback or increasing returns to adoption. Another term is hyper-selection (Ayres, 1991), which suggests a clear evolutionary basis of the concept, namely that it derives from repeated selection in favour of one option or technology. The reasons for this phenomenon include:

- *Conventional economies of scale*: By reducing per unit production costs these allow for better price competition.
- *Learning by doing*: More experience translates into lower costs of production and better performance of products, also known as learning by using (Rosenberg, 1982).
- *Dynamic efficiency*: If a product or technology is more adopted then additional resources will be available for its development and perfection, which in turn will make it relatively attractive for potential adopters. Moreover, emergence of a new technology that competes with a dominant incumbent technology often stimulates innovation in the latter with the aim to maintain its competitive position. This is known as the 'sailing ship effect' (Ward, 1967).
- *Imitation or bandwagon effect*: This demand-side effect means that the more consumers have already adopted a product, the more likely it is that newcomers will buy it as well. The latter tend to be motivated by imitation to avoid search costs or reduce the risk of dissatisfaction with the product.

- *Network externalities*: Being connected to a larger (social) network has an advantage in terms of potential interactions with other members. This is relevant for traditional and mobile phones, computers and software (e.g. Microsoft versus Apple), social network and other websites on the internet (notably for consumer-to-consumer sales).
- Related to the two previous effects are *informational increasing returns*: If a product is more adopted and hence better known, risk-averse individuals are more likely to buy it.
- *Technological interrelatedness*: The success of a technology depends on other, complementary technologies or infrastructures that are already mature. For example, the combustion engine has an advantage over electric automobiles because it goes along with an extensive, mature infrastructure of gasoline filling stations.

Not all these effects are equally relevant to all technologies or products. Which are depends on their particular production, design, economic and social features. Of relevance are, among others, the magnitude of economies of scale in production, the speed of learning in producer or user stages, the role of demand-side informational effects, the existence of network features, and the role played by complementary technologies.

The presence of increasing returns means there is no perfect competition in terms of basic, intrinsic features (quality, price) of products or services. Instead, history matters. Due to increasing returns, a product that initially obtains a larger market share, by coincidence or 'historical accident', has an advantage and can keep competitors at a distance. Increasing returns imply multiple potential equilibria. This causes path dependence as then the path taken will determine the final equilibrium. Adopting an evolutionary viewpoint, one can say that the dynamics of diversity of options and their market shares affect the ultimate outcome. This evolutionary angle on competition differs essentially from the atemporal market equilibrium thinking in standard microeconomic textbooks. Evolutionary economics – including both evolutionary game theory and agent-based models – typically study the dynamic paths to an equilibrium by assuming some kind of evolutionary mechanism of diffusion of a new technology, idea, product or service. The traditional view in economics is that multiple equilibria are undesirable and should be avoided as much as possible by making adequate assumptions. But this does not recognise that equilibria may be unstable, unreachable, sensitive to starting conditions or dependent on the historical path. Just adding mechanistic dynamics is not sufficient to fully grasp increasing returns to adoption, as they tend to result from large numbers of agents or technologies, suggesting the relevance of a population approach. Not only innovation economics, but also network economics (Economides, 1996), economic growth theory (Romer, 1994) and modern trade theory (Helpman and Krugman, 1985) have recognised the importance of increasing returns and multiple equilibria. However, evolutionary approaches to deal with them are mainly used in innovation economics. Arthur (1994, Chapter 2) shows that for constant and diminishing returns the system does not get locked-in and therefore is predictable: in the long run, on average equal market shares hold.

Path dependence means that the adoption process towards the final state of the system is non-ergodic, i.e. it depends on the way adoptions are built up. Alternatively, in an ergodic or stationary (in statistical terminology) process, the system repeats its behaviour, even if the cycle may sometimes be long. Instead, non-ergodic systems look more irregular. Evolutionary systems are generally non-ergodic. In line with this, evolutionary processes are unstable and sensitive to initial events, sometimes referred to as 'historical accidents', which Arthur compares to 'founder effects' in biology (Section 3.5). This means that small initial events have large consequences, consistent with the idea of multiple potential equilibria. This explains why evolution is connected to indeterminacy and unpredictability.

Path dependence can further be interpreted as temporally remote events having a significant or dominant impact on the present. It suggests that the logic of the world can be understood only by uncovering how it got this way (David, 1985). Note that path dependence is fundamentally related to the evolutionary nature of systems: large diversity that changes over time due to selection and innovation makes it extremely unlikely that an exact earlier state will be revisited. This becomes clear from probabilistic reasoning about large numbers as are characteristic of real-world evolutionary systems – irrespective of whether we are considering populations with a huge internal diversity of molecules, organisms, firms, consumers or products.

It should be noted that distinct authors emphasise particular aspects of path dependence and consider it more or less deterministic and restrictive, while further diversity is due to applying the notion to products, technologies or institutions. For a more complete entry into the broad literature, see Arrow (2000) and contributions in Magnusson and Ottosson (2009). While acknowledging the fundamental contributions by Paul David and Brian Arthur, Arrow notes that path dependence was already implicit in ideas expressed by Schumpeter, Veblen and Cournot (the latter writing in the first half of the nineteenth century).

Path dependence can result in a situation of lock-in, meaning that a dynamic pattern of competing technologies, habits or institutions, owing to increasing returns to scale, ends up in a situation with one being dominant or lock-in – and the alternative options being locked-out. This can be seen as consistent with the phenomenon 'the winner takes it all'. How this precisely works is discussed in Box 10.2. Unless the dominant technology becomes obsolete, a much cheaper alternative appears or a large-scale coordinated effort is undertaken to change the technology, lock-in will be extremely durable. This can go along with considerable social costs in the case of a locked-in alternative that is economically inefficient, or creating negative social or environmental effects. The next section will examine policies to encourage escape from such undesirable lock-in.

There is ample empirical support for lock-in and path dependence driven by increasing returns to adoption. The QWERTY keyboard is perhaps the most famous example. Although an ergonomically better alternative seems to be Dvorak, QWERTY dominates due to its application to typewriter keyboards, with the purpose of avoiding keys/hammers sticking. David (1985, p. 333) says about this: 'From the inventor's trial and error rearrangements of the original model's alphabetical key ordering, in an effort

Box 10.2 Arthur's evolutionary model of technological path dependence and lock-in

A theoretical model explaining path dependence driven by increasing returns to adoption with lock-in as an outcome is shown in Table B10.2.1. It generates the pattern shown in Figure B10.2.1. Intrinsically, R-agents prefer technology A (i.e. $a_R > b_R$ for intrinsic preference parameters), while S-agents prefer B ($a_S < b_S$). Returns of using A and B to R- and S-agents are, however, not fixed but change over time under the influence of adoptions of A and B. This means increasing returns to scale are at work, such as a bandwagon effect (imitation) or informational externalities. As a result, if one technology is relatively much adopted, the preference for that technology by both R- and S-agents will be higher. When the balance between adoptions is extremely shifted towards one technology, both R- and S-agents will adopt that technology. The shifting of balance may start with a random element, namely the order or path of R- and S-agents making choices: If, as shown in the figure, during some period many S-agents appear, then B will become more attractive for both types of agents.

Note: a_X and b_X denote intrinsic preferences of agent X ($X = R,S$); r and s are external returns to agents R and S, respectively, due to previous adopters; and $n_A(t)$ and $n_B(t)$ denote previous adopters at time t of A and B, respectively.

Table B10.2.1 Returns to adopting technology A or B, given n_A and n_B previous adopters, respectively

Type of agent (diversity)	Returns to adopting technology A	Returns to adopting technology B
R-agent	$a_R + r \cdot n_A(t)$	$b_R + r \cdot n_B(t)$
S-agent	$a_S + s \cdot n_A(t)$	$b_S + s \cdot n_B(t)$

Source: Arthur (1989, Table 1, p. 118). © 1989 Royal Economic Society. By permission from John Wiley & Sons Ltd.

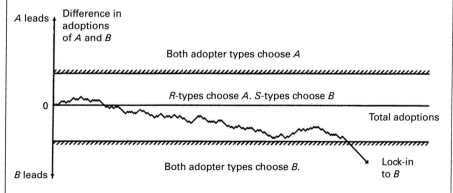

Figure B10.2.1 Illustrating path dependence with two adopters.
Source: Arthur (1989, Figure 1, p. 120). © 1989 Royal Economic Society. By permission from John Wiley & Sons Ltd.

to reduce the frequency of type bar clashes, there emerged a four-row, upper case keyboard approaching the modern QWERTY standard.' He also suggests a 'sales gimmick' was involved in the design, as all the letters present in one row allowed a salesman to impress his customers by rapidly typing 'TYPE WRITER'.

Other cases of lock-in are as follows (e.g. Liebowitz and Margolis, 1995):

- VHS video systems – an alternative that performed equally well if not better was Betamax.
- Alternating current (electricity) – an alternative that may function better under certain circumstances is direct current.
- The fossil fuel engine in automobiles – alternatives with various attractive features, notably having fewer local and global emissions, are hybrid and electric engines.
- Microsoft DOS and later Windows operating systems for personal computers – a more stable alternative is Linux.
- Social network websites attract new members or visitors more easily when they already have many. This is a main explanation for the dominance of early founded sites such as *Facebook*, *LinkedIn*, *Twitter* and *YouTube*.

Also in a spatial context, path dependence and lock-in play a role, through agglomeration effects, which are a special type of increasing returns to adoption. This may lead to one or multiple locations dominating the geography of economic activity. As in the case of technological lock-in, this can only be understood by taking dynamics into account. Through path dependence, agglomeration builds up gradually in several locations. A particular location can become locked-in, often depending on a historical accident. Arthur (1994, Chapters 4 and 6) shows that history matters for how cities develop, due to chance and the interaction of geographical factors ('necessity'), such as topology, resource availability and natural infrastructure. Infrastructure is relevant to geographical lock-in, as it functions as a complementary technology and causes increasing returns to scale through network effects (see Section 8.7).

Software and digital information are technologies that are amenable to lock-in. The dominance of initially DOS, and later Microsoft, in operating systems of personal computers illustrates this well. Since they are characterised by an almost zero cost of reproduction, producers enjoy significant scale economies. Moreover, users can, if the technology is unprotected, make free (illegal) copies (software, data, music, images). The combination of high fixed costs and low marginal variable costs means that marginal cost pricing is non-optimal and, instead value or strategic pricing, aimed at dynamic increasing returns is used (Shapiro and Varian, 1999). Modern technologies, such as social media on the internet and mobile digital communication technology, have important implications for lock-in as well. They allow rapid and wide diffusion of information about products and their prices. Advertisement costs are low, and, moreover, direct marketing – using information about the past purchase record and profile of customers – is feasible. These allow creating network and bandwagon effects on the demand side. An example is information provided by web-shops such as 'customers who bought this product also bought . . .' to influence customers' additional purchase decisions. In this way, technologies and communication can contribute to lock-in of physical products.

Finally, lock-in raises questions about the use of standards, which increase both costs and benefits. Lock-in of a suboptimal technology means a lack of fair competition among alternative designs, raising costs for consumers. In the presence of a technology standard, exchange of lessons by consumers or producers is easier, reducing costs. In addition, a standard leads to fewer resources being wasted on technologies that ultimately will become extinct. However, the evolution of technology unavoidably requires such waste, so that the relevant question is at what time during a life cycle of a technology one should implement a standard and, as a result, reduce technological diversity. The evaluation of the costs and benefits of standards is difficult because of the many uncertainties involved. In practice, standardisation is not based on some net benefit optimisation but simply the outcome of agreements among firms that have developed competing designs. In Section 10.7 we will consider the related notion of optimal diversity in more detail.

Standardisation through mass production became significant during the rise of industrialisation, accelerated through electrification and associated automated processes, and was perfected in the period after World War II to maximise economies of scale. More recently, one can see a new trend. In the computer and automobile industries, some products are now only assembled once buyers have expressed an interest in them. The internet has given rise to further important innovations in this respect, such as tailor-made solutions for specific consumer requests. This may result in flexible manufacturing systems in which economies of scope are more relevant than economies of scale associated with standardisation and mass production. One might interpret it as creating diversity in production to match diversity in demand.

10.6   Managing Technological Innovations

Evolutionary economics offers various suggestions for technological management and policy, related to fair market competition as through antitrust law, support of new technologies with public subsidies of some kind, stimulating innovation and entrepreneurship as in the form of patent law, or fostering entirely new activities or even sectors through deliberate industrial policy. At an abstract level, one can regard evolutionary policy as aimed at achieving desirable short- to long-run economic and technology dynamics by encouraging the right balance of diversity, selection and innovation.

According to Boschma et al. (2002, Section 8.2) traditional policy views fall short because of particular features of the evolutionary economy: bounded rationality, which means that price incentives will not always have the intended, optimal effect; heterogeneity, which causes policies to not have uniform effects across agents; and path dependence, implying that early, and possibly wrong, decisions have major long-term consequences. These points do not suggest that traditional policy insights are useless, but that they may need adaptation or precision. For instance, price incentives will work but their setting should reflect the difference between rational and boundedly rational behaviour. In addition, some evolutionary insights add to, or strengthen, traditional economic ones. For example, an evolutionary argument against market power in the form of a monopoly or oligopoly is that the associated market has limited technological and organisational diversity, in turn hampering future adaptive change.

The standard theoretical starting point for technological policy is the insight that the nature of technological R&D or knowledge innovation generates positive externalities, that is, it allows others who have not invested in the R&D to benefit from the new knowledge at low or zero cost. This implies that there is underinvestment in R&D as appropriability of its full benefits is not guaranteed. This can, in principle, be resolved by price corrections using taxes or subsidies, or by patent legislation. Both, however, go along with informational problems, among others, relating to uncertainty of R&D or innovation success. In other words, it is not possible in practice to come up with optimal policy solutions. Nelson and Winter (1982, p. 387) note that '[i]f patents prevent direct mimicking, but there is a "neighbourhood" illuminated by the innovation that is not foreclosed to other firms by patents, the externality problem remains.'. On the other hand, competition for patents may also lead to too many R&D efforts by all firms together, namely if the patent offers too much protection. Since a similar problem of overuse of common-pool resources with an open access character has been noted in the literature on environmental economics, this problem has been denoted the 'oil pool problem' (Nelson and Winter, 1982, pp. 387–388). Evidently, the externality and oil pool problem work in different directions on R&D expenditures, so that the net policy effect is generally unclear. The oil pool problem suggests that firms tend to cluster around new innovations, so that innovations concentrate on certain parts of the 'innovation landscape'. Since this is not desirable from a social welfare angle, governments should stimulate diversification in innovation efforts, possibly through subsidies.

Another issue relevant to policy is the relationship between market structure and R&D effectiveness. This has received much attention in evolutionary as well as neoclassical economics (Kamien and Schwartz, 1981; Nelson and Winter, 1982; Stoneman, 1995). Competitive markets can stimulate diversity in R&D research and thus enlarge the probability of successful inventions and innovations. A monopoly, on the other hand, will be better able to appropriate or internalise all benefits, because it can apply a new technology to a large quantity of output (products) or inputs (production technology), assuring low unitary R&D costs and considerably learning. However, it will tend to reduce diversity of innovation and may even misuse R&D for strategic pricing, by creating barriers to entry through gradual optimisation of its product and processes, and generally minimise more costly radical innovations. In view of this, many authors in both evolutionary and industrial economics regard the intermediate case of an oligopoly as a good compromise, as it means a trade-off between appropriability of R&D successes and variety of R&D efforts. This explains why such markets are commonly found in industries with rapid technological innovations. Nelson and Winter (1982, p. 390) warn, though, that an oligopolistic market may also combine the worst features of a monopoly and competition: either R&D will be defensively focused on imitating competitors, or progressive innovation by one player results in its domination and hence a monopoly. In addition, one should recognise that monopolies and oligopolies usually mean large firms with a high absolute level of R&D, guaranteeing economies of scale in innovation efforts. On the other hand, smaller companies often have a higher intensity of R&D outlays (Freeman and Soete, 1997).

Next, the government has a critical role in supporting basic, fundamental research, in large part undertaken at universities and affiliated institutes. As within science, there is competition for funds, a relevant management issue here regards the allocation of scarce research funds. A basic choice regards a fixed allocation to universities and a flexible allocation via science foundations in which individual researchers or research groups compete on the basis of research proposals or their curriculum vitae (i.e. in *ex post* evaluations or research awards). The allocation of funding between distinct sciences and disciplines – including natural, medical and social sciences and humanities – is difficult to decide upon and, as a result, it often depends on historical choices and path dependence. This reflects the evolutionary process that underlies scientific knowledge, as was discussed in Section 2.3. In addition, how much public research should be undertaken to match research by private companies is difficult to answer. Given the huge R&D budgets of large corporations, it is possible that public research receives insufficient funding. It is clear that both are needed as they are complementary. Nelson and Winter (1982, pp. 392–393) argue that R&D is surrounded with so much uncertainty that a central planning approach will not work, so that private research and decentralised public research are the best options. A barrier to public research planning is that while governments would need information about research inside firms, this is surrounded by secrecy. An optimal balance between private and public research, as well as optimal cooperation between the two, is thus complicated.

Science is often accused of generating too few practical innovations. Consequently, suggestions are made to control and direct science, on the output side, using indicators of the societal value of academic research and, on the input side, evaluating research proposals on the basis of potential contribution to urgent problems. These approaches may reflect a misunderstanding of the role of science in the innovation systems composed of private and public agents. It is, in fact, quite possible that, through such directive practices, science is hampered in its function of generating radical innovations through basic research with a long gestation period. It may be killing to require that such research becomes profitable in the short run and serving particular interests. Perhaps more than any other R&D activity, science is evolutionary, in the sense that it produces a lot of what looks like waste in the short run. But waste is inherent to evolutionary and all innovative systems and may, in the long run, turn out to have unexpected adaptive values – explaining the notion of exaptation in biology (Section 4.4). Anyone claiming that wasteful research should be eliminated does not understand human intellectual history nor evolution. Rather than asking researchers for promises in the form of fancy research proposals, one might focus university funding on *ex post* success. Requiring *ex ante* promises merely result in researchers becoming divided between those who are risk averse and those who oversell.

A challenge for technological policy is responding to a lock-in of an inefficient or otherwise undesirable technology. As we saw in the previous section, path dependence due to increasing returns to adoption implies that competition is imperfect. This, in turn, means that market outcomes cannot be relied upon as arranging socially desirable systems – there is a clear need for government intervention. In particular, governments can set clear long-term goals, such as the 'zero-emission' target of California. This contributes strongly to a stable and effective selection environment that helps in escaping from current lock-in. Next, semi-protected niches can be created for technologies that are still expensive but promising (e.g. certain types of renewable energy), while subsidies can be given for pathway technologies (e.g. energy storage technologies) as these will imply huge social benefits because of their application to many industries and technologies. Unlocking further requires a coordinated change-over in the case of network externalities. This is nowadays referred to as 'transition management' (Section 13.5). For this, not only escaping lock-in but also avoiding early lock-in of new technologies are valid goals as it is often uncertain which desirable technology can become superior. A wise strategy is then a kind of portfolio investment, i.e. supporting multiple technologies and keeping competition among them fair, until it becomes clear that a particular improved technology is the most desirable. How to choose an optimal diversity of technologies is discussed in the next section. Chapters 13 and 14 will offer more details on technology policy in the context of environmental problems and climate change.

10.7   Optimal Technological Diversity

Choices regarding technological diversity play an important role in economics and innovation management, but often remain implicit. Once made explicit, the objectives of efficiency – in terms of maximum benefits, profits or welfare – and diversity are usually posed as conflicting with one another, because efficiency generally has a positive, and diversity a negative, association with returns to scale in markets. This conflict is weakened in a dynamic setting as diversity then appears to have benefits in terms of keeping options open and contributing to spill-overs or recombinant innovation. In other words, an appropriate degree of diversity contributes to good long-term performance of systems.

Several approaches are available to address diversification of investment under irreversibility and uncertainty. Option value theory (Arrow and Fisher, 1974) suggests that, under certain conditions, irreversible developments should be avoided or postponed until more information is available. The notion of quasi-option value is derived from this approach, which reflects the value of information obtained by waiting rather than irreversible investment and development. The theory of real options (Pindyck, 1991; Dixit and Pindyck, 1994; Luehrman, 1998) derives from options pricing theory in financial economics (Black and Scholes, 1973). Whereas a financial option is defined as the right to make a financial investment decision in a certain time interval, a real option then is the right to make a concrete business decision which involves physical, tangible assets such as equipment and buildings. A number of real options can be distinguished, namely postponement of investment until market conditions improve, expanding or downsizing, cancelling an existing project or learning through R&D. Taking into account real options can significantly change the valuation of investments as derived from a traditional cost–benefit analysis or net present value approach.

The relationship between diversity and system performance has received attention from various angles. Hong and Page (2004) find that a group of randomly selected agents can outperform a group consisting of the best-performing agents. This is explained by greater individual ability of the latter being more than offset by the lack of problem-solving diversity at the group level. Weitzman (1992, 1993) proposed and perfected (Weitzman, 1998b) a theoretical framework based on genetic distance to study optimal diversity, with applications to biodiversity in mind. Baumgärtner (2004) and Van der Heide et al. (2005) extended this approach with ecological relationships, giving distinct policy recommendations. Weitzman (1998a) presents a formal model to derive that the number of new combinations is a combinatorial function of the number of existing ideas, called 'hybridisation of ideas'. He shows that if this number were the only limiting factor in knowledge production, super-exponential growth would result. Combining ideas and technologies, however, requires R&D efforts and thus scarce financial, labour and capital resources. Olsson and Frey (2002) connect Weitzman's recombinant growth idea with Schumpeter's view of the entrepreneur as innovating by combining existing ideas or technologies in a convex way. They demonstrate that the resulting combinatory process is constrained by various additional factors: convexity of the search space implies exhaustion of technological opportunities; the cost of combining increases with the distance between ideas; and social norms and the ruling technological paradigm prohibit certain combinations of ideas.

Within evolutionary economics, several models are worth mentioning here. The *NK* complexity model (Kauffman, 1993; Frenken, 2000; see also Box 8.2) stresses the combinatorial nature of design space. Here *N* denotes the number of elements with diversity of values, while *K* is the number of connections between elements, serving as a measure of complexity. A fitness or performance function can be defined over the design space. This has been used to study actual and hypothetical designs of steam engines and their performance, casting new light on the history of this core technology of the Industrial Revolution (Section 12.4). Based on the idea that transitions are best perceived as involving multiple transition steps, Alkemade et al. (2009) have recently used the *NK* model to examine how one can best move in complexity space from one design to others, taking into account irreversibility and flexibility, uncertainty and the theoretically best-performing alternative, which functions as the end goal. Their model allows one to trade-off flexibility, shortest route and end point.

A simple model of variable, endogenous diversity is presented in van den Bergh (2008) to illustrate how one can analyse the optimal balance between increasing returns to adoption (as discussed in Section 10.5) and modular or recombinant innovation. It clarifies the conditions under which each of four solutions is optimal: complete specialisation regardless in which option, complete specialisation in one specific option, symmetric diversity (i.e. perfect balance) and asymmetric diversity. The model can be considered as combining neoclassical and evolutionary economic elements. It describes a process of spill-over, such as in the case of thin-film technology being used in solar photovoltaic technology to convert solar radiation into electricity. One can also see it as illustrating that innovation is a modular process, in the sense that the original options are joined in a new technology. An analytical elaboration is offered by Zeppini and van den Bergh (2013) and an extension by Safarzynska and van den Bergh (2011a). A number of policy insights derive from these studies. Governments employ public investments to assure that diversity of innovations represents an acceptable trade-off between scale and diversity benefits for society. This resembles, but is not identical to, the classic compromise between exploration and exploitation, or between flexibility and efficiency (March, 1991). A more fundamental role of the government is to manipulate the innovation parameters through adequate policies. In particular, it could stimulate the development of modular technologies which allow many innovative combinations. Furthermore, in deciding about setting or encouraging technology standards, several factors need to be considered: not only the benefits in terms of increasing returns to scale, but also the risk of lock-in and limitations for recombinant innovation. Policy might also stimulate radical innovations by raising the disparity between technological options, notably by directing public R&D at 'deviant' technologies, and by funding risky R&D. Finally, coordination of technological policies among various countries can abate the antagonism between scale and diversity.

Part V

Evolutionary Cultural History

11 Prehistory Until the Rise of Agriculture

11.1 Social Science Palaeontology

What can we learn about human cultural history by looking through an evolutionary set of binoculars? This involves drawing on core elements of previous chapters, dealing with biological evolution, social evolution, gene–culture coevolution, economic evolution and evolution of organisations and technologies. The basic idea is that selection of existing, and generation of new, variety are crucial if not dominant forces in socioeconomic-technological history. Evolutionary reasoning places a strong emphasis on the interaction between chance – due to innovation, creativity and events external to human societies or even the biosphere – and necessity – through ontogeny and selection. It gives rise to a set of causal mechanisms on the basis of which one can clearly separate between ultimate and proximate causes as well as proximate and ultimate consequences, a distinction that is often blurred in non-evolutionary approaches.

By adopting an evolutionary angle we can link historically important events and patterns in a single theory. It stresses the role of diversity – of groups, industries, firms, technologies, geographies, natural and energy resources, ecosystems and climates. An 'evolutionary law', in this respect, is that selection, i.e. some element of the socioeconomic diversity disappearing, and innovation, i.e. a new element appearing, changes the system permanently. What emerges is history as a multi-population process, giving rise to an alternation of trend-like patterns and moments with structural or qualitative change. When taking a closer look, such moments always come down to a relatively rapid positive or negative change in diversity within populations of activities, ideas or technologies, going along with a redistribution of power and income among certain groups in the human population. Hence, we will pay much attention to generalised populations of diversity in various socioeconomic subsystems as a possible key to understanding the particular histories. Changing populations with large numbers of elements or agents make it unlikely that an earlier historical situation is revisited. In other words, the (co)evolution of populations implies irreversibility and path dependence, or genuine history.

An impressive synthetic study of grand history, which inserts certain evolutionary angles, is Diamond (1997a). It stresses the role of biological diversity of animals and plants in the emergence of sedentary agriculture. However, unlike this chapter, Diamond's work does not make explicit use of theories of social, economic or technological evolution. Here we will devote attention to a combination of biological and cultural evolutionary arguments to explain human cultural history. For reasons of limited space, the treatment here cannot claim any historical completeness. Its aim is merely to illustrate the potential value of evolutionary thinking for historical enquiry.

11.2   Human Origins

The emergence of humans on the planet was preceded by a long evolution of consistent physiological, mental and behavioural traits in human ancestors. This is studied by human biology, genetics, anthropology, archaeology, behaviour ecology, ethology, linguistics, neurology and even economics. It is impossible to address all the relevant insights delivered by these disciplines. The following is, therefore, inevitably a sketchy and incomplete account of human evolutionary history. Moreover, the reader is warned that dating of first appearances of various species is rough and indefinite. Publications differ on exact dates and new evidence arrives while writing. The uncertainty range of the time period of human speciation is as wide as 80 000 to 400 000 years ago. This broad range is due to two factors. First, the border between humans and apes is fuzzy. Moreover, the full extent of mental evolution and associated human behaviour is imperfectly visible in the fossil and archaeological records. On the other hand, some of the other primates provide a rough model of how our ancestors behaved and what they looked like.

Simon Conway Morris (1998) thinks that an intelligent human-like being was a likely, or almost inevitable, outcome of a planet-wide process of organic evolution. Others, such as Stephen Jay Gould (1989), have expressed the belief that we are a coincidental and unlikely outcome of evolution, and that were evolution to be repeated, with slightly different starting conditions, what he called 'rewind the tape', it would not necessarily result in a world with organisms possessing human-like intelligence. Which particular conditions gave rise to humans? Climate change has been argued as having played a pivotal role. It changed the landscape from one dominated by a continuous cover with dense forests to a savannah landscape with patchy forests. Possibly this gave rise to isolated areas or 'vegetation islands', fostering the emergence of a new species.

Another suggestion is that pre-humans have gone through a water-rich phase (Hardy, 1960; Morgan, 1997; Finlayson, 2014). This so-called 'water-ape hypothesis' is motivated by a number of considerations: humans evolved in the tropical savannah near the great lakes in Africa; they share with sea mammals a naked skin; human nostrils are downward oriented; humans have a diving reflex; humans can stop breathing as well as rapidly inhale air; the human brain is consistent with a fish-rich diet; and walking upright could have started off in shallow waters as it stabilises the body. The relevance of an aquatic evolutionary phase to fully explain humans is not taken very seriously by most biologists, but neither is it completely rejected. Sceptics will argue that for most of the mentioned features there are better explanations. For example, voluntary breath control is probably a side effect of speech. Roberts and Maslin (2016) summarise the theory as: 'both far too extravagant and too simple an explanation. It attempts to provide a single rationale for a huge range of adaptations – which we know arose at different times in the course of human evolution'.

Many earlier ecological changes may have created the necessary conditions for the emergence of humans. For example, flowering plants with fruits appeared before mammals evolved around 65 million years ago. These coevolved, through spreading of seeds in the fruits, assisted by certain mammals, including primates. Potts (1996) suggests that (pre-)human adaptation and evolution can best be understood as a response to continuous changes in the environment in which pre-humans lived – what he calls 'variability selection'. Consistent with this is that humans are versatile and flexible, allowing adaptation to a range of habitats with distinct environmental conditions.

Humans as Primates

Observing that their anatomy is almost identical, the father of pre-phylogenetic taxonomy, Linnaeus, classified humans and apes in his *Systema Naturae* (first edition 1735) as a single category. While initially calling this *Anthropomorpha* (human-like), in the 10th edition of 1758 he renamed it *Primates*. Obviously, this inclusion of humans in zoological nomenclature created resistance as it effectively downgraded humans to animals. At the time, it was not yet known that humans and modern apes evolved from a common ancestor that lived some 4 to 6 million years ago, i.e. about 200 000 to 300 000 generations ago.

Humanity's closest relatives are the apes, which include gibbons and the great apes (chimpanzees, bonobos, gorillas and orangutans). Gibbons and orangutans live mainly in trees, while gorillas, chimpanzees and bonobos move more on the ground or in lower parts of trees. This likely relates to the fact that early hominids were mainly terrestrial. Currently, fewer than 500 000 great apes are alive, distributed among four subspecies of gorilla (about 100 000 individuals), three subspecies of orangutan (about 60 000), four subspecies of chimpanzee (about 170 000–300 000), and one species of bonobo (about 30 000–50 000). The evolutionary connection between these species is shown in Figure 11.1. Humans share many genes with extant species of apes and monkeys,

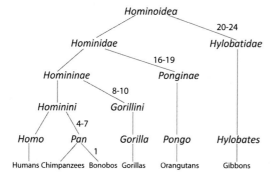

Figure 11.1 Evolutionary sub-tree of apes/hominoids.
Note: Numbers indicate approximate time of splitting in millions of years ago.

specifically 98.8 per cent with chimpanzees and bonobos; 98.4 per cent with gorillas, 96.9 per cent with orangutans, and about 93 per cent with monkeys. This particular decreasing order makes sense given the time elapsed since separation and the similarities/differences one can observe between the species. These numbers should be interpreted with care as, for instance, 2 per cent difference could mean that 20 per cent of all genes have a small but essential change, namely in 10 per cent of their molecular/information content. This could have enormous consequences for the build-up of the organisms. Still, the mentioned percentage overlaps are definitely useful in reconstructing family trees. Genetic distances to related primate species also reflect geographical distance. Humans are closest to African species, notably chimpanzees and bonobos (formerly also referred to as pygmy chimpanzees), at further distance from Asian species (orangutans), and furthest from American species. In fact, the genetic distance between humans and African species such as chimpanzees, bonobos and gorillas is smaller than that between these apes and non-African apes. Similarly, the genetic distance between humans and chimpanzees is smaller than that between chimpanzees and gorillas (Jones et al., 1992). This has given rise to the 'Out of Africa' hypothesis, now widely supported by both palaeontological and molecular findings (Ke et al., 2001).

Our next closest kin is formed by monkeys. The main division here is New World (Central and South America) and Old World (Africa and Asia) monkeys, owing to the geographical isolation and independent evolution over an extremely long period of time. The species in each group share a number of evolved features that are not found in the other group. An example is a flat nose with nostrils open to the side in New World monkeys versus a downward facing nose with nostrils closer together in Old World monkeys. Genetic and palaeontological evidence shows that the two groups of monkey species have the same origin, but separated about 40 million years ago. This has always puzzled researchers, as the ancient Gondwana, the southern part of the Pangea supercontinent, broke apart into what later would become Africa and South America much earlier, approximately 130 to 110 million years ago. Proposed hypotheses to explain the common origin come in four types: monkeys migrating to South America on a raft of vegetation, the presence of a temporary land bridge between the two continents, a floating island with monkeys on it, and monkeys hopping across a range of islands serving as stepping stones. To believe any of these explanations, one has to take into account that 40 million years ago the distance between Africa and South America was about two-thirds of the current 3000 km. The explanation taken most seriously now combines elements of two hypotheses: it involves a fall in the sea level in the Oligocene, which might have created a complete connection over land or a series of mid-Atlantic islands that made island hopping feasible (Oliveira et al., 2009).

Humans, apes and monkeys form together the simians or anthropoids. At larger genetic distance, we find more primitive species called 'half-apes' or prosimians, which include lemurs, lorises and tarsiers. Prosimians evolved about 55 million years ago, while monkeys evolved about 10 million years later and apes another 10 million years later. The geographical distribution of the 236 primate living species (note this number is debated) can be summarised as follows: 30 in Madagascar, 63 in mainland Africa,

Table 11.1 Main similarities and differences among apes, humans, other primates and other animals

---

Primates have a number of unique features relative to other animals:

- Ability to move the four limbs in various directions.
- Grasping power of the hands and feet (and independent mobility of the digits).
- Nails instead of claws.
- Well-developed hand–eye motor combination.
- Flexibility of the spine to allow twisting and turning.
- Vertical pointing body, with centre of gravity positioned close to the hind legs.
- Overlap of visual fields of eyes realising precise three-dimensional views (stereoscopic view).
- Teeth and digestive system adapted to an omnivorous diet.
- A large and complex brain compared to body size.
- A long postnatal growth and learning period before adulthood is reached.
- A small number of offspring.

Humans and apes differ from other primates in the following ways:

- They have no tail.
- Their shoulders, elbows and wrists are extremely flexible.
- Their body and limb positions and facial expressions share many similarities.
- In walking, the heel touches first to the ground.
- They communicate in subtle ways: vocally, visually (facial expressions) and through touching (holding hands, embracing, kissing, patting).
- They show empathy.

Chimpanzees and bonobos possess a variety of linguistic and mathematical abilities:

- Use words as symbols.
- Create combinations of words (new words).
- Refer to activities distant in time and place.
- Use words to deceive, gain information and comment on the environment.
- Understand the logic of cause and effect.
- Categorise new objects into groups (fruit, tool).
- Can count.

Humans differ from other primates in the following ways:

- They walk straight up on two legs (bipedalism).
- They have modified teeth and jaws.
- They have little body hair.
- Large brain (3× expected size if compared with other primates)
- They speak (other primates generate sounds through the nose).
- They are conscious or self-aware.
- They laugh.
- Through bipedalism and subtle cooperation they are better at predating and avoiding predators.
- They can carry things while walking and frequently use a variety of tools.
- They show very complex social interaction.

Based on information from Strickberger (1996, Chapter 19), de Waal (1996) and Jones et al. (1992).

63 in (south) Asia, and 80 in Central and South America. The majority of these live in rainforests, and a few in temperate zones (Jones et al., 1992, Section 1.3). Table 11.1 lists some important similarities and differences among primates, apes and humans. Most likely due to competition with their more intelligent simian cousins for food and

space in overlapping habitats, lemurs have not survived on the mainland of Africa. They are currently only found on the island of Madagascar, where apes are absent.

The diversity of primates is high compared with other mammal orders, but it might rapidly decline in coming decades. Many species already have small populations, and several face extinction in the absence of specific conservation measures. The threats to their survival are many: habitat loss through (il)legal deforestation; mortal diseases such as the Ebola virus, particularly affecting chimpanzees and gorillas; spreading of human diseases, notably viral ones, to other primates; and illegal hunting, for live capture and especially for the bushmeat trade.

Evolution from Early Hominids to Modern Humans

Why did intelligent humans evolve through the ape lineage? Pinker (1997) offers four reasons. First, primates are animals with stereoscopic vision, while most other mammals keep their head down much of the time to sniff at the ground. Primates were thus better able to evolve an understanding of the world as a three-dimensional space with other living organisms moving in it. Second, primates live in groups, which means that intelligence pays off as one can better keep track of the complex web of social relationships and the information flowing between individuals. Third, primate hands are unique in that they are capable of many tasks, especially when aided by an intelligent human mind. This involved a co-adaptation of precision in thinking and manual activities, which could only happen as primates are semi-bipedal and hence are not (all the time) using their hands to walk. Finally, hunting was a catalysing factor, on the one hand, as it stimulated intelligent cooperation and, on the other hand, as it was a great source of carbohydrates, fats and proteins, as well as all 20 amino acids, allowing the brain to grow in size.

The first phase of human evolution runs from the earliest hominid[1] type of ape *Aegyptopithecus*, some 30 million years ago, to *Proconsul*, about 20 million years ago. Somewhere between 20 and 5 million years ago the lineages of humans and the great apes split. If bipedalism is a distinguishing feature of humans versus apes – next to, of course, a large brain – then the oldest specimen of which fossils have been found is *Ardipithecus ramidus kadabba*, discovered in 2001 in Ethiopia and estimated to have lived about 5.5 million years ago. This human predecessor differed from our currently closest family members, the chimpanzee and bonobo, in that it had long fingers that were unsuitable for walking on its hands or jumping between trees. The same region provided the previously oldest human fossil of *Australopithecus ramidus* (*ramidus*), which is 4.4 million years old, and the famous Lucy – inspired by the equally famous Beatles' song 'Lucy in the Sky with Diamonds' that was repeatedly played in the

---

[1] The popular term 'hominid', a member of the family 'Hominidae', is used in various ways, many of which are not consistent with the scheme in Figure 11.1. Correct use of it denotes the great apes and all their ancestors back to the time when gibbon-like Hylobatidae separated from other apes (Hominoidea). While the term 'hominid' is frequently employed to denote all bipedal ancestors of humans, a more precise characterisation here would be 'hominid type of ape'.

expedition camp during the evening after the discovery. This speci(wo)men belongs to the 3.5 million years old *Australopithecus afarensis* which walked on two legs. Another prominent finding is still debated, namely the 6 million year old 'Millennium Man' or *Orrorin tugenensis*. As we are not sure whether it walked on two legs it might be either the first human or the last common ancestor of humans and the great apes. The previous picture may still change, though, in response to newly found fossils. Alternatively, some of the existing fossils could turn out to be dead ends of evolution.[2]

The human lineage continues with *Homo (Australopithecus) africanus* (3 to 1 million years ago), stone tool using *Homo habilis* or 'handy man' (2.5 to 2 million years ago), *Homo erectus* or 'upright man' with a much larger brain and modern body features such as long legs and short arms (more than 1 million years ago), who spread out to Eurasia, and archaic *Homo sapiens* or 'wise man' who appeared more than 300 000 years ago.[3] Earlier, between 800 000 and 400 000 years ago, Neanderthals (*Homo neanderthalensis*) had split from the line leading to modern humans. DNA research shows that Neanderthals are genetically closer to humans than any of the extant apes.

Some mixing between Neanderthals and *Homo sapiens* in Europe, Cro-Magnons, occurred (humorously represented in Figure 11.2). This explains why 1 to 4 per cent of DNA in Europeans and Asians comes from Neanderthals. It is quite certain that Neanderthals were not simply assimilated in Cro-Magnon populations. Owing to being less intelligent, they likely lost the competition for food resources and became extinct some 30 000 years ago. It is even possible that the last ones were killed by Cro-Magnons.

A third species, Denisovans, lived simultaneously with Cro-Magnons and Neanderthals until some 40 000 years ago. There is little known about it except that it split from human ancestors about 1 million years ago, and interbred with the others species. The genome of Melanesians (e.g. from Papua New Guinean) contains most Denisovan origins, namely 4 to 6 per cent. A wild suggestion by Rogers Ackermann et al. (2016) is that the success of *Homo sapiens sapiens* is due to being the outcome of a hybridisation – often acting as a creative evolutionary force – among the three species. For example, *Neanderthal* genes protected better against certain diseases (Temme et al., 2014). In addition, frequent contact between groups of different species might have stimulated diffusion and innovation of cultural habits (e.g. Neanderthals used stone tools), thus contributing to the emergence of modern culture.

---

[2] For example, in 2013, fossil skeletons of 15 specimens of extinct *Homo naledi* were found in a cave near Johannesburg in South Africa. The specimens combine primitive and modern characteristics: their fingers indicate a climbing ape, while the length of their legs and certain features of their feet, wrists and hands suggest similarity to early *Homo* species. Dating techniques show that the fossils are 'only' 236 000 to 335 000 years old. It is still a mystery how so many skeletons ended up in the cave. After eliminating common explanations, ritual mortuary of the dead appears as a serious possibility.

[3] While it was long believed that *Homo sapiens* first appeared some 200 000 years ago in Eastern sub-Saharan Africa (what is now Ethiopia), a recent discovery in Morocco of skull, face and jaw bones from various early *Homo sapiens* specimens questions shifts the date 115 000 years back (Callaway, 2017). This finding suggests that around 300 000 years ago *Homo sapiens* may have already been dispersed across the entire African continent.

Figure 11.2 Humorous account of mixing between Cro-Magnons and Neanderthals
Source: http://bizarrocomics.com. © Dan Piraro.

Historical economic modelling has been used to examine why the competition for food between Neanderthals and early *Homo sapiens* worked out positively for the latter and led to extinction of the former (Horan et al., 2005). This is often attributed to a better evolutionary adaptation of *Homo sapiens* to its natural environment. The question is, however, 'what determined this adaptation?' The authors suggest that *Homo sapiens* was less capable of competing through physical force – in terms of hunting, gathering food and even fighting – which stimulated the development of specialisation and trade. This allowed more efficient handling of various tasks, such as hunting, food gathering, producing clothes and constructing shelter, with positive impacts on survival and reproduction. Ofek (2001) suggests that Neanderthals lived in smaller groups with less labour division and did not undertake trade. He further notes that specialisation may have allowed *Homo sapiens* to even outcompete on average physically stronger Neanderthals in hunting, since only the most capable people might have been given the task to hunt, whereas less-able hunters would perform other tasks in the small economy. In addition, Neanderthals probably did not run as fast as humans given that their equilibrium organ was smaller (Spoor et al., 2003).

It has been hypothesised that relatively recently in human evolutionary history a population bottleneck in human evolution occurred caused by one of the Earth's largest volcanic eruptions some 70 000 years ago, at the site of present-day Lake Toba in

Sumatra, Indonesia (Rampino and Self, 1993). This event may have caused a global volcanic winter that lasted for several years as well as a centuries-long cooling period. It disrupted ecosystems globally and caused droughts in the tropical rainforest belt. This population bottleneck theory is supported by genetic evidence suggesting that human populations sharply decreased to 1000 to 15 000 surviving individuals. Allowing for genetic drift in a small population, this event may have contributed to rapid change of the human species (Section 4.2). This would mean that genetic differences among modern humans reflect disproportionately the effects of evolutionary change during the last 70 000 years. Similar population bottlenecks with a consistent time scale have been genetically identified for other mammal species, including several ape species. Several studies, though, cast serious doubt on the human population bottleneck due to the Toba supereruption (Anderson, 2007; Lane et al., 2013).

Geographical Spread of Modern Humans

Two competitive explanations of the final evolution and spread of modern humans are available. The first is known as the Out of Africa hypothesis in which a completely evolved human species spread to the rest of the world. A particular version of this single origin is the replacement model, which states that modern humans substituted – i.e. filled up the niches occupied by – other archaic human species outside Africa. These two explanations are consistent with the notion of 'Mitochondrial Eve' (Cann et al., 1987). The relevance of mitochondria is that they form part of the metabolic machinery which mothers transmit through the egg cell to their offspring. Analyses of mitochondrial DNA and male-specific Y chromosomes from people around the world hint at a common ancestor from Africa approximately 120 000 years ago (Poznik et al., 2013).

The alternative explanation is a multiregional model in which gene flow among regional populations took place, due to interbreeding, before human evolution was completed (Boyd and Silk, 2002, Chapter 14). This has been confused with a third theory, polygenism, according to which 'human races' are of different origins. While this is clearly an incorrect theory, the multiregional model receives some, though relatively little, support among biologists and palaeontologists. Instead, a weak version of the multiregional model is broadly considered as the most realistic. It reflects that humans have a single origin (in Africa) but later slightly evolved further in distinct directions in different parts of the world owing to some degree of isolation – while still allowing for modest gene flow. Although this might be objected as being inconsistent with evidence for Mitochondrial Eve, the debate continues as supporters of multiregionalism have identified flaws in the Mitochondrial Eve theory. Among others, the reliability of the molecular clock used to date Eve has been questioned, and it has been argued that archaic human mitochondrial strains were lost through genetic drift or a 'selective sweep' (Thorne and Wolpoff, 2003).

Because of natural barriers, humans spread intermittently across the planet. A first migration of early humans (*Homo erectus*) out of Africa to Eurasia took place about 1.9 million years ago. Humans ultimately reached China about 1.66 million years ago and Western Europe a little later (1.2 million years ago). A second species migrating to Europe was semi-modern *Homo antecessor* (800 000 years ago) and a third one *Homo heidelbergensis*, the likely ancestor of both modern humans and Neanderthals (600 000 years ago). Final waves of migrations, by modern man (*Homo sapiens*), involve those to Asia 80 000 to 120 000 years ago and to Europe and Asia 40 000 to 60 000 years ago. Finally, Australia was reached 40 000 to 50 000 years ago, the New World (North America) 16 000 to 20 000 years ago and South America less than 13 000 years ago (Callaway, 2015).

It was perhaps inevitable that closely related human species went extinct as modern humans did not leave any niche for them, due to human intellectual and derived social advantages, as well as overlapping food habits and living areas. For example, it was recently suggested that *Homo floresiensis*, nicknamed 'the hobbit' because of its small size, disappeared after modern humans appeared on the scene (Callaway, 2016). *Homo floresiensis* lived on the island of Flores, currently part of Indonesia, and descended from *Homo erectus* or *Homo habilis*. It may have become extinct around 50 000 years ago, at the same time as modern humans reached South East Asia and Australia.

Bipedalism

Humans are bipedal: they walk and run in an upright position. It is thought that the evolution from ancestral apes to humans went through apes that could walk but not run bipedally. In fact, the main connecting feature of hominids is habitual bipedalism. Several explanations have been proposed for the evolution of ancestor apes towards bipedalism. One often-mentioned theory states that apes left the forest under the influence of climate change and lived under new ecological conditions in grasslands which served as new selection pressure that, in turn, stimulated new adaptations. Bipedalism meant improved vision over high grasses and in the distance, allowing better perception of dangerous predators, free hands to carry things and make tools. In addition, grasslands hosted many larger herbivores, creating ample opportunity for effective hunting. Walking upright had other advantages. It meant less solar radiation reaching the body surface and better cooling by breezes, more relevant in a savannah landscape than a dense tropical forest from where human predecessors originate. It further allowed for a wider vertical range for collecting berries and fruits from bushes. These features made travel over long distances easier, increasing the chance of finding areas well suited for hunting and gathering. Bipedalism likely stimulated labour division, because men could walk and at the same time carry hunted food, while women could care for children. This in turn contributed to sexual selection for males who were good at hunting, indicated by carrying much food home. The resulting advantages of bipedalism have outweighed any initial disadvantages, such as being possibly slower than by moving on four legs.

Bipedalism was not an easy and isolated change. Its evolution involved various alterations in the physical structure of primates, in feet, knees, thighbone, pelvis and spine. Along with these, other new physical features arose in proto-humans, notably a larger brain and more capable hands, which – freed from moving on the ground – allowed for grasping and carrying things. Bipedalism helped in understanding the environment and making good use of it; and grasping environmental change and responding to it by controlling or influencing it (think of burning vegetation) or moving away from it. It also stimulated the use and making of tools, which started with *Homo habilis*. This was possible due to standing up and advanced hand–eye coordination. It, in turn, stimulated a further upright position. This allowed the creation of hunting gear, access to a greater variety of foods in the diet and the creation of better living conditions in terms of clothing and shelter. Through these activities, early humans had a great propensity to adapt to changing environmental circumstances, what Lewontin (1957) has called 'creative homeostasis'. In fact, their adaptive range far exceeded that of any other species, as human intelligence allowed for extended cooperation and, in due time, cultural-technological progress. As a result, humans can be found in any climate. They know how to create a suitable microclimate through clothes, housing and fire. This explains why early humans spread across the planet. In turn, it explains why the size of the global human population is much larger than one would expect for any organism its size.

Expensive Tissue and Cooking

Another important feature of humans is a highly varied diet over the last 40 000 years, involving many different animals, fish and plants. This was supported by a combination of cooperation in hunting, food sharing and use of tools and, not to forget, cooking. The human gastrointestinal tract (gut) is adapted to this omnivorous diet. While many now would judge cooking as an innocent pastime, primatologist Wrangham (2009) considers it to be the main distinguishing feature of humans among animals. He thinks that if pre-humans had not cooked, we might not have evolved into modern humans. Cooking may indeed explain why the human digestive system is small relative to body size, possibly only 60 per cent of what one would expect from comparison with other primates (Figure 11.3). This facilitated walking and running upright. Moreover, according to the 'expensive tissue hypothesis' of Aiello and Wheeler (1995), humans probably could not have evolved their energy-expensive brain without a smaller, energy-cheap gut, especially not in tropical areas as there animal meat contains little fat – as opposed to meat from animals in colder, Nordic regions.

The human gut resembles that of a carnivore animal. This does not mean humans shifted their diet completely to protein-rich meat. They always maintained great dietary variety, as this had particular advantages in terms of health, as well as quick access to energy. This variety was only possible as, due to human brain increase and associated dexterity of the hands, humans were able to externalise functions of the gut through manual processing and chemical modification of food. This involves such activities as cleaning, peeling, grinding and cooking. As a result, the digestibility and density of food was improved, saving the human gut from too much energy-intensive mechanical work, overload of substances and presence of toxins. An additional factor explaining food diversity is mercantile exchange (Section 4.8). Cooking, and later trade, substituted thus for the human incapacity – owing to short bowels as well as small teeth – to digest raw food, whether plants or animal meat. The invention of cooking may have

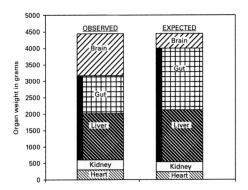

Figure 11.3 Observed and expected organ mass for a standard 65 kg human.
Source: Aiello and Wheeler (1995, Figure 3, p. 204). © 1995 The University of Chicago Press.

been difficult, but Wrangham suggests that all animals living with humans like cooked food a lot, so it is likely that once intelligent pre-humans stumbled on cooked food, they never lost interest in it. He also suggests that this introduced labour division between men and women, which may have reinforced their pair bonding. Arguably, women originally mainly cooked for offspring with males acting as free riders. This is still reflected by women cooking for children and men in most cultures around the world.

On an evolutionary or geological time scale, one must consider the human species as relatively young. This, and the fact that all humans probably descend from one ancestral group, living some 100 000 years ago, explains why genetic variation among humans is small compared to other primates, which are older. Moreover, more than 80 per cent of all human genetic variation is already found among individuals living within a single culture, while on average humans are more than 99 per cent similar to any other humans in terms of their DNA composition. Most visible differences in body features in human populations originated during the last 20 000 years. An obvious one, skin colour differences, are explained by an originally pigmented skin that conflicted with vitamin D production under the influence of ultraviolet light in temperate zones with less sunlight.

11.3   Brain–Mind Evolution

Despite much research and a huge body of literature, the evolution of the human brain still poses many unresolved puzzles. Here we list the main ones with several candidate answers, culminating in a systems view on how all factors interact and add up to explain the unusual intelligence of humans. To arrive there, we first adopt a basic engineering perspective on the brain's physical and morphological features. According to it:

- the brain approximates a spherical shape to allow best protection (smallest surface – skull) and to provide the shortest access to different parts;

- brain cell interactions rely on electric charges which travel fast through insulated axons, cable-like extensions of the cell bodies; one neurone cell links directly to a limited number of other cells;
- the interactions among neurones are a sort of democratic voting system;
- and neurone clusters occur based on similarity of functions generated by their circuits.

In humans, the part of the brain that is responsible for higher cognition, namely the cortex, makes up more than 80 per cent of total brain mass. Contrary to what is often believed, in terms of the number of neurones relative to brain size, the human brain is not so exceptional but simply a primate brain linearly scaled up (Herculano-Houzel, 2009). Next to neurones, glial cells, long thought to be merely support cells of the brain tissue, are suspected to play a role in shaping the brain and the related capacity for phenotypic plasticity. Human brains seem to be more plastic, i.e. less genetically determined, than those of other primates, which partly explains our intelligence. Support for this comes from studies that show less variety in primate than human cortex organisation (Gómez-Robles et al., 2015). The plasticity may be due to human brains being less developed at birth, allowing for a longer gestation period during which the plasticity can be employed to learn and adapt better to the particular environmental conditions. This plasticity served as a fundamental basis for the evolution of human cultures.

Brain–mind dualism is an outdated notion. The mind is the expression of the brain at work. The complexity and subtlety of both indicates consistency and association. Once I made a back-of-the-envelope calculation of how many active (moving or reactive) molecules there are in a human brain, and arrived at an estimate of 10^{22}. This is an extremely large number, but not surprising in view of the immense complexity of the mind. The encephalisation quotient (EQ) is considered a good measure of relative brain size. It is defined as the ratio of actual and predicted brain mass for an animal (notably mammal) of a given size, which is expected to highly correlate with intelligence as it corrects for allometric, i.e. nonlinear shape-to-size effects (such as a sphere changing with a power 3 to changes in the length of the radius). On this measure, humans score the highest of all mammals, followed by two species of dolphins, orcas and chimpanzees, rhesus monkeys, elephants and dogs. Another measure would be the brain's relative energy and oxygen use. In humans this equals about 20 per cent of the body's use, whereas for other vertebrates it falls in the range 2–8%, and for some other primates in the range 11–13% (Mink et al., 1981). Energy use by the human brain is equivalent to 300 kcal per day, or about 15 watts (joules/sec), so fairly efficient, namely equivalent to two modern, low-energy light bulbs. Surprisingly, this rate of energy use is rather constant over time and not much affected by the specific mental and motoric activities undertaken. During sleep the brain consumes only about 10 per cent less energy. The suspected reason is that a high baseline metabolic activity of the brain is needed to maintain the responsiveness of neurones and the functionality of brain areas, including memory (Raichle and Gusnard, 2002).

Human brains carry an animal evolutionary history with all its tendencies of instincts, automatisms, passions and emotions, as studied by evolutionary psychology and human genetics (Section 5.4). A rough, and admittedly overly simplistic, distinction of parts of the brain, reflecting its evolutionary history, is between the old reptilian part (responsible for decisions such as fight or flight), a mammalian part (focusing on care-taking as in mother–child relationships), and the recent cortex (creating consciousness and allowing deliberation and 'rational accounting'). Emotions and automatisms often have an advantage over deliberation because they are quicker and use less energy, which contributes to survival. Emotions are an evolutionary logical strategy in the face of urgent or complex decisions, such as environmental hazards. Instinct is sometimes referred to as a primary emotion while so-called secondary emotions reflect personal experiences. Both are hardwired, but unlike the secondary the primary ones are preprogrammed by our genes, due to evolutionary selection, as they relate directly to survival and reproduction. They include responses to loud noises, sudden appearance of large shapes, snake-like movements and stimuli for sexual excitement.

Conscious and Unconscious Processes in the Brain

Most activity in the brain is permanently unconscious to us, notably the autonomous nerve system automatically regulating the heart, lungs, metabolic processes, instincts and reflexes that keep the body alive and in good shape. These processes are under the control of genetically programmed autopilots in the brain and spinal cord, the neurone circuits of which are our inheritance from hundreds of millions of years of vertebrate evolution prior to the origin of human consciousness. The conscious mind can easily mislead us, namely by suggesting that the way we solve certain problems is easier than it seems, as if determined only by factors of which we are aware. But the required complicated neural circuitry remains hidden from the conscious experience so that many factors playing a role in decisions we make about complex, multifaceted issues will not be evident to us. Sleeping is an extended state of unconsciousness. Whereas plants can eat and reproduce without ever being awake, sleeping animals will not find food or a mate. A popular account of being awake and conscious is that it serves four Fs: fighting, fleeing, food gathering and fucking. Perhaps the most significant consequence of consciousness is that it leads to thinking about thinking. This in turn boosts creativity and learning. Learning operates in a different manner in conscious humans than in unconscious animals. It involves a kind of simulation of possible scenarios in the mind, which allows for more rapidly and more broadly exploring a range of possible realities than through trial and error, i.e. learning by doing. Consciousness means also that one remembers well the history of events, on the basis of which cause–effect mechanisms can be derived.

The information storing and processing capacity of the conscious mind is limited. As an indication, take the well-known distinction between short- and long-term memory. The latter is unconscious, although some of its information can move to the conscious domain. The short-term memory, however, is (close to) conscious but limited in capacity: it can hold, on average, seven symbols or words. This is one reason why, through practice, information is transferred from the conscious to the unconscious: we say that some learning has become automatic. Learning to play a musical instrument works like this: turning conscious training through repetition into automatisms. In this way, the unconscious part of the brain relieves pressure from the conscious part, to avoid memory buffer overflows. Although we know that long-term memory can be accessed by our consciousness, it remains largely a mystery how the conscious and unconscious parts of the brain interact. This is understandable in view of the well-known paradox that the human brain is both too complex to be understood by itself and too simple to understand itself.

The brain involves many automatic processes that are genetically determined, not influenced by earlier experiences. E.O. Wilson (1998) calls these 'primary epigenetic rules', as opposed to secondary epigenetic rules due to storing information in the brain, through categories such as memory, perception, emotion and cognition. The primary rules include, among others, certain reflexes, certain facial expressions, our ability to see the continuous wavelengths of visible light as discrete colours, and our ability to distinguish between noise and tone, or differentiate between acid, bitter, salty and sweet tastes. A basic property of many primary rules is that virtually continuous information (light, sound, food) is translated into a limited, discrete number of categories. This includes our tendency to employ two-part classifications in social communication: man/woman, child/adult, kin/non-kin, married/single, good/evil, in-/out-group, etc. Such biological insights about our visual, acoustic, smell and taste abilities are relevant to social science as they form the basis of our communication with other humans, through which cultures have been shaped.

The Illusion of Free Will

Most people take for granted that we have 'free will'. If not, then self-determinism, autonomy, liberty, free choice and even democracy are supposedly at stake. Among scientists and philosophers who have thought about this, a considerable number are unconvinced that humans, or for that matter, any other organisms on Earth, have free will.

In view of so many brain processes being unconscious to us and continuously interacting with our consciousness, one should doubt that free will as independent and completely conscious decision-making exists. E.O. Wilson (1998, p. 119) finds it difficult to see any sign of free will if one defines it as 'freedom from the constraints imposed by the physiochemical states of one's own body and mind'. Although the lack of free will may not charm everyone, it is undeniable that we make decisions for reasons we rarely fully understand. As Arthur Schopenhauer said, 'Der Mensch kann tun was er will; er kann aber nicht wollen was er will' ('A human being can do what it wants; but it cannot want what it wants'). Indeed, one's 'wants' are generally the result of a combination of nature and nurture, that is, ones' genes and environmental influences, and rarely or never of any previous conscious decision about which 'wants' to adopt.

Experimental findings from cognitive science and psychology show that many decisions, which human participants think they take consciously and deliberately, are already being prepared by the brain before the participants were aware of them. Soon et al. (2008) find that the time between preparation and awareness may be as much as

10 seconds. One explanation is that a network of countless external and unconscious internal stimuli contributes to brain processes taking off a certain time before arriving at a decision. The complexity associated with this network makes it impossible for people to perceive a simple cause–effect chain behind the resulting decision that undercuts free will. This allows for persistence of the comforting feeling that we have a free will.

E. O. Wilson regards confidence in free will as a biologically adaptive trait of conscious humans: it helps conscious humans to avoid falling into the trap of fatalism and depression – moods which do not guarantee high chances of survival and replication. A strong awareness of having no free will creates a moral vacuum and has implications for how we think about crime and punishment. For who is to blame for a person's misconduct: the genes, the brain, the circumstances, the past environment, the upbringing or perhaps all of these? Posing a free will makes things much easier: the 'I feeling' emerging from the brain with the supposed, unproven, free will is fully responsible. It is difficult to move away from the traditional law system as evolution has firmly anchored morality and punishment in our brain. We have evolved to punish free riders, and criminals are the worst category of all free riders. Only in exceptional cases are we willing to accept that someone was temporarily insane and thus not accountable for a crime. Swaab (2014) thinks this is one reason why criminal law has not progressed beyond the level of medicine 100 years ago. It has not developed much knowledge about the effectiveness of psychological or medical treatments of criminals because it lacks an evidence-based research approach which makes use of controlled experiments. As a result we keep punishing in old-fashioned ways without guarantees of much effect. Duntley and Shackelford (2008) propose a similar insight from the angle of evolutionary psychology. They note how natural selection has shaped behaviours that represent crimes in modern societies, such as murder, assault, rape and theft, and separate between crimes due to conflict over status, material resources and 'mating resources'. Their work suggests an 'evolutionary forensic psychology' which contributes to improvements in legal and judicial systems, in terms of, among others, jury selection, eye-witness testimony and judges' views of human nature.

Two distinct, well-informed and eloquently defended views on free will are expressed by Daniel Dennett (1984, 2003) and Sam Harris (2012). Harris argues that free will does not exist and that, by accepting its implications for individual responsibility for actions, we can create a better society. Dennett defends the view that free will and determinism are compatible ideas, as humans have freedom to act according to their own motivation, without being coerced or constrained. In line with Harris' view, I think humans are definitely constrained — in ways unclear to them — in terms of both motives and actions by body-internal, notably unconscious brain, processes as well as by uncontrolled, body-external influences.

Rather than the 'I feeling' behind free will, Marvin Minsky (1986) suggests that one should recognise a 'society of mind' at work (e.g. Minsky, 1986). With this he means that many potential ideas compete for attention, that is, appearance in the conscious part of the mind. This suggests an evolutionary metaphor. Indeed, various internal and external selection forces jointly influence which idea among a diversity wins.

For instance, when while reading you suddenly develop stomach pain, this will act as an internal selection force to attract attention from your consciousness, so that your mind can shift from reading to appropriate action to resolve the pain. When you are notice-ably dreamy when a friend is telling you something, she will make an expression or use a louder voice to get your attention, implying an external selection force that reaches the mind through the sense organs like eyes and ears. This challenges the 'I feeling' that supposedly is in control and making decisions all the time.

In view of these reflections, it is difficult to avoid the conclusion that free will is an illusion and that the feeling of free will merely reflects our conscious sensation of indeterminacy. But this, of course, will not stop the debate about whether free will is for real.

Evolution of Human Intelligence: Multiple Factors and Positive Feedbacks

How could the human brain evolve? The foregoing discussion shows that many factors and processes played a role, including intelligence payoff, benefits of communication in social groups, cooperation in hunting, and sexual selection. Together they create an impressive system of interactive positive feedback mechanisms (Figure 11.4).

To understand the net benefits of a large brain, one has first to assess its costs, in terms of energy use and risks. In adults, 20 per cent of energy intake is consumed by the brain, compared with the brain making up only 2 per cent of the body weight; and in new-borns even 60 per cent of food energy is used by the brain. An evolution of a larger brain was thus only feasible through more efficient food collection and processing, aided by bipedalism (locomotion), and alterations in teeth and jaws (Jones et al., 1992, Part 3). Other costs of a larger brain include a longer pregnancy, a higher infant dependency on parents, a higher risk of complications during pregnancy to birth – affecting the child as well as the mother – and generally a higher vulnerability to head injuries, both as an infant and adult.

In explaining the evolution of the large brain, an important role is assigned nowadays to the interaction between sexual partners, men and women. It is believed that sexual selection may have caused a runaway evolutionary process, characterised by positive feedback. In evolutionary biology this is often referred to as 'Red Queen effect', and in evolutionary ecology as an 'arms race' (see Section 4.6). Sexual selection and mutual predator–prey selection (coevolution) are both consistent with positive feedback, or a runaway process. In line with this, Wills (1993) has coined the term 'runaway brain'. This term refers to runaway dynamics, which is common in mutual predator–prey selection, where it has contributed to larger brains in both predators and prey, since catching prey as well as avoiding capture by predators both benefit from higher intelli-gence. Similarly, conflict between the sexes may have contributed to the evolution of larger brains. The idea here is that when sexual conflict is intense, as when males aggressively try to inseminate females, the intelligent ones among the latter are better able to avoid such males and copulate only with the fittest ones, while they also may have more time for feeding. It has been shown that this can contribute to larger female brains being selected in certain fish (Buechel et al., 2016). But probably sexual selection

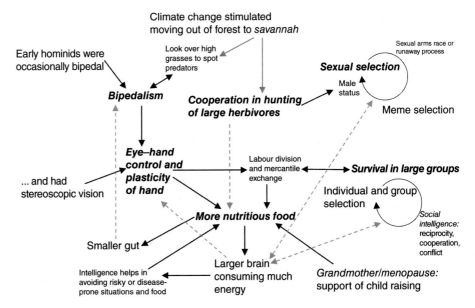

Figure 11.4 Systems dynamics representation of human brain–mind evolution.
Note: Bold, italics terms denote factors emphasised in the literature. Purple items are more specific factors. Some arrows are broken to avoid confusion about the direction of intersecting arrows.
(*A black and white version of this figure will appear in some formats. For the colour version, please see the plate section.*)

in humans worked differently. It arguably favoured interaction among relatively intelligent partners, as they are more creative and effective in gathering food, hunting and – not to forget – seduction. Miller (2000b) thinks this is the main credible explanation for the evolution of a brain that wastes so much energy. He argues that language was the main direct function served by the evolving human brain and that it was triggered by sexual selection. According to meme theory (Section 7.6), which stresses the role of imitation, sexual selection and meme selection could also have reinforced each other, as mating with good imitators, i.e. intelligent individuals, would have been attractive and produced offspring with the best imitation capacity. Blackmore (1999, p. 79) even thinks that meme selection through imitation capacity can explain better than other theories why sexual selection started to seriously affect brain size and intelligence. While sexual selection often results in male–female dimorphisms (Section 4.7), the foregoing considerations may explain why the male brain is not larger or more capable than the female one. Another reason is that many other factors (Figure 11.4) played a role in brain evolution, which affected both males and females.

Human cooperation in hunting is often mentioned as an early factor allowing selection for a larger brain. Since hunting effectiveness depends on intelligence, this may have contributed to increase in brain size. Cooperative hunting offered much food for everyone, including non-hunters, providing sufficient nutrients for children's brains to grow. Hunting may have, moreover, contributed to competition among males to be successful in hunting, so as to acquire status, which could have created positive synergy with sexual selection effects on the brain.

Another popular explanation is the 'social brain hypothesis'. The idea here is that intelligence is needed to survive, reproduce, cooperate and participate in reciprocal relationships with a large number of people, up to 150 according to Dunbar (1996). It requires an extended capacity to understand what others are thinking, and to remember their positive and negative actions towards oneself (for reciprocity) or even towards other group members (indirect reciprocity). This may have had a runaway effect, since more intelligent beings are more difficult to understand, and can play strategic games, which requires more intelligence of others to interpret or even outsmart them. Finally, one might add competition and conflict between groups as forces that might have reinforced the process towards larger human groups composed of effectively cooperating individuals (Chapter 6). In turn, this can have functioned as a catalyst for stimulating selection of increased brain size.

Early development of mercantile (i.e. non-kin) exchange, or trade, in humans, associated with living in larger groups, may have further selected for intelligence (Section 4.8). The reason is that it requires skills such as communication, negotiation and quantification, which depend on linguistic and facial expression as well as memory capacities. Moreover, such exchange presumably contributed positively to survival and reproduction of individuals partaking in it, as it allowed for variety in consumption, labour division in production and improved hunting, food gathering and child upbringing.

Less frequently mentioned is the role of the grandmother. She arguably increased the survival chances of her grandchildren through physical support, food gathering, education and social bonding. A life-long experience in raising children and a complete devotion to her offspring's offspring are critical in this respect. The value of family support by grandmothers may partly explain why humans can live relatively long in comparison with apes. The menopause, which allows women to shift the attention from their own children to grandchildren, is considered evidence for this theory, as it is exceptional in animals. According to anthropologist Kristen Hawkes (2004), the menopause evolved before the large human brain and, indirectly, made the latter's evolution possible. The reason is that the grandmother's family support allowed children to have a longer developmental period during which they depended on adults, essential for the large brain to come to maturation. A particular finding is that offspring survival applies especially to maternal grandmothers, while paternal grandmothers merely increase birth rates (Hawkes and Smith, 2009).

To summarise, the unusual intelligence of humans can probably only be satisfactorily explained by a process of evolution that involved manifold factors: bipedalism, stereoscopic vision, cooperation in hunting, sexual selection, surviving in a large social group, mercantile exchange, meme selection, and the role of the grandmother. Moreover, some of these factors also made it possible to offer ample nutritious food to children allowing the brain to grow large and consume much energy. It has further been suggested that intelligence helps in avoiding risky or disease-prone situations and food. As illustrated in Figure 11.4, these factors together created a complex evolutionary system, with various positive feedbacks and synergistic effects, which explain why the unusual brain and intelligence of humans could evolve.

Consistent with the complexity of the evolution of the human brain–mind is the long gestation period of the brain. During a crucial development phase, known as adolescence, the brain drastically alters in structure. As discussed in Box 11.1, this has received various evolutionary explanations. Finally, to illustrate that evolution can bring about considerable and rapid changes in intelligence, consider the case of Ashkenazi Jews. They are relatively intelligent, notably in terms of language and mathematical abilities. They have the highest average IQ of any ethnic group and are strongly overrepresented in occupations requiring elevated cognitive skills. As an indication, during the twentieth century, although making up only about 3 per cent of the US population, they won 27 per cent of US Nobel science prizes and accounted for more than half of the world chess champions. Their high intelligence is likely the outcome of combined cultural-genetic selection during the period 800–1700. This was driven by a combination of closed groups – limiting inward gene flow – and intellectually demanding work – due to forming an urban occupational caste of initially merchants and later money-lenders and business managers. In contrast, the majority of families in other ethnic groups worked as peasant farmers. Possibly Jewish learning and scholarship habits played a reinforcing role. Those who fared well in their occupation had, on average, more children surviving to adulthood as they could offer a prosperous livelihood. Such relatively recent selection for intellectual capacity explains why Ashkenazi Jews are susceptible to neurological diseases, such as Tay–Sachs and Gaucher. In fact, their genetic disorders relate to DNA-change mechanisms that may have contributed to raising intelligence (Cochran et al., 2006).

Limited Individual but Extreme Collective Intelligence

Despite the amazing human capacity for language and social learning, we should not overestimate human intelligence. Our capacity for abstract thinking and other activities uncommon during the Pleistocene – such as understanding probabilities and statistics, playing a musical instrument, or objective unprejudiced argumentation – is limited and requires extensive training. The brain limits are even a cause of joy: when we watch a movie on television or in the cinema, we are not hindered by the fact that we are only looking at a two-dimensional image of real-world-like events. Our intellectual capacity seems larger than it is because each of us benefits from living in a culture that has accumulated knowledge, life lessons, practices, tools and institutions through learning by doing and formal education over many generations. This is the 'cumulative culture' thesis of Joseph Henrich (2015), who stresses that the main difference between the apes and us is not orientation in space, quantification or grasping causality, but the capacity for social learning. This connects with meme theory (Section 7.6), which poses pure imitation capacity as the main distinctive feature of humans among primates. Henrich suggests that humans, put in a natural environment unknown to them, would, without their cultural baggage, have a difficult time surviving for long. The evolution of human culture has given rise to a kind of 'distributed human intelligence' or 'collective brain'

Box 11.1 Evolutionary accounts of teenager brain transformation

The brain undergoes a major restructuring during adolescence which is the cause of unorthodox and risky behaviour often observed in adolescents. The ultimate reasons for this can be identified by adopting an evolutionary angle.

Teenager behaviour is characterised by sensation seeking and relatively little impulse control, leading to so-called 'maturational imbalance'. This has been suggested to be adaptive as it prepares adolescents for future sexual relationships. Such an explanation seems consistent with adolescents, unlike younger children, moving frequently in mixed-sex social groups. Arguably, risky behaviour is part of the adolescent strategy aimed at attracting the attention of potential mates, resulting in risks being rewarded by reproductive success. A related but distinct explanation is that adolescent behaviour functions as a practice for mate competition in later life – analogous to how young mammal predators learn fighting in a playful context to help them become good hunters in adult life.

According to Jensen (2016) the evolutionarily adaptive function of teenager behaviour is likely to be a different one, namely experimenting with cultural ideas obtained from people other than parents, notably peers. This increases the diversity of teenagers' cultural baggage, going beyond the limited cultural lessons obtained from the parents and other family. In this way, life strategies and thus the chance of long-term survival and reproduction of adolescents are improved. A kind of combination of the previous explanations is that a relatively high willingness to take risks by teenagers will make it easier to leave their family and start an independent life. Another explanation is that young people taking risks offsets risk-averse or even conservative tendencies encountered in adults, which can help groups to deal better with new challenges. Perhaps this account better fits the historical evolution of humans in extended kin groups, where the majority of offspring never left their family.

The fact that adolescent behaviour frequently involves serious risks has prompted some to label it as maladaptive. In the modern world, risks are magnified by three factors. First, adolescents have easy access to hazardous technologies, such as alcohol, drugs, arms and motorised vehicles. Second, adolescents living in a liberal society, and nowadays with the internet, are bombarded by a vast variety of cultural ideas that may trigger them to participate in unorthodox behaviours and perilous activities. Third, adolescents spend much time with other adolescents without being supervised by adults, which may lead to excessive behaviours. During most of human evolutionary history, risks of these types were modest due to a limited variety of cultural ideas and technologies, and adolescents always moving in mixed groups with adults.

supporting the survival of all individuals, not unsimilar to how individually unintelligent ants need the 'intelligent ant colony' to survive (Box 5.1). This view is supported by experiments with young chimpanzees, orangutans and humans, which suggest that the main intellectual difference between them is not in the areas of spatial orientation, assessing quantities or deducing causality. It is that humans possess a much great ability for social learning and imitation (Herrmann et al., 2007). Without it we would lack spoken and written language, technology and accumulation of problem-solving knowledge beyond generations — and likely be close in intellectual ability to our ape cousins.

11.4   Pre-agricultural, Pleistocene Humans

The traditional view in history, anthropology and archaeology is that the pre-agricultural or Pleistocene human era of nomadic foraging, involving hunting and gathering, is best characterised as stasis. This means it entailed little or no change in social and cultural characteristics during a period possibly lasting from several hundreds of thousands of years until about 8000–11 000 years ago. But it is probable that hunter-gatherer groups evolved towards more complex societies already before the rise of agriculture. This would mean that social-cultural evolution started before agriculture. This involved a number of stylised facts (Jones et al., 1992; Boyd and Silk, 2002).

Human population density slowly increased over time. This is the result of, on the one hand, more efficient hunting and gathering due to human intelligence and cooperation, and, on the other hand – as for a moving band it was difficult to carry and raise children – splitting and migration of groups, leading to the occupation of new territory. The geographical spread of humans over the world has been possible because of their physiological plasticity (tolerance of temperature range, diet tolerance), cultural-technological adaptation (clothing, housing, cooking, eating animal fats, social support) and natural selection leading to genetic adaptation (body size, skin pigmentation, heat or cold tolerance and altitude tolerance). The spatial pattern of diffusion strongly correlates to genetic distances between human populations (Africans, Caucasians, Asians and American Indians), with patterns of linguistic evolution (Cavalli-Sforza and Feldman, 2003) and with animal species' extinction waves in Australia and the New World.

In some areas, particularly those that were rich in naturally available foods, humans may have already made the transition from mobile, nomadic groups (bands) consisting of several families to somewhat larger groups of sedentary people before the emergence of agriculture. This may have involved protecting, or even taking a kind of ownership, of fields rich in edible plants such as cereals. Switching from hunting to herding animals (pastoralism) also fits into this pattern. Both can be considered as an intermediate phase in a large-scale transition to agriculture. This may have gone along with changes in food habits. Humans have always been omnivorous and it is still debated whether hunting was generally more important than gathering in terms of delivering food calories and proteins. One argument against this is that too much meat is unhealthy for the body because of an excess of complex amino acids relative to calories.

Technological progress may have occurred at a slow pace and small scale. It involves such diverse activities as the construction and use of stone and wooden tools, cloths, hunting tools (spear, bow and arrow), simple houses (huts, tents) and watercrafts (floats or canoes). The materials used came directly from nature and included especially certain types of stones and organic materials (bone, horn, ivory, skins and shells). Certain tools became standardised, while a wider variety of tools developed over time, including more complex tools consisting of multiple components (such as the bow and arrow). Tools also became more efficient in design and use. For example, blades were slowly improved to provide much more cutting edge. A fundamental innovation has been the invention of fire, discussed in more detail in Section 11.5. Next to technological advance, organisational progress must have occurred, notably in terms of cooperation and labour division in hunting and, possibly, also in gathering. Some human groups may have moved from egalitarian organisations to stratified populations and complex organisations. Labour division started with a clear distinction of tasks between men and women, and later older and younger individuals. In contrast, in pure hunting-gathering bands or tribes, knowledge and skills are shared, which means that no individual can specialise to create an advantage over others, so that egalitarianism results and individualism – as we know it – is virtually absent.

Technological and organisational innovations in one region likely diffused to others. Diffusion was possible at the local scale and, by extension, to larger scales, since neighbouring groups would frequently exchange products and social habits. As a result, ultimately, tools, raw materials and ideas would have been transported or diffused over long distances. Because imitation is imperfect and unique innovations must have appeared in distinct regions, fairly early on in human cultural history a large spatial diversity of technologies and organisations must have existed. Another reason is that cultural development tends to be rather unique as it is influenced by isolation (islands, mountains, rivers, seas) and adaptation to unique regional climatic, environmental and resource (soil, water, and vegetation and animal presence) conditions.

Finally, the development of rituals and early art, such as decorative carving, body painting and cave painting, took place before agriculture, notably in Africa, Europe (especially France and Spain) and Australia. The oldest pigments have been found at Twin Rivers in Africa, estimated to be between 350 000 and 400 000 years old, while the oldest evidence for use in rituals and paintings is between 30 000 and 40 000 years ago (Grotte Chauvet in France, and Lake Mungo in Australia). Subtle cave paintings, with images portrayed as well as use of shading and perspective techniques, support the idea that at the time some type of human culture had evolved.

Such cultural artefacts during the late Pleistocene era are all part of social-cultural evolution within rather fixed genetic constraints. The reason is that the human gene pool has hardly changed since hunter-gatherers spread across the globe about 35 000 years ago. Nevertheless, selection pressure of natural conditions must have been higher than it is currently, as reflected by a relatively high rate of child mortality. Environmental conditions and selection pressure changed drastically since about 18 000 years ago when the shift away from a full glacial climate started.

The period of cultural evolution meant a spread of groups that were successful as well as an accumulation of culturally – and hence intergenerationally – transmitted habits, technology and knowledge. At an individual level it might have meant a small positive impact on longevity. Before the large-scale transition to agriculture, humans thus must have reached a historically relatively high level of social-cultural and technological development. Moreover, they already had occupied all continents, regions and climate zones of the world. The total number of people at that time is estimated to fall in the range of 1 to 20 million people, a ten-fold to hundred-fold increase from 1 million years before when the human population counted about 125 000 individuals (Cohen, 1995, Appendix 2; McNeill, 2000). This growth reflects an evolutionary successful species spreading in space.

11.5   Fires, Dogs and Cats

The 'Invention' of Fire

Controlled use of fire is widely considered to be the most important early human invention. It allowed people at the time to influence natural vegetation in major ways, both purposefully and unintentionally. The importance of fire is reflected in the Greek myth of Prometheus stealing fire from the gods, which changed the conditions for human life forever in a revolutionary and irreversible way. Georgescu-Roegen (1971) therefore speaks of 'Promethean inventions', among which he also includes the emergence of agriculture and the Industrial Revolution. One can debate the notion of invention and instead use the term discovery, since fire exists in nature and sort of 'diffused to culture'. Nevertheless, early techniques of making fire – pyrotechnology with stones or a wire – clearly represent human inventions. In addition, fire control comprised transporting and using fire for various purposes. Probably, ways to make fire were discovered several times before they globally diffused, due to low population density and isolation of regions with humans. Incidentally, evolution invented fire use before humans did; see the defence mechanism based on chemical ignition of the bombardier beetle.

There are several motivations for initial uses of fire by humans (Stewart, 1956): keeping dangerous animals at a distance, creating trails for travel in areas with dense vegetation, trapping hunting animals into small areas, improving grazing areas for big game, promoting a diversity of habitats, clearing edges of settlement areas to create more security against wild animals and other tribes, and setting vegetation on fire as an act of war. In addition, fire control allowed for rituals such as cremation of the dead. Perhaps the most crucial function of fire was that it allowed for cooking and baking of food (the biological evolutionary importance of this was discussed in Section 11.2). This involved the creation of pottery using fire heat. The combination of fire and pottery allowed advanced cooking which increased the variety of food that could be consumed. Thus, humans had access to complex organic molecules, notably amino acids, which otherwise could not have been digested or assimilated by their bodies. In addition, cooking killed disease germs and neutralised toxins that many plants produce. The resulting health benefits may have influenced the direction and pace of human physiological and brain evolution as well as of social-cultural evolution. Fire has further been speculated to be related to the origin of speaking by humans. Namely, by prolonging the day and creating a communal centre for the family or tribe, it stimulated communication in the evening before resting, thus possibly giving rise to early forms of speaking.

Since fires occur regularly in nature, during their lifetime, prehistoric humans were likely to experience them, providing the opportunity for some to recognise their power. Experimenting with fire must have happened somewhere after the first use of tools. Fire-making could use a variety of fuels, such as wood, crop residues or dry dung, and later on even charcoal, peat or coal. The use of fire may have been connected with using fish as food, notably shellfish, which were easy to catch but needed cooking or baking. Although no accurate dating of early fire-making is possible, it is believed now that *Homo erectus* first invented fire (Pyne, 2001). A few sites with burnt clay fragments and bones in Africa (Kenya, South Africa) suggest this happened more than a million years ago. Along with good clothing, fire allowed migration into colder areas. This suggests that the invention of fire is at least as old as the colonisation of Northern Europe and Asia. Nevertheless, habitual use of fire in Europe, by both Neanderthals and *Homo sapiens*, is thought not to have started until 300 000 to 400 000 years ago (Roebroeks and Villa, 2011).

Domestication of Dogs and Cats

The domestication of dogs and cats was possible due to the evolution of suitable ancestors of these species. Among the mammals that evolved after the extinction of the dinosaurs, some 70 million years ago, two main groups of carnivorous mammals evolved that ultimately gave rise to the canids, the larger family of the dog, and the felids, the larger family of the cat. Domestication, or controlling the reproduction, of the dog occurred before the rise of agriculture, and of the cat afterwards. Carnivorous animals are expensive for humans to keep because they require meat, cats more so than dogs. But the net benefits of dog and cat domestication were evidently positive.

Domestication of dogs goes back to at least the end of the last ice age, some 12 000 years ago, but may even have occurred up to 40 000 years ago. Genetic data analysis indicates that dogs are directly descended from several extinct wolf species tamed in different areas, presumably to aid in hunting. Domestication of the wolf produced physical changes that are common in domesticated mammals, such as a reduction in size, shrinking of the teeth, and a reduction in brain size. After farming developed, dogs spread and became more common, as they contributed to herding and protecting humans, livestock and agricultural crops. The dingo, the native wild dog of Australia, is speculated to be a domesticated dog from Asia that was brought to Australia more than 3500 years ago and subsequently became a feral species, arguably as – unlike in other continents – there was little competition from species occupying a similar habitat range. In fact, the dingo currently is the largest terrestrial apex predator – i.e. at the top of the food chain – in Australia. It is suspected that it was responsible for the decline and extinction of the Thylacine, popularly known as the Tasmanian tiger or wolf. The latter was a carnivorous marsupial that, owing to convergent evolution, resembled placental canids.

With currently more than 150 different breeds existing, variation among dog breeds exceeds that of any other domesticated animal. Darwin, consequently, suspected that dogs had descended from multiple, distinct wild species, but this has not been confirmed. Another explanation is that dogs have the longest domestication history, so that there was much time to create distinct varieties. Arguing against this is that current dog variety is mostly the outcome of a relatively recent interest in certain physical features of dogs, over the past few centuries. Ofek (2001, p. 50) suggests that, unlike most domesticated animals which are descended from gregarious species, the dog is descended from a truly social species, the wolf, which is characterised by labour division. This explains a relatively high degree of phenotypic plasticity, allowing for variation to appear under the force of artificial selection. In fact, dogs enter into labour division with humans, providing tasks such as guarding, guiding, herding, hunting and searching. While wolves perform similar tasks in groups driven by kin relations, dogs are willing to perform these for humans as they identify them as their relatives, having been born and raised among them.

We humans tend to enjoy the presence of dogs more than of other pet animals because like us they show a variety of expressive emotions, are empathic and have an endless interest in social interactions. Based on their most recent evolutionary history, as well as socialisation during development in human environments, dogs have developed an exceptional ability, even beyond that of extant nonhuman primates, to understand our non-verbal signals and spoken language. The average dog can comprehend the difference between more than 1000 words, as many as a three-year-old child, and grasp language subtleties such as intonation. In fact, it was recently found that a dog's brain handles spoken language in a manner similar to a human brain: meaningful words are processed in the left brain and intonation in the right brain (Andics et al., 2016). This could be due to convergent evolution, or the use of brain mechanisms that primates and wolves already had in common, or a combination. Contrariwise, humans generally can tell just from the bark sound of a dog its size or emotional state. These mutual understandings are the outcome of having evolved together for so many generations, which has resulted in the strongest interspecies bonding found in nature. Convergent or parallel evolution between dogs and humans has been noted also with respect to other characteristics, such as genes that are critical to starch digestion, which is due to similar selection by food resources (Wang, 2013). Finally, some have gone as far as to suggest that humans have learned from wolves to be territorial and hunt in groups for large game (Taçon and Pardoe, 2002). Regardless, it is likely that early dogs assisted humans in finding, pursuing and catching prey.

Cats were domesticated much later than dogs, in any case after the rise of agriculture. There is less variety and mixing of cat races. In general, cats are closer to the original ancestor because they have been much less subject to selection pressure from humans, having always been more independent and not having been used for concrete production purposes. Because of this, as well as due to their origins, cats are less social than dogs, as is wittily visualised in Figure 11.5. The main reason for domestication of cats was that they eat mice and other rodent pests which thrive on cereal crops. Taming of the cat is usually considered to have occurred about 4000 years ago in ancient Egypt, where they were eventually worshipped. The latter caused the spread of cats through export to other countries for some time to be forbidden. Ultimately, however, cats followed humans across the globe. As a consequence, they negatively affect nature since many of them

Figure 11.5 Distinct attitudes to humans of dogs and cats.
Source: http://bizarrocomics.com. © Dan Piraro.
(*A black and white version of this figure will appear in some formats. For the colour version, please see the plate section.*)

wander freely around, killing large numbers of birds and small mammals, stirred by their natural hunting instinct. For the United States these killings have been estimated to run in the 1–4 and 6–22 billion, respectively (Marra and Santelle, 2016). In view of this, some argue for more serious legislation and control of cats, or even cat predation and eradication programmes, to reduce the number of free-ranging housecats.

Both the domestication of cats and dogs can be seen as the result of coevolution of carnivore ancestors of these species and humans. As their living and hunting areas overlapped, contact was inevitable. Instead of competing, humans managed to control the carnivore species, creating a founder species (Section 3.5) that became isolated from the main population and thus could relatively easily evolve into a subspecies with desired features. There are now possibly half a billion housecats and about an equal number of dogs in the world. From a reproductive success angle such pet animals can be regarded as evolutionary winners as a consequence of free-riding on the human species.

11.6   Proximate Versus Ultimate Factors Behind the Rise of Agriculture

Humans were not the first on Earth to undertake agriculture. Domestication of other organisms creates so much evolutionary advantage that it evolved long before humans appeared on the evolutionary scene. The arthropods, notably ants, include thousands of species that can be regarded to undertake agricultural activities – such as hosting, controlling, cleaning and using the products of other species – and have done so for many millions of years. Humans have independently discovered agriculture. It took them, however, quite a long time — during most of their existence, they did not practise agriculture. The rise of agriculture occurred some 10 000 years ago, which coincides with the transition from the Pleistocene to the Holocene period, characterised by large-scale climate change followed by stabilisation of climate. This entailed the first successful experiments with farming of cereal grains, notably wheat and barley, and root crops. Around the same time, people may have started to domesticate wild animals, first for food purposes, beginning with goats and sheep, and later also for carrying loads and pulling ploughs and wagons.

Biologists, anthropologists and, recently, even economists have shone their light on this phenomenon and proposed mechanisms that can explain the historical transition to agriculture. Early 'agriculturalisation' has likely gone through different phases: limited food production on open spots between existing vegetation inside forests, temporary use of small non-irrigated gardens created through slash and burn, small gardens around houses, rotating crops, irrigated farming and mixed agriculture (crops and animals). Initially, farming was likely restricted to purely subsistence, whereas later surpluses stimulated labour division and new tasks, such as bureaucracy, religion, transport and trade.

The rise of agriculture was a clear case of coevolution at a grand scale. It involved domestication and related genetic change of animals as well as their (artificial, human-controlled) environmental conditions; changes in the economic strategies employed by humans, including agriculture or foraging, type of agriculture and domestication of particular plants and animals; early genetic changes in humans, such as adult lactose tolerance; later on the transfer of disease-causing microorganisms from animals to humans; and subsequently humans acquiring genetic resistance against these diseases.

Possible Proximate Causes of the Rise of Agriculture

Various proximate and ultimate explanations of the emergence of agriculture have been proposed (e.g. Claessen and Kloos, 1978; North, 1981; Jones et al., 1992; Diamond, 1997a; Richerson et al., 2001; Smith, 2003). We first consider the proximate ones:

1.   *Deliberate human choice*: The idea here is that their intelligence allowed humans to recognise that agriculture is superior to hunting and gathering, both in terms of productivity and way of life – i.e. an arduous nomadic versus a comfortable sedentary lifestyle. An argument against this is that the relevant unit at the time was a band or tribe with its habits and limiting structure instead of an intelligent individual free to choose what to do. Another one is that fully fledged productive agriculture is too complex to have been a singular, one-time decision, but must have been the outcome of a gradual, incremental process involving decisions by many groups over a longer period of time. Through experimentation humans stumbled on productive crops, which, by artificial selection, became more productive.

2.   *Agricultural technological progress*: As discussed in the previous section, late Pleistocene humans might have undergone some cultural development which

allowed them to make the shift to agriculture. In particular, accumulation of knowledge about techniques and tools for hunting and gathering, and knowledge about the living environment in general, allowed for a radical change. This accumulation occurred through learning by doing (efficiency) and experimentation (innovation). For example, the sickle may have been an important early tool in the farming of cereal grains. In addition, the knowledge about wild plants and animals was close to perfect, and any species suitable for domestication would have been recognised. The nutritional value of certain crops, notably grains and roots, could only be obtained through cooking or baking. This required controlled use of fire and the capacity to construct pottery (ceramics) or baskets. A (weak) counter-argument is that agricultural technology is likely to have developed rapidly only after agriculture reached a fairly large scale, because interaction among a larger population would stimulate a high rate of innovation and diffusion.

3. *Organisational evolution – labour division and skills*: Agriculture was probably preceded by an increase in emphasis on more intensified gathering, as this involved several skills and techniques that could be of use in agriculture. As food gathering was typically the area where women dominated, male group members might, because of prestige and power, have resisted shifting the balance of activities from hunting to gathering. Tribes more prone, however, to making such a shift would have been better capable of developing early agriculture.

4. *Property rights*: An early form of property rights, notably communal rights, may have created the necessary conditions for benefitting from the investments and efforts undertaken by a particular human group. This might have taken the form of communal exclusion of fields rich in wild cereals or herds of goats or sheep. It is consistent with pastoralism or herding of animals being an intermediate stage between nomadic hunting-gathering and settled agriculture. North (1981, p. 87), a Nobel laureate in economics, suggests that humans who stumbled upon areas with an abundance of desirable crops, for instance cereals, would slowly assign themselves a first right to use, and defend 'their property' against others. This way of demanding ownership of land is, in fact, how all private or common property must have started. This might have stimulated experimenting with weeding, irrigation and seed selection, which increased the productivity gap between agriculture and hunting and gathering.

5. *Population growth driving food scarcity*: A long history of slow population growth led to higher population densities. This may have contributed in many areas to local scarcity of wild animal and plant species serving as food. In turn, this could have stimulated hunter-gatherers to experiment with early agricultural practices. An argument against this theory is that population pressure, at the time, probably was more easily and quickly resolved by biological responses in the form of death and birth rates than by cultural adaptation in the form of agriculture. Another counter-argument is that the areas where early agriculture took place were characterised by a combination of good climates and rich indigenous vegetation with high diversity of potential crops (known as the 'oasis argument').

Under such resource-abundant conditions there was not a strong incentive to take the first steps towards agriculture, since the marginal payoff of labour effort applied to hunting and gathering is considerably higher than that applied to primitive, initial agriculture.[4]

6. *Extinction of animals due to hunting driving food scarcity*: Intensive hunting, partly driven by organisational and technological advances as well as population growth, may have contributed, along with other factors such as a changing climate, to the extinction of large game animals, known as megafauna. If gathering was unable to substitute for the lack of hunting game, the resulting food scarcity may have forced humans to search for alternatives strategies, including farming.

7. *Climate change*: Many believe it cannot be a coincidence that agriculture origins coincide with the end of the late Pleistocene ice age. Climate change caused a range of environmental and resource changes, such as sea-level rise, changes in humidity and temperature, and geographical shifting of ecological zones. In turn, it may have led to open, arid areas with less vegetation. In addition, climate change may have accelerated the extinction of animals through the loss of habitats, in combination with the pressure arising from hunting. Evidence in favour of this explanation is that climate change affected the whole world while agriculture arose independently in several regions and continents during a rather short period of time in human history: 10 000 to 9000 years ago in the Middle East, 9000 years ago in South East Asia, and 9000 to 7000 years ago in Mesoamerica. Richerson et al. (2001) argue that although agriculture developed in various regions during the Holocene, it did not develop anywhere during the late Pleistocene, even if the knowledge basis was roughly the same.

Explanations 5 and 7 seem to have received most support in the literature. Several of these find support in radiocarbon dating in combination with archaeological data on pollen, seeds, remains of plants, animal bones, faeces, tools used in farming and even early drawings. Moreover, some explanations suggest a possible interaction with other explanations. For example, deliberate choice (explanation 1) would have benefited from technological, organisational and institutional progress (explanations 2–4), while considering alternative options is more likely to happen under the pressure of food scarcity resulting from population growth, hunting and climate change (explanations 5–7). It seems, therefore, likely that instead of a simple, singular mechanism, a complex system of feedbacks underlies the emergence of agriculture, which included (co)evolutionary ecological and cultural factors.

---

[4] Early agriculture seems to have involved an increase in malnutrition, humans getting smaller, more infectious diseases owing to increasing population densities and people living shorter lives. In contrast, it is widely believed that a hunting and gathering lifestyle generated affluence (Gowdy, 1998). According to Robson (2003), initially agriculture may have reduced health (quality) which was compensated by an increase in fertility (quantity), explaining population growth. Galor and Moav (2002) apply a related argument but with reverse reasoning – namely, more quality and less quantity – to explain industrialisation (Section 12.1).

More on Climate Conditions

Climate change had a variety of implications that deserve more attention. During the last glacial period, most local climates were dry, local weather conditions were strongly fluctuating on a time scale of decades and the atmospheric concentration of CO_2 was low, arguably not providing favourable atmospheric fertilising conditions for cultivating crops. This all must have prevented the emergence of early agriculture. Indicative of this is that less extreme climate variation during the Little Ice Age (1300–1850) greatly affected agricultural performance, giving rise to famines (Richerson et al., 2001, p. 394). Climate and associated environmental change following the retreat of glacial ice caused an increase in the atmospheric concentration of CO_2, sea-level rise that redefined coastal lines and zones, and rainfall patterns that transformed grasslands into forests. Response strategies by humans may have included migration to rich habitats and using a wider variety of plants and animals, through hunting and gathering. In addition, in various regions, the result was a combination of warm and wet local climates, creating 'oasis' types of conditions. Successful experimentation and diffusion of farming required a sufficiently stable climate. This was indeed the final outcome of climate change. Climate variability during the early Holocene was, in fact, limited compared with that during the late Pleistocene. This allowed experiments with plants to have relevance over a longer time. In other words, both climate change and a stable climate appear to be critical factors in an overall explanation of the origin of agriculture.

Climate change and withdrawal of the glaciers may have caused the migration or extinction of the large ice age mammals, which in some regions were highly important for human hunters. Several hundred species became extinct. This was partly the result of increasing hunting pressure owing to growing human populations. The extent to which extinction is attributable to climate change and human hunting pressure probably differs between animals. For example, overexploitation may have been critical for local extinction of smaller animals such as the gazelle, while climate change may have resulted in the extinction of the largest ice age animals such as the mastodon, the mammoth, the Megaloceros (the largest deer in history), the Megatherium, the Megistotherium, the Paraceratherium and the Uintatherium.[5] Cohen (1995, pp. 36–37) notes that a warmer climate might have contributed positively to the survival of many other animals as well as plants. The net effect on food availability for humans and, in turn, on human population growth, would then be uncertain. In addition, conditions may have differed between continents. Notably, in North America extinction of large animals is likely to have been driven to a large extent by human hunting, because animals had evolved without humans and therefore had not adapted to evade them. Horan et al. (2003), though, use an economic model to argue that progress in hunting was more relevant than 'naivety' of animals.

---

[5] For information about, and images of, these animals see: http://www.enchantedlearning.com/subjects/mammals/Iceagemammals.shtml.

Ultimate Factors Behind the Rise of Agriculture

The problem with the explanations for the rise of agriculture discussed so far is that they do not clearly distinguish between proximate and ultimate factors. In an ambitious, integrative analysis of why different parts of the world have such distinct development levels, Diamond (1997a) has separated these two out (Figure 11.6). He argues that the regions of early agriculture had a path-dependent effect on the geography of later developments. The ultimate factors responsible for this included environmental features that are fixed over longer periods of time, namely climate, ecological and geographical features of continents or sub-continental regions.

Although hard to believe given its current arid landscape, the Fertile Crescent – encapsulating former Mesopotamia and currently overlapping with Turkey, Israel,

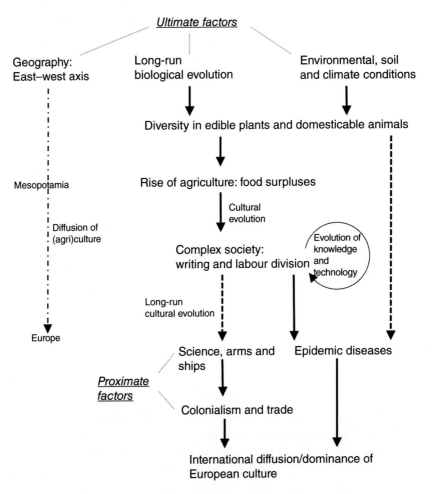

Figure 11.6 Proximate to ultimate factors behind cultural dominance of Europe.
Source: Elaboration of Diamond (1997a, Figure 4.1, p. 87).
(*A black and white version of this figure will appear in some formats. For the colour version, please see the plate section.*)

Syria, Iraq and Egypt – served as the cradle of agriculture and hence modern civilisation. Some 10 000 years ago it was an area with good climate and soil conditions, potent rivers (Euphrates, Tigris and Nile) and a high variety of domesticable plants and animals. This provided good conditions for agriculture to emerge. Ofek (2001, Chapter 12) also draws attention to the unique regional geomorphic setting of the Fertile Crescent: surrounded by five nearby major bodies of water (Mediterranean Sea, Caspian Sea, Black Sea, Persian Gulf and Red Sea) and at great distance from oceans, contributing to a stable climate with mainly regular, seasonal weather variation and a relatively low propensity of extreme weather events (floods, tidal waves and heavy storms). It should be added that essential for the diffusion of agriculture was also the invention of irrigation methods, which allowed early Sumerian societies to overcome the naturally dry climate conditions in areas close to rivers. In other words, Mesopotamia was not an opulent natural oasis such as a tropical rainforest – probably under such conditions agriculture would never have emerged. Instead, it was dominated by marshlands. Few of these are surviving now due to damming of rivers and climate change.

The most important geographical factor, according to Diamond (1997a, p. 177, notably Figure 10.1), has been the main axis of continents: north–south or vertical in the Americas and Africa, versus west–east or horizontal in Eurasia. The horizontal axis has the great advantage that, when moving away from a specific region in horizontal direction or diffusing ideas in that same direction, one remains at the same latitude and thus in approximately the same climate zone. This means that the spread of crops, adapted to particular climate zones, from their geographical origin in Eurasia in eastern or western direction was fairly easy. This allowed for a large area of diffusion and experimentation of early farming. In contrast, in Mesoamerica, agriculture could only diffuse in a vertical direction, shifting to latitudes with a distinct climate, ultimately limiting the spread of crops. Hence, the area and human population involved in farming experiments was considerably smaller, limiting progress. Statistical evidence for Diamond's theory is provided by Olsson and Hibbs (2005) and Ashraf and Galor (2011).

The rise of agriculture in Western Asia caused the first civilisations to be located in that area. Through food surpluses, agriculture allowed for the development of stratified societies with bureaucracy, language, armies, philosophers and scientists. The early agricultural area, Mesopotamia, was able to take advantage of its head start. In the modern jargon of innovation economics: Mesopotamia had a 'first-mover advantage', which it later transferred to Europe, in broad brushes, via the Fertile Crescent states, Egypt, the Persian Empire, the Greek city states and the Roman Empire to Western Europe. This (agri)cultural diffusion pattern explains, along with additional factors, Eurasian and European dominance up to the present time (accounting for the fact that United States dominance is the outcome of European colonialism).This is Diamond's theory in a nutshell.

The spread of agriculture occurred along different routes. Farmer populations grew more rapidly than those of hunters-gatherers due to ultimately – though not initially – better nourishment. This allowed farmers to support larger families. Migration of farmers, pushed by population pressure, contributed to intergroup competition, effectively among different types of farming systems that would result in the selection of the most effective ones. Spatial diffusion of knowledge about farming eventually would lead to the adoption of farming strategies by hunter-gatherers in geographically remote areas. Diffusion may further have been promoted by exchange of agricultural for hunted or gathered products. The geographic distribution of human genetic features shows that the process involved mixing of originally separate populations. Cavalli-Sforza and Feldman (1981) argue that the spread to the north and west occurred at an average speed of about 1 km per year. In other words, the diffusion of agriculture from a limited number of centres, notably the Fertile Crescent, to other regions would have taken several thousands of years.

Some historians have criticised Diamond's theory for covering too wide a time span, or insufficiently recognising autonomous social-cultural dynamics (McNeill, 1997; Mokyr, 1998). These criticisms overlook the relevance of distinguishing between proximate and ultimate factors and giving attention to geographical and biological factors, in order to explain long-term human cultural history. In addition, some have levied racist types of arguments against Diamond's theory, notably that genetic differences are sufficient to explain variance in success of distinct cultures around the world. However, most biologists nowadays agree that genetic differences are so small, and that intermixing of populations over recent evolutionary history has been so pervasive, that the notion of races finds weak support in human genetics (Chapter 9 in Cavalli-Sforza and Cavalli-Sforza, 1995). This does not deny the existence of certain genetically based differences in phenotypes among broad groups, such as white/Caucasian, Mongoloid/Asian, Negroid/Black, and Australoid. But average differences in intelligence among human populations are so small that they cannot explain distinct development levels between populations or countries.[6]

Diversity of Domesticable Plant Species in the Fertile Crescent

A great diversity of domesticable animals and plants provided an essential condition for the origin of the earliest agriculture in Western Asia, the Fertile Crescent. This allowed for food rich in carbohydrates, proteins and fats. Diamond (1997a, Table 8.1, p. 140) suggests that 33 out of the 56 species of grass with the largest seeds were located in this area. Moreover, many species included a high percentage of hermaphroditic plants, which facilitated reproduction under artificial farming conditions. In addition, wheat species had the advantage that they contained relatively many proteins compared, for example, with American cereals like rice and corn. Diamond considers as the eight 'founder crops': the cereals emmer wheat, einkorn wheat and barley; the pulses lentils, peas, chickpeas and bitter vetch; and the fibre crop flax. Only two of these, barley and

---

[6] This is not to deny any gene-based differences in intelligence; see the discussion of Ashkenazi Jews at the end of Section 11.3.

flax, were found widely outside the Fertile Crescent. Note that the success of these species is indicated by the fact that they are now grown in all continents.[7]

Different types of agriculture can be identified in terms of stability, productivity, staple food production or domestication of animals (Smith, 2003, lecture 9). Cultivation versus domestication is an important distinction, as the first can do without the second, but not vice versa. Seed versus root versus tree crops is another basic distinction. The first include cereal grains (maize, rice, wheat), which can be productive but unstable in the sense that constant human intervention, such as weeding, is needed in cultivation. Root and tree crops, known also as vegeculture (such as yams and potatoes), are more complex, in an ecological sense, and therefore less productive, even though more stable. Much is known about seed-crop domestication because its archaeological preservation is better given that it took place in arid regions, whereas root crops were more common in wet areas. Both seed and root crops provide staple foods, meaning that they can be stored for a long time, thus allowing surpluses to be accumulated over time. Root crops especially can withstand long periods of drought and cold. However, both their cultivation and their nutritional values differ. While grains are rich in proteins, root crops need a complementary diet, such as meat, fish, grains, vegetables, olives or nuts. Tree-grown fruits and nuts were easily available in subtropical and tropical regions. It is likely that early farmers still undertook some gathering of such foods, although, at a certain point in time, exchange or sharing may have allowed complete specialisation in farming, hunting and gathering.

According to McElreath (2010), over many generations of farming, a transfer of information about the environment to the genes of the domesticated species has occurred. This includes information about their energetic value, success under particular soil and climate conditions, and ease of harvesting. The reason is that generations of farmers have selected seeds from the species that performed well on these criteria. In return, some domesticated species do not perform so well in certain aspects, such as germination, which are typically complemented and compensated by intensive human management and technology. Since the latter result from cultural and technological evolution, we can say that domestication of species in agriculture is part of a grander process of gene–culture coevolution (Section 7.4). This holds equally for farmed plants and animals. To the latter we turn now.

Domestication of Mammalian Herbivores

Unlike other areas with a Mediterranean climate, such as California, Chile, South Africa and Southwest Australia, the Fertile Crescent hosted a unique diversity of herbivore animals that could be domesticated. These included four of the five currently most important domesticated species, namely the cow, goat, pig and sheep. Diamond (1997a, Chapter 9, p. 156), in analogy with Tolstoy's opening statement in Anna Karenina – 'Happy families are all alike;

---

[7] Currently, about 3000 species of plants are cultivated for food, of which some 20 dominate world supply. No more than 12 of these are responsible for about 80 per cent of the current annual tonnage of all crops consumed, including wheat, rice, corn (maize), barley, potato, sweet potato and cassava.

every unhappy family is unhappy in its own way' – proposes 'The Anna Karenina principle", which he invokes various times in his book. One version is: 'Domesticable animals are all alike; every undomesticable animal is undomesticable in its own way.' The principle conveys that a positive outcome requires multiple factors to have the right sign, whereas one or more failing factors will produce a negative outcome.

Domestication of mammalian herbivores has involved several steps, such as taming, herding and spread (Jones et al., 1992; Diamond, 1997a). Some animals are more difficult to tame than others. Notably social carnivores, such as wolves, can be more easily tamed than solitary hunters such as leopards. Sociality is not a sufficient condition though. Taming of social herbivores – herd animals – crucially depends on the nature of spacing and agility of animals. Buffalo, sheep and goats are used to staying close together, as opposed to deer and, especially, antelopes, gazelles and zebra, making the latter more difficult to tame. Taming of solitary, territorial animals, which requires much space, is generally difficult if not impossible. The only example is the cat. The foregoing illustrates that the potential for integration, through artificial selection, of a species in cultural evolution is limited by its biological evolutionary history.

Herding, or pastoralism, can be regarded as an intermediate phase between hunting and domestication. It involves herd animals that can be tamed – caprines, like goat and sheep, being the most common. First efforts to herd animals probably occurred in marginal areas, as in other areas hunting would have been a much more energy-efficient way of obtaining meat and other animal products. Probably, early on already domesticated dogs were used to aid in herding. Motivations for a transition to pastoralism might be that a domesticated stock of animals effectively functions as a permanent refrigerator, serving demand when it arises. In addition, herds of animals can be fairly easily moved to market places for sale or trade. Domesticated animals have changed, generally more in physical than behavioural features. Artificial selection for easily manageable animals and lack of food may have prompted on average smaller animals than are naturally found, also as the key to successful domestication is retaining juvenile characteristics in an adult. This process is known as neoteny and results from selecting for tameness. Table 11.2 lists early domesticated animals.

In conclusion, the Fertile Crescent offered a diversity of domesticable animals and plants that could support a complete, healthy and varied diet. In various other regions where subsistence farming was introduced by diffusion from other areas, it failed to provide such a healthy diet because local circumstances did not allow the production of the necessary variety of foods. Diamond (1997a, p. 149) illustrates this with the example of New Guinea highlanders having a structural deficit of proteins, which he suggests to have been a probable cause of cannibalism.

11.7   Between Neolithic and Industrial Revolutions

Between the rise of agriculture and industrialisation lies a period of about 10 000 years. It is impossible to summarise it in sufficient detail in the context of a single section. Merely a few general trends over long-term history can be identified here.

Table 11.2 A selection of early domesticated animals

Domesticated animal	Wild progenitor	Principal region of origin	Approximate date of domestication (BP = years before present)
Dog	Wolf	Western Asia	>12000 BP
Goat	Bezoar goat	Western Asia	9000 BP
Sheep	Asiatic mouflon	Western Asia	9000 BP
Cattle	Aurochs	Western Asia	8000 BP
Pig	Boar	Western Asia	8000 BP
Llama	Guanaco	South America	7000 BP
Horse	Horse	Central Asia	6000 BP
Donkey	Ass	Arabia, North Africa	6000 BP
Chicken	Jungle Fowl	Southern Asia	6000 BP
Turkey	Turkey	Mesoamerica	5500 BP
Dromedary	Camel	Arabia	5000 BP
Camel	Camel	Central Asia	5000 BP
Yak	Yak	Himalayas	5000 BP
Reindeer	Reindeer	Northern Eurasia	5000 BP

Source: Information based on Jones et al. (1992, p. 384) and Diamond (1997a, p. 100).

The Appearance of Complex Societies

The transition to agriculture and settlements ultimately gave rise to food surpluses. This allowed the development of larger societies with extended labour division and complex organisation. As already discussed several times in this book, extended labour division is a phenomenon found generally in evolutionary systems. Human groups thus became bigger, which led to the evolution of hierarchical organisation and related bureaucracy, giving rise to early states. Larger religions developed that supported the ruler and forced people to accept a certain degree of inequality. Systems for redistribution of social production and finance of public goods developed as well. These included taxes, initially collected in the same kind as the product (crops) over which it was levied. So this crucial role of the state, even today, emerged early on in history. According to North (1981, Chapter 8, p. 94) it should be seen as the most fundamental achievement of the ancient world. Different political decision-making systems evolved over time, including dictatorship (Pharaohs of Egypt), monarchy (Persia/Macedonia), aristocracy (Greek city states) and oligarchy (Roman Empire). The development of military technology was important in this respect, as it allowed for larger connected regions and states to emerge: from city states in Mesopotamia and ancient Greece to the huge Persian and Roman Empires. Communal property rights were altered, leading to private and state property regimes. This contributed to more skewed income and wealth distributions. Ultimately, extended legislation developed to arrange and control ownership and contracts, notably in the Greek city states and the Roman Empire.

The food surplus caused labour to be allocated to non-agricultural (non-food production) activities. Among others, it allowed thinkers and specialists to emerge, which contributed to technological progress. Somatic energy derived from the metabolism of human bodies dominated until about 8000 BC. Agriculture increased energy contained in food crops, notably through animal muscle power used in ploughing, haulage and transport. The transition to agriculture thus led to a 10- to 100-fold increase in available energy relative to the hunting-gathering era. In addition, new materials were used, initially bronze and later iron. Technology developed in line with these. All these phenomena can be understood as part of a positive feedback loop: more available energy, a higher productivity of labour, more labour division, more technological progress, more access to materials and energy, and back to the beginning of the cycle. Organisation and distribution of the output of agriculture was a social and bureaucratic activity, which stimulated the invention of early writing. This increased the efficiency of management and control of production, transport and distribution of food. In addition, writing fostered precise reasoning as well as the accumulation of ideas, which meant a transition from prehistory to history.

Humans started to live in more organised spatial structures as well, resulting in villages, and later in early cities and civilisations (from the Latin 'civitas', which means city) extending over large regions. Urban areas with a dense network of activities, markets and trade developed, in some places growing to considerable size. Markets allowed the exchange of goods, tools, land and human labour. The slow transition to cities and civilisations involved a change in group size, fostered by intercity trade. This allowed specialisation and labour division, as well as barter trade and mercantile exchange through markets over larger areas. In addition, it changed land use and planning, such as central urban areas with markets surrounded by subsequent zones of gardens with perishable crops, orchards with fruit, cropland with more durable crops and pasture with grazing animals. Ofek (2001, Chapter 13) notes that this geographical structure is well explained by the classic Thünen model of optimal land use around a city, taking into account yields, production costs and market prices of all agricultural products as well as transport distances and costs.

Fuelled by these production and trading processes, early civilisations arose. The six areas where this happened were Western Asia (starting around 3500 BC, with Sumer, Babylonia, Assyria, Phoenicia, Troy, Persia), Egypt (roughly 3000 BC), India (2600 BC, with Harrapan/Indus Valley, Dravidian, Vedic, Shramana, Magadha, etc.), Europe (1000 BC, with Minoans, Mycenaeans, Greece, Macedon, Carthage, Sicily, Etruscans and the Roman Empire), China (2000 BC, with Xia, Shang, Zhou, and Imperial China) and the Americas (1000 BC, with Olmecs, Teotihuacán, Toltecs, Maya, Aztecs and Incas). Among significant relics are the remains of cities and impressive palaces, such as the Ziggurat at Ur (Sumer, Mesopotamia), the palace of Minos at Knossos (Crete), the Acropolis in Athens, the temples and pyramids of Teotihuacán (Mexico), and the Colosseum in Rome.

Agriculture became the dominant activity around the world, with a minority of tribes continuing to practise hunting and gathering, for a variety of reasons: an insufficient diversity of animals (aboriginal people of Australia), arid zones unsuitable for farming (nomads in the Sahel), or an oversupply of natural foods (tribes in tropical forests in the

Amazon). The diversity of early villages, cities and activities meant a competition, diffusion of acquired knowledge and selection which points at an evolutionary mechanism at the level of groups, settlements, villages, cities, civilisations, as well as at organisational and spatial scales.

Group Size and Demography

Speed and scope of cultural evolution are influenced by the size and composition of unitary human groups. Agriculture caused an unprecedented growth of the human population, which went along with increasing population density and spatial concentration in certain areas, ultimately giving rise to early cities. Diamond (1997a, Chapter 14) identifies four main stages of social structures over the last 13 000 years. All of these still exist in some part of the planet:

1. The nomadic family or band: This covers 5 to 80 people and is kin based, consisting of extended families or several related ones. Everyone knows each other, allowing for informal conflict resolution through arbitrating and mediating.
2. The tribe living in a village: This entails hundreds of people, usually kin-based clans. Also here everyone knows each other so that there is informal conflict resolution. The ruler is supported by a family network and reciprocal acts.
3. A chiefdom comprising one ethnicity and one or more villages: This involves thousands of people organised in kinship groups. Here not everyone knows each other personally and, as a result, there is centralised conflict resolution. Groups have a chief (generally an inheritable position) which together form a kind of council to govern the chiefdom.
4. A state containing many villages and cities, with one or more ethnicities: This generally includes at least tens of thousands of people and is guided by laws, judges and police. Key factors behind the transition to states were population pressure, war or threat thereof (selecting for strong leadership), and a food surplus from agriculture — allowing for fulltime bureaucrats or an expensive ideology like a religion.

When moving through the social structures in the above order, non-kin relationships increase in importance. Furthermore, increase in group size tends to go along with many other changes, such as inequality in the distribution of power and wealth, more formal ways of conflict resolution and more complex organisation involving hierarchy and horizontal diversification. This process has not finished – witness that at the supranational level we lack a government. Increased international cooperation and agreements, as well as rapid long-distance travel and communication, indicate the human population is currently moving in the direction of a fifth stage, namely of a globally connected and governed population. But this transition is far from complete and we may well never finish it. The United Nations is an effort in this direction, but with limited success. Indicative is that we do not seem to be able to eradicate human conflict owing to religious, ethnic or political-ideological differences.

Contagious Human Diseases

The change through agriculture and food surpluses gave rise to large sedimentary societies with a high population density and people living close to domesticated animals. This, in turn, allowed for epidemic human diseases to spread and survive, especially as many of these evolved from germs of domesticated animals. Wolfe et al. (2007, p. 281) report that 8 of the 15 temperate human diseases likely originated from domestic animals (diphtheria, influenza A, measles, mumps, pertussis, rotavirus, smallpox, tuberculosis); three very likely came from apes (hepatitis B) or rodents (plague, typhus); and the sources of the other four (rubella, syphilis, tetanus, typhoid) are yet unknown. Most likely, measles, tuberculosis and smallpox originate from cattle, flu from pigs and ducks, malaria from birds (possibly chickens or ducks), and pertussis (whooping or 100-day cough) from pigs or dogs. In addition, many more tropical diseases arose in the Old World (Africa and Asia) than elsewhere due to the genetic distance between humans and monkeys/apes there being the shortest. Moreover, much more evolutionary time was available for disease to transfer from animals to humans there. It is suspected that many more diseases derive from pathogens originally carried by animals, including Alzheimer's, certain types of cancer, diabetes, multiple sclerosis, schizophrenia and heart failure (Ewald, 2002). This needs to be clarified by future research.

Contagion by animals would have been unlikely in hunter-gatherer cultures, as these had too low population densities, did not live together with animals, and were generally not controlled by diseases, but instead by predators, competition with other groups, and famines (McNeill, 1976). Incidentally, many groups applied conscious population controls through cultural habits and rituals – related to frequency of coitus, the probability of conception and the survival of offspring. This included such diverse practices as prolonged lactation, use of herbal and animal contraceptives, abortion, post-birth sex taboos (abstention), ritual genital mutilation, sterilisation, setting a minimum age for regular sex, polygamy, and even infanticide and putting widows to death (Hayden, 1972).

A long-term consequence of selection of disease resistance throughout medieval times occurred during the beginning of the colonial period, when West European countries sought expansion in other parts of the world. Namely, their inhabitants took microorganisms – against which they themselves had genetic resistance – into areas with people without such resistance. As a result, many of the latter were killed, not by guns but by germs. For instance, in Hawaii, there were once 1.5 million aboriginal people, whereas only 80 000 survived the first disease triggered by European invaders. When Cortez invaded Latin America, he fought against a much larger population than his, but due to smallpox, this population was extenuated which facilitated a final victory.

A Shift in Dominance from the Mediterranean to West Europe

Despite many ancient civilisations having followed a pattern of rise and decline (Tainter, 1988), a lot of their accumulated knowledge regarding organisation and technology was, in many cases, not completely lost but preserved and diffused in some form. This has contributed to a long-term, slow trend of progress. The first half of the

Middle Ages (500–1000 AD), referred to as the Dark Ages, is often regarded as an exception to this, but as we will see below this is probably too much of a caricature. The Catholic Church was the crucial institution during the Middle Ages, in terms of knowledge, literacy, property and generally material wealth. In addition, it possessed power over a large portion of the population and in the political arena, and had the capacity to define basic norms and rules for individual behaviour and interpersonal relations. It provided a bridge from the classical world, through the Roman times, to the modern era. Hence, it was the binding and highest element in human and spatial organisation in Europe. Contributing to this was its solid hierarchy with a single leader, the pope, on top of what may be considered the first supranational organisation.

The first era of rapid human cultural and urban development took place in the Mediterranean basin and west Asia: first Mesopotamia (the Fertile Crescent), then Egypt, next the Persian Empire, Greek city states and, finally, the Roman Empire. The Roman Empire disintegrated in the fifth century as its large state no longer supported military advantages and protection of property rights, and because taxes to finance these public goods were rising over time. In North Africa it allowed Muslims to spread from the Arab peninsula. In Northern Europe the Roman Empire was followed by a collection of smaller states with a political-economic system known as feudalism. This was confined to localities that were part of a hierarchical nature, with at the top a local castle with a lord and knights, providing land-use rights in exchange for military service and taxes. The system had its roots in north-west Europe and, although it was strongly influenced by Nordic-German culture and institutions, it shared some characteristics with that of the Roman Empire, such as the serf versus slave. The relationship between the lord and the serf was central, and has been regarded as a contract enforced mainly through power by the lord or manor (North, 1981, Chapter 10). The only characteristic which gave some bargaining power to the serf was the scarcity of labour, resulting in competition among lords for serfs. The spatial features of manors, notably the scattered strips of agricultural lands, may have functioned as insurance against risk of concentrated land holdings, or by allowing for more effective monitoring of production output. A third explanation is evolutionary, specifically that certain strategies have evolved due to being selected through competition among neighbouring manors with distinct strategies. Learning through imitation may further explain the diffusion of more successful strategies.

Although feudalism is often seen in a negative way, it created order in Europe, allowing for population and income growth, diffusion of technological innovations, and trade among regions specialised in different, complementary products – such as fish, metal, timber, wine or wool. Towns developed and grew, leading to local legislation, along with guilds to support property rights of, and standards employed by, producers. Trade increased, notably in Flanders and North Italy, supported by states providing protection and privileges. Owing to population growth and plagues, however, Europe fell back into Malthusian economic and population crises, with people living at subsistence level and meat being replaced by cereals. As a result, labour became scarce, which completely turned around the balance of power in agriculture: from lords to tenants, including freemen and serfs. The legal consequences were that in some cases property and user rights became less clear or uncertain, while in most regions they were guaranteed no longer by lords but by the highest jurisdiction, namely the king. North (1981, p. 136) notes that the growth of the money economy meant that the interaction between king and knights and lord and servants on the basis of exchange of user rights for military service lost value and ultimately viability.

After the demise of the Roman Empire, Mediterranean cultures still dominated, namely through Portugal and Spain, followed by North Italy, after which economic dominance shifted towards Western Europe. The Netherlands and England were the most successful, largely because they had imposed a set of private property rights that reduced transaction costs of production and commerce. This helped to respond flexibly to eventual scarcity of production factors, stimulate technological change and foster national competition as well as participation in international trade. Other countries, such as Spain and France, had inherited outdated institutions and property rights, dominated by local jurisdictions and privileges to certain groups (Cameron, 1989, Chapter 5). This contributed to Malthusian economic crises in these countries.

Development as Dependent on Ancestral and Geographical Distance

A subtle treatment of the role of a combination of genetic and cultural factors in clarifying long-term cultural history is offered by Spolaore and Wacziarg (2016). They argue that certain intergenerationally transmitted traits, such as habits and language, can act as barriers to the diffusion of technologies and institutions because they make populations dissimilar, hampering communication, trust and mutual learning (Guiso et al., 2009). In particular, a larger ancestral distance with a frontier society, i.e. with the most advanced technologies or institutions, suggests a stronger barrier for a current society to have adopted these. The degree of ancestral distance is generally associated with a longer historical separation time and geographical distance between the two societies. The first hinders relevant vertical combined cultural-genetic transmission and the second relevant horizontal cultural transmission. The latter connects to Diamond's view (Section 11.6) on the role of geographical factors in the diffusion of technologies.

An earlier econometric study by the authors (Spolaore and Wacziarg, 2013) indicates that four geographical variables account for 44 per cent of the statistical variation in per capita incomes in 2005, specifically absolute latitude (largest effect), the percentage of land area in a tropical climate, being landlocked (i.e. no direct access to seas) and being an island. In fact, the majority of the 48 landlocked nations, hosting about 7 per cent of the world population, has a low level of development and includes many of the poorest countries in the world. The importance of access to the sea for economic development is widely recognised: witness the United Nations Convention on the Law of the Sea which assigns landlocked countries the right to transport goods through neighbouring countries without having to pay taxes.

The role of ancestry is further analysed in a number of other quantitative studies. Putterman and Weil (2010) construct, for each country in the world, an ancestry weights matrix by calculating the share of the population originating, since 1500, from other countries. Using these weights, they find that two 'deep history' variables, viz. years since the agricultural transition and experience with centralised governance, are statistically strong predictors of contemporary per capita income. This approach shows that institutionally young countries, such as the United States and Canada, have benefited indirectly from transfer of insights and habits favourable to prosperous development due to their populations originating from countries with longer histories of national institutions. Comin et al. (2010) apply a similar set of ancestry weights to indicators of technological advancement, finding that long-term history positively matters for current development level. Michalopoulos and Papaioannou (2014) examine the relation between precolonial ancestral traits and economic development in sub-Saharan Africa, accounting for national boundaries in Africa as a recent outcome of colonialism. They discover that groups which share common precolonial histories and cultures, but live in different countries and are thus subject to distinct national institutions, display similar levels of economic development. In an earlier quantitative study (Michalopoulos and Papaioannou, 2013), measures of precolonial political centralisation were found to have much influence on current development. Underneath the insignificant average effect of current national institutions on the development of ethnic groups, however, lies considerable heterogeneity. The influence of national institutions turns out to be stronger the closer groups are to capital cities.

12 Industrialisation and Technological History

12.1 Preconditions and Ultimate Reasons

Since 1500, the world as a whole, but mostly Europe, has faced a significant growth in average income. This fits the now common view that industrialisation did not arise suddenly but was preceded by a period of gradual change. In fact, the term 'Industrial Revolution' was not used until 1880, when British historian Arnold Toynbee introduced it in a lecture. But simply recognising that before the 'revolution' a quite complex technology, the sailing ship, was already well developed, allowing colonisation by European countries of the rest of the world, indicates the gradual process of industrialisation that has taken place over the course of centuries. Dutch historian Jan de Vries (1994) uses the label 'industrious revolution' to emphasise that a process of household-based production contributed to a transformation of the pre-industrial economy into an industrialised system driven by markets and scale economies. These annotations do not deny, though, that the pace with which almost everything changed after 1750 had no precedent in human history[1] – hence, warrants the term Industrial Revolution. To illustrate, the relative contribution of sectors to national production, income and employment altered structurally. In the Middle Ages 80 to 90 per cent of the work force was in agriculture, which had dropped to about 50 per cent at the end of the nineteenth century, while currently it is below 10 per cent in industrialised countries.

The period 1870 to 1913 showed a spectacular growth rate, which was much higher than in the preceding period, as well as in much of the subsequent one. The war period 1914 to 1945 repressed growth considerably. But in spite of, or perhaps due to, the two world wars, the 20th century showed exceptional economic growth. In addition, an enormous influence was exerted by humans on their natural environment, both locally and globally, equally unprecedented in history. Only natural phenomena like extreme climate change and meteorites have had a similar or larger impact. The highest growth rates were realised between 1950 and 1973, which marks the beginning of the first oil crisis. Average income is now about nine times that of 1500. From an evolutionary angle, however, diversity and distribution matter: many people in developing countries still have an income below the global average in 1500. In this sense, we have made only moderate progress.

---

[1] On the other hand, the same can be said for the changes between 1500 and 1750 as compared with the period before, and the changes between 1200 and 1500 and the period before. So history contains periods with progressive progression, in the vein of exponential growth.

Following Parker (1984), one can distinguish between Smithian (after Adam Smith), Solovian (after Nobel laureate Robert Solow) and Schumpeterian (after Joseph Schumpeter) growth. The first refers to growth caused by an increase in trade or exchange, which allows one to benefit from advantages of specialisation and labour division. Solovian growth is the outcome of investment in the capital stock, which allows labour productivity to increase. This ultimately depends on savings or abstention of consumption. Schumpeterian growth, finally, is driven by technological inventions and innovations. Evidently, other factors play a role in explaining growth, such as the scale of market and thus ultimately population size, or education and learning (human capital). Decomposing historical growth in these factors is hard. One reason is that Schumpeterian and Solovian growth are strongly interactive as new technology is embodied in capital. In other words, innovations appear through appropriate capital investments. An additional role is played by improved organisation, money and legislation, mainly responsible for economic growth in classical Greek and Roman societies.

A fundamental puzzle is why something such as the Industrial Revolution would occur anywhere? As it required major technological innovations, this raises the question 'What causes a society to be successful at technological innovation?' Mokyr (1990, pp. 11–12) mentions four conditions: a population of inventors and innovators; economic and social incentives for inventions and innovations, such as property rights, market incentives and awards; tolerance to diversity and creativity; and not overly strong vested interests that resist innovations. These conditions are unlikely to be satisfied in societies that are malnourished, superstitious or extremely conservative. Table 12.1 presents another angle on invention and innovation factors, based on identifying three key dimensions: individuals, society and natural environment. Note that this list mixes proximate and ultimate factors.

Ultimate explanations of the Industrial Revolution come in various forms. A famous one is the 'Protestant ethic', proposed by sociologist Max Weber at the beginning of the twentieth century. He wondered why Europe, instead of China, India or the Ottoman Empire, emerged as the dominant economic and technological power in the world. His explanation was that, unlike other economic systems, industrial capitalism incorporated a mechanism for accumulation of wealth, which in turn was the source of investment, including what we now would call 'venture capital'. Weber considered the Protestant ethic of merchants and industrialists – characterised by discipline, diligence and sober living – as the fundamental behavioural reason. The Calvinist puritans were most efficacious in this respect, as they believed in predestination, which inspired convictions such as 'luxury is evil', 'work is a vocation' and 'the fruits of labour are for the glory of God' (Giddens 1997, pp. 576–577). Like any theory, Weber's explanation has received criticism. For example, it has been said to fail in explaining why early capitalist developments also occurred in the northern part of Catholic Italy. Another critique is that merchants, regardless of their religious upbringing, simply opted for Protestant regions as suitable to undertake their commercial activities. Blum and Dudley (2001) have tested Weber's hypothesis using data on income levels and population growth in 316 Protestant and Catholic regions in Northern and Southern Europe. While they reject it overall, they nevertheless find that Protestantism mattered in a subtle way. International contacts

Table 12.1 Factors positively affecting the rate of technological inventions

Personal factors	• Life expectancy: learning; payoff. Did early inventors get old? Or are important inventions done at young age? • Nutrition: intelligence, energy (laziness). • Risk-taking: potential payoff higher during Industrial Revolution than before; synergy with market size.
Social factors	• Tolerance and resistance to innovation. • Pragmatic attitude: experimenting, solving problems. • Ethics and values: material wealth. • Religion. • Economic factors: costs, interest rate (investments), market size, market competition. • Demographic factors: population pressure, critical mass for innovation. • Spatial factors: economics of agglomeration: most innovations since the late Middle Ages arose in urban settings. • Scientific factors.
Environmental factors	• Geographical environment: inventions and innovations stimulated by resources; focusing factor: away from scarcity (bottlenecks), towards abundant factors (opportunities). Examples: energy sources (rivers); climate and ecosystems (agriculture); rainy parts of Northern Europe and watermills, wind and windmill; Europe had a large supply of domesticated animals: impact on innovations in transport. • Cheap and ready availability of abundant energy sources, such as peat and coal.

Summary of Mokyr (1990, Chapter 7).

between Protestants transformed the network of Northern European cities into a so-called 'small world' between 1500 and 1750, in turn creating a fruitful basis for innovation, diffusion and growth. At the root of this, the authors suggest, was an inclination of Protestants to honour contracts with strangers, including in distant regions. Arguably, this was the result of Protestants' strong fear for punishment of sins. Catholics lacked it as they could be pardoned through confession or the 'sacrament of penance'.

A more fundamental question is why economic growth took off in the first place. An evolutionary, ultimate account is offered by Galor and Moav (2002, p. 1): 'the struggle for survival that had characterised most of human existence generated an evolutionary advantage to human traits that were complementary to the growth process, triggering the take-off from an epoch of stagnation to sustained economic growth'. This view fits in the 'enhancement mode' of Durham's (1991) classification of gene–culture coevolution (Section 7.4). The hypothesis of Galor and Moav is that natural selection changed the distribution of certain parental care characteristics, notably the trade-off between quantity of offspring and quality of parental care. In modern economic growth jargon, such quality improvements are a special case of human capital investment (Becker et al., 1990). In particular, the gradual emergence of the smaller family since the rise of agriculture may have played an important role in this. Hitherto, larger groups, such as tribes built around one or more extended families, dominated human biological and cultural evolution. Galor and Moav argue that human organisation by way of smaller families fostered a strategy that shifted parental investment to quality of offspring, such as through education. This resulted in an improved capacity of humans to develop and employ advanced technologies, as well as undertake entrepreneurial activities. This, together with a sufficiently large size of the communicating human population, gave rise to major technological innovations, thus creating the impetus for the take-off of the Industrial Revolution. Galor and Moav can thus be seen to propose a kind of 'endogenous evolutionary theory' of the Industrial Revolution. Relevant selection pressure was effective during the preceding 'Malthusian era' because the majority of people were living at a subsistence consumption level. In addition, Galor and Stelios (2012) claim that the selection of entrepreneurial traits provided the conditions for entrepreneurial- and innovation-based growth, as is typical of industrialisation. Their argument is that there was an almost continuous slow improvement in the entrepreneurial attitude over a long period before the Industrial Revolution, characterised by risk-tolerant or even risk-seeking behaviour.

A related, but nevertheless distinct, mechanism, involving a mixture of genetic and cultural selection, was proposed in *A Farewell to Alms* by Gregory Clark (2007). His thesis is that England was able to escape the Malthusian trap through natural selection working upon its population. Selection was relevant as Malthusian limits were operative, notably scarce food supply. The Malthusian trap means that more food due to increased agricultural productivity was immediately compensated by population growth, so that income, in effect, would not rise. A core element of Clark's explanation is contentious, namely selection causing 'survival of the richest'. He tries to provide evidence for the idea that, in centuries preceding industrialisation, poor people in England had lower survival and birth rates than rich ones, and that rich people had a culture that differed from that of the poor. As a result, middle-class values such as peacefulness, literacy through education, patience, work discipline and family planning spread through both genetic inheritance and cultural mechanisms – learning from parents and similar members of society. To support his suggestion that genetic factors play a role, Clark (2008, pp. 190–191) notes that genetic expression differs among societies depending on their social structure and institutions. If the large majority of people are serfs, and a few rich own all the capital, the environment predicts success better than genetics. However, in a modern society with easy public access to advanced education and health care, genetics may explain economic success to a greater extent. Since the Middle Ages, England has been a society where most occupations were open to all, genetics could express itself well in terms of economic success. Clark adds that in China or Japan at the time, a similar process of rapid industrialisation did not take place. He presents data showing that their richer classes, the Samurai in Japan and the Qing dynasty in China, were much less fertile than the rich in England.

Clark's work has received critical responses.[2] As one can imagine, many commentators simply disliked his invocation of genetic mechanisms to explain social-economic

---

[2] See a long list of book reviews here: http://faculty.econ.ucdavis.edu/faculty/gclark/a_farewell_to_alms.html. A summary of the main critiques and a response to these can be found in Clark (2008).

history. Others question the evidence for differential birth rates and middle-class values. In my opinion, the critics have not disproved Clark's hypothesis. Nevertheless, it remains no more than that, as the evidence certainly is not watertight. Incidentally, Clark avows that the evidence for the aforementioned theory by Galor and Moav is weaker than for his alternative explanation. Anyway, the criticism of Clark is valuable as one can distil from it suggestions for alternative factors behind the Industrial Revolution in England. Bowles (2008b) notes that while Max Weber's Protestant ethic sounds through Clark's work, in places, cumulative innovation generated by Schumpeterian entrepreneurs is a more likely factor behind the Industrial Revolution. Others refer to cumulative knowledge and technology having passed a critical threshold, which is consistent with theoretical economic studies (Hansen and Prescott, 2002). As a sub-explanation of this, human population size is regarded critical too, not just to support large markets, but also to allow for technological innovation based on aggregate human capital, which economists defend through the so-called non-rivalry feature of technology. Kremer (1993) elaborates this through a model that combines Malthusian and Boserupian (Boserup, 1965) elements. The latter means people adopt new agricultural technologies if population pressure on land increases. He shows empirically for long-term data that regions with larger population size attained higher technology levels. Many additional explanations are offered: the scientific revolution, the English reformation to Anglicanism, access to coal, profiting from overseas colonies, relatively low interest rates, urbanisation and extensive road and water infrastructure (implying a low cost of transport) in England at the time. Further factors are mentioned in the first column of Table 12.3 (in Section 12.7), which compares candidate countries for the Industrial Revolution. Of course, a combination of these various factors is particularly powerful, especially since it is likely to involve synergistic effects. It should be realised, though, that most of the mentioned alternative factors represent proximate, rather than ultimate, explanations of the Industrial Revolution.

Effective diffusion of institutions and technologies is widely regarded as being key to long-term economic progress. Spolaore and Wacziarg (2009) test how such diffusion is affected by geographical measures of human genetic distance, which they consider to be associated with a wide range of human traits transmitted from parents to offspring, such as language, family networks, religious traditions and cultural practices. They find that relative genetic distance to the frontier (England in the nineteenth century, United States in the twentieth) strongly predicts diffusion during the period from the start of the Industrial Revolution to the beginning of World War I, and then tapers off as diffusion had many more routes available due to a high number of countries having already adopted the most advanced institutions and technologies – causing the 'barrier effect' to be weakened. In a later study, the authors test the effect of genetic and linguistic distance on the diffusion of fertility decline since the nineteenth century, arguing that it involved new social norms which initially emerged in France (Spolaore and Wacziarg, 2014). A similar argument was earlier made by Watkins (1990). The fertility transition was important as through shifting from a quantity-based to a quality-based fertility strategy human capital was enhanced, in turn encouraging economic development. Transfer of knowledge, technologies and institutions by colonisers have received attention as potentially contributing to the economic success of countries. Acemoglu and Robinson (2012) suggest this as an explanation for both positive and negative 'reversal of fortune' – compared with economic, technological and institutional circumstances around the year 1500 – as occurred in various regions colonised by Europeans. For instance, North America started off with low development and population density, but shifted to a pattern of progress. In contrast, some Latin American countries, notably Mexico and Peru, moved from a relatively high development and population density to a phase of decline.

12.2   Middle Ages and Renaissance

The period ranging from about 500 to 1150 is known as the 'Dark Ages' of the Middle Ages, owing to a restraining effect exerted by religion on art, science and economic activity, reinforced by conflicts among the upper classes. Nevertheless, in this period major inventions came about as well, notably in agriculture. This had much impact as the large majority of people were involved in farming, causing agriculture to dominate the economy. In the late Middle Ages, from 1150 to 1500, the number of inventions increased significantly.

The printing press, with mechanical movable type, is the first invention of which the name of the associated inventor is still known: Johann Gutenberg. This is explained by the importance of this invention as well as the fact that the inventor himself solved all practical problems. This happened in 1453, resulting in an invention that was immediately suitable for diffusion, something which is rare. By 1480, some 400 presses had been mounted in Europe, mainly to produce bibles. The crucial importance of the printing press, a clear 'macro-invention', is that it stimulated reading and allowed for the diffusion of books on science and engineering, thus contributing to the dissemination of productive knowledge and technology. In addition, it served as a precondition for education, emancipation and, ultimately, the emergence of early democracy.

Possibly, Laurens Coster from Holland invented a printing press in the same period, as early as 1430, 23 years before Gutenberg. While it might have influenced Gutenberg's design, definite evidence for this is lacking. History presents other examples of such synchronicity or parallel, independent inventions (Kelly, 2010, Chapter 7). They suggest that sometimes the technical and knowledge conditions, and perhaps also the social demand, are for the first time ready to make a particular invention feasible or even likely. Incidentally, the first known printing with movable type, using porcelain, occurred in China, around 1045. In any case, Gutenberg can be credited with building a complete, wooden press, using metal moulds and oil-based inks, which for the first time allowcd printing books at an unprecedented speed.[3]

Another fundamental innovation was the weight-driven mechanical clock at the end of the thirteenth century. As it was more reliable than the earlier water clocks, this

---

[3] A brief instructive demonstration of how the Gutenberg press actually worked is here: https://www.youtube .com/watch?v=ksLaBnZVRnM. It also explains the ingenuity with which Gutenberg solved the problem of aligning the front and back prints on a single paper.

technology was rapidly perfected and spread throughout Europe. It had several important impacts. First, it was exemplary of what technology entailed and could realise. Second, it served as a model for developing miniature technologies. Third, it provided modules and design schemes useful for other products and sectors, such as textiles, manufacturing machinery and transport. And finally, it created order and organisation in the production process, permitting accurate measurement of productivity. This promoted planning and efficiency in productive activities and life in general.

It is generally accepted that in the period from 700 to 1200 Muslims were culturally more open minded and technologically the most advanced in the world. They merged elements of ancient Greeks, Romans, Asia, Africa and China – a special case of recombinant innovation in evolution. Their influence on Europe occurred through the Mediterranean region, notably Spain and Sicily. They were mainly diffusers and to a lesser extent innovators, contributing to the dissemination of a number of major technologies: paper, textiles of high fabric quality (probably originating from Persia), cotton production, high-quality leather, chemical technology (glass, ceramic products, naphtha, perfume and acids), high-quality water mills and clocks, metallurgy (high-quality swords), irrigation, sugar refining, and various types of food, including cereals such as rice, fruits like oranges, lemons, bananas and water melons, and vegetables such as asparagus, artichokes, spinach and aubergines. The diffusion power of Islam was due to the extended area it covered and its strategic geographical position between Asian and African to European cultures. The latter allowed for learning and variation owing to errors in transmission, giving rise to technological improvements. Around the twelfth century, however, Islam was technologically superseded by Christian Europe which moved on a path of progressive innovation.

Between 1500 and 1750 the emphasis shifted from macro-inventions to continuous, gradual progress or micro-inventions. This increased the productivity of existing techniques in virtually all areas: agriculture, the windmill, pig iron (blast furnaces), mining, textile production, hydraulic engineering, fishing techniques, and land and water transport. In addition, colonialism led to the adoption of certain techniques and crops (notably the potato) from other (all) parts of the world. Many inventions could, however, not be operationalised due to shortage of workmanship and materials, and costliness of machines. The same held for many visionary ideas by Leonardo da Vinci and others, which were, at best, imperfect blueprints that, given the lack of specialised craftsmen and instrument technology, could not be translated into actual technical designs. They could be seen, though, as contributing to the general knowledge base that made later innovations more likely.

Mining activities were critical for the access to large amounts of materials needed for the new industries. They developed from the end of the Middle Ages and required many practical challenges to be solved. Empirical engineering dealt with underground transport by rail, use of waterpower and firefighting. Early scientists studied mining-related problems of a mechanical, hydrological and mineralogical nature. From this period, it became common for engineers to write handbooks which contributed to the diffusion of knowledge about techniques throughout Europe. Of course, this was made possible by the invention of the printing press in 1453. Furthermore, fashions spread quickly over

Europe, such as the interest in gadgets, i.e. instruments that were not always functional. Governments of early European states started to stimulate and protect technologies through subsidies, patents, pensions and even awarding monopolies. Some states also supported scientific societies.

12.3 Industrialisation After 1750

The result of all these developments was that, by 1750, Europe possessed a definite technological superiority over the rest of the world. This provided the starting point for the 'Industrial Revolution'. Compared to earlier periods such as the Renaissance and the Ancient and Classical periods, several fundamental changes are notable. Machines replaced hands as instruments. Organic materials were substituted by inorganic materials. A huge increase of energy use took place, which was based on the availability of new energy sources, such as peat, coal and later oil. Cumulative technology gave rise to factories organising complementary tasks and machines, allowing labour division and task specialisation, which stimulated increases in efficiency and productivity. As production became more roundabout, intermediate industry, such as instrument making, developed into a sizeable activity. Previously, instruments had been fabricated by individuals who were self-made craftsmen or had been apprentices of others, turning instrument making into a kind of art. After industrialisation, instruments became more standardised and were produced in larger numbers. This led to lower costs, easier diffusion and gradual improvements. In addition, inventions became increasingly dependent on collaborative efforts aimed at solving well-defined technical problems, rather than brilliant individuals stumbling on some unanticipated idea. Really difficult technical problems were resolved in distinct technological areas, often involving complex design of machinery. Such technologies could not be invented overnight but required a long, cumulative process of experimentation and learning by many individuals — described as technological evolution in Chapter 10.

During the period of rapid industrialisation after 1750, continuous gradual improvements were made in many areas of production and technology. Mokyr (1990) considers three main areas of crucial significance: power technology (notably the steam engine), metallurgy (iron and steel) and textiles (replacement of human hands by a machine in cotton production, bleaching, spinning and printing). Cotton proved to be an attractive material, suitable for dyeing, washing and ventilation. Moreover, it had an elastic supply and readily lent itself to mechanisation. As a result, it became the main growth industry. The dynamics are shown by the price of cotton cloth, which fell by 85 per cent between 1780 and 1850. Scarcity influenced innovations. When, due to the Napoleonic Wars, the continental textile industries could not receive raw cotton, the idea of mechanised wet spinning of raw flax was invented, partly stimulated by a financial reward offered by Napoleon himself. But old-fashioned linen could not catch up: in 1850 there were about 250 000 power looms weaving cotton, 42 000 weaving wool and 1000 weaving linen.

Diffusion and learning followed complex patterns. For example, a loom for automatically weaving patterns coded by holes on cards was invented by two Frenchmen – one a son of an organ maker – in the late 1720s. This was the first application of binary-coded information, which inspired Charles Babbage to design his Analytical Engine, the first mechanical equivalent of a computer. This is not the only example of expertise in one area being transmitted to others. Learning, handbooks and craftsmen moving to other areas all contributed to this. In particular, engineers and inventors with a background in instrument- and clock-making in England contributed to the development of various other technologies during the early period of industrialisation.

Iron and steel were the other main (intermediate) materials of industrialisation. The main innovation here was the transformation of blast furnaces producing brittle and hard pig iron (containing much carbon) to a combination of processes designed by Henry Cort in 1784 that produced wrought iron (almost carbonless). This involved different types of furnaces (similar to those used already in glassmaking) and rolling of heated metal with grooved rollers. This process was invented by directed search with the objective of producing wrought iron, because it allowed metal products to be shaped in forges (machine parts, nails, etc.). Although steel production technology was perfected around 1740, because of price differences and possibly a lock-in mechanism, it did not overtake iron production until about 1860. The role played by coal increased in importance. It served as a fuel and as a material input to metallurgy. The productivity of coal mining did not improve much during the nineteenth century, while progress focused almost entirely on increasing safety.

Diversity was important in many ways. Competing designs of steam engines coexisted and were all perfected during some time before a dominant design appeared. In other cases, such as iron production, the older technology was rapidly replaced by the new one. Entrepreneurs organised technology and labour, while engineers focused on maintenance, and operators of machines provided feedback to inventors. As a result, the Industrial Revolution, and solving difficult technical problems associated with it, was as much the work of geniuses as of manual workers, being a social process involving competition and collaboration between a large number of individuals.

Finally, a technological invention that does not fit in any specific category is ballooning by the Montgolfier brothers in France in 1783. It was unrelated to anything before and did not depend, as many other inventions, on past cumulative or problem-oriented inventions. This invention made visible to a broad public what amazing things technological progress could achieve.

12.4   The Steam Engine from an Evolutionary Angle

If one technology is to be regarded as essential for, and even symbolic of, the Industrial Revolution, then it is without any doubt the steam engine. According to Lovelock (2014), 'accelerated evolution' was set in motion by the invention of a sufficiently energy-efficient and powerful steam engine, as it allowed large-scale use of a new energy source, namely coal. The combination with energy-dense coal was essential as burning wood would not have delivered sufficient output, given the inefficiency of early steam engines.

The path towards an operational and profitable engine was long and associated with the names of: the Marquis of Worcester, who suggested the idea in 1663; Denis Papin, an assistant to the Dutch physicist and mathematician Christiaan Huygens, who built a prototype in 1691; Thomas Savery, who built a first working version in 1698; and Thomas Newcomen, who constructed a much perfected version between 1700 and 1710. The first economically successful engine (the Dudley Castle machine) was used in a coal mine near Wolverhampton in 1712. Just a few years later it had already spread to various other European countries. James Watt became the most famous of all steam engine engineers because in 1765 he managed to introduce a number of critical improvements in the Newcomen engine: notably, separating the condenser from the cylinder to keep the latter constantly hot; and a double-acting machine that increased fuel efficiency from 1.0 to 4.5 per cent, allowing for a wider range of applications, as well as reducing wear and tear. This advance benefited from other technologies, especially machines for accurately boring cylinders. Although initially Watt's design was more expensive and technically more difficult to make than the original Newcomen engine, it ultimately dominated, saving on fuel costs and allowing for more power.

The steam engine meant a much more flexible and more powerful machine than the next best power technology, i.e. waterpower. The latter was also improved, notably through the 'breast wheel'. But, after 1850, the steam engine outcompeted all other power technologies. In the eighteenth century, some 2500 steam engines were built, of which approximately 30 per cent by Watt himself. Around 1000 of these were used in mining, which could then extend to areas where coal prices were high, such as Cornwall. After Watt's patent expired, in 1800, Richard Trevithick increased the efficiency and reduced the size greatly by increasing the pressure to ten times the atmospheric pressure. The latter type allowed for use in boats and carriages. In 1803, Arthur Woolf built the first compounding machine, which used the steam in more than one cylinder, one of which was high pressure.

It was only after all these inventions that the theory of thermodynamics or 'energy physics', which could explain many aspects of the technology, was developed. The 'father of thermodynamics', Sadi Carnot, published his *Reflections on the Motive Power of Fire* in 1824, while the first (conservation of energy) and second (entropy increase or arrow-of-time) laws of thermodynamics were formulated after 1850 by Rudolf Clausius, William Rankine and William Thomson (Lord Kelvin). This shows that fundamental scientific knowledge does not always precede applied knowledge and technology.

In a captivating study, Frenken and Nuvolari (2004) analyse the development of the steam engine during 1760 to 1800 as an evolutionary search process in a multidimensional design space, considering all feasible designs including those that never appeared, as far as we know. They use the *NK* model of Kauffman (1993), originally intended to describe evolutionary performance over relevant parameters – referred to as a 'fitness landscape'. Yearly distributions of steam engine designs and their use are modelled as if under selection pressure from the supposed technological needs of the various sectors of applications, including coal mining, other mining, water works and canals, metal working, cotton and other textile manufacturing, as well as food and beverage production. The design space is seven-dimensional, defined by the following alternative choices for each dimension: high/low pressure, with/without separate condenser, single/double acting, (not) compounding, reciprocating/rotary/water returning, open/closed top, and single/double cylinder. Combinations of these generate a total of 192 potential designs, which include 13 engines known to have actually existed, including various Newcomen and Watt versions.

The *NK* model of technological evolution describes local adaptation, or localised technological progress, not global optimisation (see also Box 8.2). This explains why not all designs were tried out. Even the development of thermodynamics after 1850 could not overcome existing sector-specific biases in certain steam engines. By applying the model to a database of individual steam engines in the period considered, Frenken and Nuvolari draw various interesting conclusions: The Watt engine is not just a marginal improvement or as they call it 'linear successor' to the Newcomen engine, but represents a new 'technological species' that opened up new applications. This follows the suggestion by Saviotti (1996) that a new application of a technology is akin to speciation in biology. Whereas the Newcomen engine(s) was dominant in coal mining, the Watt machine(s) was seen more in other mining activities, while in food and textiles both models competed. Another interesting finding is that a hybridisation of the Newcomen and Watt designs by Symington (in 1787), which some historians have characterised as an effort to circumvent Watt's patent, is interpreted by Frenken and Nuvolari as a genuine attempt to combine the merits of both. In support of this view, they stress the fact that Newcomen machines continued to be produced and used long after Watt's machine had entered the market.

12.5   Scientific (R)Evolution

More than 2000 years ago, Greece had the monopoly on science, while at the beginning of the Industrial Revolution the three large countries of Europe – England, France and Germany – dominated scientific activity in the world. In between, during the Middle Ages, influential scientists were rare due to the repressive influence of religion-oriented scholastics, notwithstanding science enjoying international communication, aided by virtually all works being published in Latin. Most technological innovations during the medieval period were not directly influenced by science. In contrast, the period between the Middle Ages and industrialisation (1500–1750) saw rapid scientific progress. In particular, chemistry, engineering and medicine showed a strong interaction between science and applications. But the effect on technological inventions was still not large. This changed during industrialisation, as along with it institutions for diffusion of scientific knowledge appeared, such as scientific societies, schools and universities. Until the eighteenth century, the majority of engineers had been self-made.

Science is used to denote the body of knowledge about the universe, the world, organic nature, humans and societies, and sometimes also the basic technologies created by humans. The (r)evolution of science was important as it diffused the scientific method – in brief, measurement and experimentation to test hypotheses. Science opened the door to a liberal, democratic way of communication about knowledge and opinions, where evidence and argument preside over status and familiarity, and taboos are removed. It allowed participation in research of an ever larger and broader group of intelligent and motivated researchers. This caused rampant growth in fundamental and applicable knowledge.

Of the sciences, mathematics was perhaps the most effective in supporting techno-logical inventions during the Middle Ages, Renaissance and Reformation, by offering techniques for advanced calculations. Think of the Dutch scientist Christiaan Huygens, who contributed to the development of the pendulum clock and gun-powered internal combustion, or Blaise Pascal in France and Gottfried Leibniz, in Germany, who developed the first calculators in the middle of the seventeenth century. In addition, trigonometry allowed for a revolution in map making. Over time, an increasing number of inventions required a certain minimum level of technological progress and know-ledge from multiple disciplines. Francis Bacon (1214–1294) made the famous distinc-tion between inventions that depend on the state of knowledge and those that could have been made at any time. The ancient and classical periods had a relatively large propor-tion of the second type of innovation. For the first type, science needed to provide insights, stimulate rigorous thinking, undertake experimentation and decompose prob-lems into manageable parts that could be linked to controllable physical processes. Such a systematic approach was also adopted, with much success, by certain inventors. Think of Thomas Edison, who generated more inventions (and patents) than any other individual, including the electric light bulb and the phonograph.

Until far into the nineteenth century, scientific knowledge neither stimulated nor restrained much the development of technology. Science and technology had distinct purposes, namely understanding versus utility. Educated people and those working with applied technology belonged to separate classes. At the end of the nineteenth century, and certainly into the twentieth century, science became not just critical for technological inventions but even stimulated them: think of electricity, the telegraph, the radio, and later modern medicine, nuclear fission, jet aeroplanes, computers and recently genetic engineering. North (1981, Chapter 13) refers to the marriage of science and industry as the 'Second Economic Revolution'. He thinks it represents a more fundamental change in the productive potential of the economy than the Industrial Revolution a century earlier, because it meant an inexhaustible supply of new knowledge that fuelled economic growth. The conditions allowing this marriage were that universities were institutionalised, the scientific disciplines of physics and chemistry were mature, a rising share of the population received higher education and systematic communication took place between inventors, engineers and scientists. At a general level, technological innovations during this period fall in three types: automation in production that relieved intensive human labour and allowed component-wise mass production; an extended knowledge about physical and chem-ical properties of materials applied to new products; and using new energy sources (notably oil and electricity) which permitted massive power generation in production and transport.

Table 12.2 The most important scientists until Darwin

Pythagoras (582–493 BC), Greece – mathematics
Hippocrates (460–375 BC), Greece – medicine
Aristoteles (384–322 BC), Greece – biology (taxonomy)
Euclides (325–265 BC), Greece – geometry
Archimedes (287–212 BC), Sicily – mathematics and physics (mechanics)
Lucretius (98–55 BC), Rome – physics, cosmology, psychology
Ptolemaeus (87–150) Greece – astronomer and geographer (solar system)
Galenus (131–200), Greece – medicine
Avicenna (980–1037), Persia – medicine
Fibonacci (1170–1250), Pisa – introduced the use of Arabic numerals in Europe
Bacon (1214–1294), England – physics and chemistry (plea for experimentation; gunpowder)
Copernicus (1473–1543), Poland – astronomy
Vesalius (1514–1564), Belgium – medicine (anatomy)
Stevin (1545–1620), Flanders – introduced the decimal point in mathematics
Brahe (1546–1601), Denmark – astronomy
Galilei (1564–1642), Italy – physics, astronomy
Kepler (1571–1630), Germany – astronomy
Harvey (1578–1657), England – medicine (blood circulation)
Descartes (1596–1650), France – mathematician and philosopher
Malpighi (1628–1694), Italy – medicine (microscopic anatomy)
Huygens (1629–1695), the Netherlands – physics and mathematics (wave theory of light)
van Leeuwenhoek (1632–1723), the Netherlands – microbiology, perfected the microscope
Newton (1643–1727), England – physics and mathematics
Leibniz (1646–1716), Germany – mathematics
Boerhave (1668–1738), the Netherlands – medicine (clinical teaching, academic hospital)
Linnaeus (1707–1778), Sweden – biology (taxonomy)
Euler (1707–1783), Switzerland – mathematics
Von Haller (1708–1777), Switzerland – medicine (physiology)
Hutton (1726–1797), Scotland – geology
Herschel (1738–1822), Germany – astronomy
Lavoisier (1743–1794), France – chemistry
Lamarck (1744–1829), France – biology
de Laplace (1749–1827), France – mathematics and astronomy
Dalton (1766–1844), England – physics (atom theory)
Gauss (1777–1855), Germany – mathematician
Faraday (1791–1867), England – physics (electricity and magnetism)
Lyell (1797–1875), England – geology
Liebig (1803–1873), Germany – chemistry (industrial and agricultural)
Darwin (1809–1882), England – biology

Innovation in science was originally closely associated with singular individuals. This is illustrated by Table 12.2, which appropriately – given the focus of this book – ends with the initiator of modern evolutionary thinking. A more recent phenomenon is a shift to project- and goal-oriented scientific research involving large teams of researchers. Prominent technological events illustrative of this are the Manhattan project (atomic bomb, 1940–1945), the Mercury and Apollo human spaceflight programmes (1958–1972), and the development of the internet (1980–1990s). Another indication is that, over time, the average number of authors on scientific articles has been steadily increasing, notably in the experimental sciences. During the second half of the nineteenth century the large multi-factory enterprise arose. It separated functional units and R&D activity, used automated production processes and a large distribution system. They had large research departments that hired experts from science with PhD degrees or even full professors, or even established working relationships with universities. This implied a direct flow of information from academia to businesses and sometimes in the opposite direction as well.

12.6   European Origins

Evolutionary analysis devotes much attention to geographical dimensions. This includes the role of geographical isolation (founder effects), divergent development in some regions, and path dependence, possibly leading to dominance of certain regions (Section 8.7). Two key questions related to geography that require an answer in this chapter are: why did industrialisation start in Europe? (addressed here); and, why did England play a central role? (next section).

   The geographical centre of long-term economic dominance in the world has altered over time. Mesopotamia, Persia, Greece, the Roman Empire, North Italy, Spain, the Netherlands and England all had their rise and fall. Their 'successor' was always geographically connected, that is, adjacent or overlapping. As a result, it inherited many elements of the previously dominant region, which contributed to continuity and cumulative economic-technological progress over a larger geographical area over extended periods of time. To illustrate, Greece was influenced by the Persians, the Romans adopted much of the culture of the Greeks, North Italy obtained the infrastructure created by the Romans, the Netherlands benefited from commercial trade between north-west and south Europe initiated by Venetian merchants, England imitated many of the agricultural innovations from the Netherlands, and the United States inherited common law and measurement units from England.

   So why did industrialisation occur in Europe? A comparison with other regions, notably those that are climatologically different, leads to the idea that the harsh climate typical of northwest Europe, in which a reasonable life or even survival is impossible without relevant technologies, provides more incentives to develop and perfect technologies than a warm climate. The opposite view is that the latter type of climate allows people to have much leisure time which can be devoted to innovative activities. Of course, such a general, deterministic and atemporal explanation is bound to fall short. From an evolutionary-historical angle, an interesting comparison is technological progress in early China versus Europe. China was technologically ahead until the end of the Middle Ages. Already by 1700 BC, China had seen the first use of bronze, from 1700–1100 BC the lunisolar calendar, and in 100 BC the first use of paper. Many Chinese technological advances related to agriculture, notably innovations in rice cultivation. In the sixth century BC an iron plough was introduced with adjustable parts, allowing for use under different soil and weather conditions. Handbooks were written on agricultural technology as early as the sixth century AD. Blast furnaces to cast iron were already in use by 200 BC – while this did not start in Europe until the fourteenth century. Water power and the spinning wheel emerged at about the same time as in Europe (third century and thirteenth century, respectively). The clock and compass were invented earlier than in Europe (tenth century). Ships were larger and safer in earlier times. The Chinese invented paper around 100 BC, which then took more than 1000 years to reach Europe. In 1045, movable type (printing) was invented, and it is unclear if the European inventor(s) knew about it. The wheelbarrow was invented in 232 AD, about 1000 years before Europe.

Why did China then fail to continue and make the transition – possibly as the first country in the world – to industrialisation? In particular, why did technological development between 1300 and 1400 stagnate? An explanation may be a change in political power that led to centralised control and, in turn, to withdrawal of support for technological innovation, as well as cultural and economic isolation. The reason for this is debated, and ranges from a conservative ruler or state fearing culturally disruptive effects of technological change, cultural factors such as pride that prevented imitation of the West, and malnutrition (notably a shortage of proteins) owing to quick population growth and increasing dependency on rice. The state factor receives most support, as, unlike in Europe, the state in China had played a pivotal role in the generation and diffusion of innovations through appropriate institutions. In line with this, Jones (1981) holds the relatively liberal economic-political environment in large parts of Europe responsible for the 'European miracle', regarding centralised states in China, India and the Islamic world as constraining innovation and trade.

Diamond (1997a, Epilogue) mentions a potential ultimate factor. Unlike China, Europe has a coastline that is strongly indented and includes large peninsulas (Scandinavia, Spain/Portugal, Italy), and a few large islands. Such relative cultural evolutionary isolation creates the right conditions for political and institutional diversity. A similar argument was previously provided by Kennedy (1989). Incidentally, the main peninsula of China is Korea, which has never been part of China. Diamond regards such geographical features as an ultimate physical-geographical factor, which explains why unification, resulting in centralised governance, worked in China but not in Europe – despite various efforts by the Romans, Charlemagne, Napoleon and Hitler. As a consequence, Europe has always had much more internal diversity of economic activity and political strategies, making it as a whole rather flexible and adaptive, while still benefitting from its large overall size through internal trade and associated international labour division. While China has considerable cultural diversity, it has considerably more lingual and political homogeneity than Europe. Incidentally, against this background, the unification of Europe through the European Union can be judged negatively as it reduces institutional-political diversity, possibly at the cost of adaptive flexibility and innovation.

Pomeranz (2000) adds another fundamental, ultimate factor for consideration. He offers the unorthodox insight that, even as late as 1750, East Asia was, together with Europe, the most advanced region in the world, in terms of life expectancy, consumption, trade and markets. He then argues that Europe saw subsequent progress largely due to the location of coal, which substituted for timber, and allowed energy-intensive industries to grow relatively quickly. He further suggests a main role of imports of essential primary resources from the Americas, which allowed Europe to specialise in, and reallocate labour from, agriculture to manufacturing. This resembles a theory proposed by Wilkinson (1973), interpreting economic development as a response to ecological or energy scarcity. According to this, the Industrial Revolution was a reaction to the scarcity of firewood in England during the second half of the eighteenth century. At first, this scarcity was undone by extracting coal at the surface with simple techniques and against relatively low costs. When these supplies of coal were exhausted, one had to delve deeper, requiring the pumping away of groundwater. Initially, this was achieved with the help of animal energy, but soon more powerful techniques were required. This led to the first large-scale application of the steam engine, giving an enormous impetus to this technology. Subsequently, improvements in the technology and applications to other industries, notably textiles, led to its wide diffusion. This, in turn, triggered other technological innovations and diffusion processes, which could benefit from the early generated knowledge and labour division. This connection between resource scarcity and industrialisation is also supported by Wrigley (1988) and Allen (2009), who both stress the role of cheap and abundant coal. Complementary factors, as discussed in the next section, were probably relevant as well – these may have been absent earlier in China, preventing ample coal availability there to prompt industrialisation. A related but more basic explanation was recently offered by Pezzey and Stern (2017). Using a model of directed technical change (Section 8.6), they show that a British Industrial Revolution was possible as firewood was scarce and could be substituted for by coal, which triggered innovation in coal-using technologies rather than in efficiency improvement of firewood-using technologies. Their analysis suggests that with less firewood scarcity or limited substitution potential, an industrial revolution in Britain would have occurred later or never.

12.7 England Versus the Low Countries

Why did industrialisation occur first in England (or Britain), and not in France, the Low Countries or Germany? The Dutch Republic, together with the north of Belgium (Flanders), jointly known as the Low Countries, seems to have been the best candidate next to England (Box 12.1). The Low Countries was the first modern economy and the most advanced nation of Europe shortly before 1800 (de Vries, 2000). As can be seen in Figure 12.1, it was the first European country that had started to grow before the Industrial Revolution, giving rise to the 'Dutch Golden Age' during the seventeenth century. According to Mokyr (2000), this was old growth rather than modern economic growth fuelled by industrialisation and associated technology. This is open to debate, though. As Mokyr himself notes (p. 509), the Golden Age involved sophisticated manufacturing, highly developed shipping and advanced engineering, notably hydraulics. De Vries' (1994) term 'industrious revolution' signals that the pre-industrial economy already contained a fair amount of industrial elements.

Box 12.1 Why the Low Countries did not spawn the Industrial Revolution

The economic hegemony of the Low Countries goes back as far as the twelfth century. At that time Brabant and Flanders were, together with North Italy, the most prosperous regions in Europe, largely due to specialisation in production of woollen textiles and dyeing. In addition, there was much trade in food – notably cereals, wine and fish – with France, England and the Baltic, in wool with England and later Spain, and in dyeing materials with northern Italy. A fundamental reason for the high productivity of the Low Countries was that it transformed agriculture through the application of hydraulic engineering techniques to regulate water flows and storage. This resulted in a complex water network with canals, channels, dams and bridges. This development was a logical, if not inevitable, response to the specific water conditions caused by the geography of the Low Countries. These resulted from the integrated delta of the rivers Meuse, Rhine and Scheldt, complemented by a dense network of smaller, derived rivers. Driven by high productivity agriculture and early industrialisation, between 1400 and 1700 per capita income in Europe grew the fastest in the Low Countries, while its level was the highest from halfway through the sixteenth century to the 1820s (Figure 12.1).

Institutional integration contributed to economic success. The seven Nordic provinces – Zeeland, Holland, Utrecht, Gelderland, Overijssel, Friesland and Groningen – formed together the Dutch Republic, which became independent in 1579. Owing to its water-rich delta environment – one of the largest and certainly the most densely populated in Western Europe – the Republic functioned like an island. It was bounded to the west and north by the North Sea, to the south by the Meuse and the Rhine branches Waal and Nederrijn (with further sub-branches), and to the east by the river (Gelderse) Ijssel – also a branch of the Rhine – and to some extent also by the more longitudal parts of Meuse and Rhine. The resulting relative institutional isolation was supportive of independent and innovative developments. At the same time, the Low Countries were economically well connected to other regions, notably overseas – witness the Hanseatic League during the thirteenth to seventeenth centuries.

From an evolutionary angle, it is interesting to note the large diversity of successful economic activities in the relatively small area that the Low Countries comprised, namely the water-rich and flat north and the urbanised south (Flanders), all with access to the sea. In fact, the diversity of activities in the north and south Low Countries was unknown for any region of similar size at the time. One reason was one of the densest networks of natural and constructed internal waterways in Europe, which allowed for transport of both freight and passengers. This contributed to a well-connected economy with considerable labour division and trade. The diversity of activities resulted in trade advantages through low transport costs, fostered learning and diffusion across sectors and adjacent regions, and created the capacity to flexibly adapt to changes in the rest of the world. In the provinces of Zeeland, Holland and Friesland, the Dutch were stimulated to find solutions to water risks and

Box 12.1 (*cont.*)

scarcity of land. Dutch engineers had already acquired much expertise about fighting the water in earlier centuries, and this had turned them into the leading hydraulic engineers of the time. This permitted the development of unique knowledge about land reclamation (including using large pumps), water defences, canal construction and ship building (Maddison, 2001).

Organisations and institutions in the Low Countries were strongly influenced by water control. Since the thirteenth century, farmers had united in water boards ('waterschap' or 'hoogheemraadschap'). These guaranteed water quality and safety, introduced taxes for water-related works and maintenance, and charged penalties to offences such as water pollution and damage to dikes. These boards are among the first democratic institutions worldwide and are active to this day. They contributed to farmers being freer than in the rest of Europe, where feudal systems dominated. Individual freedom caused agriculture to innovate and diversify, over time focusing on industrial crops such as hops, flax, tobacco and tulip bulbs, and the development of a considerable horticulture sector.

In the seventeenth century, just before the period of industrialisation, the Dutch Republic had its Golden Age. Apart from the above conditions, additional reasons are as follows. The windmill was in widespread use, with various applications, including driving sawmills, which contributed to Dutch shipyards producing the best ships in Europe. Many specialised river and canal boats were built. Another reason was that peat offered the first large-scale use of stored-up solar energy, serving as a predecessor of coal. It supported heating by households and large-scale energy use in production, such as in bakeries and brick factories. According to Wrigley (1988), though, peat was a 'renewable area resource' which created less agglomeration advantages than a 'localised non-renewable resource' such as coal, since the latter allowed greater localised concentration of industries.

Another relevant factor was the 'Hollander beater', a device for papermaking, which allowed the Dutch to produce the best quality paper at the time, and much faster than with previous technologies. Of additional importance was the emigration of many intellectuals, merchants and bankers from Flanders to Holland, in response to the struggle against the Spanish regime. Subsequently, the old University of Leiden was founded, which became one of the important centres of intellectual thought in Europe. This further boosted innovation and activity in the Dutch Republic.

Whereas the Dutch Republic possessed the most advanced political institutions and transport waterways, it perhaps suffered from what one might characterise as the 'law of the handicap of a head start' (from the Dutch phrase: 'de wet van de remmende voorsprong'). This conveys a 'first-mover disadvantage': if one is the first to realise a dominant innovation this may impose a restriction on subsequent technological or economic progress. In particular, the Dutch Republic was so proficient in agriculture that it did not have an incentive to leap far into industry. In addition, it was perhaps too small, in terms of internal market, labour population

Box 12.1 (*cont.*)

and network of entrepreneurs, engineers, scientists and inventors, especially compared with England/Britain. The Dutch economy stagnated in the eighteenth century after it lost monopolistic trading privileges due to conflicts with France and England (Maddison, 2001, p. 82). Craft guilds and local regulations may have played an additional role, as they restricted creativity, innovation and diffusion to other sectors.

Incidentally, the Netherlands has remained among the most productive in agriculture up to the present day. This may relate to the Dutch being the tallest humans in the world. Nutrition, notably dairy products, and health care are often proposed as proximate explanations. A recent study of ultimate factors suggests an additional role for sexual selection (Stulp et al., 2015). Over three generations, larger parents had more offspring, which may be due to taller men, *ceteris paribus*, being more attractive as partners. Since women have overlapping genes, sexual selection for taller men will make women taller as well. That this happened in the Netherlands is possibly because favourable food and health conditions allowed the genetic capacity for male body growth to be fully explored. This augmented phenotypic variation upon which sexual selection could then operate with notable effect.

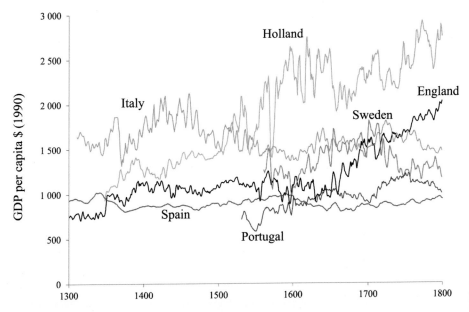

Figure 12.1 Gross domestic product (GDP) per capita in selected European economies, 1300–1800.
Note: Curves represent a 3-year average and for Spain an 11-year average.
Source: Fouquet and Broadberry (2015, Figure 1, p. 230). © 2015 American Economic Association.
(*A black and white version of this figure will appear in some formats. For the colour version, please see the plate section.*)

Table 12.3 Comparing countries' capacities to spur an industrial revolution at the end of the eighteenth century

Feature	England	The Low Countries	France	Germany
Population size	high	low	high	medium
Population density	high	high	low	medium
Education (literacy and progressive universities)	high	medium	medium	medium
Number of scientist and engineers	high	low	medium	medium
Communication and physical infrastructure	high	high	low	low
Monetary institutions (investment, trust)	high	high	medium	medium
Political institutions (trust)	high	high	medium	medium
Protection of inventions (patent law)	high	low	medium	medium
Colonial activity	high	high	medium	low
Class society	high	low	high	medium
Energy availability (peat, coal, wind, steam engine)	high	medium	medium	medium

For the other countries, one should account for the fact that France consisted of rather independent regions in terms of markets and infrastructure, while Germany did not yet function politically as one country. Table 12.3 schematically compares the four most likely candidates for early industrialisation along nine criteria. It shows that England scores highly on all criteria, the Low Countries combines 'high' and 'low' scores, and France and Germany display a medium performance on the majority of criteria. It is perhaps no surprise then that the Industrial Revolution started in England. Much can be said about the factors and their synergies. For example, Beckert (2014) considers the interaction between colonialism, slavery and the cotton industry as a key mechanism underlying industrialisation. The textile industry, and indirectly England, benefited from a resource, cotton, produced in the United States using cheap labour – through slavery – and cheap land – expropriated from native people. In a similar way, the sugar trade with the West Indies (Caribbean Basin) contributed to British industrialisation. In addition, many early investments in cotton factories were financed with wealth obtained through the interconnected trade of cotton, sugar and slaves in Africa and the West Indies.

A relatively rich and highly educated population in England is another important factor for early industrialisation (Allen, 2009). It allowed extended labour division before the Industrial Revolution, generating a large number of technicians and crafts-men – the focal point of the first elaborate economic theory by Adam Smith. The middle class of literate and well-fed people hosted most entrepreneurs, engineers and inventors. The land-owing social elite did not resist the Industrial Revolution, while it funded many industrial and technical innovation projects. In addition, England had taken the lead in clock- and watch-making during the seventeenth century. This concerned a technology that provided many lessons, insights and experts for industrial technologies. Next to population size, population density matters. It relates to economies of scale in markets and transport as well as to scarcity of fuelwood and land, while it enhances literacy given it is more worthwhile to set up schools in more densely populated areas. In addition, England and the Low Countries possessed the best transport infrastructure at the time, in terms of roads as well as canals. This contributed to free competition among firms and rapid diffusion of the best technologies. As a result, England had a unified and large market area. In France, for instance, local markets dominated.

Engineers and technicians interacted through various mechanisms, including a variety of learned societies. In particular, there was an active Royal Society of Scientists, which may have had a positive effect on diffusion and directed innovation. Moreover, inventors and engineers of machine-tool industries were part of closely connected families and communication networks. Some inventors were thus able to contribute to the improvement of a wide range of technologies. Moreover, many technical problems that were obvious stimulated organised, guided searches to solve them. In the Low Countries, the main other candidate for the Industrial Revolution, such a close cooperation between natural scientists, philosophers, engineers and entrepreneurs, and moreover with a critical mass, was missing.

Britain had possessed patent law since 1624, serving as a critical institutional factor. France implemented such a law only in 1791, while other European countries did not follow until the early nineteenth century (Box 10.1). Between 1770 and 1850, 11 962 patents were granted in Britain. Patents are particularly relevant for stimulating radical or macro-inventions. Somewhat related is that British inventors were, on average, much better rewarded than those in other countries, which could have acted as a positive incentive.

The invention and perfection of the steam engine – widely regarded as the quintessential invention of the Industrial Revolution – was a purely English undertaking since Newcomen (Section 12.4). The development of tall and hot blast furnaces to smelt iron ore using coke rather than more scarce charcoal, in order to produce cast or pig and ultimately wrought iron, also happened mainly in England. Witness the Bristol Iron Company of the Darby family. Another factor is that England had coal supplies close to the surface, which guaranteed ample availability against low prices (see the previous section). Inventors from other European countries, notably France, equally influenced textile, glass, paper, ceramics and chemical technologies. In addition, British science was experimental in approach as compared to French science, which was more mathematical and deductive. The first may have provided a better basis for inventing industrial technologies.

Perhaps the fact that Great Britain is an island acted as a more ultimate factor in prompting the Industrial Revolution. This is an overlooked issue. From an evolutionary angle, islands represent the clearest example of spatial isolation, explaining their often unique features in terms of organic life as well as culture. Great Britain is the largest island of Europe, allowing for a big population and internal market, but also relative isolation from, and protection against, external influences. This may have contributed to cultural, economic and technological evolution in a rather independent and original way. In particular, spatial isolation may have protected England, as well as Scotland and Wales, from competition by older institutions and technologies from mainland Europe.

Being an island also stimulated extensive trade over sea, in turn contributing to Britain ultimately becoming the dominant colonial power. Interestingly, similar spatial isolation might earlier have been an ultimate factor leading up to a Golden Age in the Low Countries (see Box 12.1).

After such a long list of arguments, one can only wonder how the Industrial Revolution could not have occurred in England. According to Mokyr (2000, p. 516), an industrial revolution would have happened anyway in continental Europe in the absence of a British one, even though it would have been later and slower. Just like the economic hegemony of the Netherlands was taken over by England in the eighteenth century, towards the end of the nineteenth century, England lost momentum and Germany, the United States and France took the lead in inventions, while Germany and the United States paved the way for further industrialisation. This second wave of innovations and industrialisation will not be dealt with here in detail as it involves too many particular innovations. It was briefly discussed in the context of long waves in Section 10.4.

12.8　The Rise and Fall of Population Growth

In an evolutionary treatment of the socioeconomic history of humankind, it is almost essential to zoom in on population and more generally demographic issues. According to Harari (2014), the ecological dominance of *Homo sapiens* is due to a unique capacity to cooperate in very large groups, which, through positive feedback loops, is associated with the major cognitive, agricultural and scientific-industrial transitions. Increases in agricultural productivity fuelled population growth by creating food surpluses. This explains historical civilisations with high population density in river valleys (the Euphrates/Tigris, Nile, Indus and Huang He), the Golden Ages of the Greeks (ninth to fifth centuries BC), Mediterranean prosperity (first century BC to second century AD), and economic growth in the Low Countries (since the seventeenth century). Population density has an intricate relationship with growth. Economic growth often drives population growth, which then leads to increases in population density. This in turn generates economies of scale and urbanisation, through larger labour markets, wider technological diffusion and shorter transport distances, all fuelling further growth. In initial development phases, economic growth affects welfare, which influences birth and death rates. The second usually falls faster, due to application of medical knowledge and education, thus triggering population growth. A change in birth rates can go along with cultural or religious shifts altering values attached to children and families, further affecting demographic patterns.

According to McNeill (2000), at the time of the transition to agriculture, approximately 8000 BC, humans were still outnumbered by other primates, notably baboons. Since then population growth has been large, due to the power over other species through hunting, the invention of fire, the rise of agriculture and the Industrial Revolution. Demographic patterns in the twentieth century are exceptional in both historical and likely future context: population skyrocketed due to the combined consequences of quick dissemination of health care, reducing death rates in the developing world, and continuation of poverty, keeping birth rates high. McNeill notes that 'Although the twentieth century accounts for only 0.00025 of human history (100 out of 4 million years), it has hosted about a fifth of all human-years.' This is based on estimates that 80 billion hominids were born in the past 4 million years, which have lived a total life of about 2160 billion years. Another enlightening perspective is that the current human population is about 100 000 times as large as that of other species with a similar body size and position in the food web. Cohen (1995, Chapter 2) discusses different phases of world population growth in more detail. The first occurred when humans acquired the capacity to make and use tools, possibly more than 100 000 years ago; the second one, when agriculture spread and human cities emerged, between 8000 and 4000 BC; the third one, as a result of globalisation of agriculture between 1650 and 1850, in line with the period of industrialisation and a sharp increase in international trade; the fourth due to improvements in sanitation and public health, which reduced the death rate of adults and children considerably (since World War II, still continuing); and the fifth which involves a fall in the fertility rate due to higher living standards (since the nineteenth century for Western countries and still in progress for developing countries). The population doubling time over this entire period went down from more than 40 000 to 34 years. Whereas economic growth and population growth initially proceeded at the same pace, they decoupled after 1820. In view of this, the question has been raised of how many people the biosphere can support. Many studies have tried to assess a hard limit, using notions like optimum population, carrying capacity and limits to growth. These reflect that for a decent life human individuals need minimum amounts of certain natural resources – notably land, fresh water, energy and (a)biotic resources. In addition, these studies recognise that necessary resources are finite in supply. The two elements taken together suggest that there is a limit to the size of the human population. In a statistical meta-analysis of 69 such studies, van den Bergh and Rietveld (2004) arrive at 7.7 billion people as the best point estimate of this limit.

Although most of us are quick to accept that continued population growth is abnormal, unsustainable and undesirable in itself, many of us want to believe that economic growth is a normal phenomenon that can, and even should, continue indefinitely. But even the first century of industrialisation (1750–1850) delivered only limited income growth compared to a later one, known as the 'second Industrial Revolution' (1850–1970) and, since then, the growth rate has fallen, despite the third, digital revolution, which only temporarily, during roughly 10 years (1994–2004), realised a growth spurt. Robert Gordon (2016) attributes this pattern to several factors. Two important ones are mass education and a flood of major technological inventions during the second revolution, such as the light bulb, electricity, automobile with combustion engine, flush toilet and sewage systems, phonograph (record player), radio, television, electric kitchen equipment and aeroplane. These technologies increased health and life expectancy, causing investments in education to be sustained and creating high returns. Once mass education was realised, and major innovations were exhausted, the days of high growth were over. As one may expect, not everyone agrees with this interpretation. Teulings and Baldwin (2014) collect distinct views by renowned economists.

Part VI

Evolutionary Environmental and Policy Sciences

13 Survival of the Greenest

13.1 Human Maladaptation

The relationship between the environment and the economy has been studied in a variety of disciplines that make up the broader field of 'social environmental sciences'. Among these, environmental and ecological economics have offered the most elaborate views on the causes of, and solutions to, environmental problems. Both use insights from biology, in particular population ecology, in studying ecosystem management and sustainable exploitation of renewable resources. This suggests methodological overlap with evolutionary thinking. But as we will see, the connections are actually rather patchy and offer much room for further development. Our understanding of the relationship between the economy, natural resources and the environment can benefit from evolutionary approaches. These can provide a better understanding of behavioural responses to external influences and policies. With regard to behaviour, they can well assess aspects whose analysis require a population context, such as imitation, conformism, status-seeking and diffusion of information, such as about environmental problems or less environmentally damaging technologies and lifestyles. This in turn will allow for the formulation of more effective public policies aimed at limiting environmental degradation caused by humans. Conversely, the study of certain themes within the evolutionary social sciences could benefit from giving more explicit attention to the fundamental role of environmental and resource factors. This opens the door to coevolutionary analysis of human culture and the natural environment.

An informative article on the meaning of evolutionary thinking for the environmental sciences examines 'why humans are ecologically destructive, overpopulate, overconsume, exhaust common-pool resources, discount the future, and respond maladaptively to modern environmental hazards' (Penn, 2003, p. 1). The author notes that evolutionary biology provides fundamental insights into various environmentally relevant aspects of human behaviour, such as cooperation, morality and reproduction. Regarding the latter, it is undeniable that population pressure is a key aspect of human overexploitation of the natural environment, not only in particular regions characterised by extreme food and water shortages, but increasingly so at the global scale. A general insight arising from evolutionary thinking is that human-induced environmental degradation can be interpreted as the human species being maladapted to, i.e. showing evolutionary mismatch with, its current natural environment. An explanation for this was suggested by E.O. Wilson (2009) in an interview: 'Humans have Palaeolithic

Figure 13.1 Devolution – evolution of environmental problems.
Source: http://bizarrocomics.com. © Dan Piraro.
(A black and white version of this figure will appear in some formats. For the colour version, please see the plate section.)

emotions, medieval institutions and god-like technology.' Maladaptation is likely to worsen since the environment is not exogenous but continues to be transformed by humans, at an unprecedented rate (Figure 13.1). Humans did not evolve to perceive, avoid or solve environmental challenges, especially those of a long-term and global nature. And as Stephen Hawking put it in an interview: 'We now have the technology to destroy the planet on which we live, but have not yet developed the ability to escape it.'[1] Technological change has up till now predominantly served private, commercial interests rather than relieving human pressures on the environment. This might change, of course, with adequate public policies and strategies. We will examine how these various issues can be enlightened from the angle of evolutionary social sciences.

13.2   Evolution in Environmental Social Studies

Environmental and ecological economics address the allocation, exploration and management of energy, mineral and renewable (water, forestry and fisheries) energy resources, including ecosystems. In the case of energy, the environmental and resource dimensions come together, as scarce fossil fuel resource issues are connected to climate change through greenhouse gas, notably CO_2, emissions. An integrated treatment of environment and resources has become more pressing since the notion of sustainable development was widely adopted during the 1990s. It urged for a systemic viewpoint in

---

[1] https://www.theguardian.com/commentisfree/2016/dec/01/stephen-hawking-dangerous-time-planet-inequality?CMP=share_btn_tw.

order to arrive at genuinely sustainable solutions and policies. This involves accounting for the full set of dynamic mechanisms in the overall economic-environmental system, including evolutionary processes, whether of a biological, social-cultural, economic or technological nature.

Environmental economics emerged during the 1960s. Originally, it was under the strong influence of agricultural and resource economics, especially in the United States. Its early development focused on three research themes. First, the technique of cost–benefit analysis was applied to evaluate investment projects with significant environmental impacts. Many of these related to water quality and flooding by rivers. The focus on monetary accounting to undertake cost–benefit analysis stimulated the development and application of monetary valuation techniques to value environmental changes, damages, projects, and policy scenarios. This started with revealed preference techniques like hedonic pricing and travel cost methods, followed in the late 1970s by stated preference techniques such as contingent valuation. Parallel to this, environmental policy theory was developed, with the aim of evaluating, comparing and designing instruments of regulatory policy. This was mainly motivated by achieving efficient outcomes, in the sense of welfare-maximising or cost-minimising solutions. In addition, economic growth and resource scarcity received much attention in theoretical and empirical studies. On the one hand, this was part of a wider debate on environment versus growth that involved other disciplines within the environmental sciences as well. On the other hand, it had strong ties with a rapidly growing literature on resource economics, which analyses optimal resource extraction and tests resource scarcity using a range of indicators reflecting physical conditions, extraction costs and market prices.

Traditional environmental and resource economics is strongly dominated by neoclassical microeconomics. This is most clearly exemplified by the theories of monetary valuation (Johansson, 1987) and environmental policy (Baumol and Oates, 1988), both of which assume rational agents and market equilibrium. Evolutionary approaches relieving these assumptions did not receive any systematic attention before the 1990s. In this period, along with an increasing focus on sustainable development, ecological economics was founded as a new field of research (Costanza, 1991), representing a more pluralistic and multidisciplinary approach, integrating elements of economics, ecology, geography, political science, thermodynamics and ethics. Its methodological approaches are well embedded in the use of physical-biological indicators and comprehensive systems analysis. Table 13.1 schematically compares evolutionary economics with environmental and ecological economics. Ecological economics appears to be methodologically closer to evolutionary economics than environmental economics.

Fundamental criticism on the assumptions and approaches of mainstream economics (e.g. Daly and Cobb, 1989) has inspired the emergence of ecological economics, bringing together many heterodox approaches to social science. Ecological economics additionally offers a platform for multidisciplinary research with strong social science and economic dimensions (van den Bergh 2001). It is close in spirit to evolutionary economics, which is not surprising, as the latter was also motivated by dissatisfaction with mainstream economics, notably how it coped theoretically with innovation and structural change. Indicative is that many influential 'ecological economists' have

Table 13.1 Differences in emphasis among evolutionary, ecological and environmental economics

Evolutionary economics	Ecological economics	Environmental economics
Evolutionary potential	Optimal scale	Optimal allocation
Agent, technique, product diversity	Biodiversity	Representative agents
Innovation – recombination/ mutation	Divergent views on innovation	Optimal R&D
Fitness	Equity (intra/ intergenerational)	Efficiency, cost-effectiveness
Evolutionary stability	Resilience	Sustainable macro growth
Adaptive limits	Limits to growth	Growth of limits
Path dependence	Ecological irreversibility	Economic irreversibility
Varying time scales	Medium/long run	Short/medium run
Population/distribution indicators	Physical/ biological indicators	Monetary indicators
Bounded rationality and selection	Myopic behaviour	Rational behaviour
Functional morality (fitness)	Environmental ethics	Utilitarianism
Adaptive individuals and systems	Causal processes	Equilibrium, comparative statics

recognised cultural, non-genetic evolutionary theory as a powerful framework to improve our understanding of the complex, interactive dynamics of human and natural systems (Norgaard, 1984; Faber and Proops, 1990; Costanza et al., 1993; Ayres, 1994; Gowdy, 1994, 1999). More recently, evolutionary thinking and modelling have contributed to an emerging research area dealing with management of sociotechnical systems innovations and sustainability transitions (Geels, 2002; van den Bergh et al., 2011). Evolutionary theory provides a range of concepts and mechanisms that are useful in theoretical and applied studies of so-called 'sustainability transitions' (Safarzynska et al., 2012). Section 13.5 will address this issue, while the next chapter will zoom in on the peculiarities of a transition to a low-carbon economy.

How should one evaluate the approaches and insights of environmental and ecological economics on the basis of evolutionary criteria? Resource economics, dealing among others with renewable resources such as fisheries and forestry (Johansson and Löfgren, 1985; Clark, 1990), tends to search for optimal exploitation without taking into account evolutionary or selection effects of resource exploitation. Examples of relevant selection pressure due to resource use are mesh size and season in fisheries (selection for small fish) and the use of pesticides in agriculture (selection for pest resistance). As another example, environmental macroeconomics, in particular the extension of growth theory with environmental and resource factors to address sustainability issues (Toman et al., 1995), does not take the heterogeneity of firms and technologies into account as does evolutionary growth theory (Section 8.6). In addition, it does not allow for much policy detail in terms of addressing innovations, selection environment and lock-in. Finally, economic valuation and evaluation assume perfectly rational and selfish agents that are individually efficient (Johansson, 1987; Freeman, 1993; Hanley and Spash, 1993). Evolutionary insights, however, underpin that individuals are boundedly rational as well as other-regarding, explaining phenomena like 'protest bidders', 'warm glow' and 'altruism' (Andreoni, 1995).

Some early economic studies applied evolutionary thinking to environmental problems. Nicholas Georgescu-Roegen (1971) presented the notion of exosomatic instruments as an almost natural extension of endosomatic capabilities of humans – i.e. those which are part of the human body. In line with this, he regarded human development and cultural evolution as the result of innovating in the exosomatic sphere. Boulding (1981) emphasised the analogy between ecology, evolutionary biology and economics, considering concepts such as homeostasis and population, and the distinction between genotype and phenotype. Norgaard (1984) was the first to discuss seriously the application of the concept of coevolution to the interaction between economic and natural systems. This dealt with the coevolution of pests, pesticides and environmental policy in the United States (e.g. Norgaard, 1994). Although a complex picture results, it arguably does not sharply distinguish between mechanistic interaction among subsystems ('co-dynamics') and strict coevolution as interacting populations with internal diversity which impose mutual selection pressure (Winder et al., 2005). Faber and Proops (1990) have combined elements of the neo-Austrian approach – with its emphasis on the temporal and roundabout features of economic processes – and evolutionary thinking. Their approach is developed around an analogy of the genotype/phenotype distinction, as initiated by Boulding. They also abstain from explicitly modelling populations and their internal dynamics, hence the notion of evolution is debatable. Munro (1997) offered a rigorous model analysis of optimal agricultural crop cultivation when pesticide use causes selective pressure on the genetic composition of a pest population. His approach can be seen to combine neoclassical and evolutionary economic elements. Sethi and Somanathan (1996), among many others (e.g. Noailly et al., 2006, 2009), have considered the endogenous character of norms in the context of a common-pool renewable resource using evolutionary game theory. The research on biodiversity loss includes consideration of genetic evolutionary aspects as well (Weitzman, 1998b). Some of the previously mentioned studies are discussed in more detail in subsequent sections. Finally, one of the first multi-agent approaches to evolutionary social science, the 'Sugarscape' model of Epstein and Axtell (1996; see also Section 8.5), gives a core role to a renewable resource, namely sugar. *Ceteris paribus*, individuals with more access to sugar can better survive, while some lucky ones even accumulate sugar reserves. Luck also plays a role in a genetic way, as individuals are selected based on their metabolic efficiency and vision: less need for sugar and less myopia contribute to survival. In one model version, pollution is included which provides an incentive to move to other locations.

13.3   Managing Evolutionary Resources and Ecosystems

Both ecological and environmental economics deal with ecosystem and resource management issues. The dynamics of ecosystems and natural resources can be seen to involve reversible and irreversible dynamics. The reversible dynamics relate to population growth and ecosystem succession, while the irreversible changes are partly the result of evolutionary changes, in the short run, especially covering selection processes. Irreversibility also relates to systems being able to move to alternative equilibria.

A relevant ecological notion in this context is 'resilience', a sort of extended stability concept. It has two alternative interpretations, namely the time necessary for a disturbed system to return to its original state (Pimm, 1984), and the amount of disturbance a system can absorb before moving to another state (Holling, 1973, 1986). In line with the latter interpretation, resilience has been phrased 'Holling sustainability' at the ecosystem level in ecological economics, as an alternative to 'Solow-Hartwick sustainability' at the macroeconomic level in environmental economics (Common and Perrings, 1992). Some authors have tried out analogies of resilience in socioeconomic systems. These clarify, among others, the limits of bureaucracy in adapting to changing external circumstances, the differences between social and market economy systems, and the shortcomings of policies aimed at tight control rather than stimulating experiments, diversity, flexibility and adaptability (Holling, 1978; Levin et al., 1998; Gunderson and Holling, 2002). The analysis of resilience is a useful alternative approach when traditional policy theory fails. Walker et al. (2002) mention a number of reasons for this to happen. Important ones are that economic or economic-environmental systems contain thresholds, exhibit hysteresis and show irreversible change. In traditional, neoclassical economics' terminology this means that systems are not convex.

An analogy of resource management with an immune system was proposed by Allen (2001) and Janssen (2001). This aids in understanding the way in which resilience of economic-environmental systems can be maintained in the face of external perturbations. Both types of system are hierarchical and complex, based on internally diverse components that operate individually and are involved in local interactions, while both include some type of selection. System resilience fails when an external influence is too novel, too fast, or too abundant, when the system has invested too much in the wrong defence, or when it lacks internal diversity. In terms of management advice, Allen (2001) concludes:

As with the human body, we must be wary of too much remedial intervention in the face of invasions. Overmedicating trivial illnesses leads to hardier pathogens and weakened populations of 'friendly' bacteria, leaving the body more vulnerable to progressively worse infections. In ecosystems, too much intervention can have an analogous effect.

Resilience is often linked to biodiversity. Generally, more diversity enhances the stability of systems, although this is not necessarily true for unstable environments as then robust, simple systems may be most resilient (Holling et al. 1994). Biodiversity is currently being lost at an unprecedented, even worrisome, rate. The most influential factors are the occupation and fragmentation of space by humans, the harvesting of biotic resources, the introduction of exotic species in ecosystems, the emission of toxic substances, climate change, and hunting by humans. Not surprisingly, research on biodiversity loss and conservation has received considerable attention in environmental economics (Perrings et al., 1995). An influential study by Weitzman (1998b) studied the problem of protecting biodiversity under a limited budget constraint, deriving a concrete criterion for setting priorities among biodiversity-protecting projects. This is based on the assumption that the loss of biodiversity due to the extinction of a species is exactly equivalent to the distinctiveness of that species, defined as the genetic distance between species. The resulting criterion implies that in a situation with two species, with identical costs of protecting each species and identical changes in survival probabilities, it is optimal to protect the species which is the most distinctive. Baumgärtner (2004) and van der Heide et al. (2005), however, show that distinct policy advice may result when in addition to genetic diversity ecological relationships are taken into account. The conservation problem then becomes more difficult as more variables and relationships have to be considered, including the multilevel connection between biodiversity, species and ecosystems.

Human use of ecosystems and populations of living organisms creates selective pressure that depends on the particular managing and harvesting practices. The general nature of the problem is that resource harvesting not only affects the quantity of the resource but also its quality or composition in genetic terms. Examples can be found in agriculture (use of pesticides, herbicides, fungicides, monocultures), fisheries (mesh size, season of fishing), ecosystem management (groundwater control, fire protection) and health care (use of antibiotics). Selection of resistant individuals in a population is the main evolutionary mechanism to be addressed in agriculture. Resistance usually has a (small) cost, otherwise the resistant individuals would be well represented even without selection pressure from pesticides and the like. In other words, the resistant individual is relatively susceptible to other stress or selection factors. Withdrawal of the selection pressure (e.g. removing pesticides) will then usually lead to a decline in the proportion of resistant individuals in the population. Nevertheless, selection pressure often reduces (bio)diversity in the population irrevocably. This may be regarded as an intertemporal externality. A relevant question, then, is whether the costs of overcoming resistance are increasing over time. Learning effects, such as developing new antibiotics, may temper the rise in costs. But a renewed introduction of the selection pressure will still quickly achieve widespread resistance due to the relatively large proportion of resistance-affecting genes in the population. A specific problem here is cross-resistance, i.e. resistance to one type of insecticide creating resistance to others. The risk of cross-resistance depends on the type of mechanism – metabolic, biological or chemical – through which individuals achieve resistance. This can include excretion of the unabsorbed insecticide, changes in behaviour or reduced penetration of the chemical in the body.

An elegant way to address these various issues was proposed by Alistair Munro (1997). He extends the standard, general model of optimal use of a renewable or agricultural resource with a negative intertemporal externality based on genetic selection. He illustrates this for an insecticide that raises the fitness of resistant insects relative to their susceptible competitors. Pesticide use initially reduces the pest population size, but after resistance has developed, it increases again (Figure 13.2). The traditional planning solution is myopic as it only recognises changes in the population size while it does not anticipate genetic selection. Instead, optimisation under perfect foresight about the evolutionary consequences of insecticide use gives the fully optimal plan. This involves less use of pesticides to benefit longer from resistance in the pest population, which

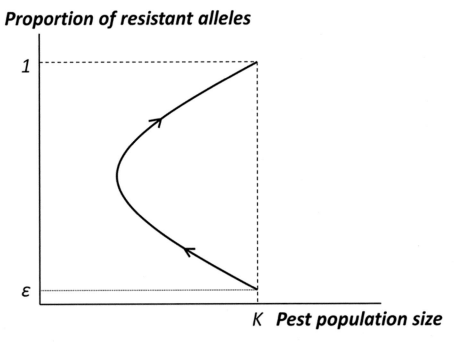

Figure 13.2 Short- and long-term effects of pesticide use on a pest population
Legend: K = carrying capacity of the pest population; ε = initial proportion of resistant alleles
(see Section 3.3) in the population, before application of pesticide.

results in moving more slowly along the curve in Figure 13.2. In this way, one enjoys lower pest population sizes and thus damage to resource or agricultural yields for an extended time. The optimal plan over time evidently depends on time preferences: for a higher discount rate, which stimulates myopic behaviour, optimal pesticide use will be higher, causing resistance to increase more rapidly. As a corollary, under myopia, investment in R&D on new pesticides will be relatively high. This model was extended by Noailly (2008), who replaced the rational intertemporally optimising farmer with a population of farmers affecting all the pests in their joint region. These are all assumed to be boundedly rational and copying other, successful strategies. The result is surprising, complex dynamics as if a stochastic factor is at play. The policy implication is that in order to achieve effective long-term pesticide use, policy-makers must direct farmers' imitation strategies in a way that the system does not traverse ecological thresholds.

Another example of selection pressure arising from human activity is illustrated by early industrialisation. In nineteenth-century industrial England, the white (actually rather grey) peppered moth was selected against because air pollution from industries put a black layer over buildings and trees. The moth lost its camouflage and became more susceptible to predating birds, causing a black variant to become dominant. This is known as melanism, the opposite of albinism. This issue was discovered in 1896 – not

Figure 13.3 White peppered moth and its black variant.
Note: In the left photo the white moth is well camouflaged, and in the right photo the black one is camouflaged.
Source: Ford (1975, plates 14 and 15). With permission of Springer.

long after Darwin's death – and has remained a classic example of observable, rapid evolution in practice. After the validity of the evidence was questioned, Michael Majerus, Professor of Evolution at the University of Cambridge, decided to undertake what was the largest experiment of this kind at the time: it lasted 7 years, from 2001–2008, and involved 4864 moths. This provided definite confirmation, which was published by others after Majerus' early death (Cook et al., 2012). Figure 13.3 shows both moths positioned on a light and dark tree. The white one is better camouflaged on the white tree, arguably as it had much more time to adapt than the black one. By the way, the process reversed owing to air pollution legislation in the 1950s which gave rise to a cleaner environment, causing the rare white variant to increase.

Environmental chemicals may also affect humans in an evolutionary sense. Pillarisetti (2016) illustrates this by showing that sex ratios may be influenced by environmental toxins. This is particular relevant to developing countries as here polluted local environments are common. Certain types of pollution tends to lower the ratio of sexually competent males to sexually competent females as ontogenetic formation of males is more sensitive to relevant substances (Cummings et al., 2007; Bergman et al., 2013). As this may counteract cultural forces, the net effect is uncertain and likely to vary between countries and cultures. Many developing countries face skewed sex ratios due to interplay of genetic, cultural, economic, environmental and socio-historical processes. With regard to cultural factors, a strong preference for sons can be found in many low-income countries. A special case is China, where such a preference was reinforced by its one-child-per-couple regulation.

While biodiversity loss receives a lot of attention in environmental science, the evolutionary dimensions of this are surprisingly neglected. An important cause of future biodiversity loss will be climate change. It will have impacts that are expected to last for many millennia (Clark et al., 2016). This means that environmental conditions will be different, and increasingly so, for an extended period of time. In other words, the selection environment will alter for all species on the planet. This is bound to have evolutionary consequences, as already indicated by major changes in biodiversity. In turn, it is quite certain, that biodiversity loss itself will change the direction of evolution of the biosphere. Myers and Knoll (2001) argue that the 'sixth extinction' caused by humans will generate evolutionary impacts in three stages:

1. Between one- and two-thirds of all extant species – until recently successful (=well adapted) owing to millions of years of evolution – will become extinct; alien invasions and other mixings of biotas will be common; progressive depletion and homogenisation of biotas will occur; and a reduction if not elimination of many biomes, including tropical forests, coral reefs, and wetlands, all of which served as centres of diversification in the past.
2. Gene pools will be so depleted that species may be unable to bounce back; fragmentation of species' ranges with disruption of gene flow will occur; effective population sizes will decline; and not only species but also biotas will enter into new areas.
3. A disruption of food chains and webs will occur; a dominance and outburst of speciation of species thriving in human-dominated ecosystems will be observed; fewer species will evolve in tropical forests and other biomes; all large mammals are likely to go extinct; and speciation of large vertebrates will come to an end; even the largest protected areas will be too small for further speciation of animals like elephants, rhinoceroses, apes, bears and big cats.

Previous mass extinctions – even the worst one at the end of the Permian 250 million years ago, with an over 70 per cent loss of all species – were followed by a bounce back of biodiversity once the cause of the extinction was gone or altered. However, in the case of a human-dominated biosphere it is unlikely that the pressures caused by the huge human population, in terms of land use and infrastructure, pollution and waste, can be reduced to pre-industrial or even prehistoric levels. This means that biodiversity bounce back is unlikely and that evolutionary processes will be quite different from how they were before humans dominated the biosphere. Future developments in this respect depend to a great extent on how humans will fare over evolutionary time scales. This is elaborated in Chapter 16.

A complementary view on the evolutionary consequences of human dominance in the biosphere is offered by van der Voet et al. (2000). They argue that impacts are heterogeneous for different types of living organisms. In evolutionary terms, one can classify influences as changing the selection environment (predation, nutrients, habitats, and chemical pollutants), introducing new mutations (exotic species, chemical pollutants), affecting spatial isolation (fragmentation through infrastructure) and influencing reproduction advantages (pesticides). Ultimate consequences are influences on speciation or extinction, thus affecting the direction of natural evolution. Predators and prey species are especially sensitive to this, for different reasons. Predators live in small numbers, follow an ecological K-strategy (Section 4.6), require a large habitat and often have no, a relatively low or even negative economic value. They are especially sensitive to land use, toxic substances – through increased concentration of these higher up in food chain – and hunting. Moreover, due to a low reproduction rate, they are not very adaptable. Prey have a positive economic value, and remain part of a pre-agricultural hunting culture. Harvesting pressure is the most serious threat to them. Most herbivore prey species that were initially predated by humans (cattle, sheep, goats, chickens) have been domesticated. The last major group of 'free prey species', namely fish populations, are being seriously threatened by overfishing, and increasingly domesticated as well, resulting in aquaculture or fish farms. This might be considered as a sort of final stage in cultural-natural coevolution.

Some studies suggest that loss of predators can negatively affect the productivity (and stability) of the entire ecosystem. As a short-term effect, lack of predators means that nutrients are not rapidly and efficiently re-used. As a long-term effect, it reduces selection pressure on prey, which may as a result become slower and more prone to disease. Illustrative of this is a study by Wardle et al. (2004). It shows that forests in different parts of the world, namely Alaska, Australia, Hawaii, New Zealand and Sweden, have low productivity, characterised by on average smaller trees and less fertile soil. Analysis indicated that the common factor was a lack of top predators, in most cases owing to the presence of humans.

Given these insights, the study of multispecies systems in resource economics (Hoekstra and van den Bergh, 2005) might be extended with evolutionary elements to consider the indirect impacts of, for example, fishing or harvesting one species upon another species (or population) through a one-directional or even mutual (coevolutionary) selection mechanism. This would mean a shift from 'mechanistic, stock sustainability' to 'evolutionary and ecological sustainability', which provides a more satisfactory, though analytically more difficult, approach to sustainability analysis.

Since the classic article 'The tragedy of the commons' by Hardin (1968), much attention has been given to the risk of overexploitation of common-pool resources (Clark, 1990; Ostrom, 1990). Although common property is often confused with open access, where exploitation is more likely, in common property resources the risk of overexploitation is also serious. It critically depends on the type of common property regime that prevails, such as public or club good, and may therefore differ from situation to situation. A relevant question is whether preventing resource conflicts and overuse requires strict regulatory policies set by higher-level governments, or whether instead one should rely on endogenous formation of use regimes. This has been tackled using an evolutionary framework (Axelrod, 1986). A main finding is that externally imposed rules and monitoring can reduce and destabilise cooperation or even completely destroy it (Ostrom, 2000). Stimulating norms through communication may be more desirable. This is certainly the case when hierarchical monitoring is imperfect. Instability in the evolutionary equilibrium can arise when certain parameters change, for instance, when sanctions decline, harvesting technology becomes more productive due to technical progress, or the price of the resource increases. Other risks to evolved norms are migration of resource users, external economic and political disturbances, and natural disasters influencing the state of the resource. These issues have been examined using a wide range of approaches, including empirical field studies and laboratory experiments (Martin, 1992). Evolutionary modelling methods used to study such issues are discussed in Box 13.1.

13.4   Coevolution and Evolutionary Growth

This section will illustrate the usefulness of evolutionary thinking for understanding long-term growth in relation to environment and resources. This relationship has received much attention from environmental economists. The so-called environment versus growth debate is relevant here – where growth commonly refers to a rise in gross domestic product (GDP), or of GDP per capita. This debate can be characterised by three core questions (van den Bergh and de Mooij, 1999): (1) Is growth desirable: are there social limits to growth? (2) Is growth feasible: are there environmental or physical limits? (3) Can growth be controlled or steered: are there governance limits? Climate change research is an area where (optimal) growth models have actually been applied (Nordhaus, 1994), leading to considerable criticism (e.g. Azar, 1998; van den Bergh, 2010). In addition, climate change is seen to revive the limits to growth debate (Antal and van den Bergh, 2016).

Evolutionary economics offers an alternative to traditional aggregate growth theory, namely so-called 'differential growth' theory (Section 8.6). This reflects a change in the frequencies of all possible individual (firm) characteristics. A central feature of it is the lack of an aggregate production function. Instead, a micro-approach is adopted in which a population of heterogeneous firms is described. Based on such an approach, various lines of research could be developed as an alternative to neoclassical growth theory in environmental economics. This would require extending evolutionary growth theories with environmental and resource variables, as well as applying coevolutionary theory to deal with long-run interaction of resources, behaviour, institutions and technologies to overcome these shortcomings (van den Bergh, 2005). Thus one could study the impact of alternative evolutionary assumptions regarding bounded rationality, diversity and interactions of agents, selection and innovation, as well as different types of policies, notably environmental regulation, investment in R&D and other policies affecting the economic selection environment. In addition, extensions of technological innovation models with environmental resource considerations might be used to study the long-run implications of innovation strategies, major innovations, increasing returns to adoption and long waves, as well as the impact of resource scarcity and environmental regulation on innovation opportunities and tempo. In these latter type of models, innovation,

Box 13.1 Economic evolution and renewable resource dynamics

Studies combining economic evolution and renewable resource dynamics in the context of common-pool resources are rare. They step away from the traditional assumption of rational harvesters and assume instead some distribution of low and high harvesting strategies which changes according to replicator dynamics. A basic model was developed by Sethi and Somanathan (1996). It combines evolutionary (replicator) dynamics and renewable resource dynamics (further elaborated in Noailly et al., 2003). The evolutionary model is set up as a two-stage game, involving a resource use strategy and the decision to punish or not, with punishment being costly. There are then three strategies, namely defecting (a high effort); enforcing (a low effort and sanctioning of defectors) and cooperating (a low effort and no sanctioning of defectors). The game can be seen as a mix of prisoner's dilemma and coordination games. The profits under each strategy are variable and depend on the distribution of strategies. Dynamics following from the selection of highest payoff strategies are modelled as replicator dynamics. The result is only pure equilibria: defecting or restraint (cooperating or enforcing). In other words, in the end, nobody or everybody cooperates. Once defector equilibrium is reached, recovery of cooperative behaviour is difficult. Note, however, that this is not the case in spatial interaction games, in which the spatial structure of agents (e.g. local interactions) is explicitly modelled (Noailly et al., 2006). This gives rise to clusters or zones of cooperation protected by enforcers against intruders, i.e. defectors.

Other evolutionary models also employ multi-agent evolutionary models with a spatial disaggregate (grid, cellular automata or network) structure. These can be quite easily built on ecological foundations, as spatial models are common in applied ecological modelling (Grimm, 1999; Janssen and Ostrom, 2005). The integration of economics and ecology has been hampered, though, by a lack of space in economic theories and models, motivated by most economic processes operating at higher scales. The most sophisticated models employ cellular automata. These were originally used to model process-based predator–prey interactions, changes in surface water quality, and fire propagation problems. They allow for an explicitly spatial process approach, making the state of a cell (partly) dependent on neighbouring cells. Whereas, in physical and biological systems, such immediate influences in space dominate, this is less the case in social and economic systems. Here, 'spaceless' or long-distance information (e.g. in the form of books, databases, websites) plays an (increasingly) important role, as witnessed by trade and globalisation processes. This means that many spatial interactions in an economic context extend beyond the scope of neighbouring cells. The application of cellular automata in social science, and hence its linkage to ecological models, could be further developed by allowing for direct interactions between non-bordering cells (Couclelis, 1985).

selection, path dependence and lock-in rather than capital accumulation only would play a central role. Attention needs to be given to two specific issues: evolutionary modelling of supply–demand chains, and of technological complementarities, especially with an eye to 'hybridisation' of technologies. Obtained insights may inform policy aimed at large-scale transitions, such as from fossil fuels to renewable energy: intermediate phases might involve hybrid technologies, such as mixing fossil and biomass fuels, or combining fuel cells running on hydrogen with combustion engines in 'hybrid cars' (Zeppini and van den Bergh, 2011). Policy lessons can be derived from this, such as which instruments – regulation through environmental levies or innovation incentives such as subsidies – to emphasise in which stage of the innovation cycle: research, development, demonstration and deployment.

A neo-Austrian approach with evolutionary elements to growth and environment was proposed by Faber and Proops (1990). The neo-Austrian approach emphasises the role of time and roundaboutness of economic production. Roundaboutness denotes the presence of various intermediate stages and multiple technologies. The approach allows for uncertainty, novelty and irreversibility of changes in the sector structure of the economy, and for a teleological sequence of production activities. The long-term relation between environment, technology and development is characterised by three elements: use of non-renewable resources is irreversible, so that a technology based on it must ultimately cease to be viable; inventions and subsequent innovations lead to both more efficient use of presently used resources and substitution by resources previously not used; innovation requires that a stock of capital goods with certain characteristics is built up. Based on these foundations, a multisector model is constructed with a production side formulated in terms of activity analysis. The model is capable of replicating transitions from simple to more complex or roundabout production activities. This is then used to simulate economic and environmental history from a pre-industrial agricultural society to an industrial society, using fossil fuels and capital. This may be considered as a formal elaboration of earlier ideas by Wilkinson discussed in Section 12.6.

13.5   Evolutionary Transition to a Sustainable Economy

Recent years have shown a rising interest in the study of large-scale sociotechnical transitions to an environmentally sustainable economy, in brief 'sustainability transitions'. This relates to key sectors such as transportation, electricity generation, agriculture, water, fisheries and tourism. The need for major transitions derives from the persistence of structural problems in these sectors – related to resource scarcity, oil dependency and contribution to local as well as global environmental problems, notably climate change. Transitions management has been proposed as a broader approach than traditional environmental policy analysis, deemed necessary to foster sustainability transitions (Geels, 2002; Grin et al., 2010).

Transitions can be seen as fundamental system changes which entail, or even require, escaping lock-in of dominant technologies, introducing major technical innovations, and changing prevailing social practices and structures. To achieve transitions, various barriers have to be removed or weakened, associated with vested interests, myopia and other aspects of human bounded rationality, lock-in of undesirable technologies, and slow diffusion of new technologies. Transition management and policy are needed for this. They can be framed as some combination of environmental regulation, innovation policy and measures to escape lock-in. One can see the search for transition policy as a kind of third best policy approach, which tries to be add realism about the complex political and socioeconomic system in which policies need to be implemented and to function. In fact, transition thinking can be considered to give more attention to the policy evaluation criterion 'social-political feasibility', additional to the usual criteria of effectiveness and efficiency. Four main theoretical approaches can be identified in the current research on sustainability transitions (Table 13.2). While the fourth one is explicitly evolutionary in nature, the other three employ many evolutionary notions, as indicated by the second column.

Transitions thinking and evolutionary thinking have many features in common. This is not surprising, given that the notion of 'major transitions' is central to evolutionary thinking (Section 3.6). Another reason is that evolutionary theories and models have been regularly employed in innovation studies, which forms one intellectual foundation of transition studies. In reviewing the young literature on evolutionary modelling of sustainability transitions, Safarzynska et al. (2012) identify five model categories: (i) multilevel modelling including approaches using an evolutionary micro–meso–macro framework (Dopfer et al., 2004); (ii) multiphase modelling such as S-curve, life cycle and multistage approaches; (iii) coevolutionary modelling; (iv) modelling of learning processes; and (v) broader systems models of transitions. Regarding category (iii), many approaches can be seen to build upon some coevolutionary core. This can apply to technological coevolution where innovations in one technological population (say, cars) affect the value of technologies in another technological population (say, traffic systems). A second type is demand–supply coevolution which depicts interactions between technologies in a population of producers and preferences in a population of consumers. This has received much attention in the context of transition studies (e.g. Windrum and Birchenhall, 2005; Oltra and Saint Jean, 2009; Windrum et al., 2009; Malerba et al., 2008; and Safarzynska and van den Bergh, 2010b). Models of this type capture that producers innovate to compete and interact with various consumers groups who learn by experimenting with new products. To understand transitions and effective policies, a good formulation is needed of the fundamental mechanisms underlying demand–supply coevolution, such as consumer behaviour in the form of imitation/conformism, status-seeking, habits or individual learning. The relevance of these depends on the particular good or service under investigation. For instance, consumption of clothes is very sensitive to imitation, while that of cars seems more driven by status-seeking.

Multi-agent models dominate evolutionary transition modelling. This is understandable as more compact representations through difference or differential equations, e.g. along the lines of replicator dynamics, tend to considerably simplify the complexity of the system in transition. However, this shortcoming can be overcome by extending replicator dynamics with recombination and mutation, namely through adapting

Table 13.2 Approaches to research on sustainability transitions

Approach	Key concepts	Policy view	Publication offering a brief introduction
Innovation systems (IS)	System failures, functions, innovation, diffusion, institutions, national and sector systems, supply chain, social, political and learning networks, user-supplier networks and industry-academia networks	Identify system failures and correct these with environmental regulation (correcting prices) and technology-specific policies	Jacobsson and Bergek (2011)
Multilevel perspective (MLP)	Multiple (competing) technologies, structural change, multiple levels (niche, regime, landscape), multiple phases, coevolution, networks, transformation, reconfiguration, technological substitution, de-alignment and re-alignment	Align technologies and user practices Strategic niche management (SNM) - reflexive management of real-world experiments	Geels (2011)
Complex systems	Variation, selection, attractors, feedback, emergence, coevolution, dissipative structures, punctuated equilibrium and self-organisation	Transition management: transition experiments, focus on frontrunners, envisioning for sustainable futures	Rotmans and Loorbach (2009)
Evolutionary systems	Population, diversity (variety, balance, disparity), cumulative change, multiple selection factors, recombinant innovation, adaptation, group (multilevel) selection, path dependence and lock-in, coevolution, selection as social learning, social and demand-side networks	Account for all selection forces (market, institutions, norms, regulation), foster status character of green products, optimal diversity, stimulate recombinant and deviant innovations, and policies to escape lock-in	Safarzynska et al. (2012)

Source: van den Bergh et al. (2011, Table 1, p. 9).

existing models in biology (Safarzynska and van den Bergh, 2011a). Such an 'innovation-selection dynamics' approach has recently seen a stylistic application to energy transitions, based on a model that captures the interaction between selection and innovation in a population of boundedly rational investors who decide each period on the allocation of investment capital among different technological options (Safarzynska and van den Bergh, 2013). Investors tend to invest in below-average cost technologies, just as under replicator dynamics. In addition, they direct a constant fraction of investments, captured by mutation and recombination rates, to alternative technologies and research on recombinant innovation. As opposed to most previous studies, mutation and recombination are defined here as conceptual variables with a concrete behavioural interpretation, namely describing the decision rules (heuristics) of investors. The authors derive the optimal diversity of technologies in terms of minimising total cost of investments over time for different selection environments. Diversifying a research portfolio turns out to be an important source of innovation due to novel combinations of incumbent technologies. In the long run, it may also ensure the lowest cost of maintaining a specific technology mix.

A model in the aforementioned category (v) by Safarzynska and van den Bergh (2011b) studies historical change from coal to gas in electricity production in the UK by considering the determinants of investments in installed capacity in power plants. An innovative feature of this study is that it compares what happens under conditions of rational and boundedly rational investors. The electricity industry is composed here of heterogeneous plants producing electricity using different energy technologies. Three energy technologies are considered, namely coal, nuclear and combined combustion gas turbine (CCGT). Demand is described by a static function, motivated by the fact that demand for electricity is inelastic, while consumers have limited opportunities to choose the electricity source. Plants set their production given a reaction function of others. A model version with rational investors is capable of replicating well core features of the history of electricity in the UK. This includes a rapid diffusion of gas in electricity production, the evolution of the average size of newly installed plants, and a high percentage of electricity sales covered by forward contracts-for-difference. In this model setting, nuclear and renewable energies have no chance to diffuse on the market. In the model version with boundedly rational investors, nuclear power typically dominates electricity production as investors focus more on short- than long-run profits, while the short-term profits are positive as the government promotes diversity of technologies. The conditions under which investors behave as fully rational or myopic is worth studying. It is possible, for example, that early entrants are more rational than late-comers who may act more like followers. Alternatively, policy instruments might affect the type of behaviour. For instance, subsidies to cover the installation cost of new power plants might encourage more myopic behaviour of investors. These issues require further analysis, and possibly a connection with insights from fields like industrial and behavioural economics.

13.6 Diversity to Avoid or Escape Lock-in

A starting point for policy thinking from an evolutionary angle is that environmental problems often go along with a loss of diversity of options. Not only is biological diversity lost but also cultural diversity is sometimes at stake. The latter is the outcome of dominant life styles, the emergence of immense transnational companies, diffusion of a standard business model and production method (e.g. in agriculture), similarity of educational systems, and integration of countries in unions with coordinated institutions and policies. One might conceptualise this issue as lock-in triggered by repeated selection. Environmentally relevant lock-ins can be found in many places. The most significant example is the complete dependence of modern economies on fossil fuels, which can be traced back to the Industrial Revolution. It has gone along with a widespread diffusion of engines to industry and transport that combust fossil fuels (coal, oil, gas) as well as with a strong dependence on fossil-fuel-based electricity generation.

General policies to deal with lock-in were discussed in Section 10.6. This included the evolutionary policy approach of optimising diversity and adaptive flexibility, so as to reduce risk of undesirable long-run lock-in (Rammel and van den Bergh, 2003), as elaborated in Section 10.7. This may be regarded as an evolutionary interpretation of the much adhered to precautionary principle. A complementary strategy is to create what Unruh (2002, p. 323) calls 'a countervailing critical mass or social consensus for policy action' through education of the general public. He suggests that providing information on extreme climate events (disasters) may create support for strict climate policies and changes in current institutions. This is supported by Janssen and de Vries (1998), who incorporated evolutionary elements into climate modelling by allowing adaptive agents to change their behavioural strategies. These agents respond to persistent surprises in global climate, as represented by the global mean temperature of the atmosphere. The distribution of strategies changes according to a selection process modelled as a replicator equation. Agent fitness is a function of the difference between expected temperature change and actual temperature change. The occurrence of a climate surprise that cannot be made consistent with the initial perspective of agents stimulates adaptations in their strategies.

The potential contribution of evolutionary thinking to environmental policy-making to stimulate innovations in energy technology is studied by van den Bergh et al. (2006, 2007). The study uses the core concepts of 'bounded rationality', 'diversity', 'innovation', 'selection', 'path dependence and lock-in', and 'coevolution' to formulate guidelines for designing energy innovation policies. Extant Dutch policies, and associated policy documents, aimed at affecting energy technologies and innovation are screened against this background. Developments in the areas of three particular energy technologies are considered, namely fuel cells, nuclear fusion and solar photovoltaic cells. Existing Dutch energy innovation policies are found to concentrate on notions like cooperation, education, future projections and demonstration projects, but neglect or even hamper the uptake of evolutionary aspects like recombination, spill-overs, serendipity and multidimensional selection environment. It is concluded that in order to incorporate relevant evolutionary notions in policy, governmental technology policies should focus more on the diversity of technologies, strategies and businesses, rather than on economic efficiency as the key goal. Moreover, it is suggested that one needs to refrain from 'picking winners' because complexity and uncertainty mean one cannot know beforehand who will be the winners in terms of economic, environmental or social benefits. Instead, policy-makers should create the conditions, notably selective pressures, under which evolutionary processes will lead to socially desirable 'survival of the greenest'. In particular, an 'extended level playing field' is needed to unlock undesirable fossil fuel technologies. This covers four elements: (1) assure that prices reflect all external costs generated by activities and products; (2) provide direct technological support in the form of niche creation or public subsidies for technologies that are low on the learning curve but are expected to have much long-term sustainability potential; (3) prevent or compensate coincidental increasing returns to adoption for early cost-effective technology that are not certain to be the environmentally best-performing ones in the long run; and (4) expose different technological options as much as possible to similar selection mechanisms to stimulate fair competition. Analysis of three energy technologies – fuel cells, nuclear fusion and solar photovoltaic energy – shows that the notion of an extended level playing field is not consistent with active policies (Table 13.3).

Table 13.3 Characteristics of three energy technologies and related policies

	Fuel cells	Nuclear fusion	Photovoltaics
Diversity	High (applications, types, fuels)	Limited	High (applications, types)
Innovation	Strong *interaction* between different sectors (incl. chemicals, energy companies, auto-producers). *Niche markets*: aerospace, vehicles.	Expertise concentrated within a small global village; much internal *cooperation* but little external interaction. No *niche markets*.	*Serendipity* and cross-fertilisation were keys to development (e.g. thin-film technology). *Niche markets*: aerospace, off-grid applications.
Cumulative R&D in IEA countries 1974–98 (in billion US$)	Unknown	26.8	5.2
Selection environment	Liberalisation can present opportunities (decentralisation). Environmental policy is the key (e.g. zero-emission laws).	Not mature enough for market introduction. Viability will depend partly on stringent CO_2 policy.	Market power has played a role. Government policy (esp. subsidies) is an important factor.
Bounded rationality	Application calls for break with existing routines. Imitation behaviour in auto-industry.	Private investors not interested due to long time horizon. Not possible to build on existing routines.	Technology involves high capital investments. Long time horizon.
Path dependence and lock-in	Limited economies of scale in applications: → fits well into decentralised systems. *Lock-in* in existing technology (e.g. combustion engine) is a significant barrier.	Heavy path dependence, economies of scale are important: → difficult to fit into decentralised systems.	Limited economies of scale in applications: → fits well into decentralised systems.
Coevolution	Interaction with other components of energy system (including fuel infrastructure) is important.	Little interaction with other energy technologies. Internal complementarity exists (e.g. plasma physics and materials science).	Implications for other components of the energy system (e.g. due to fluctuations in supply of sunlight).

Source: van den Bergh et al. (2007, Table 5.4, pp. 138–139).

Development of fuel cells was mainly stimulated by a high degree of diversity of economic agents, techniques and products and a niche market of zero-emission cars. Nuclear fusion has the disadvantage of a lack of diversity because a minimum scale is required. The Dutch history of solar photovoltaics indicates the important role of serendipity, cross-fertilisation and niche markets for technological development.

13.7   Environment in Evolutionary Social Science

I have argued that evolutionary approaches to social science research, in particular within economics, provide useful theoretical and methodological inputs to provide better answers to a number of pressing environmental problems. This involves themes like management of ecosystems subject to selection pressure, spatial modelling of local self-regulation in common-pool resources, the impact of evolutionary growth on resources and environment, and the formulation of environmental policies taking into account biological and technological diversity as well as lock-in. Evolutionary thinking was used here to address sometimes biological and in other cases economic phenomena. Certain topics in resource economics, such as pesticide use in agriculture, benefit from the inclusion of biological-genetic evolutionary mechanisms next to more short-term mechanistic processes. Non-genetic evolutionary processes, covering market, technological, institutional and organisational aspects, are relevant to virtually all themes within environmental economics.

Evolutionary economics has hardly considered environmental and resource dimensions of economic systems. The neo-Schumpeterian school in evolutionary economics has been occupied with technological innovations that occur over relatively short time horizons – at least from environmental and biological evolutionary angles. The other component of evolutionary economics, evolutionary game theory, has dealt with simple, analytical models that represent the essence of selection dynamics, thereby completely ignoring the long-term impact of innovation on the adaptive capacity of systems. More generally, the neglect of environmental and resource dynamics in evolutionary social sciences may lead to an incomplete model of reality. The reason is that major phases of economic history have been subject to strong influences from environmental and resource factors. This, in any case, holds for such major transitions as the development of agriculture and human settlements, and the Industrial Revolution, as argued in the previous chapters. Likewise, energy scarcity and climate change are likely to act as triggering factors of any future economic transitions.

It might seem odd that environmental problems are neglected in evolutionary social sciences, given that the notion of 'environment' is an integral part of evolutionary thinking. Indeed, in evolutionary biology genotype and environment 'produce' jointly a phenotype. Moreover, evolution means that living organisms are during their lifetime affected by environmental selection. This relates to the fact that organisms are part of ecosystems and sometimes also of social groups. Hence, their environment consists of both biotic and abiotic factors, while the biotic factors can be separated into social and other ones, or in intra- and multispecies interactions. Eagleman (2012) proposes German terminology to identify distinct meanings of 'environment', because of a supposed lack of accurate wording in English language: 'umwelt' (similar to environment or milieu) versus 'umgebung' (similar to surroundings or environs). With the first, he means the subset of environmental signals that organisms can pick up with their particular senses: sight, smell, taste, touch and hearing. The ranges of what these senses can detect differ among species, causing them to experience distinct, species-specific 'umwelts'. This holds as much for animals as for humans. Indeed, behavioural research shows that in undertaking cognitive tasks or making economic choices, humans are influenced by environmental conditions, including by weather conditions and such basic signals as colours of clothes worn by people with whom we interact (Alter, 2012). These considerations suggest there is much room for further research on the boundary of environmental social sciences and evolutionary thinking.

One topic that deserves attention in research is the evolution of recycling. In biological systems nutrient cycling is highly developed as over millions of years complex terrestrial and aquatic ecosystems have evolved into food webs involving an incredible amount of microorganisms (bacteria, fungi, etc.) which are capable of degrading organic waste and reusing ensuing nutrients. In comparison, the economy – complicated though it may seem at first sight – is a fairly simple system with only a few actors. Therefore, it produces more than it decomposes and reprocesses. We cannot expect recycling and reuse in the economy to become very refined in this century or perhaps ever, unless a similar subtle evolutionary process to generate relevant new microscopic activities can be set in motion. One can compare the challenge of constructing a circular economy to that of achieving artificial intelligence. Computers are also still very humble compared to the evolved human brain. Biological evolution takes a lot of time perhaps, but it has come up with sophisticated and effective designs that human planning and technology have not matched by far. Notwithstanding, there exists a field of research called 'industrial ecology' which is motivated by drawing lessons from nature for the economy, at the firm to the global level (Graedel and Allenby, 1995; van den Bergh and Janssen, 2005). A traditional economic policy perspective is still useful though as well. It stresses that instruments like pricing the environment for negative externalities will make virgin materials and waste dumping more expensive and thus recycling and reuse more attractive. In addition, pricing can limit rebound effects resulting from considerable secondary recycling activity being added to existing primary activities (Zink and Geyer, 2017). This suggests that materials and substance pricing is an essential ingredient to stimulating organisational and technological evolution towards a genuinely circular economy.

Another area of research that deserves mention is 'environmental aesthetics', which suggests that people directly derive happiness from observing nature. Environmental economists have tried to capture this by using monetary valuation techniques. E.O. Wilson's (1984) notion of 'biophilia' is relevant here, as it conveys the message that humans have instinctive, genetically based, preferences for natural environments and other species with certain features. In particular, we seem to like the savannah type of landscapes, as these represent the habitat to which we adapted most recently and fully during our evolutionary history. Possibly, such basic evolutionary insights can aid in devising strategies to increase democratic support for environmental protection. Richard Dawkins (2001) is less optimistic, arguing that sustainability doesn't come naturally. We would not face the challenge of solving unsustainability if sustainable behaviour had been built into us humans by natural selection. Instead, evolution has created in us an instinctive interest in the survival of our genes (according to individual selection theories) and possibly also in the survival of the coherent group to which we belong (according to multilevel or group selection theories). But it has not stirred an inborn concern for the survival of our species or the larger ecosystem of which we form part. In line with this, natural selection has placed in us an interest in short-term benefits, whereas sustainability is all about long-term social benefits, often requiring short-term individual sacrifices. Dawkins further draws attention to the fact that humans are eager to create wealth and possess conspicuous goods — crucial factors behind the size and contamination of the global economy — because of what has become a maladaptation, namely our instinctive status-seeking which originally was selected for reasons of achieving reproductive success. He thinks that humans are not more selfish than other animals, just more effective and hence destructive. His positive message is: 'If any species in the history of life has the possibility of breaking away from short-term Darwinian selfishness and of planning for the distant future, it is our species.' He notes that we accepted contraception and taxes, so why not sustainability. Georgescu-Roegen (1972, p. 35) is less optimistic: 'Perhaps, the destiny of man is to have a short, but fiery, exciting and extravagant life rather than a long, uneventful and vegetative existence.' As sustainability is not in our genes, one can only hope 'sustainability memes' will diffuse quickly and widely.

14 Evolving Solutions for Climate Change

14.1   Consumption, Production and Population Externalities

According to the Intergovernmental Panel on Climate Change (IPCC), there is broad scientific agreement on the climate changing as a result of anthropogenic emissions. The mechanisms of this process, as well as its physical, biological and socioeconomic impacts, have been mapped out in detail, as summarised in IPCC (2014). A precautionary perspective emanating from the concern about climate change is that we should avoid a rise in the global average temperature beyond 2°C and, if possible, even beyond 1.5°C. This poses an enormous challenge for the design and implementation of policies that steer our economies away from high-carbon production, consumption, transportation and electricity generation. Studies show we need to reduce the carbon intensity of output by more 80 per cent (Jackson, 2009; PWC, 2012; Antal and van den Bergh, 2016). As Table 14.1 illustrates, this depends on scenarios of economic and population growth, which jointly determine the scale of the economy in GDP terms and, through the carbon intensity of output (e.g. $ or €), CO_2 and other greenhouse gas (GHG) emissions.

We have seen how climate change in the past likely contributed to the rise of agriculture (Section 11.6). Ironically, the resulting complex society is now undermining the climate conditions essential to its continuity. In other words, social and climate systems are involved in a long-term mutual feedback cycle. It is probably not the first time that human presence is affecting the Earth's climate. Some studies suggest that the Black Death during the mid-fourteenth century contributed to subsequent global cooling as the pest pandemic reduced the population of Europe by 30 to 60 per cent, causing agricultural lands to become unused. Hence, an increasing amount of land became covered by dense natural vegetation, involving considerable uptake of CO_2 from the atmosphere. In turn, this triggered the Little Ice Age, starting in the sixteenth century and lasting until about 1850 (van Hoof et al., 2006).

Humans have also altered local climates. An extreme example is the Sahara in North Africa. Until some 6000 years ago it was a fertile, green area with lots of species, humidity and lakes, fed by considerably more rainfall than currently. It is still debated how the Sahara turned into a desert. One explanation is that 'de-vegetation' started by nomadic herders guiding their domesticated animals to ecologically sensitive areas, which reduced grasslands and increased fires. While wild grazers avoid eroded lands, as they offer less edible vegetation and the associated open landscapes make them more

Table 14.1 The challenge of climate goals quantified as carbon intensity reduction

Reduction in carbon intensity of output to meet 2°C target with 66% probability	GDP increase	
	1.5%	0%
Total reduction 2013–2050	81%	67%
Annual reduction during 2013–2050	4.4%	2.9%
Historical annual reduction 1970–2013	<1% for CO_2 and <1.5% for all greenhouse gases	

Based on data and calculations in Antal and van den Bergh (2016).

susceptible to predators, herded animals were guided by humans to overcome such instinctive behaviour. As a result, eroded landscapes had no opportunity to recover well (Wright, 2017). Possibly, humans also cleared lands by deliberate burning, to make them more suitable for herd animals. In addition, global climate trends at the time due to orbital changes of the Earth might have contributed to make ecosystems more fragile. Altogether, these factors increased albedo effects and reduced evapotranspiration and hence rainfall. When the extent of eroded lands crossed a threshold, desertification became expansive and irreversible. Similar phenomena occurred in other areas, but much later, such as in western North America and New Zealand during the early nineteenth century. These historical cases should be taken as a warning for current dryland management in times of rapid climate change.

The dominant economic view on climate policy is that we need to regulate the emission of CO_2 and other greenhouse gases (GHGs) through market mechanisms such as carbon taxes or emissions trading. The strongest support for this comes from environmental economics, which conceptualises enhanced global warming as a market failure, namely an environmental externality caused by GHG emissions. The notion of externality means that decisions by the GHG emitters, including consumers and producers, affect the production or welfare of others, in the case of climate change in both near and distant futures. The underlying policy goal is to avoid such harmful impacts on others by correcting the market failure. The aim is not to minimise the level of environmental externalities. Indeed, since externalities are pervasive, the latter would imply reducing the size of the market economy up to the point of eliminating it. The central policy insight of this externality-welfare approach is that regulating pollution through quota, technical standards or prices (carbon taxes or emissions trading) will stimulate emitters to reduce emissions to a safe level (Sterner, 2002). Another element of climate policy is some implicit or explicit form of subsidising technological R&D. This is motivated by positive knowledge externalities associated with R&D, meaning that others can copy the innovation without having undertaken any R&D. As a result, innovators can only appropriate a part of the innovation benefits, which discourages them to undertake R&D (Section 10.2).

From a population-evolutionary angle, one can observe that whereas regulation of climate externalities tends to focus on the carbon content of consumption and production, possibly the worst decision one can make in terms of climate change externalities is not to buy a product or service but to have a child (Harford, 1998). The reason is that it implies additional emissions during the child's lifetime, that is, decades into the future — or even longer if one includes the probable descendants of the child. With an increasing number of people on the planet, the carbon budget associated with a safe climate is quickly filled up. This has urged some to propose that next to a fixed emission tax on the carbon content of goods and services, additional birth taxes – or in case of a regional or national approach, immigration taxes – are needed (Kennedy, 1995). Alternatively, a system of pollution rights with a fixed ceiling consistent with the prevailing climate goal would automatically accommodate additional emissions through a higher market price of carbon, regardless of whether these emissions come from more consumption per capita or more people (i.e. births). One reason why the decision of having a child should be priced, or otherwise regulated, is that parents make this decision while arguably only accounting for their own welfare effects and neglecting any social or environmental costs in the future. Things are more complex, though, because humans must be expected to be generally showing limited rationality with regard to decisions about having children. For example, the desired number of children is often influenced by the culture and religion to which parents belong. Moreover, parents are insufficiently rational to perceive all future private costs of raising children until adulthood. Under good climate policy, such costs would include those of carbon pricing. Parents will thus be unable to respond rationally or optimally to such policy-induced private costs, which suggests that birth regulation is required as well to assure that climate goals can be reached. As an indication of the significance of such birth regulation, Bohn and Stuart (2015) calculate that an optimal child tax equals 21.1 per cent of per capita income during a generation's time (set at 30 years).

A system of tradable birth rights is an alternative solution for the previous problem. It was initially proposed by Boulding (1964), and elaborated by Daly (1977) and others (De La Croix and Gosseries, 2009). One can characterise it best as a combination of regulation and market, or effective social control and maximum individual liberty. The system would start with a population limit derived from insights about the maximum number of people the Earth can carry in view of limited natural resource availability and environmental carrying capacities (Cohen, 1995; van den Bergh and Rietveld, 2004). Next, this limit would translate into birth rights, to be distributed among adult women and, possibly, also adult men. Then a market would be created, and those with an excess of birth-right permits or credits would sell to those with insufficient ones. For instance, if everyone had 0.9 birth rights to begin with, then those wanting a child would have to buy a 0.1 right. One could give rights to both women and men or only to men, as they cannot give birth. Boulding mentioned, as an additional advantage of the plan, that in the long run the income distribution would become more equal as the rich would have relatively many children and become poorer while the poor would have fewer children and become richer. Of course, the current situation in most countries goes completely against these policy suggestions, being characterised by direct or indirect subsidisation of births or children. The notable exception is China, which at the end of the 1970s installed a system of non-tradable birth permits allowing for only one child per couple.

Since October 2015, two children are, in principle, allowed although explicit permission from the Chinese government is required for the second child.

14.2   Evolutionary Analysis and Design of Climate Policies

Does evolutionary thinking alter the standard picture of climate policies to make a rapid transition to a low-carbon economy? Here are some suggestions based on applying insights from previous chapters.

Escaping and avoiding carbon lock-in (Unruh, 2002) represent challenges for low-carbon transitions. They can be clarified by adopting an evolutionary approach. Lock-in of undesirable technologies results from an extended history of investing, as well as innovating, in fossil-fuel-based technologies, such as coal- and gas-based electricity generation and combustion engines. Increasing returns to scale on demand as well as supply sides of markets have created significant net benefits for associated products beyond their intrinsic benefit, merely because of scale effects. A head start along with scale advantages affects long-run outcomes, also known as path dependence. Increasing returns to scale come in many forms, such as traditional economies of scale, informational externalities, learning and network effects, and even interactions between complementary technologies such as infrastructures or batteries with vehicles. Particular policies may reduce or limit increasing returns to scale and hence discourage a path to, or encourage escape from, carbon lock-in. For details, see Section 14.5.

An evolutionary angle further draws attention to the importance of optimal technological diversity, such as in low-carbon transport, battery production or renewable energy. This requires a trade-off between the aforementioned increasing returns to scale associated with concentrating on a limited number of technologies, and the benefits of assuring a minimal technological diversity. The latter include enjoying potential spill-overs, recombinant innovation (van den Bergh, 2008), and keeping options open in the face of uncertain market and institutional developments. As discussed in Section 10.7, this means merging aspects of conventional economics (optimal choices) and evolutionary economics (diversity and recombination) under evolutionary uncertainty. To complicate things, diversity is a concept that covers at least three dimensions, namely variety or the number of distinct types in a population, balance or distribution of types, and disparity as a measure of distance between types (Stirling, 2007). If policy-makers aim to stimulate an optimal diversity of energy technologies, they need to adequately take this triple dimensionality into account (Section 10.7).

Drawing on earlier research on the collapse of ancient civilisations, Tainter (2011) argues that solutions to energy scarcity tend to create more system complexity and associated indirect energy use. An advantage of an evolutionary approach to climate policy and energy technologies is that by reckoning with ultimate effects it can help to avoid unnecessary increases in system complexity and thus improve the effectiveness of policies and strategies. To this end, one has to be aware of, and account for, various indirect effects, such as carbon leakage, energy rebound, green paradox and environmental problem shifting (van den Bergh et al., 2015). Here, carbon leakage points to shifts in carbon dioxide emissions across countries following a relocation of polluting industries and changes in international trade patterns, encouraged by differences in the stringency of climate policies among countries. Energy, and consequently carbon rebound, denotes that energy conservation stimulates more indirect energy use due to higher intensity of use of more efficient machines (e.g. cars), re-spending of monetary savings following energy conservation or diffusion of energy-efficient technologies. The green paradox refers to undesirable oil market responses to renewable energy subsidies, namely oil price drops that will stimulate demand for oil and associated emissions. An example of environmental problem shifting is using biofuels to reduce CO_2 emissions, which then creates more use of land, water, pesticides and herbicides in agriculture.

These various indirect effects result from a too narrow policy focus, lacking a dynamic systems perspective. To overcome this, evolutionary economic models have been developed to study policy packages that aim to stimulate a low-carbon transition (surveyed in Safarzynska et al., 2012). Instead of equilibrium markets, they describe coevolutionary populations of firms and consumers. This allows the integration of bounded rationality of consumers, habits, social interactions, such as imitation, and status-seeking and firm competition, through both prices and innovation. Because of these features, and as multiple market and institutional selection factors are included, they can examine a broad range of policies. A particular strength of such models, in comparison with traditional market equilibrium models, is that they allow analysis of price incentives and information provision in a single framework (Safarzynska and van den Bergh, 2010b; Nannen and van den Bergh, 2010). Such instruments are currently studied by distinct disciplines (economics, sociology, social psychology) using incoherent frameworks and theories, which complicates a systematic and fair comparison of them.

Climate policy faces considerable resistance because many politicians and citizens fear it will harm economic growth or employment (Antal, 2014). For this reason it is useful to consider the potential macroeconomic impact of climate policy proposals. Elsewhere, I have argued that adhering to the idea of 'agrowth', denoting less preoccupation with GDP growth in rich countries, where average income increases do not guarantee welfare improvements, might facilitate social-political acceptance of serious climate policy (van den Bergh, 2017). Another idea that fits under the heading of macroeconomics is the distinction between top-down and bottom-up mechanisms. This relates to the choice between local and national climate strategies and policies. With regard to bottom-up, the idea of local 'transition experiments', triggered by prosocial behaviour and community initiatives, as a potential source of learning has received considerable attention (Grin et al., 2010). The power of an evolutionary approach is that it can systematically analyse the combination of top-down climate policies and bottom-up initiatives in terms of overall effectiveness. Last but not least, an evolutionary framework with diversity of economic agents (firms or consumers) is capable of studying inequitable impacts of climate policies, whether in terms of employment or income distribution, with high population resolution. Reliable knowledge about inequitable consequences is relevant as the perception of inequity is likely to hamper social-political support for implementation of climate policies.

Evolutionary game theory models have been employed to illuminate effective negotiations for a climate agreement. Compared with traditional games assuming rational agents, they add realism by imposing bounded rationality, limited knowledge and learning. Subsequently, they analyse how this influences the success of negotiations and characteristics of agreements (Courtois and Tazdaït, 2014; Smead et al., 2014). One finding is that prior commitments can be a way to build up an agreement. In particular, large players should try to make bilateral or multilateral agreements as these are easier to accomplish and facilitate later to reach a broader global agreement. This could take the form of so-called climate clubs as some authors have called them (Weischer et al., 2012). These could levy carbon tariffs at their joint border to avoid unfair competition from non-members with weak or no climate policy (Nordhaus, 2015), and even return the tariff revenues to the countries of origin to create a carrot-and-stick effect (van den Bergh, 2016).

So far we have been talking about reducing GHG emissions, sometimes referred to as climate mitigation. A complementary area of policies and strategies concerns adaptation to inevitable climate change. Biologists have accumulated considerable evidence that many natural populations of animals and plants are already responding to climate change by shifting their geographical spread, altering their reproduction cycle and plasticity, and by genetic change. Indeed, individuals with a tolerance to a wider temperature range have a selective advantage by being better adapted to climate change. Only by using all these options for adaptation can populations adequately and rapidly match the speed and magnitude of climate change impacts on their habitats. Those that cannot, or run into insuperable barriers such as mountain ranges or human settlements, run the risk of extinction. Likewise, humans have to adapt to a changing climate – unfortunately, not all regions will be able to do this with success, witness Pacific islands threatened by sea-level rise such as the Maldives, Micronesia, the Seychelles and the Solomon Islands. The term adaptation is common in both evolutionary and climate change studies, with meanings that are not identical but nevertheless overlap. Climate adaptation denotes measures that help a system resist or absorb changes that have occurred or are expected to occur. The latter distinguishes it clearly from evolutionary adaptation, which, unlike humans and human institutions, cannot anticipate changes in selection pressures' potential impacts. In the literature on climate adaptation, special attention is given to economic and geographic vulnerability, referring to people likely to experience negative impacts of climate change – because of a combination of exposure and sensitivity – and are not very capable of developing robust adaptation responses in terms of social organisation, infrastructure or finance. This evidently involves many people in poor countries but also relatively poor people in rich countries who sometimes live in marginal areas that are prone to climate risks. Box 14.1 provides some illustrations of how socioeconomic and evolutionary adaptation might help in reducing negative impacts of climate change on humans and the biosphere.

At a more abstract level, James Lovelock (2014, p. 61) suggests that the biosphere may adapt to climate change in ways that temper it. He refers to the hypothesis that the 'rate of evolution is the greatest when need is greatest', and suggests that evolution of new plants in tropical and arctic regions may stabilise ecosystems, which then feeds

---

Box 14.1 Evolutionary ecological-economic adaptation to climate change

Even if we are able to quickly implement serious climate policies, the climate will further alter for a while. This means humans will have to adapt to a changing climate, which can take various forms: create flood areas to buffer seasonal changes in river flows and rainfall, build dikes and other protective infrastructures, remove environmental pressures that add to climate change pressures but are more easily controllable (e.g. local air or marine pollution), grow drought-resistant crops, invest in water storage, construct waterproof houses, avoid new buildings in risk-prone areas, or create incentives and insurances to reduce risks and damages.

Resilience is a notion often used in the context of climate adaptation, reflecting the idea that local social-economic systems need to be redesigned so that they maintain their basic functions by resisting or absorbing stresses associated with climate change, such as droughts, floods and heatwaves. A diversity of adaptive management strategies is useful as we have little experience and can learn from experimentation and comparison. Special attention is needed for climate resilience of urban areas. Their population density assures that climate disasters such as floods cause many casualties and damages, while the distribution of risks and adaptive capacity among inhabitants is often skewed.

Understanding biological evolutionary processes contributes to effective biodiversity management strategies under climate change. Hoffmann and Sgrò (2011) provide a list of suggestions: give special attention, in conservation programmes, to species that are particularly susceptible to climate change because of limited adaptation potential, due to little genetic variation or lack of phenotypic plasticity; improve connectedness of threatened populations in fragmented habitats with other, similar populations; identify populations and species that might be threatened in the future by using criteria for evolutionary potential, such as local adaptation to climate gradients and probable levels of genetic variance in the future; and use seeds and pollen of key species originating from climatically diverse areas to stabilise ecosystems in areas subject to rapid climate change, while accounting for risks such as disease transmission.

Creation of biodiversity reserves is a crucial additional strategy. In designating areas as reserves, one should prioritise those containing genetic variation across multiple species, with environmental heterogeneity and sharp climatic gradients over small geographical areas – which together create varied selection conditions. Moreover, reserves could be designed such that connecting areas enable gene flow and allow for continuous adaptation to relevant selection pressures, such as fires, droughts and species interactions, all of which are affected by climate change. This is relevant as species in a community or ecosystem are likely respond differently to climate change, which can destabilise the overall system. In other words, the elements of an ecosystem do not shift nicely in a coordinated way – as if one superorganism – through space to find the adequate new climate conditions.

Another aspect of evolutionarily adaptive management, in the context of climate change, is ecological engineering. For example, the stability of many coral reefs around

Box 14.1 (*cont.*)

the world is seriously threatened by climate change. Some have realised that if we can regrow forests, why not coral reefs? This is the motivation for the Coral Restoration Foundation. It grows corals that are offspring of wild patches, which survived mass die-offs, in nurseries. This means they are selected for being more robust to environmental conditions, notably bleaching and adjustment to warmer waters. They are transferred to coral on the reef where they need several months to firmly establish and stabilise. Using this procedure, the foundation has already planted more than 18 000 corals in the Molasses Reef in the Florida Keys, with a survival rate of 80 per cent after 2 years. The goal is to thus contribute to genetic diversity, making the reef maximally resilient to climate change. Another example concerns coral reefs in the American Samoa Islands, east of Australia, which are adapted to very high water temperatures, namely up to 35°C. These are transferred to the Great Barrier Reef, to provoke adequate epigenetic changes in the populations to make them more resilient to climate change. In addition, recent studies examine artificial selection and breeding of corals in laboratories that simulate climate change.[1]

back to local up to global climates. This self-regulation or homeostasis of the biosphere, resisting extreme climate change, can be seen to extrapolate his original Gaia hypothesis (Section 4.5). He formally elaborated this in the Daisyworld model (Watson and Lovelock, 1983), which can be regarded as as an evolutionary model describing the dynamics of a population of black and white daisies. When the Sun emits more energy, selection causes the white daisies, reflecting light, to increase their share in the population while the black ones, absorbing light, become rarer. As a result, the surface temperature of Daisyworld remains fairly stable. In an extension, Lovelock (1992) added trophic levels (herbivores and carnivores) and grey daisies to the model, showing that this contributed to homeostasis. Lenton and Lovelock (2001) respond to various criticisms on the model.

The previous suggestions are not meant to imply that traditional climate policy approaches, in particular within environmental and ecological economics, completely neglect evolutionary thinking. Among others, diversity is part of the core policy insight of environmental economics as price incentives are seen to minimise total abatement costs by equalising marginal abatement costs among heterogeneous polluters. Admittedly, though, this represents a static and exogenous view of heterogeneity (diversity), which considerably differs from how it is conceptualised in evolutionary approaches. In addition, environmental economics now has well-integrated insights from innovation studies (e.g. Jaffe et al., 2005), which to some extent use evolutionary methods, and fully supports the idea that we need a policy package consisting of carbon pricing and subsidising promising but still expensive technologies with favourable environmental

---

[1] http://grist.org/science/coral-reefs-are-in-trouble-meet-the-people-trying-to-rebuild-them/#.

features. A main conclusion to be drawn is that an evolutionary perspective on climate policy, rather than suggesting a radical change in policy design, mostly refines and complements current insights about particular instruments. Some of the foregoing issues are elaborated in subsequent sections.

14.3   Evolution of Low-carbon Technologies

This section examines the potential contribution of, and limits to, technological evolution in combating global warming. Although there is widespread optimism about the contribution of technological innovation in solving environmental and climate problems, notions such as 'green technology' and 'environmental innovation' are not unproblematic. In many cases, it is virtually impossible to know for sure that a new technology will ultimately create a net environmental benefit over its lifetime, after all indirect and unintended effects have materialised. This is not just because of general uncertainty about the future, but owing to unintended and avoidable indirect effects of environmental innovations being ubiquitous. An effective climate policy package is needed to assure that new technologies are low carbon net of counterproductive systemic effects. It will be argued that this requires a balance of environmental regulation and technology policy instruments.

Environmental innovations differ in several ways from other more general innovations. They are usually factor-saving rather than quality improving: climate innovations save on use of the factor energy, or reduce the carbon intensity, of products and services. But this does not mean improved product functionality for users. In addition, solving environmental problems tends to go along with diffuse public benefits and concentrated private costs. Hence, the gainers from a certain environmental innovation are not so easily organised while the losers can effectively pose resistance. A crucial environmental innovation for a low-carbon economy is renewable energy. The desired transition to it is logical from an environmental but not from an economic angle, unlike energy transitions in the past: food energy $\rightarrow$ animal power and firewood $\rightarrow$ coal $\rightarrow$ oil, gas, electricity. All these historical transitions were economically rational owing to cost advantages, quality improvement or new functions. But renewable energy does not offer any of these. Its large-scale diffusion will, therefore, not occur quickly without regulation of energy markets to give fair competitive advantage over fossil fuel technologies.

In other words, moving to a low-carbon world requires more than funding new technologies. Mowery et al. (2010, p. 1022) think that the popular analogies of the Manhattan and Apollo projects are inaccurate or even misleading for design of public R&D programmes on climate innovation. Unlike these ambitious political projects, where public R&D was guiding and the government was effectively the sole consumer, climate innovation involves a multitude of decentralised solutions associated with different technologies, involving many types of energy conservation to various types of renewable energy, a combination of private and public R&D projects and diffusion of technologies in markets with many producers and consumers. The dynamics of this system are evolutionary in nature, unsuitable for planning but in need of repeated interaction between cumulative innovation and selection, guided by steady public policies. Perrow (2010) argues that the climate strategy of carbon capture and storage (CCS) has features that may endorse an Apollo/Manhattan type of research approach. This strategy is, however, seen by many as, at best, a transitionary rather than definite solution to control excess CO_2 in the atmosphere.

The foregoing arguments mean that climate innovation represents a tough problem. It will not be simply solved through big money funding by the Musks, Gates and Zuckerbergs of this world. Incidentally, Elon Musk seems to share this view, having stated that instead of subsidies to renewable energy companies, governments should price carbon as it would considerably reduce the transition time to 100 per cent renewable energy.[2] As we will see later, however, innovation and diffusion subsidies serve a complementary role to carbon pricing, meaning that innovation policy is no substitute for environmental regulation. Furthermore, innovation is not the only or major climate solution in the coming two or three decades – the time frame we have to reduce GHG emissions to close to zero to avoid dangerous climate change. During that period, behavioural change of consumers, notably in rich countries, and application of existing low-carbon technologies in production are just as important as new technologies. It is true, though, that affordable breakthrough technologies, such as for renewable energy, will play a crucial role 30 years from now. The period until then should be used to maximally stimulate these technologies, through regulation with carbon pricing and innovation subsidies. This approach is elaborated and further motivated below.

Innovation and Knowledge Are a Part of the Solution

When reading the literature on innovation and environment, one easily gets the impression that innovations are always good and signify a relatively easy and cheap solution to pressing environmental problems, including climate change. Given the limited time frame for stopping further global warming, the question is not whether we can find innovative solutions sometime in the future but whether we can realise these with sufficient rapidity. There is no doubt that, with enough time available, we will be able to have a full and spontaneous transition to renewable energy in one or two centuries. History shows that in specific countries, the full realisation of energy transitions, such as to coal, to oil and to electrification, took some 200, 85 and 65 years, respectively (Huberty and Zysman, 2010). This should come as no surprise given that technological evolution needs sufficient generations of technologies and inventors to achieve its triumphs. In addition, going from invention to innovation requires a considerable monetary investment in R&D over an extended period of time. This may be truer of environmental than commercially oriented innovations for various reasons: environmental innovations often require progress at the level of fundamental knowledge, such as on new materials with a combination of desired features; they do not easily find a

---

[2] See http://www.theguardian.com/technology/2015/dec/03/elon-musk-calls-carbon-price-halve-transition-time-clean-energy.

large market as their benefits are social rather than private; and they often involve processes rather than products. For example, in the user phase 'green electricity' is not different from 'grey electricity' produced with fossil fuels. Both make electric equipment run in the same manner. Therefore, green electricity has no special attraction for users from the viewpoint of functionality, commercial value or provision of status.

Virtually all economic studies, irrespective of the particular model and assumptions used, show that the major part of the reduction of greenhouse gas emissions in the coming decades is unlikely to be realised through technological innovations only (Jorgenson, 2009; Hedenus et al., 2006; Fischer and Newell, 2008; Goulder, 2004; Nordhaus, 2002; IEA, 2010). Instead, it requires environmental regulation that alters relevant consumer and producer behaviour. This in turn will cause changes in the production input structure and in the sector and demand structure of national economies. Undoubtedly, this will involve the adoption of already existing technologies, possibly with minor improvements, but it will not depend on extended innovation patterns and major technological breakthroughs. One study, using energy-related patents as a proxy for energy innovation, finds that a third of the overall response of energy use to prices was the outcome of induced innovation and two-thirds was of factor substitution (Popp, 2001). So, innovation and innovation policy cannot be used as an excuse for avoiding regulatory policies. Only the latter guarantee a rapid redirection of consumer and producer decisions towards considerably lower GHG emissions.

At a more general level, one can ask how important knowledge is as a production factor. The World Bank (2006, Table 1.1) calculated the following composition of capital (for 2000) based on monetary valuation of different types of capital: for low-income countries total capital consists of 58 per cent intangible (human capital and informal institutions), 16 per cent produced and 26 per cent natural capital; for the OECD this division is 81, 17 and 2 per cent, respectively. This suggests a minor role for natural resources such as energy. However, Ayres and Warr (2005) present empirical evidence that energy is a much more crucial factor of historical economic growth. They argue that although the cost of energy at a national level for most countries has been in the range of 7 to 10 per cent over the past decades, this does not reflect well the relative importance of energy. Using an alternative thermodynamic modelling approach based on accounting for useful physical work, called 'exergy services',[3] they accurately project historical economic growth in the United States on the basis of empirical estimates of improvements in exergy conversion for physical work. Stern (2011) reviewed a related literature and found that substitution of low- for high-quality energy carriers, such as a moving from coal to electricity, has been critical to economic growth in the past. He added that future gains of this type are more limited.

Some have questioned the environmental and social benefits of ever more knowledge. For instance, Ehrlich et al. (1999) point out that there is much production of

---

[3] Whereas energy is conserved, 'exergy' is the thermodynamic term used to denote non-conserved, available or useful energy that is able to perform mechanical, chemical or thermal work, given relevant environmental boundary conditions. Exergy is a less abstract basis for discussing thermodynamics than the commonly used notion of entropy (Ayres and Massini, 2004).

irrelevant knowledge or even disinformation – fake or false information, notably through commercial advertisement. A great deal of potential knowledge is irreversibly foregone owing to both biological and cultural diversity loss. Moreover, new knowledge and technological innovations have many unwanted effects that were unforeseen and unintended. Certain policies may counter these negative effects, such as public production of knowledge, subsidising the dissemination of new knowledge through good education, and reducing incentives for the production and distribution of disinformation, such as through regulating advertisement. From an evolutionary cultural angle this comes down to creating selection factors that steer knowledge development and diffusion in a societally beneficial direction.

Renewable Energy Technologies Have a Low EROEI

Next to energy conservation, a main strategy to combat CO_2 emissions is renewable energy. A relevant question is how efficient renewable energy technologies can become in terms of their use of labour, capital as well as energy itself. One composite measure proposed to capture this is the Energy Return on (Energy) Investment or ERO(E)I (Murphy and Hall, 2010). It is defined as the energy output obtained in a process divided by the energy input or cost needed to extract, produce, deliver and use the output. If renewable energy needs, indirectly, many energy and labour inputs, and technological advance cannot reduce these inputs considerably, then the energy surplus or net energy available, calculated as the output minus the input, will remain small. An economy running entirely on renewable energy would then have to devote a disproportionally large share of labour and capital to provide intermediate services to the renewable energy sector.

For developed countries, Giampietro et al. (1997) calculate that complete dependence on biofuels would imply that – depending on the country and climate zone – 20 to 40 per cent of the working force would be employed in the energy sector, and an even larger portion if polluting emissions and soil erosion problems associated with large-scale biofuel production were to be countered. In general, a serious investment in biofuels will lead to a high demand for land and water, leading to negative effects on other uses of these resources, notably food production. Massive adoption of biofuels would further reverse the historical trend of less labour use per unit of energy obtained.

More generally, one can rank combinations of resources and technologies (and climate zones) in terms of their EROEI performance (Murphy and Hall, 2010). Historical oil exploration was characterised by easily reachable fields with good concentrations and thus scores the best with an EROEI of over 100, followed by coal, which has had a constant EROEI of about 80 since the 1950s. More recently discovered oil and gas fields perform less well. Note that the global average EROEI of oil went down from more than 100 to less than 40 today, despite technological progress in exploration, drilling and transport technologies. The reason is that effort has shifted to less easily accessible reserves and reserves with lower concentrations. Next, the EROEI of nuclear fission varies between 5 and 15, that of hydropower is above 100 (but has limited application), of wind 18, of solar photovoltaics about 7, of solar (flat plate) thermal collectors 1.9, of solar concentrated heat power 1.6, and of biodiesel 1.3. None of these numbers is absolute as there is uncertainty about estimates, variation between regions and climate zones, as well as uncertain aggregation to a global scale. But the wide range indicates that a transition to renewable energy would, unlike historical transitions to coal and oil, mean moving to a less concentrated, and from that perspective, less attractive source of energy. This underpins the challenge of such a transition. It is, moreover, not to be expected that the EROEI of renewables will easily and rapidly increase in the near future without significant public funding of basic R&D and subsidy-like incentives for private R&D.

Variation in and Motivations of Eco-innovators

As we have seen in Chapter 10, technological change is driven by a process of technological evolution, involving not only distinct designs but also a variety of actors who, in the present context, may be called eco-innovators and eco-adopters. For the formulation of environmental, innovation and diffusion policies, it is important to know which incentives apply to these distinct actors. The heterogeneity of firms and industries implies a diversity of determinants of, and motivations for, innovation and adoption of climate-relevant innovations. Moreover, environmental innovations may well cover a broader set of motivations than regular innovations as they are inspired not only by market opportunities but also by health, environmental and ethical concerns. Factors positively affecting environmental innovation show a great variation (Porter and van der Linde, 1991; Lanjouw and Mody, 1996; Wagner, 2007; Horbach, 2008; Popp et al., 2010):

- Current environmental regulation, with a stronger effect on less innovative firms.
- Anticipation of, or fear for, future regulation.
- A high rate of expenditure on pollution abatement.
- Existence of niche markets, reflecting environmentally conscious or wealthy consumers.
- A first-mover advantage for firms in a national or international context.
- Public image, aimed at increasing market share or profits, reinforced by social or commercial pressure exerted by environmental organisations, customers and suppliers.
- Firm awareness of environmental pressure and of opportunities to relieve it, possibly improved by having an environmental manager and management system (notably with ISO/EMAS certification). This is associated with a high education of employees.
- Voluntary agreements or covenants involving firms, environmental organisations, unions and/or a regulator.

These factors do not apply in the same manner to all firms and industries. Governments should thus employ distinct, well-tailored policies for stimulating environmental innovations.

14.4   Ultimate Effects of Climate Innovation and Policies

Many studies suggest that energy conservation through behavioural change and use of more energy-efficient technologies can cause rebound, in terms of not only energy use but also CO_2 emissions – given that most energy generation is still based on fossil fuels. Rebound denotes unintended, indirect effects of energy conservation which cause net energy savings to be lower than the initial or direct ones (Sorrell, 2009). An extreme case is when rebound equals 100 per cent or more, known as backfire or the Jevons' paradox. In this case, net energy savings will be zero or negative, respectively. The problem of rebound is especially significant to developing countries. The reason is that their final demand is still far from saturated, so that both demand and production sectors can greatly expand, something that is stimulated by the application of more energy-efficient technologies (van den Bergh, 2011).

To understand rebound, consider the following four fundamental reasons for it. First, improvements in efficiency relieve limits of various types – time, money, scarce resources, production factors and space. Some economic activities can subsequently grow, giving rise to an increase in associated energy consumption, partly undoing any initial energy savings. Second, if general purpose technologies, such as engines, batteries or computers, become more energy efficient this can stimulate their diffusion to new activities, sectors and applications. For example, the number of appliances used in household activities – many of which were previously limited to use in industrial activities – is steadily rising. Third, solutions to environmental problems generally take the form of increasing complexity in human technologies, organisations, institutions and public regulation (Tainter, 2011). This in turn often triggers new demand for energy. For example, the number of environmental scientists has sharply increased in the last decades and so has the number of international conferences to which they all travel, the majority by plane. Fourth, the behaviour of individuals, households and firms is characterised by bounded rationality, taking the form of myopia, habits, imitation or cognitive limitations and associated decision biases. As a result, the phenomenon of rebound is generally not recognised, causing people to easily overestimate their energy savings. This is strengthened by the fact that rebound comprises many channels and mechanisms, among others (van den Bergh, 2011):

- More intensive use of equipment (e.g. cars) after its effective energy cost has fallen due to higher energy efficiency (e.g. a more fuel-efficient engine).
- Purchase of larger, heavier equipment or with more energy-using functions in response to higher energy efficiency of such equipment.
- Spending financial savings due to energy conservation on other energy-consuming goods or services. Figure 14.1 illustrates its relevance for various countries and fuels, where variation is due to differences in domestic prices and energy or carbon intensity of GDP.
- New energy-efficient devices embody energy in their production.
- Energy savings by many agents can reduce energy prices and in turn the prices of energy-intensive goods, which will stimulate demand for these.
- More energy-efficient technologies diffuse to new users and applications.

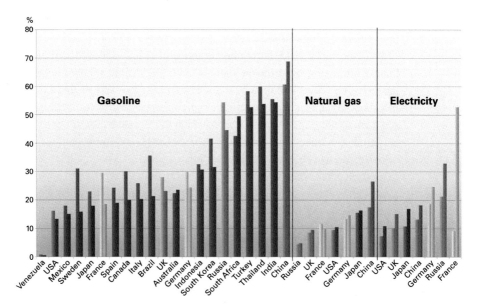

Figure 14.1 Re-spending rebound, for three energy carriers.
Notes: National averages, 2009; rebound for energy in first columns, and for carbon dioxide emissions in second columns; colours denote countries (standard blue means data was only available for gasoline).
Source: Antal and van den Bergh (2014, Figure 1, p. 586).
(*A black and white version of this figure will appear in some formats. For the colour version, please see the plate section.*)

The debate on energy conservation and rebound is old. In the 1970s and 1980s, it was motivated by the need to move away from expensive energy resources resulting from oil embargos by Organization of the Petroleum Exporting Countries (OPEC) and international conflicts in the Middle East. Currently, it is driven by the search for solutions to climate change and peak oil. It seems little has been learned, because the belief is once again widespread – among economists, environmentalists, energy experts and politicians alike – that improving energy efficiency and stimulating voluntary action offer easy solutions. Nevertheless, as Brookes (1990) argues, countering global warming will unfortunately not be that easy, as rebound will make these strategies less effective than many think.

To limit rebound of energy, or better of CO_2 emissions, conservation efforts and related technological innovation activities need best to be induced by rebound policy. Notably higher energy prices will ensure that prices of all goods and services appropriately signal fossil energy use, and thus CO_2 emissions, to consumers. In addition, overall rebound could be limited by a hard ceiling on relevant environmental (CO_2) emissions. The combination of a price mechanism and a quantity limit suggests the instrument of tradable emission permits, also known as cap and trade, as an effective 'rebound policy'. Indeed, if rebound occurs, then the ceiling for emissions will push up the permit price until rebound effects are sufficiently discouraged (van den Bergh, 2011). Consumption is then effectively redirected to low-carbon opportunities.

The December 2015 Paris climate agreement has achieved national pledges rather than coordination of stringent national policies. Afraid of harming their competitive positions, countries will then be inclined to realise the pledges through weak regulatory measures. Examples include stimulating companies and consumers to voluntarily conserve energy through information provision or covenants, imposing moderate standards on a limited number of technologies and products, and implementing low-carbon taxes or setting a low cap in an emissions trading system. All such polices will cause rebound of energy conservation and associated CO_2 emissions. To this, we should add that aviation has not been assigned any emissions reduction goals under the Paris agreement, so consumers will not be discouraged from spending any monetary savings associated with energy conservation on energy-intensive plane trips. Furthermore, as developing countries and emerging economies generally show high rebound of energy conservation, given that the Paris agreement includes their pledges, one should expect substantial rebound.

In addition, one should reckon with two other ultimate, systemic effects of the Paris agreement – carbon leakage and a green paradox (van den Bergh, 2012). The first denotes a relocation of polluting industries or shifts in international trade owing to policy differences between countries, notably rich and poor ones. This leads to a dislocation of emissions to countries with weaker policies. The green paradox means that fossil fuel markets will respond with lower prices to the threat of renewable energy when no minimum fossil fuel market prices are guaranteed through some form of carbon pricing (see Section 14.5).

A carbon price would discourage consumers from spending their saved money on carbon-intensive alternatives. The reason is that pricing accomplishes complete emissions control, since all goods and services would be priced in proportion to the emissions caused during their life cycle. Other instruments, such as technical emission standards on a subset of all technologies and products, cannot achieve this fine-tuned control and are likely to result in higher rebound. The reason is that they do not prevent substitution of the regulated technology by other, less regulated ones, which also use energy. An example is the European Union's policy since 2009 to phase out traditional light bulbs. Unintendedly, it created widespread use of unregulated light emitting diodes (LEDs), in turn reducing initial energy savings (van den Bergh, 2015). Witness the Eiffel Tower in Paris, which on many occasions since the end of 2009 was completely showered in LED light.

To go beyond the Paris agreement requires a majority of citizens to support stringent climate policies. Unfortunately, many are still disinterested or even sceptical about climate change. Ecklund et al. (2016) establish a modest association of it with scepticism about evolution (Section 2.4), which is predicted by political conservatism, a lack of confidence in science, low education and religiousness. While Evangelical Protestants show relatively strong scepticism towards both evolution and climate change, religion generally is found to have a stronger connection with scepticism about evolution.

14.5   Policy Mix for Innovation and Diffusion

Addressing Systemic Challenges

Policies to move to a low-carbon economy, and stimulate appropriate environmental innovations, include both environmental regulation and technology-specific actions. They serve different purposes and are therefore largely complementary. Unfortunately, many policy reports and political statements suggest, incorrectly, that public innovation policy can act as a substitute for environmental regulation. This view can be found on both ends of the political spectrum. For example, the Bush administration transmitted this line of thought as one reason for not wanting to ratify the Kyoto Protocol. But many supporters of ambitious renewable energy subsidy programmes, as in Germany, often express a similar stance. However, the purpose of environmental regulation is to reduce negative environmental externalities, while innovation policy intends to address positive knowledge externalities due to R&D spill-overs. This insight is now recognised in both economics and innovation studies (Jaffe et al., 2005), but still needs to diffuse to policy-makers and politicians.

Implementing only one of these policies is likely to lead to unintended and undesirable outcomes. Consider the example of renewable energy subsidies as an instrument of technology policy. In the absence of adequate environmental regulation, such as a carbon tax, such subsidies will magnify the supply of energy, notably electricity. Owing to a depressing impact on fossil fuel prices this can even stimulate a more rapid extraction of fuel resources in turn leading to an increase in GHG emissions, which aggravates global warming. This has been called a 'green paradox' (Sinn, 2008), which can be seen as a special case of rebound. The problem here is that policies such as renewable energy subsidies, but also strategies such as energy conservation through more energy-efficient technologies, reduce the demand for fossil fuels while forgetting about potential responses from the supply side of fossil fuel markets. Lower demand will reduce prices of fossil fuels, through both market competition and technology effects, which will stimulate their pace of extraction and supply as their reserves in stocks underground will become less profitable over time, further depressing prices and, in turn, encouraging demand. In other words, climate solutions cannot be seen separately from oil markets. Incidentally, a similar story can be told for investing large scale in nuclear fission or fusion without carbon pricing. They also will trigger fossil fuel market responses that can frustrate the intention of such investment strategies by increasing fossil fuel use and GHG emissions. To avoid a green paradox, one has to simultaneously make fossil fuels more expensive through environmental regulation (van der Ploeg, 2011).

On the other hand, one should also expect unintended and undesirable outcomes if only environmental regulation is implemented, that is, without complementary innovation policy. Market selection pressure will then favour currently cost-effective technologies which may lead to an early lock-in of these at the disadvantage of technological alternatives that are more desirable from a long-run angle. The problem here is that cost-reducing potential of not yet cost-effective technologies is neglected.

Adding innovation or technology-specific policies will help to influence the speed and direction of technological change, and keep promising, but still expensive, options open.

More policy pressure is needed to unlock the current fossil-fuel-based energy and transport systems, unless the price incentive of regulation is sufficiently strong, but political support for high carbon prices is still fictional. A kind of separate 'unlocking policy' can compensate for increasing returns to scale, on demand and supply sides of markets, which are fundamental to lock-in. A range of specific policies to escape lock-in has been proposed in the past (Section 10.6): price subsidies to consumers to stimulate adoption, and thus diffusion, of a new technology; setting a clear future goal (e.g. California's ZEV or Zero Emissions Vehicles programme), creates semi-protected niches (e.g. with public subsidies), and public procurement (governments buying green products or investing in eco-efficient public buildings to help to create new markets). Additional, potential 'unlocking strategies' that have received less attention include (van den Bergh et al. 2007): restricting advertising (e.g. of automobiles with oversized combustion engines), stimulating status-seeking (e.g. of electric cars or solar roof top panels), and discouraging innovation in undesirable dominant technologies (e.g. in the internal combustion engine by automobile manufacturers). Furthermore, hybrid solutions are empirically suggested to possibly offer a way out of a lock-in situation (e.g. hybrid cars as a transition technology from fossil fuel combustion to fully electric engines). This is supported by evolutionary model studies showing that a transition to a low-carbon economy can occur along a path starting with hybrid, recombinant innovations (Zeppini and van den Bergh 2011).

Finally, it worth noting that, from an evolutionary angle, mainstream economics' emphasis on carbon pricing as an essential instrument of climate policy is not wrong. In fact, a complex evolutionary system invites an incentive structure characterised by decentralisation, such as through price regulation, rather than hierarchical control. One might say that, whereas mainstream economics recognises the need for pricing solutions because of the presence of many agents or targets, an evolutionary viewpoint will do so stressing that the targets are moving in unpredictable directions. In other words, an evolutionary approach confirms or reinforces certain existing policy insights, indicating their robustness.

Accommodating Basic and Applied Research as Well as Diffusion

A fundamental and unresolved issue regarding policy targeted at environmental innovation is when to focus on fundamental innovations, mainly in the public sector, but to some extent in private firms, and when to focus on applied R&D and diffusion. Machlup (1962) suggests that fundamental or basic research is aimed at discoveries, whereas applied research deals with inventions. Some applied inventions require that certain enabling basic discoveries have already been made.

Mowery et al. (2010) conclude from research in agricultural, biomedical and information technology that a focus on major, radical innovations and deviant technologies is needed. They suggest that considerable technological diversity is required.

An important trade-off is between the benefits of increasing returns to scale and the benefits of keeping options open and allowing for spill-overs and recombinant innovation (Section 10.7). To ameliorate the conflict between these, one might encourage an international coordination of diversity where different countries specialise more or less in technological R&D that matches well their history (university research and industries), expertise (labour market, education), and geographical and climate conditions (wind, sun, land). Then diversity can be combined with a critical scale in R&D for each technology or technological area (van den Bergh et al., 2006).

To illustrate what this can mean for evaluation of concrete technologies, consider nuclear fusion, suggested as a possible solution to climate change. Its main shortcoming from an evolutionary angle is that it is characterised by a lack of diversity, which does not promise a smooth and rapid innovation until successful performance. Indeed, because of the huge cost, only one large-scale experiment is currently undertaken worldwide, namely the ITER reactor constructed at Cadarache in southern France. Its cost is covered by a consortium of countries: the EU, China, India, Japan, Russia, South Korea and the United States The project has been characterised by overshoot of time deadlines and budget. Originally the cost was estimated to be €5 billion, but already some €14 billion have been spent, while the current overall cost estimate has risen to €20 billion. ITER started in 1988 and although originally planned to finish in 2018, the current deadline is predicted to be 2035. And this is just an experiment to test if net energy can be generated by a large-scale fusion reactor. An initial productive electricity plant would be the next step, likely to be ready at the earliest sometime after 2050.

Clearly, lack of diversity is a serious weakness of the innovation trajectory of nuclear fusion. Moore's law of exponential growth in performance or density applying to solar photovoltaics or computer chips, means their cost per unit of output has dropped. No such phenomenon has been observed for nuclear fusion: its performance has improved, but its complexity and cost have increased in the process. This relates to technologies such as wind turbines and solar photovoltaic electricity being characterised by much diversity in types, applications and countries, which has also gone along with considerable variety in institutional and policy conditions, all providing a better guarantee for successful and rapid innovation, spill-overs and learning. It would, thus, be advisable to quickly shift the remaining ITER 'fusion funding' to innovation research on such alternatives. The theme of technological diversity merits more research, where attention should be given to the different dimensions, benefits and costs of diversity (Skea, 2010; Stirling, 2010; van den Heuvel and van den Bergh, 2009; Shemelev and van den Bergh, 2016). Some suggest that given the limited time offered by rapid climate change we can afford the luxury of 'wasting scarce funds' on supporting diversity. Indeed, we have no alternative as major innovations to allow for a transition after 2040/2050 will require a certain degree of diversity of options and experiments. Until that time, we have to reduce CO_2 emissions mainly by environmental regulation, notably CO_2 pricing, of firm and consumer behaviour, involving application of current technologies. Other aspects of the timing of climate policy are discussed in Box 14.2.

A general framework to study allocation of scarce funds for governmental subsidies to market and R&D support is offered by Koseoglu et al. (2013). It is used to compare

Box 14.2 Appropriate timing of climate policy

A critical feature of climate policy is its timing. Evolutionary economic approaches have been employed to identify phases in which distinct policies or 'time strategies' are needed (Nill and Kemp, 2009; Sartorius and Zundel, 2005). This involves rather abstract notions such as 'utilisation of time windows of opportunities' and temporal phases such as R&D, applied R&D, demonstration and experiments, and market diffusion (or deployment). Each phase is confronted with specific combinations of uncertainty, learning and market opportunities, with associated cost reduction potential and lock-in tendencies. For wind energy, this approach suggests that upscaling is relevant, while solar photovoltaics requires a continued focus on fundamental R&D.

Another aspect of time strategies is the duration of policy pressure. Popp et al. (2010) argue that environmental regulation will be permanently needed for the adoption and diffusion of pollution control technologies as there is little private benefit to pollution control. On the other hand, in the case of energy-using technologies consumers and firms benefit from adopting more efficient alternatives as this reduces their energy bill. Regulation is still needed, however, since not all benefits are private, i.e. external costs are involved. An additional issue is when to shift the attention from R&D to large-scale diffusion and, associated with this, moving from fundamental, mainly public to applied, private R&D. Because of the interaction and feedback between the various stages of RDD&D (the 'nonlinear model' of Research, Development, Demonstration and Deployment) one should stimulate market development before a technology is cost-effective.

Recently, Acemoglu et al. (2012) have undercut the dominant view in climate economics (e.g. Nordhaus, 2002), which suggests that policy intervention should be modest and implemented gradually. Instead, they suggest that any delay in policy intervention will turn out to be unnecessarily costly. The sooner and the stronger the policy response, the shorter the harsh transition phase. Stavins (2009) presents a similar reasoning, arguing that we should not wait to implement stringent price-based climate policy because tomorrow's world will be wealthier and therefore better able to carry the policy costs. We should progressively introduce policy and make it more stringent over time to allow producers, consumers, investors and innovators to adapt. This will support the quickest transition to a sustainable economy and will reduce the long-run policy costs. Delaying necessary price corrections will mean either having to accept high damage costs of climate change, or having to rapidly introduce much more stringent policies at a future date, which will force the economy to make a more radical change with associated excessive adaptation costs and risks of economic crises.

the strategies of Germany, with more focus on market instruments, of California, with a greater emphasis of R&D, and of China, characterised by central planning. Demand subsidies aimed at stimulating diffusion may have adverse effects, through energy rebound and oil market responses (see the previous section). Germany's heavy focus on demand subsidies for renewable energy in past decades has also seen criticism for being a non-effective way to lower prices of renewable electricity structurally and to mean a redistribution of money from poor to rich households (Frondel et al., 2008, 2010; Andor et al., 2015). Other evidence goes in the same direction. Based on a study for Denmark, Germany and the UK, Klaassen et al. (2005) provide evidence that public R&D contributed considerably more than market-related learning to cost reduction of wind turbine technologies, namely on average 5.4 per cent for learning by doing and 12.6 per cent for R&D based 'learning-by-searching'. One would expect an even larger difference given that the learning curve of solar photovoltaics shows a higher progress rate than that of wind, which has been largely attributed to R&D investments (van der Zwaan and Rabl, 2003). Another study indeed confirms that for organic photovoltaics demand support reduces costs less than R&D support (Nemet and Baker, 2009). Ultimately, though, this issue is not about choosing between the two subsidy types but balancing them. According to the 'nonlinear invention–innovation–diffusion model', innovation without diffusion does not work well because of feedback from diffusion to innovation. Indeed, renewable energy innovations such as solar photovoltaic panels need to be integrated in buildings, requiring expertise of architects, intermediaries and constructors, as well as complementary innovations in construction.

In discussing clean energy, Jeffrey Funk (2013, Chapter 10) suggests that we are currently focusing too much on demand-based or diffusion subsidies, which have encouraged the implementation of existing technology that is quickly outdated, rather than the development of new types of solar cells, batteries and wind turbines with better performance. Funk calls the current demand-oriented policies, such as in Germany's 'Energiewende', ineffective and 'feel-good' approaches. Stimulating demand through subsidies when the price gap between fossil and solar photovoltaic electricity is so large is also an expensive way to reduce CO_2 emissions. In order to realise the ambitious goal of a close-to-zero carbon economy in three decades we need dramatic reductions in the cost of clean energy.

Demand-based subsidies have primarily encouraged the implementation of single or polysilicon-based solar cells, while it is now clear that newer forms of so-called thin-film solar cells have much lower costs. Funk makes similar arguments for wind turbines and electric vehicles. For wind turbines, we need materials with higher strength-to-weight ratios in order to build large wind turbines in areas with high wind speeds. For electric vehicles that can eventually compete with traditional vehicles, we need to find materials for batteries that have sufficiently high energy and power densities. Funk argues that all these fundamental innovations are more critical to making a large-scale transition to clean energy in a few decades than in the short term raising the cumulative production (and consumption) of solar cells, wind turbines and electric vehicles in factories. Witness in this respect the role played by long-term research on semiconductors in altering the cost and performance of computers. This was accomplished by steady and rapid improvements in the performance of integrated circuits, as captured by Moore's law, namely exponential growth of chip capacity of about 40 per cent per annum. This, in turn, reduced the cost of computers, which increased demand for them. No demand-side policy was needed. Of course, one cannot hope to achieve such a favourable progress rate for renewable energy technologies, certainly not without public support of basic research. One has to bear in mind that environmental innovations do not offer quality improvements or new functions that attract users, as do computers.

15 Evolutionary Policy and Politics

15.1 Towards an Evolutionary Policy Theory

Evolutionary thinking in the social sciences can offer refreshing views on public policy, resulting in what one might call 'Darwinian policy'. We have already illustrated this in previous chapters for environmental and climate challenges. Here, I will try to offer a more general and foundational account by integrating elements from public policy theories and evolutionary social sciences. An early informative discussion of evolutionary public policy was provided in the last chapter of Nelson and Winter (1982). More recently, a biologist and an economist made a strong joint plea for using insights from evolutionary sciences in the formulation of public policies (Wilson and Gowdy, 2013). Other studies, mentioned later on, tend to adopt a more narrow perspective, usually focusing on technology policy.

Ulrich Witt (2003) considers evolutionary policy from three angles: the political economy of actual policy-making, the analysis of effective policy instruments in view of given aims and the choice of such aims. He argues that both positive and normative dynamic aspects of evolutionary approaches improve upon the comparatively static interpretations of policy as in public choice theory, economic policy theory and social philosophy. According to Moreau (2004), maintaining the flexibility and diversity of the economic system is the main purpose of evolutionary policy. This is achieved through four specific aims: prevent lock-in of undesirable trajectories, stimulate incremental innovations in a world governed by radical uncertainty, create incentives for explorative activities, and establish institutional structures in response to what has been learned. In addition, he notes Hayek's 'Impossibility Theorem' as a shortcoming of evolutionary policy:

The state has no privileged knowledge, therefore there is no reason that it should be better informed than private agents on the optimal nature of market outcomes. The state cannot be considered as more efficient than the market process for solving the coordination problem among individual decisions with the information disseminated among market participants. (Moreau, 2004, p. 872)

For a critical view on this, including the conditions under which intervention can still be successful, see Wegner (1997).

The current chapter aims to discuss principles of positive and normative approaches to evolutionary policy. This involves addressing, among others, welfare and happiness, distributional issues, consumption, environmental problems, technology policy and responses to economic crises. We will pay attention to policy criteria and goals from an evolutionary slant, and critically examine the related issue of 'evolutionary progress'. Two different but complementary angles will be adopted, namely policy design informed by evolutionary thinking (Section 15.3) and policy-making as an evolutionary process (Section 15.4). The former treats policy as an independent variable that can be chosen by policy-makers or politicians. Deliberately chosen policies can shape the evolution of the economy, in which case normative criteria are needed for the selection of these policies. The second angle means considering policy as dependent on many factors in society and politics, notably as a part of the evolutionary dynamics of the economy, society and political system. This, too, yields policy insights in so far as a better understanding of the mechanisms of political change can inform the process of institutional-constitutional reforms. It is important to take both dimensions of evolutionary policy into account because without understanding how policy changes – the positive dimension – normatively inspired proposals are likely to be unrealistic, ineffective or even misguided. It is true that the two sub-approaches have developed quite independently, so that the treatment here can be seen as a modest effort to offer a more complete evolutionary account of public policy.

15.2   Policy Criteria and Evolutionary Progress

First, one needs to understand which policy criteria derive from adopting an evolutionary viewpoint. Witt (2003), Moreau (2004) and Schubert (2012) are earlier efforts in this vein. They offer interesting ideas but do not satisfactorily resolve the problem of designing a normative framework. A complementary set of thoughts is presented here. A starting point is the relationship between evolution and progress, about which there is much misunderstanding. According to Maynard Smith and Szathmáry (1995, p. 4), 'the notion of progress has a bad name among evolutionary biologists'. They suggest that the history of life is best depicted as a branching tree rather than progress on a linear scale.[1] The term 'evolutionism' is often used to denote the false idea that evolution is consistent, if not synonymous, with progress. Notwithstanding, many still regard evolution as generally leading to progress, witness expressions like 'climbing Mount Improbable' (Dawkins, 1996), 'lifting in design space through ever larger cranes' (Dennett, 1995) or 'evolution uses its own products to climb upon' (Blackmore, 1999). These authors will be the first to admit, though, that evolutionary progress is not inevitable or steady. Box 15.1 lists eight reasons.

The insights in Box 15.1 are endorsed by recent evolutionary theory in the social sciences. It has, therefore, little in common with old Social Darwinism, a pejorative term to denote an ideology which placed individuals and races on an evolutionary

---

[1] Nevertheless, certain authors, notably Conway Morris (2003) and Dawkins (1997), maintain that the broad strokes of evolution are almost inevitable, including the emergence of intelligent organisms such as humans. See also the discussion of 'convergence' in Section 4.4.

Box 15.1 Reasons for evolution not necessarily coinciding with progress

Both genetic and non-genetic evolution do not necessarily give rise to progress, nor to a continuous increase of complexity – for various reasons:

1. Selection is a local search process, which leads, at best, to a local optimum (not 'survival of the fittest', as first coined by Herbert Spencer, but survival of the fitter or relatively fit).

2. Adaptations are often compromises between different objectives, being stimulated by a multitude of selection forces. This suggests that evolution is better regarded as multicriteria evaluation than as single-objective optimisation.

3. 'Historical constraints' cause strong resistance against changes, whether in ideas, institutions, behaviours or technologies. In biology, this is referred to as 'bauplan limits', 'development constraint' (Gould and Lewontin, 1979) or phylogenetic inertia (E.O. Wilson, 1975, p. 20). In economics, it is treated under the headings of increasing returns to scale, path dependence and lock-in (Arthur, 1989). In organisation theory, it has been called 'structural inertia' and 'imprinting' (Hannan and Freeman, 1989, p. 70 and p. 205, respectively).

4. Not all evolution is adaptive microevolution: e.g. randomness, genetic (molecular) or cultural drift and coincidental founder effects play an important role. In addition, macroevolution creates boundary conditions for adaptation and may destroy outcomes of microevolution, in a way as to set back time ('initialise').

5. Selection can only 'capture' variations that exist. The process of creation of variation is limited and partly random.

6. Agents explore only a minor range or subset of the opportunity space, which is reflected by the notion of bounded rationality. This takes the form of myopia, habits, satisficing and various decision biases triggered by cognitive limits.

7. Coevolution and niche construction mean that the environment is not constant and exogenous to the individual species' evolution, but influenced by it (Section 4.6). Coevolution means adaptation to an adaptive environment. All straightforward notions of static or dynamic optimisation within fixed constraints are lost then.

8. Sen (1993) notes that evolution as improving species or their fitness does not imply improving the welfare or quality of life of each individual. The reason is that a higher fitness and survival do not necessarily imply a happier or more pleasant life. In fact, evolutionary models often show that inequality easily arises in evolutionary systems, suggesting that part of the population is relatively unhappy (e.g. Epstein and Axtell, 1996). One aim of evolutionary policy could thus be to counter tendencies of such inequality arising due to evolutionary forces.

'ladder of progress' and proposed that the strongest or fittest should survive, thus justifying inequality of opportunities and outcomes (Hodgson, 2004). Modern evolutionary social sciences recognise that evolutionary outcomes are not necessarily beneficial in every particular context and may require compensatory action encouraged by public policies (Witt, 2003). In particular, certain types of cultural or technological innovation may conflict with social goals such as equity or environmental sustainability. For more discussion of this issue, see Sections 1.4 and 7.2.

Schubert (2012) critically reviews the literature which suggests that policy should focus on the encouragement of innovation and learning. He notes that novelty and evolutionary change have an unclear normative dimension. Mainstream evaluation of welfare is difficult, if not impossible, as it assumes fixed preferences, which is inconsistent with learning and novelty. He proposes an alternative 'procedural welfare approach' in which biological evolutionary insights about how humans respond to novelties play an important role. Their responses are diverse and involve both positive and saturation elements. Individuals should be able to learn new preferences, requiring public policies that do not constrain their opportunity sets. This is consistent with 'libertarian paternalism' informed by behavioural economics, as proposed by Thaler and Sunstein (2008). It is fair to note another obstacle to fixed, given preferences, namely derived from the (debated) modularity view of the brain (Section 5.4): if there are distinct decision modules, or actually multiple agents, in the brain, then it is not always evident what is good for the individual – and by derivation for society. This does not deny, of course, that for many consumption goods or services, it is rather clear what is bad for an individual or a society, or what adds insignificantly to either survival or long-term happiness.

Many evolutionary economic studies have constrained themselves to a positive analysis (Verspagen, 2009). An agnostic position is maintained then with respect to the criteria and valuation schemes according to which these alternative states can be compared. This is unsatisfactory, however, as one implicitly always adheres to a normative position if one performs an analysis to enlighten or inform policy. A normative theory of social welfare and public decision-making that takes account of evolutionary dynamics is not well developed (Witt, 2003). This could take evolutionary theories of ethics as a starting point. For example, Kenneth Binmore's (2005) evolutionary approach to morality argues that moral rules are a cultural system that evolved historically with the advent of modern humans and their generally egalitarian societies. Another possible theory, discussed in Kallis and Norgaard (2010), is the philosophy of pragmatism (Dewey, 1943; Rorty, 1991). It is based on a notion of thought (and knowledge) as evolutionary processes of diversity-generating innovation and selection.

Without making a choice for the moment, we can see a number of partly consistent practical entry points to resolve this issue and advance normative evolutionary policy analysis. First, although evolutionary processes – apart from artificial selection of animals and plants – do not assume any goal or target, normative elements can be added by policy-makers. In this sense evolutionary policy analysis allows for the same distinction between positive and normative as neoclassical economic policy theory:

the normative goals can be the same, namely maximum social welfare. Even though one may adopt the same normative principle, the positive approaches evidently differ between evolutionary and neoclassical economics, namely evolutionary processes versus market equilibrium processes, respectively. As a result, the outcomes of the confrontation of normative goal and positive model may, and are likely to, differ between the two approaches.

A second consideration is that evolutionary theory poses a fundamental challenge to the neoclassical economic approach to policy evaluation, notably to the notions of social welfare and consumer sovereignty. Evolutionary thinking suggests that the famous Arrow (1950, 1951) Paradox, stating that aggregation of individual rational preferences into internally consistent ('rational') social preferences is impossible, even understates the problem of constructing a social welfare function. The reason is that preferences reflect bounded rationality and are evolving (Norton et al., 1998; Witt, 2003). As a result, there cannot be any constant static social welfare function. The practical implication might be to entirely avoid the use of social welfare notions.

A third relevant consideration is that a definition of evolutionary progress can provide explicit normative guidance. Even though such progress is not always occurring in evolutionary systems, one can still aim to use policies to bring about evolutionary progress. When choosing this route, a problem is that there are different interpretations or criteria of evolutionary progress in use (e.g. Gould, 1988; Gowdy, 1994, Chapter 8; Potts, 2000): increasing diversity and increasing complexity, emergence of new levels due to evolution (see item 'Major transitions' in Section 3.6), extended division of labour, new ways of transmitting information, population growth, being better adapted to the environment, greater control of the environment and increasing efficiency of energy capture and transformation (Schneider and Kay, 1994). Taking several of these into account while assuming or testing for a degree of consistency would be a way to go forward. When different indicators sketch different pictures one can add weights or make a clear choice or trade-off.

A fourth consideration is that evolution introduces unpredictability to some degree because of system complexity, novelty and surprises (but see the discussion of convergent and recurrent evolution in Section 4.4). This suggests two possible directions for choosing a policy goal: (1) a focus on effectiveness: better to inefficiently reach a target than efficiently miss it; or (2) a focus on risk-aversion, such as by adopting a precautionary principle. From an evolutionary angle, cost-effectiveness would be broadened to include the cost, over time, of maintaining sufficient diversity to improve selection outcomes, avoid lock-in and keep options open. Real options theory already goes some way in this direction, compared with traditional cost–benefit analysis, by accounting for uncertainty and irreversibility and building in flexibility through keeping options open or allowing a switch between development paths (Dixit and Pindyck, 1994), but it neglects endogenous diversity, which results from selection and innovation (van den Bergh, 2008). Among the costs over time of policy are those associated with the distributional consequences of creative destruction, which is inherent to evolutionary change. The fact that the gainers of innovations do not compensate losers makes the whole process of innovation possible and is an important reason for the dynamism of the capitalist economy – as well as its disasters, fluctuating inequalities and human tragedies, such as are common during periods of economic crisis.

We do not aim to settle the difficult issue of evolutionary policy criteria here. The first consideration is attractive, but hampered by the fourth one. Nevertheless, a clear distinction between normative and positive dimensions of an evolutionary approach is desirable. In addition, normative goals could be multifold, including ones independent of evolution, such as measures of subjective well-being or happiness, and ones more closely related to evolutionary processes, such as suggested under the third consideration.

15.3   Evolutionary Studies of Policy Design

Notwithstanding the lack of a general evolutionary policy framework, many studies adopting an evolutionary approach have provided valuable policy insights. This holds especially for policies associated with technological innovation. In fact, many analyses of innovation and technological change make implicit or explicit use of evolutionary notions such as selection, diversity and mutation, resulting in what is known as neo-Schumpeterian, or evolutionary, economics (Dosi et al.. 1988; Witt, 1993; Metcalfe, 1998). The policy lessons of this literature are summarised by Cantner and Pyka (2001), Fagerberg (2003) and Dosi et al. (2006).

Fagerberg (2003) emphasises that evolutionary economics provides a distinct perspective on policy from neoclassical economics built around the notion of market failures and public goods. The evolutionary approach, in his view, downplays this public-good aspect and associated policy insights. In addition, he stresses that, from an evolutionary angle, the notion of optimal growth is dubious. He sees the main role of an evolutionary approach to inform policies that can increase the economic system's ability to generate new variety. This implies the need for policies such as subsidising R&D into new types of activities and policies encouraging uptake of innovations to overcome inertia or the tendency to lock-in. Azar and Sandén (2011) argue that we have to make political choices regarding future technologies in order to escape lock-in of undesirable technologies, instead of striving for technologically neutral policies. As a consequence, technology policy should really be a hot topic in politics, which it isn't.

An evolutionary policy view stresses bounded rationality in the invention, innovation and diffusion phases. It regards the evolution of technology as happening through local interactions and recombinant innovation. In line with this, Metcalfe (1998) argues that asymmetry of knowledge and information due to heterogeneity of firms (potential innovators) is at the core of the evolutionary economic approach to technological innovation. Given a lack of knowledge or uncertainty about the distribution and dynamics of this information, policy-makers cannot optimise technical change from a social welfare angle. Instead, the two main strategies open to them are moulding the selection environment to direct selection processes rather than select the optimal technology or technological path themselves, and to create favourable conditions for innovation, using notions such as 'innovation systems' and 'niche management'.

Of course, such strategies conflict with Azar and Sandén's (2011) above-mentioned advice. Innovation system analyses are capable of providing information about system failures and connecting points for effective policy within a national or regional innovation system. They consider a broad set of innovating actors, such as governments, research institutes, private companies or non-governmental organisations, as well as their networks (Edquist, 2005). A recent branch of this literature emphasises the various 'functions' of innovation systems (Carlsson et al., 2002; Hekkert et al., 2007). This approach has recently seen much application to problems associated with environmental innovations and sustainability transitions (Jacobsson and Bergek, 2011).

Recently, much attention has been given in transition studies to niche management as an evolutionary strategy to foster innovations (Kemp et al., 1998; Schot and Geels, 2007). It involves processes of social experimentation and learning, the creation and exploitation of technological niches, increasing variety by niche branching, and assuring stabilisation through implementing rules and accepted practices. Policy can try to guide this process by stimulating diversity, extending the network of actors and stakeholders involved and by destabilising the dominant sociotechnical regime. From an evolutionary angle, stimulation of niches is a specific case of a general condition that stimulates major innovations, namely isolation. It involves regional or local experimentation, such as decentralised, small-scale renewable energy production. A diversity of such initiatives can lead to a proliferation and healthy competition of local institutions and supplier–user interactions. In fact, this characterised the initial stage of electrification at the end of the nineteenth century.

Other issues have already been discussed before in this book. One is optimal technological diversity (Sections 10.7 and 13.6), which requires an optimal balance between increasing returns to scale (i.e. short-term benefits of little diversity) and recombinant innovation (i.e. long-term benefits of much diversity). The government can stimulate a wide range of technologies in terms of variety, disparity and balance. This includes the development of modular technologies and information exchange to increase the likelihood of cross-fertilisation and modular innovations. Policy might further stimulate radical innovations by raising the disparity between technological options, notably by directing public R&D at 'deviant technologies' and by funding risky R&D. Another issue is lock-in of undesirable technologies, institutions and behaviours (Arthur, 1989). From an evolutionary angle, the path-dependent trajectory to a state of lock-in involves systematic loss of diversity. Destabilising the dominant regime and enhancing diversity are key to escaping a state of lock-in. Policy suggestions were discussed in Sections 10.6 and 14.5.

Consumer regulation is another area where evolutionary policy thinking could serve a useful role. This is relevant for public health (e.g. contagious diseases, smoking, drinking), safety (insurance, traffic) and environmental risks. Evolutionary approaches combine various aspects of bounded rationality and other-regarding preferences with populations of interacting and heterogeneous agents. This considers social interactions such as imitation, reputation and status-seeking as well as habitual behaviour, decisions under uncertainty and myopia. Evolutionary thought can help here in assessing effective as well as efficient policies or even suggest new instruments next to standards, taxes and subsidies, such as specific tools of information provision. An example is provided by Nannen and van den Bergh (2010), who translate an evolutionary view of market and nonmarket interactions between individuals or firms into policy recommendations such as 'prizes' and 'advertisement'. These have the advantage that they can attract attention to exemplary, role-model behaviour and activities, which stimulate imitation through social network interactions. This may also be helpful as a strategy to escape a lock-in situation. Note that prizes and advertising as instruments of public policy have received no serious attention in conventional policy modelling as the aggregate nature of the description there neglects population and network structures. Penn (2003) argues that, from an evolutionary angle, individual incentives can be an effective way of environmental policy. This seems to be consistent with the main policy message of environmental economics (Baumol and Oates, 1988), but Penn adds the provision that these incentives go beyond narrow economic, monetary interests. They should, for instance, also take social interactions, such as reputation effects, into account. Examples are features that determine status and for which individuals show rivalry behaviour, such as positional goods. But social phenomena can be linked to a wider set of other-regarding preferences, namely reciprocal fairness, inequity aversion, pure altruism, altruistic punishment and spite or envy (Fehr and Fischbacher, 2002).

Consumers cannot be seen in isolation from producers, investors and innovators as they interact in markets. There is a small literature of models describing coevolving consumers and producers (Janssen and Jager, 2002; Windrum and Birchenhall, 2005; Safarzynska and van den Bergh, 2010b; Witt, 2011). One policy-relevant insight is that status-seeking can be exploited by policy-makers to stimulate the emergence of product niches that can counter lock-in or dominant regimes. The reason is that the niche product, by definition, may have status features as it is scarce. This can then be exploited to make the market grow to a critical level, beyond which increasing returns to scale – imitation, information externalities, economies of scale and learning in production – can stimulate continued growth of the market beyond the niche stage. In addition, distinct sub-groups, such as poor and rich, with different social strategies or peer groups, may require specific policy strategies. In other words, the diversity of consumers may translate into a diversity of focused policies.

In general, lessons about public policy can be based on evolutionary insights into brain modules. This relates to an idea from behavioural economics (Thaler and Sunstein, 2008), namely to 'nudge' individual behaviour. This entails changing relevant institutional and other environmental conditions, reframing information that supports decisions, or showing behaviour by others. This can trigger other types of brain modules and decision processes which then allow one to move away from bad automatisms or habits, such as eating sugar- or fat-rich food or buying goods one does not really need and will hardly use. Moreover, by stimulating awareness of the environmental or social impact of one's decisions and of the comparison with consumption choices by other people, nudges can foster individual and social learning that, in turn, will guide individually or socially desirable changes in beliefs and preferences.

Finally, evolutionary policy presents a distinct way of responding to financial-economic crises. Many traditional macroeconomists neglect diversity that is central to evolutionary thinking, and instead tend to think in terms of aggregate variables and representative agents. Characteristics of crises include social phenomena such as mass behaviour, loss of trust and distributional inequity: some families lose a lot, while others even gain. To fully understand the dynamics of macroeconomics one has to pay more attention to the underlying heterogeneity. In this respect, lessons can be drawn from behavioural finance (Section 8.4). An evolutionary view on macroeconomics and growth (Mulder et al., 2001) allows one to better understand the real microeconomic mechanisms underlying the many positive feedbacks within and between financial markets, housing markets and labour markets. Such feedbacks are responsible for both periods of boom and bust. An evolutionary angle can identify solutions that reduce the volatility of markets instead of trying just to move in a crisis period back to the boom part of the cycle. The reason is that it recognises the importance of imitation and diffusion as critical forces. These need to be balanced by counteractive policies, such as information provision, rewarding examples of good behaviour, and punishment of risky or opportunistic behaviour – notably by investors and banks. A disaggregated, population view might help to identify policies tailored to the particularities of different social and income groups – instead of resorting to general austerity policies as applied in most European countries – and assess the systemic consequences of these policies. An evolutionary perspective further suggests policy diversity, learning and adaptation, by allowing for distinct policy experiments in different regions or countries (Kerber, 2005, 2011).

15.4   Evolutionary Political Dynamics and Policy Change

The previous section considered studies that treat policy as a choice variable, independent of the evolutionary social-economic system. However, policy-making itself can also be considered an evolutionary social process, with political actors, and possibly also scientist advisors, acting as agents of influence. Indeed, evolutionary theory can offer insights about the dynamics of policy change, resulting in an evolutionary type of public choice theory. This comes down to emphasising the role of time, the sequence of events and, more generally, the historical pattern of policy change (Mahoney, 2000). It involves consideration of the bounded rationality of agents in the policy process (Witt, 2003), for political opinion formation and knowledge creation (Wohlgemuth, 2002), for diversity of policy ideas and their selection in a multidimensional environment (Kerr, 2002), and for path dependence of policy processes and policy change (Pierson, 2000). By clarifying these various issues, an evolutionary study of policy-making and policy dynamics can inform policy design and yield recommendations for institutional-constitutional reforms that change the rules of the policy-making process.

Owing to bounded rationality, political actors are heavily selective in their learning and the way they filter information. Perceptions and attention are affected by preferences in a recursive manner that binds together similar-minded political actors into social networks and communities with highly intensive internal communication but relatively insulated from external influences (Witt, 2003). Discourses about policy ideas, i.e. shared understandings of explanations of problems or specific solutions, result in simplification of complex information and construction of mental maps, which join individual actors into 'discourse communities' (Hajer, 1995). Only a limited number of policy problems can garner sufficient public and policy attention. This produces an uneven temporality of policy evolution as the early agenda-setting stage becomes critical for subsequent outcomes and is heavily attended by political actors. In line with this, political scientists have invoked an analogy to the idea proposed by Eldredge and Gould (1972), namely 'punctuated equilibrium' – controversial in evolutionary biology (Section 4.4) – which denotes that policy is for extended periods in a state of stasis interjected by phases of rapid or radical policy change ('punctuations') where the pace and extent of transformation is significantly accelerated (Somit and Peterson, 1992; Baumgartner and Jones, 1993). This has been illustrated for environmental policy (Baumgartner, 2006) and tobacco regulation (Givel, 2006). An alternative model is that policy activity exhibits an S-shaped curve of phases of intense activity followed by stability as indicated by public budget spending, media coverage or parliament activity.

At the heart of theories of policy evolution is the notion that political and economic environments impose selective pressures upon alternative political strategies and that political actors, through processes of trial, error and learning, continuously adapt their strategies to this selection environment (Kerr, 2002; Ward, 2003; Lewis and Steinmo, 2010). Of course, since this environment is changing and also evolutionary, consisting of populations of producers, consumers and other stakeholders, the notion of coevolution may be relevant as well. At any given moment, there are competing ideas and solutions to policy problems, which are replicated through imitation and survive through competition in political arenas (Kingdon, 1995; John, 1999). Political actors have multidimensional motives and interests which assume meaning by being expressed in specific ideas. Actors struggle in the policy arenas to see their ideas chosen and implemented. The constraining selection environment is multidimensional – influenced by media, elections, public opinion and vested interests with power – as well as context specific and changing over time (John, 1999; Kerr, 2002). Inspired by Kuhn, Hay (1999) introduces the concept of 'policy paradigms', perceived as high-level packages of affiliated ideas which selectively constrain policy change to stay within limited paths. During economic crises, one can often witness a clash of paradigms, such as 'austerians' and 'new Keynesians' demanding distinct policies.

As the notion of 'policy paradigm' already hints, increasing returns and path dependence, central concepts in evolutionary theories, are prominent in political environments. The high start-up and exit costs that characterise collective action and the development of institutions, as well as the positive feedbacks involved in the social interpretation of information and the attraction of a critical mass of communicators, further create path dependence (Pierson, 2000; Witt, 2003). Once an institution develops, it empowers certain groups at the expense of others, which in turn alters the institution to favour the interests of the first (Mahoney, 2000). Understanding policy evolution as a path-dependent process shifts the emphasis in analysis to earlier historical events, which by design or accident may have locked-in institutional development to a certain path (Mahoney, 2000). Not all studies mentioned are explicitly evolutionary, but they nevertheless breathe an evolutionary spirit. In addition, it should be said that the term 'evolution' is also frequently misused in political science. Dowding (2000) and John (2003) offer critical accounts of the literature, accentuating that one should clearly distinguish between learning as an intentional process and evolution as an unintentional process.

To predict or guide political and policy dynamics, it is pertinent to understand how people interact in the public sphere. For this reason, insight into the origins of morality and motivations of people in their different roles as private consumer versus public citizen is useful. Box 15.2 provides some insights from a gene–culture evolutionary angle.

Box 15.2 Gene–culture coevolution of early public persona and politics

The emergence of culture in early groups with intelligent humans created the need for primitive politics. This involved interactions between individuals showing their public persona, generally associated with other-regarding values and preferences. Depending on the context, individuals would thus shift between private and public personas, or alternate between a focus on self-regarding goals such as material wealth, status and leisure and on other-regarding goals such as altruism, fairness, equity and more generally concern for the welfare of others. Some individuals were, at times, guided by 'universalist' or deontological goals, notably character virtues such as honesty, zest for work, loyalty and religiousness. This all suggests that there exist multiple personas within an individual human brain, something which is consistent with (debated) modular brain theory (Section 5.4).

Evolution of the public persona's morality occurred in small bands where politics were closely connected to private life. As a result, in their public persona role, individuals could expect to consider concrete consequences of their public actions, such as focused on mediating conflicts or achieving cooperation, for their own personal situation. This is quite different from modern-day politics in societies with large numbers of citizens, where the private and public spheres of politically active individuals are strongly separated. In such circumstances, other non-consequentialist motives drive people, such as universalist goals ('voting and participating in collect-ive action are good'). Alternatively, politicians or voters may fool themselves and think that their actions have significant consequences for themselves or others ('perceived consequentialism'). Again other people act as a public persona driven entirely by self-regarding goals, which Burnham et al. (2016, Table 8.1) call 'homo autisticus'. Or, people belonging to a particular social group act altruistically towards other members of it, but do not extend such altruism to people from other groups or even show outright hostility towards outsiders. Such 'parochialism' is an important mechanism driving group selection (Section 6.5). Indeed, intergroup competition and conflict over resources have likely been key selection forces behind public or political morality. Various group-internal processes have played a role as well, notably socialisation of children, punishment of free riders and public personas jointly creating a set of moral rules that improved the stability and functioning of their group.

Box 15.2 (*cont.*)

Group politics based on public personas is not found in nonhuman species, which raises the question how the public persona could evolve in the first place. One hypothesis emphasises the role of lethal hunting weapons. Their appearance not only transformed hunting, but arguably also undermined the social hierarchy in hominin groups as traditionally based on physical force and aptitude of dominant individuals. Any supremacy of leaders would be moderated by other members of the group through threat or actual use of lethal weapons. Hence, an egalitarian political system emerged that was characterised by consensual decision-making.[2] This was aided by already evolved human communication skills. In turn, the increased fitness of non-authoritarian leadership skills likely invited for the selection of improved cognition, linguistic abilities and social talents. This egalitarian political system was only overturned much later when (agri)cultural changes allowed for the accumulation of material wealth, which could be used to support social hierarchies involving authoritarian leaders with access to, or even a monopoly on, armed forces.

Based on Gintis et al. (2015) and Gintis (2016).

It argues that human morality, cognition and affection are the result of an interaction between genetic and cultural evolution. This means they have been reshaped throughout cultural settings in which humans have lived: from early hominin groups, through hunter-gatherer bands and early agricultural civilisations to larger and increasingly complex societies.

Darwinian evolutionary thinking has been used in various ways to illuminate the political dynamics of the 2016 presidential elections in the United States that ended in the victory of Donald Trump. One argument starts from the idea that political conservatism puts the interests of the family first and considers the family as a 'child-rearing team' (Newson and Richerson, 2016). In line with this, marriage is seen as a partnership to raise children. In contrast, liberals tend to consider marriage in the first place as a means for adults to get happy, and understand that limiting the number of children is a logical consequence of this goal. People living in urban areas, a relatively recent development from an evolutionary time perspective, are likely to be more liberal as they have been exposed to a broader range of cultural habits and political opinions. Rural isolation, especially when combined with orthodox religion, hampers liberal thinking and keeps humans in a conservative mode. The latter dominated human cultures until modernisation allowed Westerners – initially Europeans – to break away from it. The large majority of the rural population in the United States has not fully caught up with modernisation and voted for Trump, arguably triggered by his conservative messages of 'family values' and 'own first'. An alternative evolutionary

---

[2] This echoes the finding of Boehm (2001) that in some primate groups egalitarianism replaced hierarchy due to physically relatively weak individuals joining forces in order to resist or even dominate strong ones.

explanation for Trump's success comes from 'political evolutionary psychology' (van Rueden, 2016). It states that as a product of our history of living in hunter-gatherer groups with frequent internal and external conflicts, many people, notably men – who in large majority voted for Trump – intuitively trust authoritarian, physically dominant leaders to protect their group against existential threats. In fact, 70 per cent of historical US presidential campaigns between Democratic and Republican candidates were won by the taller one (Sun, 2016). Van Vugt and Ahuja (2011) regard our leadership preferences as a mismatch and suggest this explains why we so frequently see leaders failing. Indeed, in modern business and political contexts, reasoning, processing information and communication skills matter more than physical dominance. Then again, it should not be discounted that Trump was able to effectively communicate big claims and grave accusations to millions of followers through his Twitter account, what is more, without any quality or fact checking. Some call this spreading fake news or 'false memes', pointing to a cultural evolutionary explanation of Trump's success (Section 7.6). Dennett refers to Twitter and Google as 'meme-spreading machines'.[3]

15.5   Differences with Conventional Policy Approaches

Here we compare an evolutionary angle on policy with two alternative approaches, namely neoclassical economic policy theory and public choice theory. The latter have dominated academic views on public policy for a long time. Neoclassical economics weighs heavily in debates about economic, health and environmental policy, while public choice has influenced institutional reforms such as privatisation. They both build upon a theoretical foundation of rationally optimising agents. We will argue that evolutionary thinking offers a distinctive, and in several ways improved, approach to public policy and its change.

Evolutionary Versus Neoclassical-Economic Views on Policy

The traditional economic theory of public policy is the result of applying neoclassical welfare theory, which comes down to connecting a competitive equilibrium and a social welfare optimum, or less ambitiously, a Pareto efficient situation.[4] The positive part of this approach, the competitive equilibrium supported by representative or average, rational consumers and producers, is contested by evolutionary economics. Instead, it describes a population of diverse agents, bounded rationality, social (nonmarket) interaction, resulting in persistent disequilibrium, multiple attractors and path dependence. This generally conveys a less optimistic view on policy as effectiveness is not

---

[3] https://www.theguardian.com/science/2017/feb/12/daniel-dennett-politics-bacteria-bach-back-dawkins-trump-interview.

[4] Pareto efficiency, or Pareto optimality, denotes an allocation of scarce productive or consumptive resources that is not amenable for improvement, in the sense that the welfare of at least one individual increases and that of no other individual decreases.

guaranteed. In addition, as discussed in Section 15.2, the normative part of neoclassical economics reflected in social welfare and efficiency may be contested from an evolutionary perspective. Disputing both the normative and positive basis of neoclassical economic policy analysis means that the correspondence between market equilibrium and a social welfare optimum is lost. In turn, the two fundamental theorems of welfare economics no longer hold.[5] As a result, neither central planning nor market solutions can be guaranteed to achieve socially optimal outcomes.

We use here as an illustration the economic theory of environmental policy, which offers an approach to evaluate, compare and design instruments of regulatory policy. It has been approached from neoclassical and evolutionary economics' angles (Section 13.2), allowing for a comparison. The first focuses attention on the efficiency (either welfare maximisation or cost minimisation) of policy instruments, such as environmental levies and standards (Baumol and Oates, 1988). It assumes rational producers and consumers and clearing of markets. The approach neglects selection other than through market competition, as well as irreversibility due to path dependence. Neoclassical economic policy analysis addresses certain aspects of diversity, namely heterogeneity of abatement options and related costs of polluters, and shows that if firms are cost-minimising total, national or sectoral abatement costs will be minimised if regulation employs price-based instruments, such as taxes, charges or levies. These will make sure that marginal abatement costs are equal among polluters, and that those with costly abatement options will reduce pollution less than those with cheaper options. This approach assumes only static heterogeneity and does not account for dynamic diversity or selection effects of resource exploitation (Munro, 1997). Neither does mainstream resource economics give much attention to the evolution of norms, for example, through local interaction of resource users. Ostrom (1990) has considered the endogenous character of norms in a common-pool resource context using evolutionary thinking and concludes that centralised, top-down policy may erode the evolved basis for such norms. Finally, economic growth theory does not take the heterogeneity of firms and technologies into account, as is the case in evolutionary (differential) growth theory. In addition, it does not allow for much policy detail in terms of innovations, selection environment and lock-in. Finally, cost–benefit analysis is often used to evaluate public investment projects. It requires the application of monetary valuation techniques, both revealed and stated preference techniques, to assess nonmarket effects such as environmental damage and health impacts in monetary terms. Valuation depends on assumptions such as rational behaviour of agents and market equilibrium. Evolutionary insights, instead, are consistent with individuals being boundedly rational as well as other-regarding. As a result, they can incorporate phenomena such as 'protest bidders', 'warm glow', 'cold prickle', and 'altruism' (Andreoni, 1995).

---

[5] The first theorem states that in the absence of market failures, any (Walrasian) competitive market equilibrium gives rise to a socially optimal or (Pareto) efficient allocation of resources. According to the second theorem, any efficient allocation can be realised by a certain competitive equilibrium. In the case of market failures, regulation by a government needs to assure that market outcome and social optimum coincide.

Evolutionary Policy Versus Public Choice Theory

There are several political science theories that explain policy change, most notably Kingdon's (1995) multiple streams model, Sabatier's (1988) advocacy coalition approach, various policy network approaches such as corporatism and iron triangles, political economy models and adaptive policy theories (Kerr, 2002). An exhaustive comparison of evolutionary with other theories of policy is beyond the purposes of this section. To demonstrate the usefulness of evolutionary theory in explaining political change and deriving policy-relevant results thereof, we compare it with public choice theory. It regards the government or public sector as composed of various groups of agents or stakeholders who act in a self-interested manner. At the highest level, this includes voters, civil servants and politicians.

This approach is more descriptive or positive than, and thus complementary to, neoclassical economic policy theory, which combines normative and positive elements. While a dominant part of public choice theory transfers the utilitarian axioms of the rational, maximising agent of the neoclassical economic model to political science and institutional change, one can find also approaches, such as applications of game theory, that invoke lighter assumptions. Public choice theory seeks to compare the costs and benefits of distinct policies for stakeholders. This involves predicting outcomes on the basis of the power that particular actors have, accounting for how this power is wielded in the political process (Hahn, 1990). Competitive market pressures in the public and economic sphere are expected to increase efficiency since they reduce rent-seeking wastage by politicians who are lobbied by special interests and by bureaucrats who have to satisfy politicians' desires to keep their positions (Buchanan, 1986). As an implication for institutional design, the main role of the government is the establishment or redefinition of property rights. In line with such insights, the privatisation of public management and the introduction of competition in administration have been advocated on the basis of public choice theories.

The positive theory of evolutionary policy change (Section 15.4) shares with public choice theory the assumption of actors in the policy arena intentionally trying to promote their interests. But the two theories differ in important ways. The attention given by evolutionary theory to bounded rationality and social preferences contrasts with the assumptions of public choice theory of agents being rational and selfish. Evolutionary theory highlights the importance of learning, imitation, other-regarding behaviour (altruism) and communication between actors, factors often underplayed in public choice models (Witt, 2003). Political ideas and discourses assume a role beyond representing well-articulated interests. They are the glue that articulates such imperfectly formed interests. This view allows more optimism about political mediation, compromises and collaboration than the zero-sum competitive view of public choice. It fits well with insights from the adaptive management literature, which stresses the critical role of information brokers and boundary organisations that bring together disparate sources of information. This helps to create new discourses around conflicting problems, which can result in new political alliances and solutions (Feldman et al., 2006).

Some evolutionary theorists share the Hayekian anti-interventionist view (see Section 8.4): policy arrangements and institutional norms have evolved over time under the pressure of various selection mechanisms, which a central coordinating agency should, as much as possible, avoid to change – given its limitations in information processing that hamper recognising the full value of selected and adapted institutions (Hayek, 1967; Pelikan and Wegner, 2003). Public choice theory added fuel to Hayekian arguments by breaking down the state into particular interests aimed at directing and capturing the benefits of new policies to their advantage. Hayek's and Buchanan's theories underpinned 1980s political programmes that sought to reduce state interventionism and privatise its competencies. Contrary to this view, evolutionary thinking argues that path dependence provides a counter force to the functionalist mode of explanation that underlies Hayek's evolutionary view as well as public choice theory, which consider existing political arrangements as an inevitable equilibrium outcome. Increasing returns to adoption or scale causing path dependence, however, mean that present institutions were not inevitable or are not necessarily superior in some functional sense (Kerr, 2002). Political path dependence means that the capacity to change public policy is more limited than conventionally thought (Pierson, 2000).

More generally, there is a tension between positive understandings of evolution at a higher level as a complex outcome of the struggle between different interests, and normative evolutionary policy recommendations that assume benign policy-makers with the capacity to control the policy system. One way out of this tension is to see evolutionary policy as being about designing institutional procedures rather than choosing instruments, which create a multifaceted selection environment and provide incentives for experimentation, learning and adaptive policy-making (Witt, 2003). The recent literature on transition management and policy (Section 13.5) reflects this insight, and comes up with proposals to reconceptualise policy problems and reform policy practice.

A second tension revolves around what Witt (1992, 2003) calls the 'endogenisation of the theorist-analyst' within evolutionary accounts of policy change. Public choice theory is somewhat contradictory since its own recommendations, such as outsourcing and privatisation, are to be implemented by the bureaucrats and politicians themselves. This means bureaucrats are seen as self-interested except when it comes to applying the policy proposals of public choice theory. An evolutionary approach resolves this tension by perceiving the theorist-analyst as a participant in the policy process, who generates information and seeks to attract attention (Witt, 2003), i.e. a partisan in the struggle of ideas for survival. A political process that facilitates a plurality of contributions enhances scientific contestability and is more likely to lead to better outcomes. In this sense, policy design recommendations from an evolutionary angle are not infallible, but rather tentative ideas to experiment with. Table 15.1 summarises main differences between the policy theories.

15.6 Towards Evolutionary Socioeconomic Policy

We have developed the idea of how public policy is shaped from the angle of evolutionary thinking. This involves a framework composed of the notions of diversity, populations of agents, selection, inheritance, innovation, coevolution, path dependence

Table 15.1 A comparison of evolutionary and other policy theories

Feature	Neoclassical Economics	Public Choice Theory	Evolutionary Theory
Agents and behaviour	Representative agent Rational – utility or profit maximising	Multiple stakeholders Rational choice Rent-seeking	Population of diverse and interacting agents Bounded rationality Searching, learning
Policy criteria/goals	Optimal allocation Efficiency Cost-effectiveness	Efficiency Individual freedom of choice	Agnostic, or Diversity Evolutionary potential Precautionary principle
Emphasis in policy instruments	Price instruments such as taxes and tradable permits	Property rights	Information diffusion Maintaining or stimulating diversity
Examples of real-world policy proposals	Full cost pricing Cost–benefit evaluation of policies Pricing of externalities	Privatisation Identify stakeholder conflicts Control rent-seeking	Protected niches for new technologies Stimulating deviant technologies Taxing conspicuous consumption Creating a level playing field to avoid or escape lock-in
Policy analyst	External to policy	External to policy	Generator of ideas/diversity Participant in struggle of policy ideas

and lock-in. We argued that this framework offers useful policy advice and distinctive insights about the impact of policy as well as the dynamics of policy change. Policy advice from such a standpoint emphasises, among others, innovation and learning, diversity management, stimulating recombinant innovation, exploiting status-seeking behaviour of consumers to stimulate escape from lock-in and protected niches to nurture new technological variants. An understanding of policy-making as an evolutionary process highlights the possibility of technological, institutional and even political lock-in. It, moreover, regards the policy arena as a field where different ideas struggle for survival while the power of vested interests determines their success. Evolutionary thinking about policy can thus be seen as providing a micro-level basis for observed aggregate political dynamics.

Policy evaluation becomes more complicated, though, if norms and values are believed to be evolving. Evolutionary theory often adopts – implicitly or explicitly – an approach close to that of a philosophical pragmatism that places a priori normative primacy on diversity, experimentation, learning and democratic deliberation. Still, in political practice, it is hard to defend experimentation or diversity at all costs, and some balance will always need to be struck with other goals or benefits, notably returns to scale. One could say that the recognition of socioeconomic evolution means a shift to a lower weight given to traditional short-term efficiency and effectiveness, and a larger weight given to experimentation and diversity. The latter have the ultimate aim of contributing to long-term efficiency, which is, however, surrounded by much uncertainty due to system complexity and novelty. This raises the question, largely ignored in traditional policy theories, about what is the desirable level of technological and institutional diversity at different points in time.

There are still many challenges ahead. A systematic assessment of linkages between existing dynamic theories of policy and politics with evolutionary thinking could generate insights enriching both theories. Some of the existing theories use concepts similar to those found in evolutionary theories, such as path dependency, but they are not evolutionary in a strict sense. Imposing more explicit evolutionary mechanisms could make lessons more precise. A specific issue is to use insights from group or multilevel selection theories (Chapter 6) to describe political conflicts between distinct stakeholders or interest groups, each composed of a variety of individuals with particular preferences. Another theme is the development of a normative framework in which to perform evolutionary policy analysis. One might investigate policy advice for different criteria or weights of these and see how robust insights are. A third issue is the design of policy instruments that specifically employ the population features of the system to be managed, such as awards, stimulation of networks and reduction of status-seeking. Only with evolutionary methods of analysis can the implications of such instruments be fully understood and analysed. Finally, it would be great if policy experiments and in-depth *ex post* studies of policy could be undertaken to add more empirical weight to the often abstract discussions on evolutionary policy.

This chapter has not said much about how evolutionary thinking can support or undercut political ideologies. Earlier parts of the book addressed problematic associations with Social Darwinism and eugenics (see Sections 1.4 and 7.2, respectively). Among others, it was argued there that any theory or technology developed by humans can be used with good or bad intentions or for worthy or lamentable goals. In addition, we saw that evolutionary reasoning and insights can support equally left- and right-wing political views. With regard to policy advice, it is important to strive for scientific rather than political correctness or, in Pinker's (1997, p. 48) words, to 'expose whatever ends are harmful and whatever ideas are false, and not confuse the two'.

16 Evolutionary Futures

16.1 Science and Technology

How will biotic and cultural spheres on this planet fare in coming decades, centuries, millennia and even far beyond that? Evolution, whether biological or cultural, is like a blind watchmaker without goals or foresight, while random external events may alter its course. Add to that the complexity of the 'anthroposphere' evolving at high speed, and foreseeing evolution's future best resembles gazing into a crystal ball. Rather than making systematic predictions, I will sketch here odds and dead ends, based on adopting biological, cultural, economic and technological evolutionary outlooks. Arguably, evolutionary thinking offers a better way to conjecture the future of complex systems than traditional aggregate and linear approaches. As Ayres (2000, p. 81) put it, 'Forecasters normally concentrate mainly on trends – the smoother the better. Yet this may be a case of looking for a lost coin under a streetlight.' Evolutionary thinking has proven able to cast light on major system transitions, as discussed for biology in Section 3.6 and for social-cultural settings in Sections 10.4 and 13.5. But explanation is insufficient for prediction. At best, it helps to anticipate patterns of change or, if relevant, deliberate on corrective steering.

One thing is rather certain, hence provides a good starting point. The tree of technology and knowledge (Chapter 10) is likely to grow larger in the future and probably will never cease spawning new branches. Recombination is the magic word here, as most, if not all, innovations can be understood as combining existing modules of knowledge or technologies. This will certainly hold true for a future with hyper-connected social networks, permitting an accelerating rate of information exchange and recombinant innovations, assisted by a dense web of highly educated individuals across the globe.

Some have suggested that technology can still undergo significant improvement, as many inventions contribute to the emergence or diffusion of others. Illustrative of this are the printing press (accurate replication of information), the light bulb (lengthening the time for reading, learning and research) and modern transport and communications technologies (reducing distance and barriers between minds). We can become still more effective in exploring the potential recombinant space of ideas and technologies. Advanced observation and experimental research technologies are helping us in many ways, aided by the internet and communication technologies. Despite a rapid amplification of information in past decades, such technologies have facilitated tracing what is already out there in terms of ideas or technological modules that might fuel recombinant innovations.

In addition, new technologies often have synergistic effects with existing technologies, making the latter more effective. For example, the computer has provided great help in uncovering the human genome, whereas knowledge about the latter has aided in the development of medical treatments and technologies, and nanotechnology has brought about cost reductions in manufacturing solar photovoltaic cells. Surely, not all combinations and positive feedbacks have been exploited yet, and we can expect technology to deliver more productivity improvements in the near and distant future.

Energy drives the modern economy and material welfare. Hence, a major challenge for human ingenuity and invention is making low-entropy energy available to a growing number of humans, as well as transforming this energy in the most efficient manner for production and consumption purposes. Depending on the right combination of policies, in coming decades solar photovoltaics will advance to such an extent that energy becomes abundant and cheap, as well as considerably less harmful to the environment. Other crucial issues that necessitate human creativity are health care, water supply and food production. Health care should involve a greater role for 'evolutionary medicine' (Section 1.3) as this helps in identifying ultimate causes of physical and mental problems and designing effective solutions. Better health will mean, though, a longer average lifespan, the downside of which is more pressure on the environment. Many regions in the world suffer from water scarcity. Technological solutions here include solar energy to desalinate seawater or growing genetically modified crops tolerant to drought conditions, because of improved water efficiency, root performance or collection of water by leaves. Although there still is some resistance to GM crops, this is likely to dwindle in the near future as the technology becomes widespread and turns out to be devoid of the negative consequences as once guaranteed by its opponents (van den Bergh and Holley, 2002; NASEM, 2016). Noteworthy in this respect is that on 29 June 2016 about a third of living Nobel laureates signed an open letter supporting 'precision agriculture'. They claimed that strong opposition by Greenpeace and other organisations was based on misrepresenting both risks and benefits of GM crops.[1]

The evolution of technology (and the economy) will depend a great deal on incentive structures to be developed over the next decade, not in the least on post-Paris climate agreement refinements. If strict and consistent climate and technology policies are implemented around the world, a smooth transition to a low-carbon economy is feasible. But this is far from a done deal. In its current design, the Paris agreement is likely to be ineffective, due to a number of hidden systemic effects – such as rebound of energy conservation, carbon leakage between countries, and fossil fuel market responses to protect oil and coal interests. Fixing these problems requires global policy coordination around a carbon price (Sections 14.2 and 14.5).

Fundamental technological breakthroughs are still possible in materials technology, nanotechnology, biotechnology and genetic engineering, mobility and perhaps even

---

[1] https://www.theguardian.com/environment/2016/jun/30/nobel-winners-slam-greenpeace-for-anti-gm-cam paign.

food. But let us not be too optimistic. History is full of mistaken technological promises. Recent ones that are yet to deliver include: biofuels, the fuel cell, the hydrogen economy, the circular or 'cradle-to-cradle' economy, nanotechnology, intelligent robots and nuclear fusion.

To properly understand the reach of technological innovation, one can draw insights from evolutionary theories of technical change, as inspired by Schumpeter, and Nelson and Winter. This involves impacts and phenomena at multiple levels in the economy: technological innovation at the firm level, diffusion at the market and sector level, economic growth and long waves at the macro level, and trade and technological spill-overs at the international level (Chapter 10). A deep understanding of these issues is imperative if we want to predict the long-term future of society.

A phenomenon that has been on the table for a while, but the ramifications of which are still not well grasped, is path dependence. It occurs when there are increasing returns to scale at demand or supply sides of markets. Examples are learning by doing, imitation, network externalities (such as telecommunication, computer software), informational increasing returns (if more adopted, then better known), and technological interrelatedness or complementarity (such as fuel infrastructure and vehicles with combustion engines). Increasing returns are critical to the outcome of competition among alternative technologies. A company or technology which coincidentally manages to obtain a large market share early on is likely to realise a lasting cost advantage. In turn, its market share can then continue to grow relatively quickly resulting in lock-in or market dominance. As shown in Section 10.5, this can be cast in evolutionary terms by conceptualising a population with an unequal distribution of features, whose dynamics are affected by imitation, network externalities and other increasing returns to adoption.

Representative-agent theory cannot capture this phenomenon well, which explains why the problem of path dependence was discovered relatively late and is still not a central focus of microeconomics. Economics conceptualises it as the existence of multiple equilibria. To decide which one is most likely reached, the potential paths towards them must be analysed, which typically requires evolutionary rather than equilibrium economics. The existence of multiple potential technology paths implies a role for public policy. With the right combination of carrots and sticks, it can make a socially and environmentally desirable path more likely.

16.2 Cultures and Religions

Cultural evolutionary theories stress the role of horizontal transmission mechanisms. These distinguish cultural from biological evolution (Chapter 7). Such theories can help to explain why one can expect less diversity and more uniformity of cultural habits worldwide. This is a likely consequence of multiple factors: increased international communication, English being the accepted global lingua franca; increased access to the internet; commercial advertising trying to create mass markets; and higher education and rising incomes in non-Western countries. This will go hand in hand with a global spread of life styles, and associated similarities in clothing, eating, entertainment, travel, music and television, as we have already seen happening in developed countries. As an unintended side effect, environmental pressure will continue to build up, unless public policies stimulate sufficient counteractive actions.

When humans made the transition from hunter-gatherers to agrarians, they started living in settlements and larger societies. This created opportunities for wealth accumulation, conspicuous consumption and material status-seeking. While initially limited to the happy few at the top of the power pyramid, urbanisation caused larger populations to become more affluent. With industrialisation and mass production, this exploded to virtually entire populations. As a result, extraordinary product diversity allowed economies of scale linked to mass consumption to be combined with distinction and individuality (Frank, 1999).

This historical process can be seen as a maladaptation of humans, driven by our evolved preference for status, ultimately caused by sexual selection (Section 4.7). One could say that the search for individual accomplishment and prestige, in combination with an increasing diversity of goods and services over time, has prompted many people to shift their investments of effort and money away from children to luxury and overt consumption. While in the past status served a function of biological reproduction, it increasingly contributes to cultural reproduction. In other words, for many people cultural fitness seems to have replaced genetic fitness as a guide. One may wonder whether future investment in social status by humans will continue to increase or, instead, will stabilise or even diminish. Some have suggested that the only way to tone down consumption, for environmental reasons, is to weaken or redirect our social status behaviour (Brekke and Howarth, 2002). Nonetheless, having a child is likely the worst decision humans can make in terms of environmental impacts generated over time (Section 14.1). It is, therefore, not evident that focusing on cultural reproduction at the cost of biological reproduction is such a bad thing to do.

Many cultures are intertwined with religious institutions and practices. Moreover, religions are bound to remain a dividing force in societies around the world. Section 7.7 showed that the existence, features and dynamics of religions can be well explained by adopting an evolutionary V-S-I-R approach. Religions are driven, on the one hand, by internal homogeneity and, on the other, by rivalry and conflict with other religions or world views. Whereas competition among religions was typical in the past, nowadays interaction with atheism is gaining relevance, ranging from violent attacks by religious zealots to intellectual, democratic exchange of ideas.

To understand the future of religions as powerful institutions operating from local to global scales, theories of organisational evolution, as discussed in Chapter 9, may be useful. They help to understand the corporate organisational structure of churches, their dynamics and their competition within a wider population of religious and non-religious organisations. For example, the Catholic Church resembles a multinational in the way it tries to increase its 'market share' and defend its interests on all continents. This involves lobbying politicians to obtain preferential treatment. Islam is less centralised, which could spell more uncertainty about its future: will it be increasingly dominated by extremism, or gradually become more moderate as has been the path followed by the

Catholic Church. For the moment, the first seems more likely, in light of its deep internal divisions along Sunni, Shi'ah and regional lines, and the lack of genuine democracy in many countries where Islam dominates. This will likely translate into continued population growth and a negative influence on education and development. What we still lack is cultural evolutionary analysis and predictions of the ultimate outcome of competition between atheism/agnosticism and religions.

It was suggested in Section 7.7 that we can improve our understanding of religions by applying different evolutionary theories, including group selection and memetic thinking. Both reflect on the idea that social learning and, in particular, imitation are key to social evolution, including evolution of religions. Memetic thinking stresses that human evolution is guided by a combination of gene and meme selection. Currently, memetic forces appear to dominate. While meme theory is generally seen as anecdotal or a mere analogy of gene theory, many now think that it deserves serious attention in both human evolutionary studies and the social sciences. Horizontal meme-type transmission is common nowadays due to the influence of mass media, new social media on the internet and mobile communications technologies. As illustrated in Section 7.6, memetic thinking explains certain aspects of humans and human culture surprisingly well, and draws attention to understudied topics such as distinct meme pressures leading to differential cultural evolution in cities and rural areas.

Meme thinking raises a general issue relevant to evolutionary social science, namely what the exact distinction is between evolution and learning. This was also discussed in Section 7.5, where we considered the classic opposition between Darwinian and Lamarckian interpretations of social evolution. One can find authors who defend either or both interpretations. However, the term 'Lamarckism' requires a clear distinction between genotype and phenotype as, unlike Darwinism, it suggests that phenotype affects genotype.

Darwinian-style cultural evolution involves a V-S-I-R algorithm (Section 2.1) of interactions between the components of variation (V), selection (S), innovation (I) and replication (R). This book has stressed that in cultural, economic and technological settings the V, S and R components differ from their counterparts in biology. One should note, though, that many differences are already found within biology, that is, between distinct biological contexts and species. In this respect, cultural evolution does not represent a sharp break with, but continues a pattern already observed in, biology. Arguably, Lamarckian social evolution differs from Darwinian versions in that it reflects the capacity of humans and their organisations to generate knowledge through individual or social, partly conscious and deliberate, learning. The reality is more complex, though, as both processes are active. Moreover, learning by human individuals, groups and organisations evolves through repeated interaction between cultural-economic-technological selection processes and – intended and unintended – innovation. The Baldwin effect is conceptually relevant here, as it explains how evolutionary selection can give rise to improved learning capacity. This effect may have served as a foundation for the emergence of human intelligence and the meme replicator. An analogous process goes some way to explain the sustained improvement of learning and research capacities of firms and institutions.

16.3   Economic and Political Systems

What will happen to the economy depends equally on political systems, technological change and, in some countries, on religious factors. Technology will affect communication and trade. The role of digital communication and the internet will, without doubt, continue to gain importance over the coming decades. Preserving cultural, economic, technological and even political, diversity should be a political aim as it contributes to global learning and resourcefulness. Unfortunately, consumption pattern and production methods appear to become more similar across the globe. This poses a risk as less diversity translates into limited insurance and resilience during times when change and adaptation are needed.

Such issues can be studied through an evolutionary lens discussed in Chapters 8 and 13. Evolutionary economics represents what one might call a 'real micro-level approach', more so than traditional microeconomics, because representative agents are replaced by populations of agents. This approach is ideally suited to incorporate insights from behavioural economics – the soft revolution in economics. Indeed, a focus on multiple, similar but different, interactive agents within a population translates into an approach that is both behavioural and evolutionary in spirit. Unlike traditional microeconomics, evolutionary economics can comprehensively address bounded rationality and interactions such as social comparison (status-seeking), imitation, information exchange and diffusion of expectations and opinions. The lack of such behaviours represents a failure of mainstream economic theories. It explains, among other factors, why macroeconomics is so bad at predicting economic crises (Driscoll and Holden, 2014).

Organisational decentralisation has taken place since the 1980s when large corporations, such as General Motors and IBM, had difficulty adapting to economic crises. In response, they changed to more adaptive and flexible smaller units, both geographically and product-wise. Since then, other companies have followed suit with similar strategies. This involved the creation of informal networks to partly replace previous hierarchies, implying a shift towards bottom-up decision-making, flexible organisational structures and less clear-cut boundaries between internal and external organisation. This has gone along with slicing up global value chains. Such restructuring processes are still ongoing and may continue well into the future, aided by the internet. The latter offers the malleability of flexible work, telecommuting and small companies characterised by a high degree of adaptive capacity. Given the trend towards decentralisation, a critical question then remains: will the future be dominated by 'small is beautiful' ideals and centralised global governance, by transnational companies and nation states, or by some other combination? Certainly, companies, products and services associated with human health, notably involving use of information about the human genome, will gain in importance. Along with this, human lives will be prolonged and public and private health care boosted.

Additional fundamental questions relate to political systems, notably how they need to adapt to effectively confront current population and environmental challenges. Given that such systems result from a long history of institutional evolution, any quick fixes are improbable. Perhaps some economic, environmental or political crisis is needed to enforce the necessary change. Bill Clinton supposedly once observed that democracies are fundamentally conservative and only progress through crisis. Other questions in this respect relate to the future of capitalism versus socialism. Is China, with a centrally guided economy, better equipped to achieve climate goals than the decentralised Western economies? And was Marx correct – capitalism breaks down due to its tendency of (re-)creating inequality of power and wealth – or did Schumpeter get it right – capitalism will gradually evolve into socialism? It seems that Schumpeter holds the stronger card, as the world is dominated by varieties of a mixed capitalist–socialist system, also known as the welfare state. Communism in its purest form has not fared well, for various reasons – one being that it has been unable to compete with capitalist countries and, another, that it has tended to hamper individual freedoms and democratic values.

That was the past, but what of the future? Which nation(s) will dominate the world into the future: the United States with its capitalistic bent, the EU with its socialist face, or communist China? Perhaps this question is answered by China having already evolved into a semi-capitalistic economy, though it is still far from being democratic. It is likely that with rising education levels and access to the internet, a spill-over of democracy to non-democratic countries will occur in coming decades. Interestingly, while there is some evidence that the labels 'Capitalism' and 'Communism' are not very divisive in terms of economic policies, they do define a sharp division among social and environmental policies (Mulligan et al., 2004).

Will the future comprise continued economic (GDP) growth, or will growth rates continue to fall, to zero or even to negative levels? The latter may be likely in view of decreasing returns to education and innovation. In addition, an effective Paris climate agreement might constrain economic growth worldwide by causing cheap energy to become scarce, at least during some transition to affordable renewables. On the other hand, without effective climate policies, we will see serious economic damage owing to extreme climate change. These may depress economic growth as well, notably in countries particularly sensitive to climate change, including poor, low-lying, agriculture-dependent and tropical nations. A combination of these factors promises the worst outcomes – Bangladesh being a gloomy example. On the positive side, less growth is not necessarily a problem for the rich countries, where average incomes are so high that further welfare progress – if possible at all – depends more on full employment and a redistribution of income and wealth than on continued rises in average income. Taking an evolutionary viewpoint is helpful here, as its highly disaggregate approach allows gauging the distributional consequences up to the lowest level of individual citizens, along with their personal income, wealth, behavioural, educational and consumptive characteristics.

An evolutionary view of macroeconomics and growth (Section 8.6) should be particularly useful in understanding and modelling positive feedbacks within and between financial, housing and other markets. It may allow better predictions of changes in interest rates, investments and the resulting cyclical boom and bust patterns.

The reason is that an evolutionary approach pays close attention to the mechanisms of imitation and diffusion, which are critical to business cycles. Such an approach will further provide a basis for identifying effective policies to reduce the volatility of markets instead of trying to lurch back to the boom part of the cycle in a crisis period. In addition, a disaggregated, population-based evolutionary approach helps to identify policies aimed at maintaining or restoring equity after a crisis. It further allows analysis of the consequences of diversity of, and experimentation with, policies in different regions, rather than applying a uniform austerity approach in all countries. This can contribute to new directions in the study of macroeconomic policy responses to economic crises – which is useful given that old strategies do not work well.

Both biological and economic evolutionary processes occur in geographical settings (see item v in Section 3.5, and Section 8.7). This underpins the relevance of geographical heterogeneity, which creates opportunities and limits for economic activity. In particular, it explains the existence of regional specialisation, spatial interaction and diffusion and localised collective learning. In addition, spatial isolation is crucial as it contributes to distinct economic regimes, thus fostering different types of experimentation and innovation trajectories. A spatial interpretation of path dependence and lock-in exists as well: it is triggered by agglomeration economies and historical spatial contingencies like natural infrastructure (e.g. rivers or climate). Using such evolutionary concepts can shed new light on the old debate on convergence versus divergence of growth between rich and poor nations.

The most concerning spatial heterogeneity is without doubt the gap between rich and poor in the world. From an evolutionary perspective the role of resource scarcity is relevant here. A counterintuitive theory in this respect is known as the 'resource curse', which states that countries with considerable natural resource wealth typically experience slow economic growth. Several factors contribute to this: the government tends to be too large and costly, resource income stimulates corruption, export-led growth based on non-resource sectors is difficult to achieve because of an expensive currency and high wages, and entrepreneurial activity is crowded out (Sachs and Warner, 2001). At a more general level, Acemoglu and Robinson (2012) blame extractive institutions for underperformance of economies. If environmental policy trends continue, the contribution to environmental pressures from the North/West will decline in relative and possibly absolute terms because environmental pressures from the global South will strongly increase, due to both economic and population growth. So, whereas the North will be able to pursue a flatter environmental impact curve, the South will have to spike (Harper, 2000). This represents a severe challenge. Technology diffusion and leapfrogging may help, but since they facilitate both environmentally damaging and efficient technologies, their net effect is uncertain.

In the battle of ideological and political ideas, can we hope that the most equitable and sustainable outcome for our society will triumph? Is the political spectrum likely to narrow as the left and right move closer to one another thanks to greater awareness and respect for scientific knowledge? Or is the oscillation between left and right in politics a perpetuum mobile? Unfortunately, these questions will remain unanswered.

16.4   The Biosphere

The term 'anthropocene' unofficially denotes a new geological era dominated by humans. It was suggested in the 1970s by ecologist Eugene Stoermer and invigorated in the 1990s by Nobel laureate atmospheric chemist Paul Crutzen. James Lovelock (2014) proposes a start date of 1712, as then the first successful Newcomen steam engine appeared on the scene, which marked the beginning of accelerated technological and economic evolution. Human dominance of the biosphere stems from two factors: the current human population is some 100 000 times larger than that of any other species with a similar body size and position in the food web; and human intelligence and interaction have translated, through both cooperation and competition, into a wide diffusion of advanced technologies supporting a high average consumption of materials and energy per capita. The combination of these factors has a particularly severe impact on the biosphere. To get an indication, consider that human appropriation of net primary production by terrestrial ecosystems globally is estimated to be 24 per cent. Of this, 53 per cent is due to harvest, 40 per cent to land use, and 7 per cent to human-induced fires (Haberl et al., 2007).

Already in our lifetime, we are seeing considerable change in biological diversity. Species go extinct at a rate that is thousands of times faster than the natural process. Myers and Knoll (2001) discuss in depth what the 'sixth extinction' triggered by humans will generate in terms of evolutionary impacts in various ensuing stages (Section 13.3). They stress that while previous mass extinctions (Section 4.4) were followed by a bounce back of biodiversity, this is an unlikely scenario in a human-dominated biosphere. The reason is that the pressures from human populations and their economies will not easily return to pre-industrial or even prehistoric levels. Hence, the biosphere has started to and will continue to evolve in a direction strongly guided by anthropogenic selection factors.

Human-induced climate change is an omnipresent factor – no individual, species or ecosystem can escape it. It will make many dry regions drier and many temperate areas greener. It is likely to destabilise numerous ecosystems because their species components have distinct temperature tolerance ranges and will, when shifting in space, run into human and natural infrastructural barriers. Species may also have to adapt to conditions such as fragmented and impoverished ecosystems, or even garbage dumps, resulting, potentially, in the evolution of different snouts and digestive systems. According to Ward (2001), crows, dandelions, owls, pigs, raccoons, rats and snakes, and various others species will benefit from altered ecological conditions due to human impacts. Dixon (1981) goes further in depicting futures, by considering the evolution of new species and accounting for continental drift to substantially alter the geography of the Earth. Weisman (2007) considers how the world would alter if humans disappeared. He claims that it would take 500 years for residential neighbourhoods to convert to forest, while radioactive waste, plastics and bronze statues would for a long time serve as proof of previous human existence. Of course, the biosphere can live without humans – it has done so for the most part of its lifetime (Section 3.6), but humans may find life uncomfortable without a healthy and resilient biosphere.

Given human dominance on Earth, coevolution of humans and the biosphere takes on a new meaning. To put it differently, humans are involved in niche construction at the largest scale (Section 4.6). The evolutionary significance of this is that the selection environment is changed for both humans and other species in the biosphere. Although human niche construction generates diversity through evolutionary innovation in socio-economic and technological contexts, it also contributes to diversity loss in the organic part of the world. Of potential future biodiversity loss, 50 per cent may be triggered by climate change – the rest by a combination of overfishing, (illegal) hunting, land use, deforestation, a higher probability of fires and ecosystem fragmentation. In a stroke of optimism, despite these stark challenges, E.O. Wilson (2016) suggests that we could save current biodiversity by setting aside half the planet for biodiversity reserves, free from direct human interference. Of course, for this to be successful we would need to avert, or at least greatly reduce, the impact of climate change. Raskin (2016) speaks about the 'great transition' that is necessary to stay within planetary boundaries. In his view, it rests on global civilisation achieving a shift in values to human solidarity, ecocentrism and non-material welfare. As undesirable alternatives, he presents Barbarisation and Conventional World: the first represents a fortress world in which only a small minority would live well, while the second is characterised by markets and institutional reform. The first would arguably end in social and environmental chaos, and the second in gradually increasing environmental problems. I personally am sceptical about quickly altering fundamental human values. It will not work, or at best will go slow and reach a small part of the world population. We anyway lack effective policies to achieve such value change, and if they existed we would be unable to sell them politically. Instead, our best bet for the coming decades is to implement systemic, regulatory policies to alter daily decisions by consumers, firms and investors (Chapter 14).

While natural history shows that no species has eternal life, human extinction at short notice on a geological time scale seems unlikely. The reason is that humans are capable of adaptation to new circumstances more so than other species. Theoretically, extinction can occur with a small probability from a nuclear war, a natural disaster like a meteor collision with Earth, a pandemic, extreme climate change or a combination of such factors. One route to salvation for the planet appears unattractive – to humans at least: pathogenic microorganisms might rescue the planet from human overpopulation. To avoid this, medical science must keep up the evolutionary race with them (Pirages, 2000). However, it is more likely that humans will achieve fairly high life expectancies overall, with most deaths caused by some type of cancer – unless we manage to control the latter by adopting evolutionary treatment techniques (Box 1.2). So, at worst, global environmental change would decimate rather than eradicate the human population and, as a result, give more space for the rest of the organic world to utilise. We should thus expect humans to be around for a long time, which means that the biosphere may gradually adapt to humans.

16.5   Future Humans

Evolution is about populations, so an obvious question is: how will the human population fare? Probably some 80 billion humans have lived on this planet since

Homo sapiens evolved, and it is expected that some 15 per cent of this number will be alive by the year 2100. The precise pattern of future human population growth, and the demographic structure underlying it, depend on the aforementioned factors: the biosphere, culture, religion, economy, technology, politics, and in particular how these affect living standards, education, women's rights and social norms about the family – key factors in parents' child-rearing decisions.

Population is a key factor underlying environmental pressure. This is illustrated by the $I = P * A * T$ equation (Ehrlich and Holdren, 1971). It decomposes environmental impact I into population P, affluence A, and technology T. The latter denotes the efficiency with which resources are converted in goods suitable for consumption and with which waste is cleaned up. Human population pressure has caused resource scarcity, land scarcity, local pollution and global warming. In addition, it presents challenges for food production and water provision. A meta-analysis of 69 studies assessing a potential limit to world population indicates that we have already passed the carrying capacity of Earth, estimated to be ~7.7 billion people (van den Bergh and Rietveld, 2004). Nevertheless, some remain optimistic that technological change will relax any population limits. As Frank Herbert illustrated in his 1965 science fiction novel *Dune*, even under extreme water scarcity conditions 'stillsuits' are a technological solution to keep a sizeable population alive, as they preserve bodily fluids by recycling sweat and urine.

In the longer term, not only population but also evolutionary dynamics matter. There is uncertainty about what the driving forces of human evolution will be in the near and distant future. Should we expect technologies like artificial insemination to increase the proportion of humans with low fertility in the long run, or even complete dependence on medical intervention to guarantee healthy human babies? And will orthodontics cause a proliferation of humans with genetically mal-programmed teeth? Likewise, will contact lenses, eyeglasses and cornea-reshaping surgery in response to myopia and hyperopia guarantee that more people will have bad eyesight in the future? If so, the good news is that there will be more work for associated medical specialists. More fundamental questions can be raised. Will critical selection pressures on humans be intra-human or associated with the natural environment and competition with other species? And if the former holds, which will dominate: sexual or cultural selection? Of course, things are more complicated as certain cultural habits may influence sexual selection and vice versa (Boxes 4.1 and 12.1).

Selection by culture, notably food habits, has affected the genetic composition through gene–culture coevolution. Religion has exerted and still exerts selective pressure, as it often provides implicit or explicit values or even rules about mating, abortion and quantity of offspring. Lactose resistance and a lighter skeleton are two consequences of selection by altered food habits due to the transition from a hunter-gatherer lifestyle to one based on agricultural food production (Section 11.6). In addition, cultural selection through health care and violent (military) conflict are possibly important selection factors – also as individuals in the world, and in many countries, are differently affected by these. Possibly, the human skull and probably brain have been shrinking in response to cultural selection. Like domesticated animals that tend to be less intelligent than their wild counterparts, we are taken care of by other humans. Such 'self-domestication' of the human species allows individual humans to specialise in certain tasks or even adopt low-effort lifestyles rather than having to excel in a variety of activities, as is needed for survival in non-cultural settings (Courtiol et al., 2012). Will this trend continue?

Sexual selection is considered by many as crucial to the future evolution of humanity, as it has been in the past (Sections 4.7 and 11.3). Indeed, what matters for a large part of humanity these days is not so much survival up to reproductive age as having the opportunity to reproduce and raise a family. Preferences for potential mates matter in this respect, which are largely genetically determined, but also influenced by cultural phenomena. The latter include nowadays advertising and movies where particular facial and body shapes, as well as mating behaviours, are disproportionally present. Such information tends to reinforce rather than drastically alter the genetic disposition of mating preferences. Indeed, evolutionary psychology has confirmed preferences for mates to be consistent across cultures. Intelligence and physical attraction are greatly valued and hence highly correlated between spouses. Nevertheless, the role of culture and religion is relevant as well, evident in high spouse correlations for political and religious inclinations. This is not surprising, as living with someone who has rather distinct opinions on society and politics is likely to generate conflict. Also as expected, people negatively value emotional instability in a partner, with neuroticism as an extreme case (Buss, 1989).

It could be that widespread use of contraceptive practices in modern life is weakening the strength of sexual selection. According to Pinker (1997, p. 42), '[h]ad the Pleistocene contained trees bearing birth control pills, we might have evolved to find them as terrifying as venomous spiders'. Because this figurative tree emerged only recently, humans have no genetic tendency to dislike its low-hanging fruit. Meme theory suggests that in the future people may derive more satisfaction and status from contributing ideas to the next generation rather than babies. Influenced by culture and its memes, in fact, the majority of people have voluntarily decided to limit the number of their offspring to two or fewer. The Marxist-communist was even able to mould the entire Chinese population into the one-child family ideal.

A related issue is whether, and how, human amorous relationships will evolve. Will we move in a polyamorous or even polygamous, notably polygynous, direction, or will we stick mainly to monogamy? If wealth and technology make child-raising considerably easier then there is perhaps less need for long-term pair bonding and polygyny can become more common. In turn, this could affect sexual selection and its impact. Of course, such developments would strongly depend on various other factors, such as the presence of religions and their impact on laws and social norms.

Now that we are talking about human fundamentals, what is a likely future for human food? Habits in this respect are diverse and changing across history and the globe. It is clear that humans are currently maladapted not only in their dealings with the natural environment (Chapter 13), but also in terms of modern food culture, characterised by overly sweet, greasy, salty or fast foods, and often a combination of such health-threatening constituents. Food habits may even affect human evolution, possibly over relatively short periods of time. To illustrate, it is likely that a combination of improved food quality over centuries, excessive milk consumption and sexual selection turned the Dutch into the tallest people in the world (Box 12.1). Food contributing to short-term health benefits does not imply, though, automatic translation into long-term health benefits. According to Le (2016), we should trust in traditional cuisines, because our ancestors and their food habits have survived cultural and biological selection over centuries, if not longer.[2]

As long as humans do not migrate to outer space, it is improbable that humanity will evolve into a distinct species, even over the course of millions of years. Emergence of a new human species would require isolation of a small sub-population of humans for an extremely long time, hundreds of thousands of years at least. This is virtually impossible as modern technology and globalisation have advanced too far to allow long-term segregation of any human groups. Moreover, such a sub-population should live under particularly stable ecological or cultural conditions, so that genetic drift can have its effect. Human evolution is not easily stimulated as we do not feel any competitive pressure from other species given our total ecological dominance in the biosphere. Therefore, any future biological evolution of humans would likely be marginal and slow. This is not unusual in natural history. Other species, such as crocodiles, have altered relatively little over the course of 50 million years. They are what Darwin referred to as 'living fossils'. An additional argument to expect slow future evolution of humans is that, compared with other animals and primates, they have many DNA repair mechanisms, which is one reason for their relatively long life.

The ecological dominance of modern humans suggests that, once they appeared, it was unlikely that another similarly intelligent species would evolve – even with enough time available. In fact, it is now believed that in the past hundreds of thousands of years several intelligent *Homo* species lived simultaneously during some time with modern humans on Earth, including Denisovans in middle Asia, *Homo floresiensis* in what is now the Indonesian island Flores, *Homo naledi* and other early *Homo sapiens* species in Africa, and Neanderthals in Europe. None of these survived side by side, with modern humans descending from Cro-Magnons. In fact, humans currently have to do their utmost best – using explicit biodiversity conservation programmes – to keep alive the few species of great apes. If astronauts were to leave Earth and return to it after a long absence due to time travel, they might find modern humans had disappeared due to environmental and health disasters, while an ape lineage had evolved into a species with human-like intelligence. A scenario of this kind was elaborated by Pierre Boulle in his 1963 science fiction novel *La Planète des Singes* (*The Planet of the Apes*). It describes an ape society consisting of gorilla soldiers, orangutan administrators and chimpanzee intellectuals, who enslave a decimated population of unintelligent humans – a somewhat unrealistic scenario, to say the least.

---

[2] Katan (2016) unmasks 70 diet myths and argues that there is insufficient evidence for the widely held belief that fruits and vegetables reduce the risk of cancer and cardiovascular diseases. The reason is that people who regularly eat large amounts of them tend to lead a relatively healthy life, while disentangling the two factors is hard.

In terms of human appearances, one might presume that a mixed, brown skin colour would ultimately dominate due to the genetic mixing of distinct skin colours. However, sexual preferences and selection could prevent this from happening. Another idea is that the genes of humans who are better adapted to the urban conditions and information culture might proliferate. It is not clear, though, what it could precisely imply: more whizz-kids and nerds, or eyes better adapted to computers and videogames – or the opposite, if they were to reproduce less due to spending all their leisure time with these technologies? In any case, there seems to be a trend of substituting face-to-face human relationships with computer entertainment and contact through phone and digital message channels.

Some futurist views suggest that the gap between rich and poor is so persistent that it may translate in a 'genetic divide' (McKibben, 2003). Unlike the poor, the rich would have access to better, even genetic, medical technologies. This is a recurrent theme in science fiction – see, for example, the 1997 film *Gattaca* (its title derives from the four nucleotide bases in DNA, i.e. G, A, T and C; see Section 3.3). With access to genetic improvements, the rich would become physically and mentally superior, competing away the poor. Or the poor would revolt beforehand, as in the 2013 film *Elysium*, and thus eliminate the divide. Either way, it is unlikely that any cultural divide would persist during hundreds of thousands of years such that humans would evolve into two separate species.

According to Harari (2016, p. 1), humanity has moved beyond its three historically dominant challenges – famines, epidemics and wars – as 'more people die today from eating too much than from eating too little; more people die from old age than from infectious diseases; and more people commit suicide than are killed by soldiers, terrorists and criminals combined'. He suggests that along with it the widespread belief in a cosmic plan by a supernatural god is losing ground. Instead we are increasingly relying on 'Dataism' – god-like technologies, such as computers, algorithms, and online big data. Since it is unclear how we will manage such technologies, the future is highly uncertain. It may well be that the internet-of-all-things will come to rule the world as a benevolent dictator once it has access to all human-generated data. Or perhaps emotionless robots will take over and deal with humans as we currently do with farm animals or even annoying insects. Another wild idea is that humans might jointly develop, with the internet, a global brain via continuous interactions of mobile phones and computers. This is an old theme in science fiction, since William Gibson's 1984 *Neuromancer* about early computer hacking, continued in post-cyberpunk, such as Greg Egan's 1994 *Permutation City* about a virtual world that is a copy of the real world. Not only does it allow humans to extend their lifespan, but it also hosts a competitive module generating evolution of life from viruses to intelligent beings. The global brain might involve humans physically connected to the internet through a plug into their nerve system, similar to how a virtual reality game enters the 'gameport' of a human body in Cronenberg's creepy movie *eXistenZ* (1999).

Climate change can be seen as an inevitable consequence of human evolutionary success. Although we easily talk about a carbon-free economy, it is not easily achieved as it requires what James Lovelock (2014) calls a 'sustainable retreat' of humans, in terms of population and consumption. But he thinks our tribal nature is not consistent with effective global governance to enforce such a solution. His advice: learn to live with climate change, especially since the outcome may be more attractive than generally thought. Humans are like the cyanobacteria, which through the oxygen revolution changed the face of the Earth. We humans do it through fossil fuels and climate change, potentially preparing a transition to a biosphere with supremacy of electronic artificial intelligence driven by silicon-based energy. Such artificial life is anyway better able than carbon-based life forms to withstand a hot world. Lovelock believes, referring to Nobel laureate Jacques Monod's 1970 book *Chance and Necessity*, that humans are still able to stimulate the evolution of such artificial intelligence, out of sheer necessity – a debatable teleological interpretation of evolution. His optimism is based on parallel computers and the rapid evolution they arguably can undergo. Against this view, plead various factors: extreme climate change will likely destroy human lives and cultures, diminishing the capacity for a next technological transition; moreover, it would take artificial intelligence a very long time before it could survive and reproduce without humans, time which climate change does not allow. Artificial intelligence is anyway an overestimated and overused term. Altogether, Lovelock's ideas express a very personal interpretation of evolutionary progress.

One should further recognise that the various aforementioned projections of computer intelligence may be ill-founded. Taking an evolutionary angle, one immediately notices that critical diversity is lacking. There are no multiple competing internets, implying that a global brain cannot emerge through an evolutionary process. Similar to discussions about artificial intelligence, there is a lot of hope but limited progress. On the other hand, artificial intelligence could in principle be promoted at the level of individual machines or robots through evolutionary methods, namely by building many of them, letting them compete and having them subjected to various other selection pressures. The two main constraints here are that population numbers will not even come close to those which have been able to craft natural history, and that complex machines are unable to autonomously replicate (see further the comparison of the complexity of a human brain and a computer in Section 2.4). The best one may expect is that artificial intelligence emerges by way of evolution in computer software rather than in hardware. This will, though, not easily translate into electronic hardware species that can survive and reproduce, needed to become fully independent from their 'human parents'.

16.6   More Futurist Scenarios

Long-term thinker par excellence Stewart Brand has said that science is the only news as it alone provides the information that changes the world irreversibly (Leyden, 2001), but we certainly can also learn a lot from brainstorming about the future, as is common in futurism and science fiction. Here one comes across ideas that we will soon be able to genetically engineer ourselves to become less aggressive, more cooperative or even predominantly altruistic, less interested in affluence, less status oriented, less religious, smaller, less fertile and intolerant of meat. In this way, all kinds of problems could be solved or reduced, such as violent conflicts, consumerism, population growth, overuse of natural resources and environmental destruction. Genetically engineering a slighter and more food-efficient human body could also help humans to explore extra-terrestrial space, which some see as the obvious next phase of human cultural evolution. An 'interplanetary man' or 'interstellar humanity' (Dick, 2000) would not merely reduce the likelihood of human extinction, but increase that of developing into multiple human-like species in a remote future.

A radically opposite idea is to stimulate voluntary action to not have offspring. This would ultimately eliminate the human species altogether. Of course, this runs contrary to human or animal nature and so is unlikely to materialise. On the other hand, the Chinese already once submitted to the one-child family ideal, which is just one step away from the no-child family goal. Another alternative is to reduce overconsumption given that its social costs exceed its social benefits, as argued long ago by John Stuart Mill (1848), Thorstein Veblen (1899) and John Kenneth Galbraith (1958). When revisiting this theme with hard data analysis, Trentmann (2016) shows that the trend of increasing consumption from the fifteenth to twenty-first century is robust. This should not come as a surprise, as it is driven by the quest for self-development and status-seeking through conspicuous consumption. At best, it the outcome of the affluenza virus: bored people spending abundant incomes and leisure time on shopping. Some propose that a 'sharing economy' provides a way out. Insights so far indicate it would bring about marginal benefits as it merely adds a new layer of activity to the material-intensive economy and simply creates more consumption opportunities (Frenken et al., 2015).

Some think the only option to reduce the negative impact of humanity on the planet is to re-engineer humans up to the point where they become considerably less intelligent. This is motivated by the belief that human intelligence is key to ecological dominance of the human species and is therefore the source of all evil. Economic studies find that human skills and inventiveness are among the principle factors explaining economic growth and, indirectly, environmental pressure. Considerably less intelligent humans would mean the end of current forms of civilisation, as these require intelligence beyond a threshold – consider the Dunbar (1996) number that suggests a link between primate intelligence and group size (Section 5.1). An opposing idea is that our social and environmental problems are due to insufficient rather than too much human intelligence. The advice emanating from this line of thought is to use genetic engineering to lengthen the period of brain development during the foetal stage in the womb so as to increase the quantity of neurones in the brain. It is difficult to say which of these genetic experimental alternatives would threaten the survival of humanity more.

Such futurist ideas are nothing new. In the late nineteenth century, H.G. Wells suggested in *The Time Machine* (1895) that humanity's dependence on technology would ultimately result in the shrinking of legs and muscles, up to the point that future humans would start to look like huge heads with large brains on top of miniscule bodies. Of course, historical constraints or 'bauplan limits', referring to a degree of irreversibility, make such a scenario implausible. Wells further depicted humans evolving into a race of docile and uninvolved beings, a result of no longer having to struggle with their environment. This view is more in line with the (incorrect) 'use and disuse' principle of Lamarck than with Darwinian evolution. Another classic novel addressing such issues is Olaf Stapledon's *Last and First Men: A Story of the Near and Far Future* (1930). It follows humanity from the present to two billion years in the future, depicting 18 successive human species, including giants, dwarves, flying men and species with enormous brains. The last species is the most advanced, consisting of philosophers and artists with liberal sexual morality and, for the first time, able to avoid the cycle of progress and breakdown of civilisations.

Science fiction movies and literature offer many other extreme ideas on human evolution: mutants arising from nuclear laboratories, hydrogen bombs or radioactive waste; genes of other species mixing with those of humans; or cybernetic species arising from the integration of humans and technology. For example, the 1999 book *Darwin's Radio* by Greg Bear describes rapid mutation and even speciation in a human foetus while still in the womb, due to an 'endogenous retrovirus'. In another book from 1985, *Blood Music*, he describes how nanotechnology based on engineered human lymphocytes makes body cells evolve up to the point of becoming intelligent. In the classic comic *Spiderman*, a young man is bitten by a spider with irradiated venom, giving him spider-like powers. In *The Fly* – a tale by George Langelaan in Playboy magazine of 1957 – an inventor tries out a teleport machine, but when a fly enters the transmission booth he ends up being permanently integrated with it. Another modification with potentially huge social-economic consequences is to modify the human body so that it needs less or even no sleep, as told by Nancy Kress in the 1993 novel *Beggars in Spain*. More examples are given in Barceló and Lemkow (2016). But these ideas are likely to remain fiction. At best, one should expect cultural diffusion of body modification practices to continue, in the form of cosmetic surgery, implants, piercings and tattoos, including more extreme cases such as tongue splitting and eyeball tattooing. But unless these affect sexual selection on a grand scale, they are unlikely to alter humans genetically.

All the same, futurism has found a niche in science. The 'Millennium Project' (Glenn, 2000), based on 100 experts rating 19 potential factors of influence, identified three scenarios for the next 1000 years: (1) still alive at 3000, with environmental and health problems solved but some minor problems continuing; (2) decimation of the human population owing to nuclear wars and environmental disaster, with remaining humans living in a chaotic society; and (3) the great divide between three species: traditional humans, humans merged with technology and artificial life forms. Perhaps the most useful function of generating such visions is that they can guide future policies to reach or avoid certain (un)desirable futures. For this to work, we would have to be aware of the ultimate consequences of any policies.

16.7   Prospect for Expanded Evolutionary Thinking

This book has argued that evolutionary thinking in the social sciences is likely to see increasing application. One reason is that the V-S-I-R algorithm of evolution is such a general and powerful mechanism that its widespread application to understand the dynamics of many complex systems studied by science is only a matter of time. Adding ultimate explanations to proximate ones tends to change people's views on many social science issues that were previously taken for granted. We must therefore put aside any fears in the social sciences about thinking along biological-genetic and non-genetic evolutionary lines. Instead, we should welcome these as tools to better understand ultimate drivers of our society, economy, technology, and overuse of natural resources and the environment. Table 16.1 reviews key notions of evolutionary thinking as applied to areas of study covered in this book.

As clarified in Chapter 6, application of (cultural) group selection theories in the social sciences can generate many benefits. Indeed, assortment, a basic factor behind group selection, is common in societies where communication is cheap and easy. For example, the internet allows people to communicate with others who have similar interests, problems or opinions, even though they are anonymous or spatially distant. The explanatory power of group selection lies in its ability to connect functionality of groups or institutions, as stressed in traditional sociological approaches such as functionalism, to evolutionary cause–effect mechanisms. This leads to a well-founded and dynamic evolutionary basis for traditionally static functionalist thinking. Among others, this can be used to broaden insights to how group norms are crowded out by markets, how group phenomena play a role in common-pool dilemmas, how socioeconomic power emerges, how groups compete and end up in conflict and how individual versus group selection – together forming multilevel evolution – interact and affect complex system dynamics. Specific problems to modern society can be addressed, namely: what happens according to group selection if individuals participate in multiple, overlapping groups? Or can group selection be managed with the aim of solving tough collective action problems, such as stopping human-induced climate change? To accomplish such applications of cultural group selection, relevant insights about groups from the angles of economics, sociology and social psychology may be integrated into existing cultural group selection models to enhance their relevance and accuracy.

An old question in evolutionary biology is equally relevant to the social sciences: does evolution always mean social progress – however precisely measured? The answer is 'no', for many reasons: local search, coevolution, multiple selection factors, historical constraints or path dependence and random factors like cultural drift, etc. (Box 15.1 offers further discussion). One should thus expect evolution to be generally caught in local optima. Adequate policies may, though, move evolutionary systems in the direction of global optima, what one might call progress. But evolutionary thinking complicates things as compared with, for example, the dominant policy framework in economics. This explains why the notion of optimal policy is less common in evolutionary social science. But, as argued in Chapter 15, there is definitely a place for normative policy questions and analyses in evolutionary social science.

Related to 'evolution as progress' is the widespread fear that evolutionary thinking in the social, environmental and policy sciences will lead to a revival of Social Darwinism, notably support for harsh policies. However, the association of evolutionary social sciences with Social Darwinism lacks not just subtlety but, worse, it ignores the progress made in the evolutionary social sciences over the course of a century. The

Table 16.1 Evolutionary concepts illustrated for six areas of study

Evolutionary concepts	Area of study					
	Nature	Culture	Economy	Organisations	Technology	Policies
Population	Living organisms	Behaviours, beliefs, memes, artefacts	Goods, services, buyers, suppliers	Firms, public entities, leisure clubs, unions, NGOs, religions, etc.	Machines, tools, production lines, engines, vehicles, computers, knowledge	Regulation, laws, price incentives, norms, nudges, information (provision)
Variation	Physiology, morphology, behaviour	Habits, norms, preferences, specialisms, opinions, ideas, phrases, melodies, goods, services	Functions, designs, costs, prices, image, status-giving, satisfaction, agent goals and behaviours	Organisational structure, management system, firm size, profits, costs, revenues, workers, patents, routines	Design, productivity, cost, energy efficiency, image, reputation, blueprints	Design, efficiency, efficacy, equity, social-political acceptance, particular instruments (e.g. tax or tradable permit)
Selection	Scarcity food or space, mates (sexual selection), climate and weather factors (jointly: Natural selection)	Social interaction, imitation, peer influence, reputation, upbringing by parents and adults; also natural selection	Markets, reputation, public opinion, regulation, fashions, natural resource limits	Organisation-internal politics, mergers, takeovers, market competition, regulation	Scientific review, limited investment opportunities, learning curves, physical and chemical constraints, patent legislation	Political-democratic processes, public media and opinion, scientific evaluation, international organisations
Innovation	Genetic mutation and recombination during meiosis, sexual reproduction	Trial and error, planned change, deviant conduct, lateral thinking, social interaction, science, political modernisation, public media	Technical and social innovations, entrepreneurs, R&D institutes and departments	Search, imitation, merger, division, new employees, entrepreneurs	Factor-saving vs quality improving, R&D, learning, combining modules, serendipity, analogy, extrapolation, science-driven	Scientific input, political creativity, political coalitions and deals, imitation of successes in other regions

Table 16.1 (cont.)

Evolutionary concepts	Area of study					
	Nature	Culture	Economy	Organisations	Technology	Policies
Replication	Sexual and non-sexual reproduction, imitation of other members of one's social group	Docility, teaching, upbringing, imitating majority or those with success/prestige, information diffusion (books, digital)	Imitation, fashion, habits, information diffusion	Routines, teaching, coaching, conformist behaviour, imitation, information diffusion	Network effects, spill-overs, imitation, blueprints, books, digital information, patents	Legislation, education, coaching, public media, political parties, parental upbringing, indoctrination
Adaptation	To natural environment, or to social environment (if social species)	To scarce local natural resources and climate, to human social limits	To basic preferences, culture, scarce natural resources, technologies	To economy, market opportunities and constraints, technological conditions	To scientific and physical constraints, to economy, to market opportunities	To social, political and environmental constraints
Coevolution	Multispecies relationships	Gene–culture coevolution	Demand–supply Preferences-technologies	Firms-technologies-preferences	Complementary technologies	Politics and public opinion
Multilevel processes	Individuals and groups, assortment, competition, conflict, fusion, splitting	Cultural group selection, transmission of unique cultural information in distinct groups affecting group success and dispersal	Micro (individuals, products, technologies), meso (groups, organisations), macro (public institutions)	Firm-internal and - external selection, employees and firm routines	Complex technologies means interaction of local, national and global markets	Higher-level governance and local common-pool norms, government institutions composed of self-interested public officers with own goals
Path dependence	Bauplan limits, hyper-selection	Historical or development constraints	Increasing returns to scale on supply and demand sides of markets	History of merger, market dominance, institutional lock-in	Technological history and lock-in, increasing returns to adoption	Constitution as historical accidents, majority voting rules, lobbying

Note: List of items mentioned are not exhaustive but merely illustrative.

main problem is that accusations of Social Darwinism do not distinguish normative ('ought') from explanatory or positive ('is') approaches. As argued in Section 1.4, like Social Darwinism itself – which should really be called 'Social Lamarckism' or 'Social Spencerism' – such accusations are guilty of the 'naturalistic fallacy' or 'ought–is confusion' by making the error of deducing 'what ought to be' from 'what is'. Modern evolutionary approaches in the social sciences recognise that evolution is not identical to progress. Among others, it can give rise to extreme income or wealth inequality; witness the growing number of billionaires in the world – and it is just a matter of time before the first trillionaire appears. This is in line with the Pareto 'power law' distribution or '80/20 rule' which tends to be generated by evolutionary systems (Devezasa and Modelski, 2003): a small proportion of all people own most of the wealth; and among the richest a small subgroup owns the larger proportion; and so on. One can respond to such outcomes of social-economic-technological evolution in three ways: accept them as they are, propose corrective policies as a proximate solution, or better, guide the evolutionary process in a more socially desirable direction as an ultimate solution. Evolutionary social sciences can escape political bias by exploring a variety of normative, subjective or political goals. This is, in fact, the same challenge as faced by traditional social science theories.

This book has illustrated the value of evolutionary thinking for various types of policy: agricultural, environmental, technological and health – see Section 1.3, Chapter 10 and Chapters 13 to 15. For example, evolutionary analysis recommends lower optimal doses of pesticide use. The reason is that this analysis takes into account not only the decimation of the pest population size in the short run, but also the creation of genetic resistance in the pest population in the long run, due to selection pressure exerted by the pesticide. Another example of evolutionary policy insights comes from group selection experiments with egg production by chickens (Section 6.4). These show that unlike artificial individual-level selection of high egg production, group selection is able to achieve an optimal trade-off between productivity and aggression, which is relevant when chickens are housed in multiple-hen cages.

As argued in Chapter 14, climate policy and a transition to a low-carbon economy can benefit from evolutionary thinking. These issues concern complex systems and long-term issues that need to be informed by a good understanding of both proximate and ultimate causes and effects, as well as how these are interlinked. Evolutionary analysis helps us to uncover cultural and biological ultimate causes or effects, as well as to unravel relevant cultural-biological coevolutionary phenomena that remain usually hidden to social sciences. Notions already in use in the policy sciences, such as path dependence and lock-in, become sharper when approached from an evolutionary, population angle. Moreover, the notion of 'optimal diversity' in technologies and possibly at other levels – culture and economy – can be addressed well, from a dynamic angle using evolutionary approaches.

If the reader's interest in evolutionary social, environmental and policy sciences is not yet satisfied, may it at least have been aroused. Evolutionary thinking enriches the way we deliberate in the social sciences about many issues. It can contribute to ameliorating human maladaptation and sustaining human existence under agreeable conditions until the distant future.

References

Acemoglu, D. (2002). Directed technical change. *Review of Economic Studies* 69(4): 781–809.

Acemoglu, D., and J. A. Robinson (2012). *Why Nations Fail: The Origins of Power, Prosperity, and Poverty*. New York: Crown Publishers.

Acemoglu, D., P. Aghion, L. Bursztyn and D. Hemous (2012). The environment and directed technical change. *American Economic Review* 102(1): 131–166.

Acerbi, A., and A. Mesoudi (2015). If we are all cultural Darwinians what's the fuss about? Clarifying recent disagreements in the field of cultural evolution. *Biology and Philosophy* 30:481–503.

Aghion, P., and P. Howitt (1992). A model of growth through creative destruction. *Econometrica* 60(2): 323–351.

Aghion, P., and P. Howitt (1998). *Endogenous Growth Theory*. Cambridge, MA: MIT Press.

Aiello, L. C., and P. Wheeler (1995). The expensive tissue hypothesis: the brain and the digestive system in human and primate evolution. *Current Anthropology* 36: 199–221.

Akerlof, G. (2007). The missing motivation in macroeconomics. *American Economic Review* 97: 5–36.

Aktipis, C. A., A. M. Boddy, R. A. Gatenby, J. S. Brown and C. C. Maley (2013). Life history trade-offs in cancer evolution. *Nature Reviews Cancer* 13(12): 883–892.

Alchian, A. (1950). Uncertainty, evolution and economic theory. *Journal of Political Economy* 58: 211–222.

Alexander, R. D. (1987). *The Biology of Moral Systems*. New York: Aldine de Gruyter.

Alkemade, F., K. Frenken, M. P. Hekkert and M. Schwoon (2009). A complex systems methodology to transition management. *Journal of Evolutionary Economics* 19: 527–543.

Allen, C. R. (2001). Ecosystems and immune systems: hierarchical response provides resilience against invasions. *Conservation Ecology* 5(1):15.

Allen, R. C. (2009). *The British Industrial Revolution in Global Perspective*. Cambridge: Cambridge University Press.

Allen, W. (2005). 'Challenging Darwin': A new book argues that diversity undermines sexual selection theory. *Bioscience* 55(2): 101–105.

Alter, A. (2012). We are blind to much that shapes our mental life. *In* J. Brockman (ed.) *This Will Make You Smarter: New Scientific Concepts to Improve Your Thinking*. London: Transworld Publishers.

Andics, A., A. Gábor, M. Gácsi et al. (2016). Neural mechanisms for lexical processing in dogs. *Science* 353(6303): 1030–1032.

Andersen, E. S. (1994). *Evolutionary Economics: Post-Schumpeterian Contributions*. London: Pinter.

Anderson, K. (2007). Super-eruption: no problem? *Nature News* doi:10.1038/news070702-15

Andor, M., M. Frondel and C. Vance (2015). Installing photovoltaics in Germany: a licence to print money? *Economic Analysis and Policy* 48: 106–116.

Andreoni, J. (1995). Warm-glow versus cold-prickle: the effects of positive and negative framing of cooperation in experiments. *Quarterly Journal of Economics* 60: 1–14.

Antal, M. (2014). Green goals and full employment: are they compatible? *Ecological Economics* 107:276–286.

Antal, M., and J. C. J. M. van den Bergh (2014). Re-spending rebound: a macro-level assessment for OECD countries and emerging economies. *Energy Policy* 68: 585–590.

Antal, M., and J. C. J. M. van den Bergh (2016). Green growth and climate change: conceptual and empirical considerations. *Climate Policy* 16(2): 165–177.

Armstrong, K. (1993). *A History of God*. London: W. Heinemann Ltd.

Arrow, K. J. (1950). A difficulty in the concept of social welfare. *Journal of Political Economy* 58(4): 328–346.

Arrow, K. J. (1951). *Social Choice and Individual Values*. New York: John Wiley.

Arrow, K. J. (2000). Increasing returns: historiographic issues and path dependence. *European Journal of the History of Economic Thought* 7(2): 171–180.

Arrow, K. J., and A.C. Fisher. (1974). Environmental preservation, uncertainty, and irreversibility. *Quarterly Journal of Economics* 88(2): 312–319.

Arthur, B. (1989). Competing technologies, increasing returns, and lock-in by historical events. *Economic Journal* 99: 116–131.

Arthur, B. (1994). *Increasing Returns and Path-dependence in the Economy*. Ann Arbor, MI: University of Michigan Press.

Ashby, R. (1960). *Design for a Brain*. New York: Wiley.

Ashraf, Q., and O. Galor (2011). Dynamics and stagnation in the Malthusian epoch. *American Economic Review* 101(5): 2003–2041.

Atran, S. (2002). *In Gods we Trust*. Oxford: Oxford University Press.

Aunger, R. (ed.) (2000). *Darwinizing Culture: The Status of Memetics as a Science*. Oxford: Oxford University Press.

Aunger, R. (2002). *The Electric Meme: A New Theory of How We Think*. New York: The Free Press.

Avilés, L. (2002). Solving the freeloaders paradox: genetic associations and frequency-dependent election in the evolution of cooperation among nonrelatives. *Proceedings of the National Academy of Sciences of the USA* 99, 14 268–14 273.

Avilés, L., J. Fletcher and C. Cutter (2004). The kin composition of social groups: trading group size for degree of altruism. *American Naturalist* 164: 132–144.

Axelrod, R. (1984). *The Evolution of Cooperation*. New York: Basic Books.

Axelrod, R. (1986). An evolutionary approach to norms. *American Political Science Review* 80(4): 222–238.

Ayres, R. U. (1991). Evolutionary economics and environmental imperatives. *Structural Change and Economic Dynamics* 2(2): 255–273.

Ayres, R. U. (1994). *Information, Entropy and Progress: Economics and Evolutionary Change*. New York: AIP Press, American Institute of Physics.

Ayres, R. U. (2000). On forecasting discontinuities. *Technological Forecasting and Social Change* 65: 81–97.

Ayres, R. U., and A. Massini (2004). Exergy: Reference states and balance conditions. *In* C. Cleveland (ed.), *Encyclopedia of Energy*, Vol 2. Boston, MA: Elsevier.

Ayres, R. U., and B. Warr (2005). Accounting for growth: the role of physical work. *Structural Change and Economic Dynamics* 16(2): 181–209.

Azar, C. (1998). Are optimal CO_2 emissions really optimal? *Environmental and Resource Economics* 11: 301–315.

Azar, C., and B. A. Sandén (2011). The elusive quest for technology neutral policies. *Environmental Innovation and Societal Transitions*, *1*(1), 135–139.

Baldwin, J. M. (1896). A new factor in evolution. *The American Naturalist* 30: 441–451.

Balter, M. (2014). How birds survived the dinosaur apocalypse. *News from Science* May 6, 2014. http://www.sciencemag.org/news/2014/05/how-birds-survived-dinosaur-apocalypse

Barceló, M., and L. Lemkow (2016). New options and identities: Body enhancement in science fiction narrative. *In* A. Delgado (ed.), *Technoscience and Citizenship: Ethics and Governance in the Digital Society*. Berlin: Springer, pp. 63–82.

Barkow, J., L. Cosmides and J. Tooby (1992). *The Adapted Mind: Evolutionary Psychology and the Generation of Culture*. Oxford: Oxford University Press.

Barrett, S. (2006). Climate treaties and 'breakthrough' technologies. *American Economic Review* 96(2): 22–25.

Barro, R., and X. Sala-i-Martin (1995). *Economic Growth*. New York: McGraw-Hill.

Basalla, G. (1989). *The Evolution of Technology*. Cambridge: Cambridge University Press.

Bauer, M., C. Blattman, J. Chytilova et al. (2016). Can war foster cooperation? *Journal of Economic Perspectives* 30(3): 249–274.

Baumeister, R. F. (2007). Is there anything good about men? Invited Address, American Psychological Association. https://psy.fsu.edu/~baumeisterticelab/goodaboutmen.htm

Baumeister, R. F. (2010). *Is There Anything Good About Men? How Cultures Flourish by Exploiting Men*. Oxford: Oxford University Press.

Baumgartner, F. R. (2006). Punctuated equilibrium theory and environmental policy. *In* R. Repetto (ed.), *Punctuated Equilibrium and the Dynamics of US Environmental Policy*. New Haven, CT: Yale University Press, pp. 24–46.

Baumgartner, F. R., and B. D. Jones (1993). *Agendas and Instability in American Politics*. Chicago, IL: University of Chicago Press.

Baumgärtner, S. (2004). Optimal investment in multi-species protection: interacting species and ecosystem health. *EcoHealth* 1(1): 101–110.

Baumol, W. J. (1982). Contestable markets: an uprising in the theory of industry structure. *American Economic Review* 72(1): 1–15.

Baumol W. J., and W. E. Oates (1988) *The Theory of Environmental Policy* (2nd edn). Cambridge: Cambridge University Press.

Becker, G. S., K. M. Murphy and R. Tamura (1990). Human capital, fertility, and economic growth. *Journal of Political Economy* 98(5): S12–37.

Beckert, S. (2014). *Empire of Cotton: A Global History*. New York: Alfred A. Knopf.

Bednar, J. (2016). Robust institutional design. *In* D.S. Wilson and A. Kirman (eds.), *Complexity and Evolution: A New Synthesis for Economics*. Cambridge, MA: MIT Press, pp. 167–184.

Behe, M. J. (2006). *Darwin's Black Box: The Biochemical Challenge to Evolution*. New York: Free Press.

Benartzi, S., and R. H. Thaler (1995). Myopic loss aversion and the equity premium puzzle. *Quarterly Journal of Economics* 110(1): 73–92.

Ben-Ner, A., and L. Putterman (2000). On some implications of evolutionary psychology for the study of preferences and institutions. *Journal of Economic Behavior and Organization* 43: 91–99.

Bergman, A., J. J. Heindel, S. Jobling, K. A. Kidd and R. T. Zoeller (2013). *State of the Science of Endocrine Disrupting Chemicals: 2012*. Geneva: United Nations Environment Program (UNEP), and World Health Organization (WHO).

Bergstrom, T. C. (2002). Evolution of social behaviour: Individual and group selection. *Journal of Economic Perspectives* 16 (2): 67–88.

Bergstrom, T. C. (2003). The algebra of assortative encounters and the evolution of cooperation. *International Game Theory Review* 5: 211–228.

Bergstrom, T. C., and O. Stark (2003). How altruism can prevail in an evolutionary environment. *American Economic Review* 83: 149–155.

Binmore, K. G. (2005). *Natural Justice*. New York: Oxford University Press.

Birchenhall, C. (1995). Modular technical change and genetic algorithms. *Computational Economics* 8, 233–253.

Black, F., and M. Scholes, (1973). The pricing of options and corporate liabilities. *Journal of Political Economy* 81(3): 637–654.

Blackmore, S. (1999). *The Meme Machine*. Oxford: Oxford University Press.

Bloom, P., and D. S. Weisberg (2007). Childhood origins of adult resistance to science. *Science* 316(5827): 996–997.

Blum, U., and L. Dudley (2001). Religion and economic growth: was Weber right? *Journal of Evolutionary Economics* 11: 207–230.

Boehm, C. (2001). *Hierarchy in the Forest: The Evolution of Egalitarian Behavior*. Cambridge, MA: Harvard University Press.

Bogers, M., A. Afuah and B. Bastian (2010). Users as innovators: a review, critique, and future research directions. *Journal of Management* 36(4): 857–875.

Bohn, H., and C. Stuart (2015). Calculation of a population externality. *American Economic Journal: Economic Policy* 7(2): 61–87.

Bornstein, G., and M. Ben-Yossef (1994). Cooperation in intergroup and single group social dilemmas. *Journal of Experimental Social Psychology* 30: 52–67.

Boschma, R. A., and J. G. Lambooy (1999), Evolutionary economics and economic geography. *Journal of Evolutionary Economics* 9(4): 411–429.

Boschma, R. A., K. Frenken and L. G. Lambooy (2002). *Evolutionaire Economie: Een Inleiding. [Evolutionary Economics: An Introduction.]* Bussum Uitgeverij: Coutinho.

Boserup, E. (1965). *The Condition of Agricultural Growth: The Economies of Agrarian Change Under Population Pressure*. Chicago, IL: Aldine Press.

Boulding, K. E. (1964). *The Meaning of the Twentieth Century*. London: Allen & Unwin, Ltd.

Boulding, K. E. (1981). *Evolutionary Economics*. Beverly Hills, CA: Sage Publications.

Bowler, P. J. (2013). *Darwin Deleted: Imagining a World without Darwin*. Chicago, IL: University of Chicago Press.

Bowles, S. (2000). Economic institutions as ecological niches. *Behavioral and Brain Sciences* 23: 148–149.

Bowles, S. (2008a). Policies designed for self-interested citizens may undermine 'the moral sentiments': Evidence from economic experiments. *Science* 320: 1605–1609.

Bowles, S. (2008b). Genetically capitalist. *Science* 394: 394–396.

Bowles, S. (2009). Did warfare among ancestral hunter-gatherers affect the evolution of human social behaviours? *Science* 324(5932): 1293–1298.

Bowles, S., and H. Gintis (1998). The moral economy of communities: structured populations and the evolution of pro-social norms. *Evolution and Human Behavior* 19: 3–25.

Bowles, S., K.-J. Choi and A. Hopfensitz (2003). The co-evolution of individual behaviours and social institutions. *Journal of Theoretical Biology* 223: 135–147.

Bowles, S., A. Kirman and R. Sethi (2017). Retrospectives: Friedrich Hayek and the market algorithm. *Journal of Economic Perspectives* 31(3): 215–230.

Boyd, B. (2009). *On the Origin of Stories: Evolution, Cognition, and Fiction*. Cambridge, MA: Belknap Press of Harvard University Press.

Boyd, R., and P. J. Richerson (1985). *Culture and the Evolutionary Process*. Chicago, IL: University of Chicago Press.

Boyd, R., and Richerson, P. J. (1990). Group selection among alternative evolutionarily stable strategies. *Journal of Theoretical Biology* 145: 331–342.

Boyd, R., and J. B. Silk (2002). *How Humans Evolved* (3rd edn). New York: W.W. Norton.

Boyer, P. (2001). *Religion Explained*. New York: Basic Books.

Brekke, K. A., and R. B. Howarth (2002). *Status, Growth, and the Environment: Goods as Symbols in Applied Welfare Economics*. Cheltenham, UK: Edward Elgar.

Brekke, K. A., and O. Johansson-Stenman (2008). The behavioural economics of climate change. *Oxford Review of Economic Policy* 24(2): 280–297.

Brewer, J., M. Gelfand, J. C. Jackson et al. (2017). Grand challenges for the study of cultural evolution. *Nature Ecology and Evolution* 1(2): 1–3.

Brockman, J. (ed.) (2012). *This Will Make You Smarter: New Scientific Concepts to Improve Your Thinking*. London: Transworld Publishers.

Brodie, R. (1996). *Virus of the Mind: The New Science of the Meme*. Seattle, WA: Integral Press.

Brookes, L. (1990). The greenhouse effect: the fallacies in the energy efficiency solution. *Energy Policy* 18(2): 199–201.

Brooks, M. (2014). Why are Jamaicans so good at sprinting? *The Guardian*, 21 July 2014. http://www.theguardian.com/commentisfree/2014/jul/21/jamaicans-sprinting-athletics-commonwealth-games

Bruni, L., and Porta, P. (eds), (2005). *Economics & Happiness: Framing the Analysis*. Oxford: Oxford University Press.

Bryant, J. M. (2004). An evolutionary social science? A skeptic's brief, theoretical and substantive. *Philosophy of the Social Sciences* 34(4): 451–492

Bryson, B. (1990). *Mother Tongue: The Story of the English Language*. London: Penguin.

Buchanan, J. (1986). *Liberty, Market and the State: Political Economy in the 1980s*. Sussex, UK: Wheatsheaf.

Buechel, S. D., I. Booksmythe, A. Kotrschal, M.D. Jennions and N. Kolm (2016). Artificial selection on male genitalia length alters female brain size. *Proceedings of the Royal Society B*, http://rspb.royalsocietypublishing.org/content/283/1843/20161796.

Bunge, M. (2003). *Emergence and Convergence: Qualitative Novelty and the Unity of Knowledge*. Toronto: University of Toronto Press.

Burnet, F. M. (1976). A modification of Jerne's theory of antibody production using the concept of clonal selection. *CA: A Cancer Journal for Clinicians* 26(2): 119–121.

Burnham, T. C., S. Lea, A. Bell et al. (2016). Evolutionary behavioral economics. *In* D. S. Wilson and A. Kirman (eds), *Complexity and Evolution: A New Synthesis for Economics*. Cambridge, MA: MIT Press, pp. 113–145.

Buskes, C. (2006). *Evolutionair Denken: De Invloed van Darwin op ons Wereldbeeld. [Evolutionary Thinking: The Influence of Darwin on our Worldview.]* Amsterdam: Nieuwezijds.

Buss, D. M. (1989). Sex differences in human mate preferences: evolutionary hypotheses tested in 37 cultures. *Behavioral and Brain Sciences* 12:1–49

Buss, D. M. (1995). Evolutionary psychology. A new paradigm for psychological science. *Psychological Inquiry* 6: 1–49.

Cairns, J., J. Overbaugh and S. Miller (1988). The origin of mutants. *Nature* 335: 142–145.

Callaway, E. (2015). Teeth from China reveal early human trek out of Africa. *Nature* news, doi:10.1038/nature.2015.18566

Callaway, E. (2016). Did humans drive 'hobbit' species to extinction? *Nature* news, doi:10.1038/nature.2016.19651

Callaway, E. (2017). Oldest *Homo sapiens* fossil claim rewrites our species' history. *Nature* 546 (June): news, doi:10.1038/nature.2017.22114

Calvin, W. (1987). The brain as a Darwin machine. *Nature* 330: 33–44.

Camerer, C. F., and E. Fehr (2006). When does 'economic man' dominate social behaviour? *Science*: 311: 37–52.

Cameron, R. (1989). *A Concise Economic History of the World: From Paleolithic Times to the Present*. Oxford: Oxford University Press.

Campbell, D. T. (1960). Blind variation and selective retention in creative thought as in other knowledge processes. *Psychological Review* 67: 380–400.

Campbell, D. T. (1969). Variation and selective retention in socio-cultural evolution. *General Systems* 14: 69–85.

Campbell D. T. (1974). Evolutionary epistemology. *In* P. Schilipp (ed.), *The Philosopy of Karl Popper*. LaSalle, IL: Open Court Publishing Co.

Campbell, N. A. (1996). *Biology*, 4th edn. Menlo Park, CA: Benjamin/Cummings Publishing Company.

Cann, R. L., M. Stoneking and A. C. Wilson (1987). Mitochondrial DNA and human evolution. *Nature* 325: 31–36.

Cantner, U., and A. Pyka (2001). Classifying technology policy from an evolutionary perspective. *Research Policy* 30(5): 759–775.

Carey, G. (2002). *Human Genetics for the Social Sciences*. London: Sage.

Carlsson, B., S. Jacobsson, M. Holmén and A. Rickne (2002). Innovation systems: analytical and methodological issues. *Research Policy* 31: 233–245.

Carneiro, R. L. (1967). On the relationship between size of population and complexity of social organisation. *Southwestern Journal of Anthropology* 23: 234–243.

Carneiro, R. L. (1968). Ascertaining, testing, and interpreting sequences of cultural development. *Southwestern Journal of Anthropology* 24: 354–374.

Carneiro, R. L. (1970). Scale analysis, evolutionary sequences, and the rating of cultures. *In* R. Naroll and R. Cohen (eds), *A Handbook of Method in Cultural Anthropology*. New York: Natural History Press, pp. 834–871.

Carroll, G. R., and M. T. Hannan (2000). *The Demography of Corporations and Industries*. Princeton, NJ: Princeton University Press.

Carroll, R. L. (2000). Towards a new evolutionary synthesis. *Trends in Ecology and Evolution* 15: 27–32.

Castells, M. (1996). *The Rise of the Network Society*. Volume I of *The Information Age: Economy, Society and Culture*. Oxford: Blackwell.

Castro Dopico, X., M. Evangelou, R. C. Ferreira et al. (2015). Widespread seasonal gene expression reveals annual differences in human immunity and physiology. *Nature Communications* 6, article number 7000.

Cavalier-Smith, T. (2000). Membrane heredity and early chloroplast evolution. *Trends in Plant Science* 5: 174–182.

Cavalli-Sforza, L. L., and F. Cavalli-Sforza (1995). *The Great Human Diasporas: The History of Diversity and Evolution*. Cambridge, MA: Perseus Books.

Cavalli-Sforza, L. L., and M. W. Feldman (1973a). Models for cultural inheritance. I. Group mean and within group variation. *Theoretical Population Biology* 4: 42–55.

Cavalli-Sforza, L. L., and M. W. Feldman (1973b). Cultural versus biological inheritance: phenotypic transmission from parent to children (a theory of the effect of parental phenotypes on children's phenotype). *American Journal of Human Genetics* 25: 433–445.

Cavalli-Sforza, L. L., and M. W. Feldman (1981). *Cultural Transmission and Evolution: A Quantitative Approach*. Princeton, NJ: Princeton University Press.

Cavalli-Sforza, L. L., and M.W. Feldman (2003). The application of molecular genetic approaches to the study of human evolution. *Nature* (Genetics Supplement) 33: 266–275.

Chen, S., C. Chang and M. Wen (2014). Social networks and macroeconomic stability. *Economics: The E-Journal* 8: 1–40.

Choi, J.-K., and S. Bowles (2007). The coevolution of parochial altruism and war. *Science* 318: 636–640.

Chopra, D., and R. E. Tanzi (2015). *Super Genes: Unlock the Astonishing Power of Your DNA for Optimum Health and Well-Being*. New York: Penguin Random House.

Claessen, H., and P. Kloos (1978). *Evolutie en Evolutionism. [Evolution and Evolutionism.]* Assen: Van Gorcum.

Clark C. W. (1990). *Mathematical Bioeconomics: The Optimal Management of Renewable Resources* (2nd edn). New York: Wiley.

Clark, G. (2007). *A Farewell to Alms: A Brief Economic History of the World*. Princeton, NJ: Princeton University Press.

Clark, G. (2008). In defence of the Malthusian interpretation of history. *European Review of Economic History* 12(2): 175–199.

Clark, P. U., J. D. Shakun, S. A. Marcott et al. (2016). Consequences of twenty-first-century policy for multi-millennial climate and sea-level change. *Nature Climate Change* 6(4): 360–369.

Cloak, F. T. (1975). Is a cultural ethology possible? *Human Ecology* 3: 161–182.

Clune, J., J.-B. Mouret and H. Lipson (2013). The evolutionary origins of modularity. *Proceedings of the Royal Society B, Biological Sciences*, http://rspb.royalsocietypublishing.org/content/280/1755/20122863

Coase, R. (1937). The nature of the firm. *Economica* 4(16): 386–405.

Cochran G., J. Hardy and H. Harpending (2006). Natural history of Ashkenazi intelligence. *Journal of Biosocial Science* 38(5): 659–693.

Cohen, J. E. (1995). *How Many People can the Earth Support*. New York: Norton.

Cohen, J. (2014). Early AIDS virus may have ridden Africa's rails. *Science* 346(6205): 21–22.

Coleman, W. (2001). The strange ´laissez faire´ of Alfred Russel Wallace: the connection between natural selection and political economy reconsidered. *In* J. Laurent and J. Nightingale (eds), *Darwinism and Evolutionary Economics*. Cheltenham, UK: Edward Elgar, pp. 36–48.

Colwell, R. (1981). Group selection is implicated in the evolution of female-biased sex ratios. *Nature* 290: 401–404.

Comin, D., W. Easterly and E. Gong (2010). Was the wealth of nations determined in 1000 BC? *American Economic Journal: Macroeconomics* 2: 65–97-

Common, M., and C. Perrings (1992). Towards an ecological economics of sustainability. *Ecological Economics* 6: 7–34

Conlisk, J. (1989). An aggregate model of technical change. *Quarterly Journal of Economics* 104: 787–821.

Conway Morris, C. (1998). *The Crucible of Creation: The Burgess Shale and the Rise of Animals*. Oxford: Oxford University Press.

Conway Morris, C. (2003). *Life's Solution: Inevitable Humans in a Lonely Universe*. Cambridge: Cambridge University Press.

Cook, J. (2017). Understanding and countering climate science denial. *Journal & Proceedings of the Royal Society of New South Wales* 150(2): 207–219.

Cook, L. M., B. S. Grant, I. J. Saccheri and J. Mallet (2012). Selective bird predation on the peppered moth: the last experiment of Michael Majerus. *Biology Letters* 8(4): 609–612.

Cordes, C., P. Richerson, R. McElreath and P. Strimling (2008). A naturalistic approach to the theory of the firm: the role of cooperation and cultural evolution. *Journal of Economic Behavior and Organization* 68(1): 125–139.

Cosmides, L., and J. Tooby (1994). Better than rational: evolutionary psychology and the invisible hand. *American Economic Review* 84(2): 327–332.

Cosmides, L., and J. Tooby (1997). *Evolutionary Psychology: A Primer*. http://www.cep.ucsb.edu/primer.html

Costa, J. T. (2014). *Wallace, Darwin and the Origin of Species*. Cambridge, MA: Harvard University Press.

Costanza, R. (ed.) (1991). *Ecological Economics: The Science and Management of Sustainability*. New York: Columbia University Press.

Costanza, R., L. Wainger, C. Folke and K.-G. Maler (1993). Modeling complex ecological economic systems: towards an evolutionary, dynamic understanding of people and nature. *BioScience* 43: 545–555.

Couclelis, H. (1985). Cellular worlds: a framework for modelling micro–macro dynamics. *Environment and Planning A* 17: 585–596.

Courtiol, A., J.E. Pettayd, M. Jokelae, A. Rotkirchf and V. Lummaa (2012). Natural and sexual selection in a monogamous historical human population. *Proceedings of the National Academy of Sciences of the USA* 109(21): 8044–8049.

Courtois, P., and T. Tazdaït (2014). Bargaining over a climate deal: deadline and delay. *Annals of Operations Research* 220(4): 205–221.

Coyne, J. A. (2009). *Why Evolution is True*. Oxford: Oxford University Press.

Crawford, C., and D. Krebs (eds) (1998). *Handbook of Evolutionary Psychology*. Mahwah, NJ: Lawrence Erlbaum.

Crick, F. (1981). *Life Itself: Its Origin and Nature*. New York: Simon and Schuster.

Crispo, E. (2007). The Baldwin effect and genetic assimilation: revisiting two mechanisms of evolutionary change mediated by phenotypic plasticity. *Evolution* 61(11): 2469–2479.

Croft D. P., L. J. N. Brent, D. W. Franks and M. A. Cant (2015). The evolution of prolonged life after reproduction. *Trends in Ecology and Evolution* 30(7): 407–416.

Cullen, B. (1995). On cultural group selection. *Current Anthropology* 36: 819–820.

Cullis, R. (2007). *Patents, Inventions and the Dynamics of Innovation*. Cheltenham, UK: Edward Elgar.

Cummings, A. M., T. Stoker, and R. J. Kavlock (2007). Gender-based differences in endocrine and reproductive toxicity. *Environmental Research* 104(1): 96–107.

Currie, T., P. Turchin, J. Bednar et al. (2016). Evolution of institutions and organizations. *In* D. S. Wilson and A. Kirman (eds), *Complexity and Evolution: A New Synthesis for Economics*. Cambridge, MA: MIT Press, pp. 201–238.

Cyert, R. M., and J. G. March (1963). *A Behavioral Theory of the Firm*. New York: Prentice Hall.

Daly, H.E. (1977). *Steady-State Economics: The Political Economy of Biophysical Equilibrium and Moral Growth*. Los Angeles, CA: W.H. Freeman and Co.

Daly, H. E. (1993). The perils of free trade. *Scientific American* 269(5): 24–29.

Daly, H. E., and J. W. Cobb (1989). *For the Common Good: Redirecting the Economy Toward Community, the Environment and a Sustainable Future*. Boston, MA: Beacon Press.

Darwin, C. (1958). *The Autobiography of Charles Darwin: 1809–1882. With original omissions restored. Edited with Appendix and Notes by his grand-daughter Nora Barlow*. London and Glasgow: Collins Clear-type Press. Available at: http://darwin-online.org.uk/content/frameset? viewtype=text&itemID=F1497&pageseq=1.

Darwin, C. (1998 [1859]). *On the Origin of Species by Means of Natural Selection or, The Preservation of Favoured Races in the Struggle for Life* (with an introduction by J. Wallace, 1998). Hertfordshire, UK: Wordsworth. (See Darwin Online for other editions: http://darwin-online.org.uk/content/frameset?itemID=F381&viewtype=text&pageseq=1)

Darwin, C. (2004 [1871]). *The Descent of Man: Selection in Relation to Sex*. London: Penguin Classics.

David, P. A. (1985). Clio and the economics of QWERTY. *American Economic Review* 75(2): 332–337.

Davila Ross, M., M. J. Owren and E. Zimmermann (2010). The evolution of laughter in great apes and humans. *Communicative and Integrative Biology* 3(2): 191–194.

Dawkins, R. (1976). *The Selfish Gene*. Oxford: Oxford University Press.

Dawkins, R. (1982). *The Extended Phenotype: The Long Reach of the Gene*. Oxford: Oxford University Press.

Dawkins, R. (1986). *The Blind Watchmaker*. New York: W.W. Norton & Company.

Dawkins, R. (1993). Scientific versus theological knowledge. Letter in *The Independent*, 20 March 1993. (reproduced online with the title 'The Emptiness of Theology', http://www.independent .co.uk/voices/letter-scientific-versus-theological-knowledge-1498837.html).

Dawkins, R. (1996). *Climbing Mount Improbable*. London: Penguin.

Dawkins, R. (1997). Human chauvinism and evolutionary progress. *Evolution* 51(3): 1015–10120.

Dawkins, R. (2001). Sustainability doesn't come naturally: a Darwinian perspective on values. Inaugural lecture, the Values Platform for Sustainability, The Environment Foundation, 14 November, 2001, Royal Institution, London.

Dawkins, R. (2006). *The God Delusion*. Oxford: Oxford University Press.

Dawkins, R. (2009). *The Greatest Show on Earth: The Evidence for Evolution*. New York: The Free Press.

de la Croix, D., and A. Gosseries (2009). Population policy through tradable procreation entitlements. *International Economic Review* 50(2): 507–542.

de Haan, J. (2006). How emergence arises. *Ecological Complexity* 3(4): 293–301.

de Vries, J. (1994). The Industrial Revolution and the industrious revolution. *Journal of Economic History* 54(2): 249–270.

de Vries, J. (2000). Dutch economic growth in comparative-historical perspective, 1500–2000. *De Economist* 148(4): 443–467.

de Waal, F. (1996). *Good Natured: The Origins of Right and Wrong in Humans and Other Animals*. Cambridge, MA: Harvard University Press.

de Waal, F. (2002). Evolutionary psychology: the wheat and the chaff. *Current Directions in Psychological Science* 11(6): 187–191.

de Waal, F. (2006). *Primates and Philosophers: How Morality Evolved*. Princeton, NJ: Princeton University Press.

Delli Gatti, D., S. Desiderio, E. Gaffeo, P. Cirillo and M. Gallegati (2011). *Macroeconomics from the Bottom-up*. Berlin: Springer.

Demsetz, H. (1967). Toward a theory of property rights. *American Economic Review* 57(2): 347–359.

Dennett, D. (1984). *Elbow Room: The Varieties of Free Will Worth Wanting*. Cambridge, MA: MIT Press.

Dennett, D. (1991). *Consciousness Explained*. Boston, MA: Little Brown.

Dennett, D. (1995). *Darwin's Dangerous Idea: Evolution and the Meanings of Life*. New York: Simon and Schuster.

Dennett, D. (2003). *Freedom Evolves*. London: Penguin.

Dennett, D. C. (2006). *Breaking the Spell: Religion as a Natural Phenomenon*. London: Allen Lane.

DeScioli, P., and B.J. Wilson (2011). The territorial foundations of human property. *Evolution and Human Behavior* 32: 297–304.

Devezasa, T., and G. Modelski (2003). Power law behaviour and world system evolution: a millennial learning process. *Technological Forecasting and Social Change* 70(9): 819–859.

Dewey, J. (1943). The development of American pragmatism. *In* D. D. Runes (ed.), *Twentieth Century Philosophy*. New York: Philosophical Library, pp. 451–467.

Diamond, J. (1997a). *Guns, Germs and Steel: A Short History of Everybody for the Last 13000 Years*. London: Vintage.

Diamond, J. (1997b). *Why is Sex Fun? The Evolution of Human Sexuality*. New York: Basic Books.

Dick, S. J. (2000). Insterstellar humanity. *Futures* 32: 555–567.

Diener, E., and M. E. P. Seligman (2002). Very happy people. *Psychological Science* 13: 81–84.

Ding, N., L. Melloni, H. Zhang, X. Tian and D. Poeppel (2015). Cortical tracking of hierarchical linguistic structures in connected speech. *Nature Neuroscience* 19: 158–164.

Dixit, R. K., and R. S. Pindyck (1994). *Investment under Uncertainty*. Princeton, NJ: Princeton University Press.

Dixon, D. (1981). *After Man: A Zoology of the Future*. New York: St. Martin's Press.

Dobzhansky, T. (1973). Nothing in biology makes sense except in the light of evolution. *American Biology Teacher* 35(3): 125–129.

Doody, J.S., S. Freedberg and J. S. Keogh (2009). Communal egg-laying in reptiles and amphibians: evolutionary patterns and hypotheses. *The Quarterly Review of Biology* 84(3): 229–252.

Dooley, K. J. (1997). A complex adaptive systems model of organization change. *Nonlinear Dynamics, Psychology, and Life Sciences* 1(1): 69–97.

Dopfer, K., and J. Potts (eds) (2014). *The New Evolutionary Economics*. Cheltenham, UK: Edward Elgar.

Dopfer, K., J. Foster and J. Potts (2004). Micro–meso–macro. *Journal of Evolutionary Economics* 14: 263–279.

Dosi, G. (1982). Technological paradigms and technological trajectories. A suggested interpretation of the determinants and directions of technical change. *Research Policy* 11(3): 147–162.

Dosi, G. (1988). Sources, procedures, and microeconomic effects of innovation. *Journal of Economic Literature* 26: 1120–1171.

Dosi, G., C. Freeman, R. Nelson, G. Silverberg, and L. Soete (eds) (1988). *Technical Change and Economic Theory*. New York: Pinter Publishers.

Dosi, G., G. Fagiolo, M. Napoletano and A. Roventini (2013). Income distribution, credit and fiscal policies in an agent-based Keynesian model. *Journal of Economic Dynamics and Control* 37: 1598–1625.

Dosi, G., L. Marengo and C. Pasquali (2006). How much should society fuel the greed of innovators? On the relations between appropriability, opportunities and rates of innovation. *Research Policy* 35: 1110–1121.

Dosi, G., K. Pavitt and L. Soete (1990). *The Economics of Technological Change and International Trade*. New York: Harvester Wheatsheaf.

Dowding, K. (2000). How not to use evolutionary theory in politics: a critique of Peter John. *British Journal of Politics and International Relations* 2(1): 72–80.

Driscoll, J. C., and S. Holden (2014). Behavioral economics and macroeconomic models. *Journal of Macroeconomics* 41: 133–147.

Dunbar, R. (1996). *Grooming, Gossip and the Evolution of Language*. London: Faber and Faber.

Dunbar, R. (2016) Do online social media cut through the constraints that limit the size of offline social networks? *Royal Society Open Science* 3: 150292.

Duntley, J. D., and T.K. Shackelford (2008). Darwinian foundations of crime and law. *Aggression and Violent Behavior* 13: 373–382.

Durham, W. H. (1991). *Coevolution: Genes, Culture and Human Diversity*. Stanford, CA: Stanford University Press.

Eagleman, D. (2012). The Umwelt. *In* J. Brockman (ed.), *This Will Make You Smarter: New Scientific Concepts to Improve Your Thinking*. London: Transworld Publishers.

Earp, S. E., and D. L. Maney (2012). Birdsong: is it music to their ears? *Frontiers of Evolutionary Neuroscience* 4, article 14, https://www.frontiersin.org/articles/10.3389/fnevo.2012.00014/full

Ecklund, E. H., C. P. Scheitle, J. Peifer and D. Bolger (2016). Examining links between religion, evolution views, and climate change skepticism. *Environment and Behavior*, http://journals.sagepub.com/doi/abs/10.1177/0013916516674246.

Economides, N. (1996). The economics of networks. *International Journal of Industrial Organization* 14(6): 673–699.

Edelman, G. M. (1987). *Neural Darwinism: The Theory of Neural Group Selection*. Oxford: Oxford University Press.

Edmonds, B. (2005). The revealed poverty of the gene-meme analogy: why memetics per se has failed to produce substantive results. *Journal of Memetics - Evolutionary Models of Information Transmission*, 9. http://cfpm.org/jom-emit/2005/vol9/edmonds_b.html

Edquist, C. (2005). Systems of innovations: perspectives and challenges. *In* J. Fagerberg, D. C. Mowery, and R. R. Nelson (eds), *Oxford Handbook of Innovation*. Oxford: Oxford University Press, pp. 181–208.

Ehrlich, P. R., and J. P. Holdren (1971). Impact of population growth. *Science* 171(3977): 1212–1217.

Ehrlich, P. R., and P. H. Raven (1964). Butterflies and plants: a study in coevolution. *Evolution* 18: 568–608.

Ehrlich, P. R., G. Wolff, G. C. Daily et al. (1999). Knowledge and the environment. *Ecological Economics* 30: 267–284.

Eiben, A. E., and Smith, J. E. (2003). *Introduction to Evolutionary Computing*. Berlin: Springer.

Eibl-Eibesfeldt, I. (1989). *Human Ethology*. New York: Aldine de Gruyter.

Eigen, M., and R. Winkler (1983). *The Laws of the Game: How The Principles of Nature Govern Chance*. Princeton, NJ: Princeton University Press.

Eldredge, N. (1997). Evolution in the marketplace. *Structural Change and Economic Dynamics* 8(4): 385–398.

Eldredge, N., and S. J. Gould (1972). Punctuated equilibria: an alternative to phyletic gradualism. *In* T. J. M. Schopf (ed.), *Models in Paleobiology*. San Francisco, CA: Freeman Cooper,82–115.

Engber, D. (2012). Are humans monogamous or polygamous? *Slate*, October 9, 2012. http://www.slate.com/articles/health_and_science/human_evolution/2012/10/are_humans_monogam ous_or_polygamous_the_evolution_of_human_mating_strategies_.html

Epstein, C., and R. Axtell (1996). *Growing Artificial Societies: Social Science from the Bottom Up*. Cambridge, MA: MIT Press.

Erwin, D. H., M. Laflamme, S. M. Tweedt et al. (2011). The Cambrian conundrum: Early divergence and later ecological success in the early history of animals. *Science* 334: 1091–1097.

Eshel, I., L. Samuelson and A. Shaked (1998). Altruists, egoists and hooligans in a local interaction model. *American Economic Review* 88: 157–179.

Eshel, I., L. Samuelson and A. Shaked (1999). The emergence of kinship behaviour in structured populations of unrelated individuals. *International Journal of Game Theory* 28: 447–463.

Ethiraj, S. K., and D. A. Levinthal (2004). Modularity and innovation in complex systems. *Management Science* 50(2): 159–173.

Ewald, P. W. (2002). Controlling diseases. *In* J. Brockman (ed.), *The Next Fifty Years: Science in the First Half of the 21st Century*. New York: Vintage Books.

Faber, M., and J. L. R. Proops (1990). *Evolution, Time, Production and the Environment*. Heidelberg: Springer-Verlag.

Fagerberg, J. (1988) Why growth rates differ. *In* G. Dosi, C. Freeman, R. Nelson, G. Silverberg and L. Soete (eds), *Technical Change and Economic Theory*. London: Pinter Publishers.

Fagerberg, J. (2003). Schumpeter and the revival of evolutionary economics: an appraisal of the literature. *Journal of Evolutionary Economics* 13(2): 125–159.

Fehr, E., and U. Fischbacher (2002). Why social preferences matter: the impact of non-selfish motives on competition, cooperation and incentives. *Economic Journal* 112: C1–C33.

Fehr, E., and S. Gächter (1998). Reciprocity and economics: the economic implications of *Homo reciprocans*. *European Economic Review* 42(3–5): 845–859.

Fehr, E., and S. Gächter (2002). Altruistic punishment in humans. *Nature* 415: 137–140.

Fehr, E., and K. M. Schmidt (1999). A theory of fairness, competition, and cooperation. *Quarterly Journal of Economics* 114 (3): 817–868.

Feldman, M. S., A. M. Khademian, H. Ingram and A. S. Schneider (2006).Ways of knowing and inclusive management practices. *Public Administration Review* 66(6): 89–99.

Fine G, and J. Deegan (1996). Three principles of serendipity: insight, chance and discovery in qualitative research. *International Journal of Qualitative Studies in Education* 9(4): 434–447.

Finlayson, C. (2014). *The Improbable Primate: How Water Shaped Human Evolution*. Oxford: Oxford University Press.

Fischer, C., and R. Newell (2008). Environmental and technology policies for climate mitigation. *Journal of Environmental Economics and Management* 55(2): 142–162.

Fisher, R. A. (1930). *The Genetical Theory of Natural Selection*. Oxford: Clarendon Press.

Fodor, J. (2007). Why pigs don't have wings. *London Review of Books* 29(20): 19–22.

Fog, A. (1997). Cultural *r/K* selection. *Journal of Memetics* 1. http://cfpm.org/jom-emit/1997/vol1/fog_a.html

Ford E. B. (1975). *Ecological Genetics* (4th edn). New York: Wiley.

Forsyth, D. (2006). *Group Dynamics* (4th edn). Belmont, CA: Thomson Wadsworth.

Foss, N. J. (1993). Theories of the firm: contractual and competence perspectives. *Journal of Evolutionary Economics* 3: 127–144.

Foster, J. (1997). The self-organisation approach in economics. *Structural Change and Economic Dynamics* 8(4): 427–452.

Foster, K., T. Wenseleers and F. Ratnieks (2005). Kin selection is the key to altruism. *Trends in Ecology and Evolution* 21: 57–60.

Fouquet R., and S. Broadberry (2015). Seven centuries of European economic growth and decline. *Journal of Economic Perspectives* 29(4): 227–244.

Frank, S. (1986). Hierarchical selection theory and sex ratios. I. General solutions for structured populations. *Theoretical Population Biology* 29: 312–342.

Frank, S. (1997). The Price equation, Fisher's fundamental theorem, kin selection, and causal analysis. *Evolution* 51, 1712–1729.

Frank, S. (1998). *Foundations of Social Evolution*. Princeton, NJ: Princeton University Press.

Frank, R. (1999). *Luxury Fever*. New York: Free Press.

Freeman A. M. III (1993). *The Measurement of Environmental and Resource Values: Theory and Methods*. Baltimore, MD: Resources for the Future Press.

Freeman, C. (1987). *Technology Policy and Economic Performance*. London: Pinter.

Freeman, C. (ed.) (1990). *Introduction to Economics of Innovation*. International Library of Critical Writings in Economics. Cheltenham, UK: Edward Elgar.

Freeman, C., and L. Soete (1997). *The Economics of Industrial Innovation* (3rd edn). London: Pinter.

Freeman, L. C. (1957). *An Empirical Test of Folk-Urbanism*. Ann Arbor, MI: University of Michigan Press.

Frenken, K. (2000). A complexity approach to innovation networks. *Research Policy* 29(2): 257–272.

Frenken, K., and R. Nuvolari (2004). The early development of the steam engine: an evolutionary interpretation using complexity theory. *Industrial and Corporate Change* 13(2): 419–450.

Frenken, K., T. Meelen, M. Arets and P. van de Glind (2015). Smarter regulation for the sharing economy. *The Guardian* 20 May 2015. https://www.theguardian.com/science/political-science/2015/may/20/smarter-regulation-for-the-sharing-economy

Friedman, D. (1991). Evolutionary games in economics. *Econometrica* 59(3): 637–666.

Friedman, M. (1953). On the methodology of positive economics. *In* M. Friedman, *Essays in Positive Economics*. Chicago, IL: University of Chicago Press.

Fritz, S. (2008). Why dogs don't enjoy music. *Scientific American Mind* 19: 15.

Frondel, M., N. Ritter and C. M. Schmidt (2008). Germany's solar cell promotion: Dark clouds on the horizon. *Energy Policy* 36(4): 4198–4204.

Frondel, M., N. Ritter, C. M. Schmidt and C. Vance (2010). Economic impacts from the promotion of renewable energy technologies: the German experience. *Energy Policy* 38(8): 4048–4056.

Funk, J. L. (2013). *Technology Change and the Rise of New Industries*. Stanford, CA: Stanford University Press.

Fusco, G., and A. Minelli (2010). Phenotypic plasticity in development and evolution: facts and concepts. *Philosophical Transactions of the Royal Society B, Biological Sciences*. 365: 547–556.

Futia, C. A. (1980). Schumpeterian competition. *Quarterly Journal of Economics* 94(4): 675–695.

Futrell, R., K. Mahowald and E. Gibson (2015). Large-scale evidence for dependency length minimization in 37 languages. *Proceedings of the National Academy of Sciences of the USA* 112(33): 10 336–10 341.

Galbraith, J. K. (1958). *The Affluent Society*. New York: New American Library.

Galor, O., and O. Moav (2002). Natural selection and the origin of economic growth. *Quarterly Journal of Economics* 117(4): 1133–1191.

Galor, O., and M. Stelios (2012). Evolution and the growth process: Natural selection of entrepreneurial traits. *Journal of Economic Theory* 147(2): 759–780.

Garcia, J., and J. C. J. M. van den Bergh (2011). Evolution of parochialism by multilevel selection. *Evolution and Human Behavior* 32: 277–287.

Geels, F. W. (2002). Technological transitions as evolutionary reconfiguration processes: a multi-level perspective and a case-study. *Research Policy* 31: 1257–1274.

Geels, F. W. (2007). Analysing the breakthrough of rock 'n' roll (1930–1970). *Technological Forecasting and Social Change* 74: 1411–1431.

Geels, F. W. (2011). The multi-level perspective on sustainability transitions: Responses to seven criticisms. *Environmental Innovation and Societal Transitions* 1(1): 24–40.

Georgescu-Roegen, N. (1971). *The Entropy Law and the Economic Process*. Cambridge, MA: Harvard University Press.

Georgescu-Roegen, N. (1972). Energy and economic myths. *In* N. Georgescu-Roegen (ed.), *Energy and Economic Myths*. New York: Pergamon, pp. 3–36.

Giampietro, M., S. Ulgiati and D. Pimentel (1997). Feasibility of large-scale biofuel production: does an enlargement of scale change the picture? *BioScience* 47(9): 587–600.

Gibbs, W. W. (2001). Neuroscience: optical illusions - side splitting. *Scientific American* 284(1): 24–26.

Gibson, E. (2012). Pubic hair has a job to do – stop shaving and leave it alone. *The Guardian*, 7 August 2012. http://www.theguardian.com/commentisfree/2012/aug/07/pubic-hair-has-job-stop-shaving.

Giddens, A. (1997). *Sociology* (3rd edn). Cambridge: Polity Press.

Gintis, H. (2000). *Game Theory Evolving: A Problem-Centered Introduction to Modeling Strategic Interaction*. Princeton, NJ: Princeton University Press.

Gintis, H. (2016). A typology of human morality. *In* D.S. Wilson and A. Kirman (eds.), *Complexity and Evolution: A New Synthesis for Economics*, Cambridge, MA: MIT Press. pp. 97–112.

Gintis, H., C. van Schaik and C. Boehm (2015). Zoon politikon: the evolutionary origins of human political systems. *Current Anthropology* 56(3): 327–353.

Givel, M. (2006). Punctuated equilibrium in limbo: The tobacco lobby and US state policy making from 1990 to 2003. *Policy Studies Journal* 43(3): 405–418.

Glenn, J. C. (2000). Millennium Project's draft scenarios for the next 1000 years. *Futures* 32: 603–612.

Glimcher, P. W. (2016). Proximate mechanisms of individual decision-making behavior. *In* D.S. Wilson and A. Kirman (eds.), *Complexity and Evolution: A New Synthesis for Economics*. Cambridge, MA: MIT Press, pp. 85–96.

Godfrey L. R. (ed.) (1983). *Scientists Confront Creationism*. New York: Norton.

Goldschmidt, R. (1940). *The Material Basis of Evolution*. New Haven, CT: Yale University Press.

Goldschmidt, T. (1996). *Darwin's Dreampond: Drama in Lake Victoria*. Cambridge, MA: MIT Press.

Gómez-Robles, A., W. D. Hopkins, S. J. Schapiro and C. C. Sherwood (2015). Relaxed genetic control of cortical organization in human brains compared with chimpanzees. *Proceedings of the National Academy of Sciences of the USA* 112(48): 14 799–14 804.

Goodnight, C. (1990a). Experimental studies of community evolution. I. The response to selection at the community level. *Evolution* 44: 1614–1624.

Goodnight, C. (1990b). Experimental studies of community evolution. II. The ecological basis of the response to community selection. *Evolution* 44: 1625–1636.

Goodnight, C. (2000). Heritability at the ecosystem level. *Proceedings of the National Academy of Sciences of the USA* 97: 9365–9366.

Goodnight, C., and L. Stevens (1997). Experimental studies of group selection: what do they tell us about group selection in nature? *The American Naturalist* 150(Suppl.): S59–S79.

Goodwin, R. M. (1967). A growth cycle. *In* C. H. Feinstein (ed.). *Socialism, Capitalism and Economic Growth*. London: Macmillan.

Gordon, R. J. (2016). *The Rise and Fall of American Growth: The US Standard of Living since the Civil War*. Princeton, NJ: Princeton University Press.

Goudsblom, J. (1995). *Fire and Civilization*. New York: Viking Penguin.

Goudsblom, J. (2001). *Stof waar Honger uit Ontstond: Over Evolutie en Sociale Processen*. Amsterdam: Meulenhoff.

Goudsblom, J. (2002). Introductory overview: the expanding anthroposphere. *In* B. de Vries and J. Goudsblom (eds.). *Mappae Mundi: Humans and their Habitats in a Long-term Socio-Ecological Perspective – Myths, Maps and Models*. Amsterdam: Amsterdam University Press.

Gould, S. J. (1988). On replacing the idea of progress with an operational notion of directionality. *In* M. Nitecki (ed.). *Evolutionary Progress*. Chicago, IL: University of Chicago Press.

Gould, S. J. (1989). *Wonderful Life: The Burgess Shale and the Nature of History*. New York: W.W. Norton & Co.

Gould, S. J. (1997a). Darwinian fundamentalism. *The New York Review of Books*, June 12, 1997, pp. 34–37.

Gould, S. J. (1997b). Evolution: the pleasures of pluralism debate. *New York Review of Books*, June 26, 1997, pp. 47–52.

Gould, S. J. (2002). *Rocks of Ages: Science and Religion in the Fullness of Life*. New York: Ballantine Books.

Gould, S. J., and N. Eldredge (1993). Punctuated equilibrium comes of age. *Nature* 366: 223–227.

Gould, S. J., and R. C. Lewontin (1979). The spandrels of San Marco and the Panglossian paradigm: a critique of the adaptationist programme. *Proceedings of the Royal Society of London B* 205: 581–598.

Gould, S. J., and E. S. Vrba (1982). Exaptation: a missing term in the science of form. *Paleobiology* 8(1): 4–15.

Goulder, L. H. (2004). *Induced Technological Change and Climate Policy*. Washington DC: Pew Center on Global Climate Change.

Gowdy J. (1994). *Coevolutionary Economics: The Economy, Society and the Environment*. Dordrecht, the Netherlands: Kluwer.

Gowdy, J. (ed.) (1997). Economics and Biology. Special issue of *Structural Change and Economic Dynamics* 8(4).

Gowdy, J. (ed.) (1998). *Limited Wants, Unlimited Means: A Reader on Hunter-Gatherer Economics and the Environment*. Washington DC: Island Press.

Gowdy J. (1999). Evolution, environment and economics. *In* J. C. J. M. van den Bergh (ed.), *Handbook of Environmental and Resource Economics*. Cheltenham, UK: Edward Elgar.

Graedel, T. E., and B. R. Allenby (1995). *Industrial Ecology*. Englewood Cliffs, NJ: Prentice Hall.

Grafen, A. (1984). Natural selection, kin selection and group selection. *In* J. Krebs and N. Davies (eds), *Behavioural Ecology* (2nd edn). Oxford: Blackwell, pp. 62–84.

Grafen, A. (2007). Detecting kin selection at work using inclusive fitness. *Proceedings of the Royal Society (London)* B 274: 713–719.

Griffin, A., and West, S. (2002). Kin selection: fact and fiction. *Trends in Ecology and Evolution* 17: 15–21.

Griffiths, P. E. (2008). Ethology, sociobiology, and evolutionary psychology. *In* S. Sarkar and A. Plutynski (eds), *A Companion to the Philosophy of Biology*. Oxford: Blackwell.

Grimm, V. (1999). Ten years of individual-based modelling in ecology: what have we learned and what could we learn in the future. *Ecological Modelling* 115(2–3): 129–148.

Grin, J., J. Rotmans, J. Schot, F.W. Geels and D. Loorbach (2010). *Transitions to Sustainable Development: New Directions in the Study of Long Term Transformative Change*. London: Routledge.

Groenewegen, P. (2001). The evolutionary economics of Alfred Marshall: An overview. *In* J. Laurent and J. Nightingale (eds.), *Darwinism and Evolutionary Economics*. Cheltenham, UK: Edward Elgar, pp. 49–62.

Guiso, L., P. Sapienza and L. Zingales (2009). Cultural biases in economic exchange? *Quarterly Journal of Economics* 124: 1095–1131.

Gunderson L., and Holling C.S. (eds.) (2002). *Panarchy: Understanding Transformation in Human and Natural Systems*. Washington DC: Island Press.

Gürerk, O., B. Irlenbusch and B. Rockenbach (2006). The competitive advantage of sanctioning institutions. *Science* 312: 108.

Haberl, H., K. H. Erb, F. Krausmann et al. (2007). Quantifying and mapping the human appropriation of net primary production in Earth's terrestrial ecosystems. *Proceedings of the National Academy of Sciences of the USA* 104(31): 12 942–12 947.

Haeckel, E. (1879). *The Evolution of Man: A Popular Exposition of the Principal Points of Human Ontogeny and Phylogeny*. New York: Appleton.

Hahn, R. W. (1990). The political economy of environmental regulation: towards a unifying framework. *Public Choice*, 65, 21–47.

Hajer, M. (1995). *The Politics of Environmental Discourse: Ecological Modernization and the Policy Process*. Oxford: Clarendon.

Haken, H. (1977). *Synergetics: An Introduction*. Berlin: Springer.

Hall, C. A. S., S. Balogh, and D. J. R. Murphy (2009). What is the minimum EROI that a sustainable society must have? *Energies* 2: 25–47.

Hamilton, W. D. (1964). The genetical theory of social behaviour: I & II. *Journal of Theoretical Biology* 7: 1–16, 17–32.

Hamilton W. D. (1966). The moulding of senescence by natural selection. *Journal of Theoretical Biology* 12(1): 12–45

Hamilton, W. D. (1975). Innate social aptitudes of man: an approach from evolutionary genetics. *In* Fox, R. (ed.), *Biosocial Anthropology*. London: Malaby Press, pp. 133–153.

Hanley N., and Spash C. L. (1993). *Cost–Benefit Analysis and the Environment*. Aldershot, UK: Edward Elgar.

Hannan, M., and G. R. Carroll (1989). Density delay in the evolution of organizational populations: a model and five empirical tests. *Administrative Sciences Quarterly* 34: 11–30.

Hannan, M. T., and J. Freeman (1977). The population ecology of organizations. *American Journal of Sociology* 83 (4): 929–984.

Hannan, M. T., and J. Freeman (1989). *Organizational Ecology*. Cambridge, MA: Harvard University Press.

Hansen, G. D., and E. C. Prescott (2002). Malthus to Solow. *American Economic Review* 92(4): 1205–1217.

Harari, Y. (2014). *Sapiens: A Brief History of Humankind*. New York: Harper.

Harari, Y. (2016). *Homo Deus: A Brief History of Tomorrow*. New York: Harper.

Harcourt, G. C. (1972). *Some Cambridge Controversies in the Theory of Capital*. Cambridge: Cambridge University Press.

Hardin, G. (1968). The tragedy of the commons. *Science* 162: 1243–1248.

Hardy, A. (1960). Was man more aquatic in the past? *New Scientist* 7 (17 March): 642–645.

Harford, J. D. (1998). The ultimate externality. *American Economic Review* 88(1): 260–265.

Harper, P. (2000). The end in sight? Some speculations on environmental trends in the twenty-first century. *Futures* 32: 361–384.

Harris, M. (1977). *Cannibals and Kings: The Origins of Culture*. New York: Random House.

Harris, S. (2012). *Free Will*. New York: Free Press.

Harrison, P. (2015). *The Territories of Science and Religion*. Chicago, IL: University of Chicago Press.

Hartley, T. (2017). The evolution of first possession among animals, and communal command and titled property ownership among humans. Working paper. Bristol: Bristol University.

Hawkes, K. (2004). Human longevity: the grandmother effect. *Nature* 428 (6979): 128–129.

Hawkes, K., and K. Smith (2009). Evaluating grandmother effects. *American Journal of Physical Anthropology* 140(1): 173–176.

Hay, C. (1999). Crisis and the structural transformation of the state: Interrogating the process of change. *British Journal of Politics and International Relations* 1(3): 317–344.

Hayden, B. (1972). Population control among hunter/gatherers. *World Archaeology* 4(2): 205–221.

Hayek, F. A. (1944). *The Road to Serfdom*. Chicago, IL: University of Chicago Press.

Hayek, F. A. (1948). *Individualism and Economic Order*. Chicago, IL: University of Chicago Press.

Hayek, F. A. (ed.). (1967). Notes on the evolution of systems of rules of conduct. *In Studies in Philosophy, Politics and Economics*. Chicago, IL: University of Chicago Press, pp. 66–81.

Hayek, F. A. (1976). *Law, Legislation and Liberty. Vol. 3. The Political Order of a Free People*. London: Routledge.

Hazen, R.M. (2005). *Gen*e*sis: The Scientific Quest for Life's Origin*. Washington DC: Joseph Henry Press.

Hazen, R. M., D. Papineau, W. Bleeker et al. (2008). Mineral evolution. *American Mineralogist* 93: 1693–1720.

Heath, B., R. Hill and F. Ciarallo (2009). A survey of agent-based modeling practices (January 1998 to July 2008). *Journal of Artificial Societies and Social Simulation* 12(4): article 9, http://jasss.soc.surrey.ac.uk/12/4/9.html.

Hedenus, F., C. Azar and K. Lindgren (2006). Induced technological change in a limited foresight optimization model. *The Energy Journal* 27: 109–122.

Hekkert, M., R. Suurs, S. Negro, S. Kuhlmann and R. Smits, (2007). Functions of innovation systems: a new approach for analysing technological change. *Technological Forecasting and Social Change* 74: 413–432.

Helpman, E., and P. Krugman (1985). *Market Structure and Foreign Trade*. Cambridge, MA: MIT Press.

Henrich, J. (2004). Cultural group selection, coevolutionary processes and large-scale cooperation. *Journal Economic Behavioral Organization* 53: 3–35.

Henrich, J. (2015). *The Secret of Our Success: How Culture is Driving Human Evolution, Domesticating Our Species, and Making Us Smarter*. Princeton, NJ: Princeton University Press.

Henrich, J., and R. Boyd (2001). Why people punish defectors: weak conformist transmission can stabilize costly enforcement of norms in cooperative dilemmas. *Journal of Theoretical Biology* 208: 79–89.

Henrich, J., and R. McElreath (2003). The evolution of cultural evolution. *Evolutionary Anthropology* 12: 123–135.

Henrich, J., R. Boyd and P. J. Richerson (2008). Five misunderstandings about cultural evolution. *Human Nature* 19(2): 119–137.

Henrich, J., R. McElreath, A. Barr et al. (2006). Costly punishment across human societies. *Science* 312: 1767–1770.

Henrich, N., and J. Henrich (2007). *Why Humans Cooperate*. Oxford: Oxford University Press.

Herculano-Houzel, S. (2009). The human brain in numbers: a linearly scaled-up primate brain. *Frontiers in Human Neuroscience* 3(31), http://dx.doi.org/10.3389/neuro.09.031.2009.

Herrmann, E., J. Call, M. Hernández-Lloreda, B. Hare and M. Tomasello (2007). Humans have evolved specialized skills of social cognition: the cultural intelligence hypothesis. *Science* 317: 1360–1366.

Heylighen, F. (1996). The growth of structural and functional complexity during evolution. *In* F. Heylighen and D. Aerts (eds). *The Evolution of Complexity*. Dordrecht, the Netherlands: Kluwer Academic Publishers.

Hicks J. R. (1932). *The Theory of Wages*. London: Macmillan.

Hinde, R. A. (1999). *Why Gods Persist*. London: Routledge Press.

Hislop, A. (1959). *The Two Babylons: Or, the Papal Worship*. 1853 (2nd American edn). Loizeaux: Neptune.

Hodgson, G. M. (1988). *Economics and Institutions. A Manifesto for a Modern Institutional Economics*. Oxford: Polity Press.

Hodgson, G. M. (1993a). *Economics and Evolution: Bringing Life Back Into Economics*. Ann Arbor, MI: University of Michigan Press.

Hodgson, G. M. (1993b). The Mecca of Alfred Marshall. *Economic Journal* 103: 406–415.

Hodgson, G. M. (ed.) (1995). *Economics and Biology*. Aldershot, UK: Edward Elgar.

Hodgson, G. M. (1997). The evolutionary and non-Darwinian economics of Joseph Schumpeter. *Journal of Evolutionary Economics* 7: 131–145.

Hodgson, G. M. (2002). Darwinism in economics: from analogy to ontology. *Journal of Evolutionary Economics* 12: 259–281.

Hodgson, G. M. (2004). Social Darwinism in Anglophone academic journals: A contribution to the history of the term. *Journal of Historical Sociology* 17(4): 428–463.

Hodgson, G., and T. Knudsen (2006). Dismantling Lamarckism: why descriptions of socioeconomic evolution as Lamarckian are misleading. *Journal of Evolutionary Economics* 16(4): 343–66.

Hoekstra, R., and J. C. J. M. van den Bergh (2005) Harvesting and conservation in a predator–prey system. *Journal of Economic Dynamics Control* 29(6): 1097–1120.

Hoffmann, A. A., and C. M. Sgrò (2011). Climate change and evolutionary adaptation. *Nature* 470(7335): 479–485.

Hofstadter, D. R. (1979). *Gödel, Escher, Bach: An Eternal Golden Braid. A Metaphorical Fugue on Minds and Machines in the Spirit of Lewis Carroll*. New York: Basic Books.

Hofstadter, R. (1944). *Social Darwinism in American Thought: 1860–1915*. Philedlphia, PA: University of Pennsylvania Press.

Hofstede, G. (1980). Motivation, leadership, and organization: Do American theories apply abroad? *Organizational Dynamics* 9(1): 42–63.

Hofstede, G. (2001). *Culture's Consequences: Comparing Values, Behaviours, Institutions, and Organizations Across Nations* (2nd edn). Thousand Oaks, CA: Sage Publications.

Holland, J. H. (1975). *Adaptation in Natural and Artificial Systems.* Cambridge, MA: MIT Press.

Holland, J. H. (1998). *Emergence: From Chaos to Order.* Cambridge, MA: Perseus Books.

Hölldobler, B., and E. O. Wilson (1994). *Journey to the Ants: A Story of Scientific Exploration.* Cambridge, MA: Harvard University Press.

Holling C. S. (1973) Resilience and stability of ecological systems. *Annual Review of Ecological Systems* 4: 1–24.

Holling, C. S. (ed.) (1978). *Adaptive Environmental Assessment and Management.* New York: Wiley.

Holling C. S. (1986). The resilience of terrestrial ecosystems: Local surprise and global change. *In* W. C. Clark and R. E. Munn (eds), *Sustainable Development of the Biosphere.* Cambridge: Cambridge University Press.

Holling C. S., D. W. Schindler, B. Walker and J. Roughgarden (1994). Biodiversity in the functioning of ecosystems: an ecological primer and synthesis. *In* C. Perrings, K.-G. Mäler, C. Folke, C. S. Holling and B.-O. Jansson (eds), *Biodiversity Loss: Ecological and Economic Issues.* Cambridge: Cambridge University Press.

Hong, L., and S. E. Page (2004). Groups of diverse problem solvers can outperform groups of high-ability problem solvers. *Proceedings of the National Academy of Sciences of the USA* 101(46): 16 385–16 389.

Honing, H., and A. Ploeger (2012). Cognition and the evolution of music: pitfalls and prospects. *Topics in Cognitive Science* 4(4): 513–524.

Horan, R. D., J. F. Shogren and E. H. Bulte (2003). A Paleoeconomic theory of co-evolution and extinction of domesticable animals. *Scottish Journal of Political Economy* 50(2): 131–148.

Horan R. D., E. H. Bulte and J. F. Shogren (2005) How trade saved humanity from biological exclusion: an economic theory of Neanderthal extinction. *Journal Economic Behavior Organization* 58(1): 1–29.

Horbach, J. (2008). Determinants of environmental innovation – New evidence from German panel data sources. *Research Policy* 37: 163–173.

Hrdy, S. (2009). *Mothers and Others: The Evolutionary Origins of Mutual Understanding.* Cambridge, MA: Harvard University Press.

Huberty, M., and J. Zysman (2010). An energy system transformation: Framing research choices for the climate challenge. *Research Policy* 39(8): 1027–1029.

Hudson, M. (2016). Pablo Picasso: women are either goddesses or doormats. *The Telegraph,* 8 April 2016. http://www.telegraph.co.uk/art/artists/pablo-picasso-women-are-either-goddesses-or-doormats.

Hull, D. L. (1982). The naked meme. *In* H. C. Plotkin (ed.), *Learning, Development and Culture: Essays in Evolutionary Epistemology.* New York: New York, pp. 273–327.

Hull, D. L. (1988). *Science as a Process: An Evolutionary Account of the Social and Conceptual Development of Science.* Chicago, IL: University of Chicago Press.

Hull D. L. (2001). *Science and Selection: Essays on Biological Evolution and the Philosophy of Science.* Cambridge: Cambridge University Press.

Hume D. (1739). *A Treatise on Human Nature.* Republished 1962. New York: Macnabb.

Hunt, E.K. (2003). *Property and Prophets: The Evolution of Economic Institutions and Ideologies* (7th edn). London: Sharpe.

Huron, D. (2001). Is music an evolutionary adaptation? *Annals of the New York Academy of Sciences* June: 43–61. http://onlinelibrary.wiley.com/doi/10.1111/j.1749-6632.2001.tb05724.x/abstract

Huseynov, A., C. P. E. Zollikofer, W. Coudyzer et al. (2016). Developmental evidence for obstetric adaptation of the human female pelvis. *Proceedings of the National Academy of Sciences of the USA* 113(19): 5227–5232.

Huxley, T. H. (1863). *Evidence as to Man's Place in Nature*. London: Williams and Norgate.

Iannaccone, L. (1999). Economics of religion. *Journal of Economic Literature* 36: 1465–1496.

IEA (2010). *World Energy Outlook*. Paris: International Energy Agency.

IPCC (2014). *Fifth Assessment Report of the Intergovernmental Panel on Climate Change*. Geneva: IPCC. http://www.ipcc.ch/report/ar5

Iwai, K. (1984). Schumpeterian dynamics, part I: an evolutionary model of innovation and imitation. *Journal of Economic Behaviour and Organization* 5(2): 159–190.

Jablonka, E., and M. J. Lamb (2006). *Evolution in Four Dimensions: Genetic, Epigenetic, Behavioral and Symbolic Variation in the History of Life*. Cambridge, MA: MIT Press.

Jackson, T. (2000). *Why is ecological economics not an evolutionary science? 3rd Biennial Conference of the European Society of Ecological Economics (ESEE)*, Vienna University of Economics and Business Administration.

Jackson, T. (2009). *Prosperity without growth? The transition to a sustainable economy*. London: UK Sustainable Development Commission. http://research-repository.st-andrews.ac.uk/handle/10023/2163

Jacobi, M. N., and M. Nordahl (2006). Quasispecies and recombination. *Theoretical Population Biology* 70(4): 479–485.

Jacobsen, N. B., and S. Anderberg (2006). Understanding the evolution of industrial symbiotic networks: the case of Kalundborg. *In* J. C. J. M. van den Bergh and M. A. Janssen (eds), *Economics of Industrial Ecology: Use of Materials, Structural Change and Spatial Scales*. Cambridge, MA: MIT Press, pp. 313–316.

Jacobsson, S. B., and A. Bergek (2011). Innovation system analyses and sustainability transitions: contributions and suggestions for research. *Environmental Innovation and Societal Transitions* 1: 41–57.

Jaffe, A. B., R. G. Newell and R. N. Stavins (2005). A tale of two market failures: technology and environmental policy. *Ecological Economics* 54: 164–174.

Janssen, M. A. (1998). *Modelling Global Change: The Art of Integrated Assessment Modelling*. Cheltenham, UK: Edward Elgar.

Janssen, M. A. (2001). An immune system perspective on ecosystem management. *Conservation Ecology* 5(1), article 13, http://www.consecol.org/vol15/iss1/art13

Janssen M. A. (ed) (2002) *Complexity and Ecosystem Management: The Theory and Practice of Multi-agent Systems*. Cheltenham, UK: Edward Elgar.

Janssen M. A., and B. de Vries (1998). The battle of perspectives: a multi-agent model with adaptive responses to climate change. *Ecological Economics* 26: 43–66.

Janssen, M. A., and W. Jager (2002). Simulating diffusion of green products: Co-evolution of firms and consumers. *Journal of Evolutionary Economics* 12: 283–306.

Janssen M. A., and E. Ostrom (2005). Governing social–ecological systems. *In* K.L. Judd and L. Tesfatsion (eds) *Handbook of Computational Economics II: Agent-based Computational Economics*. Amsterdam: North-Holland.

Jensen, C. X. J. (2016). Adolescent behavior doesn't make sense (except in the light of cultural evolution). *This View of Life Magazine*, https://evolution-institute.org/article/adolescent-behavior-doesnt-make-sense-except-in-the-light-of-cultural-evolution

Johansson P.-O. (1987). *The Economic Theory and Measurement of Environmental Benefits*. Cambridge: Cambridge University Press.

Johansson P. O., and K. G. Löfgren (1985). *The Economics of Forestry and Natural Resources*. Oxford: Blackwell.

John, P. (1999). Ideas and interests; agendas and implementation: An evolutionary explanation of policy change in British local government finance. *British Journal of Politics and International Relations* 1(1): 39–62.

John, P. (2003). Is there life after policy streams, advocacy coalitions, and punctuations: Using evolutionary theory to explain policy change? *The Policy Studies Journal* 31(4): 481–498.

Johnson, D. (2015). *God is Watching You: How the Fear of God Makes us Human*. Oxford: Oxford University Press.

Jones, E. (1981). *The European Miracle: Environments, Economies and Geopolitics in the History of Europe and Asia* (3rd edn). Cambridge: Cambridge University Press.

Jones, S., R. Martin and D. Pilbeam (1992). *The Cambridge Encyclopedia of Human Evolution*. Cambridge: Cambridge University Press.

Jorgenson, D., R. Goettle, M. Sing Hoc and P. Wilcoxen (2009). Cap and trade climate policy and US economic adjustments. *Journal of Policy Modeling* 31: 362–381.

Kahneman, D., and A. Tversky (1979). Prospect theory: an analysis of decision under risk. *Econometrica* 47: 263–291.

Kallis, G., and R. Norgaard (2010). Coevolutionary ecological economics. *Ecological Economics* 69: 690–699.

Kamien, M. I., and N. L. Schwartz (1981). *Market Structure and Innovation*. Cambridge: Cambridge University Press.

Kao, A. B., and I. D. Couzin (2014). Decision accuracy in complex environments is often maximized by small group sizes. *Proceedings of the Royal Society B* 281: 20133305.

Katan, M. (2016). *Voedingsmythes – Over Valse Hoop en Nodeloze Vrees [Food Myths – About False Hope and Needless Fear]*. Amsterdam: Prometheus & Bert Bakker.

Kauffman, S. A. (1993). *The Origins of Order: Self-Organization and Selection in Evolution*. Oxford: Oxford University Press.

Kauffman, S. A. (1995). *At Home in the Universe: The Search for Laws of Self-Organization and Complexity*. Oxford: Oxford University Press.

Ke, Y., B. Su, X. Song et al. (2001). African origin of modern humans in East Asia: a tale of 12 000 Y Chromosomes. *Science* 292(5519): 1151–1153.

Kelly, K. (2010). *What Technology Wants*. New York: Viking Press.

Kemp, R., J. Schot and R. Hoogma (1998). Regime shifts to sustainability through processes of niche formation: the approach of strategic niche management. *Technological Analysis and Strategic Management* 10: 175–196.

Kendal, J., J. J. Tehrani and F. J. Odling-Smee (2011). Human niche construction in interdisciplinary focus. *Philosophical Transactions of the Royal Society of London Series B* 366: 785–792.

Kennedy, J. (1995). Changes in optimal pollution taxes as population increases. *Journal of Environmental Economics and Management*, 28(1): 19–23.

Kennedy, P. (1989). *The Rise and Fall of the Great Powers: Economic Change and Military Conflict from 1500 to 2000*. New York: Vintage Books.

Kerber, W. (2005). Applying evolutionary economics to economic policy: The example of competitive federalism. *In* K. Dopfer (ed.), *Economics, Evolution and the State: The Governance of Complexity*. Cheltenham, UK: Edward Elgar, pp. 296–324.

Kerber,W. (2011). Competition, innovation, and maintaining diversity through competition law. *In* J. Drexl, W. Kerber and R. Podszun (eds), *Economic Approaches to Competition Law: Foundations and Limitations*. Cheltenham, UK: Edward Elgar, pp. 173–201.

Kerr, P. (2002). Saved from extinction: Evolutionary theorising, politics and the state. *British Journal of Politics and International Relations* 4(2): 330–358.

Keuning, D., and D.J. Eppink (1996). *Management en Organisatie: Theorie en Toepassing [Management and Organisation: Theory and Application]*. Stenfert Kroese/Educatieve Partners, Houten.

Khalil, E. L. (1998). The five careers of the biological metaphor in economic theory. *The Journal of Socio-Economics* 27(1): 29–52.

Kimura, M. (1968). Evolutionary rate at the molecular level. *Nature* 217: 624–626.

Kimura, M. (1983). *The Neutral Theory of Molecular Evolution*. Cambridge: Cambridge University Press.

Kingdon, J. W. (1995). *Agendas, Alternatives, and Public Policies* (2nd edn). New York: Longman.

Kirk, A. (2016). Rio 2016 alternate medal table: How countries rank when we adjust for population and GDP. *The Telegraph*, 22 August 2016. http://www.telegraph.co.uk/olympics/2016/08/21/rio-2016-alternate-medal-table-how-countries-rank-when-we-adjust/.

Kirman, A.P. (1992). Whom or what does the representative individual represent? *Journal of Economic Perspectives* 6(2): 117–136.

Kitcher, P. (1982). *Abusing Science: The Case Against Creationism*. Cambridge, MA: MIT Press.

Kitcher, P. (2007). *Living with Darwin: Evolution, Design, and the Future of Faith*. Oxford: Oxford University Press.

Klaassen, G., S. Miketa, K. Larsen and T. Sundqvist, T. (2005). The impact of R&D on innovation for wind energy in Denmark, Germany and the United Kingdom. *Ecological Economics* 54: 227–240.

Klepper, S. (1997). Industry life-cycles. *Industrial and Corporate Change* 6: 145–182.

Klepper, S. (2002). Firm survival and the evolution of industry. *Rand Journal of Economics* 33(1): 37–61.

Kniffin, K.M., and Wilson, D.S. (2010). Evolutionary perspectives on workplace gossip. *Group and Organization Management* 35: 150–176.

Konner, M. (2016). *Women After All: Sex, Evolution, and the End of Male Supremacy*. New York: W.W. Norton.

Koopmans, R. (2006). Het mysterie van de naastenliefde: een evolutionair-sociologische benadering. (The mystery of charity: an evolutionary-sociological approach). *Sociologie* 2: 114–138.

Koseoglu, N. M., J. C. J. M. van den Bergh and J. Subtil Lacerda (2013). Allocating subsidies to R&D or to market applications of renewable energy? Balance and geographical relevance. *Energy for Sustainable Development* 17: 536–545.

Kremer, M. (1993). Population growth and technological change: One million BC: to 1990. *Quarterly Journal of Economics* 108(3): 681–716.

Krier, J.E. (2009). Evolutionary theory and the origin of property rights. *Cornell Law Review* 95: 139–160.

Kroeber, A.L. (1948). *Anthropology: Race, Language, Culture, Psychology, Pre-history*. New York: Harcourt Brace.

Krugman, P. (1979). A model of innovation, technology transfer, and the world distribution of income. *Journal of Political Economy* 87(2): 253–266.

Kuznets, S. (1940). Schumpeter's business cycles. *American Economic Review* 30(2): 257–271.

Laland, K. N., and G. R. Brown (2002). *Sense and Nonsense: Evolutionary Perspectives on Human Behaviour*. Oxford: Oxford University Press.

Laland, K. N., N. Boogert and C. Evans (2014). Niche construction, innovation and complexity. *Environmental Innovation and Societal Transitions* 11: 71–86.

Laland, K. N., F. J. Odling-Smee and M. W. Feldman (1996). The evolutionary consequences of niche construction. *Journal of Evolutionary Biology* 9: 293–316.

Laland, K., et al. vs. Wray, G. A. et al. (2014). Does evolutionary theory need a rethink? Yes Urgently/No, all is well. *Nature* 514: 161–164.

Land, M., and D.-E. Nilsson (2002). *Animal Eyes*. Oxford: Oxford University Press.

Landa, J. T. (2008). The bioeconomics of homogeneous middleman groups as adaptive units: theory and empirical evidence viewed from a group selection framework. *Journal of Bioeconomics* 10: 259–278.

Lane, C. S., B. T. Chorn and T. Johnson (2013). Ash from the Toba supereruption in Lake Malawi shows no volcanic winter in East Africa at 75 ka. *Proceedings of the National Academy of Sciences of the USA* 110(20): 8025–8029.

Lanjouw, J. O., and A. Mody (1996). Innovation and the international diffusion of environmentally responsive technology. *Research Policy* 25: 549–571.

Lansing, J. S., and J. N. Kremer (1993). Emergent properties of Balinese water temples. *American Anthropologist* 95(1): 97–114.

Laurent, J. (2001). Keynes and Darwinism. *In* J. Laurent and J. Nightingale (eds), *Darwinism and Evolutionary Economics*. Cheltenham, UK: Edward Elgar, pp. 63–84.

Laursen, K. (2011). User–producer interaction as a driver of innovation: costs and advantages in an open innovation model. *Science and Public Policy* 38(9): 713–723.

Le, S. (2016). *100 Million Years of Food: What Our Ancestors Ate and Why It Matters Today*. New York: Picador/St. Martin's Press.

Lea, S. E. G. (2006). Money as tool, money as drug: the biologial psychology of a strong incentive. *Behavioral Brain Sciences* 29: 161–209.

Lehmann, L., L. Keller, S. West and D. Roze, (2007). Group selection and kin selection: two concepts but one process. *Proceedings of the National Academy of Sciences of the USA* 104: 6736–6739.

Lehre, A.-C., K. P. Lehre, P. Laake and N. C. Danbolt (2009). Greater intrasex phenotype variability in males than in females is a fundamental aspect of the gender differences in humans. *Developmental Psychobiology* 51(2): 198–206.

Lengnick, M. (2013). Agent-based macroeconomics: A baseline model. *Journal of Economic Behavior and Organization* 86: 102–120.

Lens, L., S. Van Dongen, S. Kark and E. Matthysen (2002). Fluctuating asymmetry as an indicator of fitness: can we bridge the gap between studies? *Biological Reviews of the Cambridge Philosophical Society* 77(1): 27–38.

Lenton, T.M., and J.E. Lovelock (2001). Daisyworld revisited: quantifying biological effects on planetary self-regulation. *Tellus Series B* 53(3): 288–305.

Levin, S., S. Barrett, S. Aniyar et al. (1998). Resilience in natural and socioeconomic systems. *Environment and Development Economics* 3(2): 222–235.

Levinthal, D. A. (1997). Adaptation on rugged landscapes. *Management Science* 43(7): 934–950.

Levy, D. J., A. C. Thavikulwat and P. W. Glimcher (2013). State dependent valuation: The effect of deprivation on risk preferences. *PLoS One* 8: e53978.

Levy, H., M. Levy and S. Solomon (2000). *Microscopic Simulation of Financial Markets: From Investor Behavior to Phenomena*. New York: Academic Press.

Levy, M. J. Jr. (1966). *Modernization and the Structure of Societies*. Princeton, NJ: Princeton University Press.

Lewens, T. (2015). *Cultural Evolution: Conceptual Challenges*. New York: Oxford University Press.

Lewis, O., and S. Steinmo (2010). Taking evolution seriously in political science. *Theory in Biosciences* 129(2–3): 235–245.

Lewontin, R. C. (1957). The adaptation of populations to varying environment. *Cold Spring Harbor Symposia on Quantitative Biology* 22: 395–408.

Lewontin, R. C. (1974). *The Genetic Basis of Evolutionary Change*. New York: Columbia University Press.

Lewontin, R., S. Rose and L. Kamin (1984). *Not in Our Genes: Biology, Ideology and Human Nature*. London: Penguin.

Leyden, P. (2001). Taking the really long view: a GBN interview with Stewart Brand. *Global Business Network*, September 2001.

Leydesdorff, L. (2001). *A Sociological Theory of Communication: The Self-Organisation of the Knowledge-Based Society*. Amsterdam: Universal Publishers.

Li, W.-H. (2006). *Molecular Evolution*. Sunderland, MA: Sinauer Associates.

Liebowitz, S. J., and S. E. Margolis (1995). Path-dependence, lock-in, and history. *Journal of Law, Economics, and Organization* 11: 205–226.

Lim, C., and R. Putnam (2010). Religion, social networks, and life satisfaction. *American Sociological Review* 75: 914–933

Lipson, H., and J. B. Pollack (2000). Automatic design and manufacture of robotic lifeforms. *Nature* 406: 974–978.

Lomax, A., and C. M. Arensberg (1977). A worldwide evolutionary classification of cultures by subsistence systems. *Current Anthropology* 18: 659–708.

Lovelock, J. (1979). *Gaia: A New Look at Life on Earth*. Oxford: Oxford University Press.

Lovelock, J. (2014). *A Rough Guide to the Future*. Allen Lane/Penguin, London.

Lovelock, J. E. (1992). A numerical model for biodiversity. *Philosophical Transactions of the Royal Society B* 338(1286):383–391.

Lovelock, J.E., and L. Margulis (1974). Atmospheric homeostasis by and for the biosphere: the Gaia hypothesis. *Tellus Series A* 26 (1–2): 2–10.

Luehrman, T. (1998). Strategy as a portfolio of real options. *Harvard Business Review* 76(5): 89–99.

Luhmann, N. (1984). *Sociale Systeme: Grundriss einer Allgemeinen Theorie [Social Systems: Outline of a General Theory]*. Frankfurt am Main: Suhrkamp Verlag.

Lumsden, C. J., and E. O. Wilson (1981). *Genes, Mind, and Culture. The Coevolutionary Process*. Cambridge, MA: Harvard University Press.

Lumsden, C., and E. O. Wilson (1983). *Promethean Fire: Reflections on the Origin of Mind*. Cambridge, MA: Harvard University Press.

Lundvall, B.-A. (1988). Innovation as an interactive process: from user–producer interaction to the national system of innovation. *In* G. Dosi, C. Freeman, R. Nelson, G. Silverberg, and L. Soete (eds), *Technical Change and Economic Theory*. London: Pinter, pp. 61–83.

Lundvall, B.-A. (ed.) (1992). *National Systems of Innovation: Towards a Theory of Innovation and Interactive Learning*. London: Pinter.

Lundvall, B.-Å. (2007). National innovation systems: from List to Freeman. *In* H. Hanusch and A. Pyka (eds). *Elgar Companion to Neo-Schumpeterian Economics*. Cheltenham, UK: Edward Elgar.

Lynch, M. (2007). The frailty of adaptive hypotheses for the origins of organismal complexity. *Proceedings of the National Academy of Sciences of the USA* 104, Suppl. 1: 8597–8604.

MacArthur R. H., and E. O. Wilson (1967). *The Theory of Island Biogeography*. Princeton, NJ: Princeton University Press.

Machlup, F. (1962). The supply of inventors and inventions. *In* R. R. Nelson (ed.), *The Rate and Direction of Inventive Activity: Economic and Social Factors*. Princeton, NJ: Princeton University Press, pp. 14–170. (see also http://www.nber.org/chapters/c2116).

Maddison, A. (2001). *The World Economy: A Millenial Perspective*. Paris: OECD.

Magnusson, L., and J. Ottosson (eds) (2009). *The Evolution of Path Dependence*. Cheltenham, UK: Edward Elgar.

Mahoney, J. (2000). Path dependence in historical sociology. *Theory and Society* 29: 507–548.

Malerba, F., R. Nelson, L. Orsenigo and S. Winter (2008). Public policies and changing boundaries of firms in a history friendly model of the coevolution of the computer and semiconductor industries. *Journal of Economic Behaviour and Organisation* 67: 355–380.

Malthus, T. R. (2005 [1798]). *An Essay on the Principle of Population*. New York: Cosimo, Inc..

Mandeville, B. (1970 [1714]). *The Fable of The Bees*. London: Penguin.

March, J. G. (1991). Exploration and exploitation in organizational learning. *Organization Science* 2(1): 71–87.

Margulis, L. (1970). *Origin of Eukaryotic Cells*. New Haven, CT: Yale University Press.

Margulis, L., and D. Sagan (1986). *Origins of Sex: Three Billion Years of Genetic Recombination*. New Haven, CT: Yale University Press.

Marra, P. P., and C. Santella (2016). *Cat Wars: The Devastating Consequences of a Cuddly Killer*. Princeton, NJ: Princeton University Press.

Marshall, A. (1890). *Principles of Economics, Vol. 1*. London: Macmillan.

Marsili, O. (2001). *The Anatomy and Evolution of Industries: Technological Change and Industrial Dynamics*. Cheltenham, UK: Edward Elgar.

Martin. F (1992). *Common-pool Resources and Collective Action: A Bibliography*. Workshop in Political Theory and Policy Analysis. Bloomington, IN: Indiana University.

Martincorena, I., A. S. N. Seshasayee and N. M. Luscombe (2012). Evidence of non-random mutation rates suggests an evolutionary risk management strategy. *Nature* 485: 95–98.

Maturana, H. R., and F. J. Varela (1980). *Autopoiesis and Cognition: The Realization of the Living*. Dordrecht, the Netherlands: Reidel Publishing.

May, R. M. (ed.) (1976). *Theoretical Ecology: Principles and Applications* (2nd edn). Oxford: Oxford University Press.

Maynard Smith, J. (1964). Group selection and kin selection. *Nature* 201, 1145–1147.

Maynard Smith, J. (1982). *Evolution and the Theory of Games*. New York: Cambridge University Press.

Maynard Smith, J. (1989). *Evolutionary Genetics*. Oxford: Oxford University Press.

Maynard Smith, J. (1995). Genes, memes, and minds: review of D. Dennett's *Darwin's Dangerous Idea* (1995). *New York Review of Books*, November 30, 1995, pp. 46–48.

Maynard Smith, J., and G. Price (1973). The logic of animal conflicts. *Nature* 246: 15–18.

Maynard Smith, J., and E. Szathmáry (1995). *The Major Transitions in Evolution*. Oxford: Oxford University Press.

Mayr, E. (1959). The emergence of evolutionary novelties. *In* S. Tax (ed.), *The Evolution of Life: Evolution after Darwin, Vol. 1*. Chicago, IL: University of Chicago Press, pp. 349–380.

Mayr, E. (1988). *Toward a New Philosophy of Biology: Observations of an Evolutionist*. Cambridge, MA: Harvard University Press.

McClelland, E. (2009). Taking sprinting to new heights. *Slate*, August 20, 2009. http://www.slate.com/articles/news_and_politics/recycled/2009/08/taking_sprinting_to_new_heights.html.

McElreath, R. (2010). The coevolution of genes, innovation and culture in human evolution. In J. Silk and P. Kappeler (eds), *Mind The Gap: Tracing the Origins of Human Universals*. New York: Springer, pp. 451–474.

McGiffert, A.C. (1932). *A History of Christian Thought, Vol. 1*. New York: Scribner's.

McKenzie, G. J., R. S. Harris, P. L. Lee, and S. M. Rosenberg (2000). The SOS response regulates adaptive mutation. *Proceedings of the National Academy of Sciences of the USA* 97 (12): 6646–6651.

McKibben, B. (2003). *Enough: Staying Human in an Engineered Age*. New York: Times Books.

McNeill, J. R. (2000). *Something New Under the Sun: An Environmental History of the Twentieth-Century World*. New York: W.W. Norton & Co.

McNeill, W. H. (1976). *Plagues and People*. New York: Anchor Books, Random House.

McNeill, W. H. (1997). History upside down. A review of *Guns, Germs, and Steel* by J. Diamond. *The New York Review of Books* 44(8), 15 May 1997.

McWhorter, J. H. (2015). How dare you say that! The evolution of profanity. *The Wall Street Journal* 17 July 2015.

Melis, A., B. Hare and M. Tomasello (2006). Chimpanzees recruit the best colaborators. *Science* 311, 1297–1300.

Merlo, L. M. F., J. W. Pepper, B. J. Reid and C. C. Maley (2006). Cancer as an evolutionary and ecological process. *Nature Reviews Cancer* 6: 924–935.

Merton, R. K., and E. Barber (2004). *The Travels and Adventures of Serendipity: A Study in Sociological Semantics and the Sociology of Science*. Princeton, NJ: Princeton University Press.

Mesoudi, A. (2011) *Cultural Evolution: How Darwinian Theory Can Explain Human Culture and Synthesize the Social Sciences*. Chicago, IL: University of Chicago Press.

Metcalfe, J. S. (1998). *Evolutionary Economics and Creative Destruction*. London: Routledge.

Michalopoulos, S., and E. Papaioannou (2013). Pre-colonial ethnic institutions and contemporary African development. *Econometrica* 81: 113–152.

Michalopoulos, S., and E. Papaioannou (2014). National institutions and subnational development in Africa. *Quarterly Journal of Economics* 129(1): 151–213.

Mikes, G. (2000b). *English Humour for Beginners*. London: Penguin.

Miles, R. E., and C. C. Snow (1984). Fit, failure and the hall of fame. *California Management Review*, 26(3): 10–28.

Miller, G. F. (2000a). Evolution of human music through sexual selection. In N.L. Wallin, B. Merker and S. Brown (eds.). *The Origins of Music*. Cambridge, MA: MIT Press, pp. 329–360.

Miller, G. F. (2000b). *The Mating Mind: How Sexual Choice Shaped the Evolution of Human Nature*. London: Heineman.

Milo, R. (2013). What is the total number of protein molecules per cell volume? A call to rethink some published values. *Bioessays* 35(12): 1050–1055.

Mingers, J. (1996). A comparison of Maturana's autopoietic social theory and Gidden's theory of structuration. *Systems Research* 13(4): 469–482.

Mingle, M., Eppley, T., Campbell et al. (2014). Chimpanzees prefer African and Indian music over silence. *Journal of Experimental Psychology: Animal Learning and Cognition* 40(4): 502–505.

Mink J. W., R. J. Blumenschine and D. B. Adams (1981). Ratio of central nervous system to body metabolism in vertebrates: its constancy and functional basis. *American Journal of Physiology* 241(3): R203–212.

Minsky, M. (1986). *The Society of Mind*. New York: Simon and Schuster.

Miranda, E. R., and J. A. Biles (eds) (2007). *Evolutionary Computer Music*. London: Springer.

Mitchell, G., D. Roberts, S. van Sittert and J. Skinner (2013). Growth patterns and masses of the heads and necks of male and female giraffes. *Journal of Zoology* 290(1): 49–57.

Mitchell, M. (1999). Can evolution explain how the mind works? A review of the evolutionary psychology debates. *Complexity* 3(3): 17–24.

Mitchell, M., and C. E. Taylor (1999). Evolutionary computation: an overview. *Annual Review of Ecological Systems* 30: 593–616.

Mitri, T. E. (1999). The narrow path of genuine dialogue between Christians and Muslims. Inaugural lecture as full professor on the Dom Helder Camara Chair for Peace and Justice, Faculty of Theology, VU University Amsterdam, 4 March 1999.

Mitsch, W. J., and S. E. Jørgensen (2003). Ecological engineering: a field whose time has come. *Ecological Engineering* 20(5): 363–377.

Mokyr, J. (1990). *The Lever of the Riches: Technological Creativity and Economic Progress.* Oxford: Oxford University Press.

Mokyr, J. (1998). *Review of Guns, Germs, and Steel by J. Diamond. Reviews in the Humanities & Social Sciences.* EH-Net, May 1998 (http://www2.h-net.msu.edu/reviews).

Mokyr, J. (2000). The Industrial Revolution and the Netherlands: why did it not happen? *De Economist* 148(4): 503–520.

Moor, N., W. Ultee and A. Need (2009). Analogical reasoning and the content of creation stories. Quantitative comparisons of preindustrial societies. *Cross-Cultural Research* 43: 91–122.

Moreau, F. (2004). The role of the state in evolutionary microeconomics. *Cambridge Journal of Economics* 28(6): 847–874.

Morgan, E. (1997). *The Aquatic Ape Hypothesis*. London: Souvenir Press.

Morgan, L. H. (1877). *Ancient Society*. New York: World Publishing.

Morin, O. (2016). *How Traditions Live and Die*. New York: Oxford University Press.

Moser, P. (2013). Patents and innovation: evidence from economic history. *Journal of Economic Perspectives* 27(1): 23–44.

Mowery, D. C., R. R. Nelson and B. R. Martin (2010). Technology policy and global warming: why new policy models are needed (or why putting new wine in old bottles won't work). *Research Policy* 39(8): 1011–1023.

Muir, W. (1996). Group selection for adaptation to multiple-hen cages: selection program and direct responses. *Journal of Poultry Science* 75: 447–458.

Mulder, P., and J. C. J. M. van den Bergh (2001). Evolutionary economic theories of sustainable development. *Growth and Change* 32(4): 110–134.

Mulder, P., H. L. F. de Groot and M. W. Hofkes (2001). Economic growth and technological change: a comparison of insights from a neoclassical and an evolutionary perspective. *Technological Forecasting and Social Change* 68: 151–171.

Müller, G. B. (2007). Evo-devo: Extending the evolutionary synthesis. *Nature Reviews Genetics* 8: 943–949.

Mulligan, C. B., R. Gil and X. Sala-i-Martin (2004). Do democracies have different policies than nondemocracies. *Journal of Economic Perspectives* 18(1): 51–74.

Munro, A. (1997). Economics and biological evolution. *Environmental and Resource Economics* 9: 429–449.

Muramoto, O. (2004). The role of the medial prefrontal cortex in human religious activity. *Medical Hypotheses* 62(4): 479–485.

Murdock, G. P. (1945). The common denominator of cultures. *In* R. Linton (ed.), *The Science of Man in the World Crisis*. New York: Columbia University Press.

Murphy, D. J., and C. A. S. Hall (2010). EROI or energy return on (energy) invested. *Ecological Economics Reviews, Annals of the New York Academy of Sciences* 1185: 102–118.

Myers, N., and A. H. Knoll (2001). The biotic crisis and the future of evolution. *Proceedings of the National Academy of Sciences of the USA* 98(10): 5389–5392.

Nannen, V., and J. van den Bergh (2010). Policy instruments for evolution of bounded rationality: application to climate-energy problems. *Technological Forecasting and Social Change* 77: 76–93.

Naroll, R. (1956). A preliminary index of social development. *American Anthropologist* 58: 687–716.

NASEM (2016). *Genetically Engineered Crops: Experiences and Prospects.* Washington DC: The National Academies Press.

Nei, M. (2007). The new mutation theory of phenotypic evolution. *Proceedings of the National Academy of Sciences of the USA* 104: 12 235–12 242.

Nelson, R. R. (ed.) (1993). *National Innovation Systems: A Comparative Analysis.* Oxford: Oxford University Press.

Nelson, R. R. (1995). Recent evolutionary theorizing about economic change. *Journal of Economic Literature* 23(March): 48–90.

Nelson, R. R. (2001). Preface. *In* J. Laurent and J. Nightingale (eds), *Darwinism and Evolutionary Economics.* Cheltenham, UK: Edward Elgar, pp. ix–xii.

Nelson, R. R. (2007). Universal Darwinism and evolutionary social science. *Biology and Philosophy* 22: 73–94.

Nelson, R. R., and S. Winter (1982). *An Evolutionary Theory of Economic Change.* Cambridge, MA: Harvard University Press.

Nemet, G. F., and E. Baker (2009). Demand subsidies versus R&D: comparing the uncertain impacts of policy on a pre-commercial low-carbon energy technology. *The Energy Journal* 30(4): 49–80.

Nesse, R. M., and G. C. Williams (1994). *Why We Get Sick: The New Science of Darwinian Medicine.* New York: Vintage Books.

Neveu, A. (2013). Fiscal policy and business cycle characteristics in a heterogeneous agent macro model. *Journal of Economic Behavior and Organization* 92: 224–240.

Newson, L., and P. J. Richerson (2016). Darwin applied to Trump: can evolutionary theory help us understand the appeal of Donald Trump? *This View of Life Magazine,* https://evolution-institute.org/focus-article/darwin-applied-to-trump-can-evolutionary-theory-help-us-understand-the-appeal-of-donald-trump/

Nielsen, F. (1994). Sociobiology and sociology. *Annual Review of Sociology* 20: 267–303.

Nill, J., and R. Kemp (2009). Evolutionary approaches for sustainable innovation policies: From niche to paradigm? *Research Policy* 38(4): 668–680.

Nilsson, D.-E., and S. Pelger (1994). A pessimistic estimate of the time required for an eye to evolve. *Proceedings: Biological Sciences* 256(1345): 53–58.

Noailly, J. (2008). Coevolution of economic and ecological systems. *Journal of Evolutionary Economics* 18: 1–29.

Noailly J., J. C. J. M. van den Bergh and C. Withagen (2003). Evolution of harvesting strategies: replicator and resource dynamics. *Journal of Evolutionary Economics* 13(2): 183–200.

Noailly J., J. C. J. M. van den Bergh and C. Withagen (2006). Evolution of social norms in a common-pool resource game. *Environmental Resource Economics* 36(1): 113–141.

Noailly, J., J. C. J. M. van den Bergh and C. Withagen (2009). Local and global interactions in an evolutionary resource game. *Computational Economics* 33(2): 155–173.

Noë, R., and P. Hammerstein (1995). Biological markets. *Trends in Ecology and Evolution* 10: 336–339.

Noë, R., J. van Hooff and P. Hammerstein (eds) (2001). *Economics in Nature. Social Dilemmas, Mate Choice and Biological Markets*. Cambridge: Cambridge University Press.

Nonacs, P., and K. Kapheim (2007). Social heterosis and the maintenance of genetic diversity. *Journal of Evolutionary Biology* 20: 2253–2265.

Nordhaus, W. D. (1994). *Managing the Global Commons: The Economics of Climate Change*. Cambridge, MA: MIT Press.

Nordhaus, W. D. (2002). Modeling induced innovation in climate change policy. *In* A. Grubler, N. Nakicenovic and W. Nordhaus (eds), *Technological Change and the Environment*. Washington DC: Resources for the Future Press.

Nordhaus, W. D. (2015). Climate clubs: overcoming free-riding in international climate policy. *American Economic Review* 105(4): 1339–1370.

Norgaard R. B. (1984). Coevolutionary development potential. *Land Economics* 60:160–173.

Norgaard R. B. (1994). *Development Betrayed: The End of Progress and a Coevolutionary Revisioning of the Future*. London: Routledge.

North, D. C. (1981). *Structure and Change in Economic History*. New York: W.W. Norton.

North, D. C. (1997). Institutions. *Journal of Economic Perspectives* 5(1): 97–112.

Norton, B., R. Costanza and R. C. Bishop (1998). The evolution of preferences.Why 'sovereign' preferences may not lead to sustainable policies and what to do about it. *Ecological Economics* 24: 193–211.

Nowak, M. A. (2006). *Evolutionary Dynamics. Exploring the Equations of Life*. Cambridge, MA: Harvard University Press.

Nowak, M., and K. Sigmund, (2000). Games on grids. *In* U. Dieckmann, R. Law and J. Metz, (eds), *The Geometry of Ecological Interactions*. Cambridge: Cambridge University Press, pp. 135–150.

Nowak, M., and K. Sigmund (2005). Evolution of indirect reciprocity. *Nature* 437(27): 1291–1298.

Nowak, M. A., C. E. Tarnita and E. O. Wilson (2010). The evolution of eusociality. *Nature* 466: 1057–1062.

Nowak, M. A., C. E. Tarnita and E. O. Wilson (2011). Reply. *Nature* 471: E9–E10.

Numbers, R. (2006). *The Creationists: From Scientific Creationism to Intelligent Design*. Cambridge, MA: Harvard University Press.

Odling-Smee, F. J., K. N. Laland and M. W. Feldman (2003). *Niche Construction: The Neglected Process in Evolution*. Princeton University Press: Princeton, NJ.

Ofek, H. (2001). *Second Nature: Economic Origins of Human Evolution*. Cambridge: Cambridge University Press.

Ohta, T. (1992). The nearly neutral theory of molecular evolution. *Annual Review of Ecology and Systematics* 23: 263–286.

Ohtsuki, H., C. Hauert, E. Lieberman and M. Nowak (2006). A simple rule for the evolution of cooperation on graphs and social networks. *Nature* 441: 502–505.

Okulicz-Kozaryn, A. (2010). Religiosity and life satisfaction across nations. *Mental Health, Religion & Culture* 13: 155–169.

Oliveira, F.B., E.C. Molina and G. Marroig (2009). Paleogeography of the South Atlantic: a route for primates and rodents into the New World? *In* P. A. Garber, A. Estrada, J. C. Bicca-Marques, E. W. Heymann and K. B. Strier (eds), *South American Primates: Comparative Perspectives in the Study of Behavior, Ecology, and Conservation*. New York: Springer, pp. 55–68.

Olson, M. (1982). *The Rise and Decline of Nations: Economic Growth, Stagflation and Social Rigidities*. New Haven, CT: Yale University Press.

Olsson, O., and B. S. Frey (2002). Entrepreneurship as recombinant growth. *Small Business Economics* 19: 69–80.

Olsson, O., and D. Hibbs (2005). Biogeography and long-run economic development. *European Economic Review* 49(4): 909–938.

Oltra, V., and M. Saint Jean (2009). Sectoral systems of environmental innovation: an application to the French automotive industry. *Technological Forecasting and Social Change* 76 (4): 567–583.

Ostrom, E. (1990). *Governing the Commons: The Evolution of Institutions for Collective Action*. Cambridge: Cambridge University Press.

Ostrom, E. (2000). Collective action and the evolution of social norms. *Journal of Economic Perspectives* 14(3): 137–158.

Otto, S. P., and S. L. Nuismer (2004). Species interactions and the evolution of sex. *Science* 304(5673): 1018–1020.

Paley, W. (2009 [1802]). *Natural Theology: or, Evidences of the Existence and Attributes of the Deity, Collected from the Appearances of Nature*. Cambridge: Cambridge University Press.

Panchanthan, K., and R. Boyd (2005). Indirect reciprocity can stabilize cooperation without the second-order free rider problem. *Nature* 432: 499–502.

Papineau, D. (2005). Social learning and the Baldwin effect. In A. Zilhão (ed.), *Evolution, Rationality and Cognition*. London: Routledge, pp. 40–60.

Park, C. (2004). Religion and geography. *In* J. Hinnells (ed.), *Routledge Companion to the Study of Religion*. London: Routledge, pp. 439–455.

Parker, W. N. (1984). *Europe, America, and the Wider World: Essays on the Economic History of Western Capitalism*. Cambridge: Cambridge University Press.

Parsons, T. (1966). *Societies: Evolutionary and Comparative Perspectives*. Englewood Cliffs, NJ: Prentice Hall Inc..

Patterson, O. (2016). The secret of Jamaica's runners. *The New York Times*, Sunday Review, 13 August 2016.

Paul, R. A. (2015). *Mixed Messages: Cultural and Genetic Inheritance in the Evolution of Human Society*. Chicago, IL: University of Chicago Press.

Pelikan, P., and G. Wegner (eds.) (2003). *The Evolutionary Analysis of Economic Policy*. Cheltenham, UK: Edward Elgar.

Penn D. J. (2003). The evolutionary roots of our environmental problems: towards a Darwinian ecology. *Quarterly Review of Biology* 78(3): 275–301.

Pepper, J., and B. Smuts (2000). The evolution of cooperation in an ecological context: an agent-based model. *In* T. Kohler and G. Gumerman (eds), *Dynamics in Human and Primate Societies*. Oxford: Oxford University Press, pp. 45–76.

Pepper, J., and B. Smuts (2002). A mechanism for the evolution of altruism among non-kin: positive assortment through environmental feedback. *American Naturalist* 160: 205–212.

Perez, C. (2013). Unleashing a golden age after the financial collapse: Drawing lessons from history. *Environmental Innovation and Societal Transitions* 6: 9–23.

Perrings C., K.-G. Mäler, C. Folke, C. S. Holling and B.-O. Jansson (eds) (1995). *Biodiversity Loss: Economic and Ecological Issues*. Cambridge: Cambridge University Press.

Perrow, C. (2010). Comment on Mowery, Nelson and Martin. *Research Policy* 39(8): 1030–1031.

Peters, S. E., and M. Foote (2002). Determinants of extinction in the fossil record. *Nature* 416 (6879): 420–424.

Pezzey, J., and D. Stern (2017). Directed technical change and the British Industrial Revolution. CAMA Working Paper 26/2017, Crawford School of Public Policy, Australian National University, Canberra. https://econpapers.repec.org/RePEc:een:camaaa:2017–26.

Pierson, P. (2000). Increasing returns, path dependency, and the study of politics. *American Political Science Review* 94: 251–267.

Pillarisetti, J. R. (2016). Skewed sex ratio, environmental toxins and human wellbeing: the need for policies. *International Journal of Environmental Studies* 21: 692–701.

Pimm S. L. (1984). The complexity and stability of ecosystems. *Nature* 307:321–326.

Pindyck, R. S. (1991). Irreversibility, uncertainty, and investment. *Journal of Economic Literature* 29: 1110–1148.

Pindyck R. S. (2000). Irreversibilities and the timing of environmental policy. *Resource and Energy Economics* 22: 233–259.

Pinker, S. (1994). *The Language Instinct: How the Mind Creates Language*. London: Penguin.

Pinker, S. (1997). *How the Mind Works*. London: Penguin.

Pinker, S. (2002). *The Blank Slate: The Modern Denial of Human Nature*. London: Penguin.

Pirages, D. (2000). Diversity and social progress in the next millennium: An evolutionary perspective. *Futures* 32: 513–523.

Plotkin, H. C. (1982). *Learning, Development and Culture: Essays in Evolutionary Epistemology*. Chichester: Wiley.

Plotkin, J. B. (2017). No escape from the tangled bank. *Nature* 551: 42–43.

Polderman, T. J., B. Benyamin, C. A. de Leeuw et al. (2015). Meta-analysis of the heritability of human traits based on fifty years of twin studies. *Nature Genetics* 47: 702–709.

Pomeranz, K. (2000). *The Great Divergence: Europe, China, and the Making of the Modern World Economy*. Princeton, NJ: Princeton University Press.

Popp, D. (2001). The effect of new technology on energy consumption. *Resource and Energy Economics* 23: 215–239.

Popp, D., R. G. Newell and A. B. Jaffe (2010). Energy, the environment and technological change. *In* B. Hall and N. Rosenberg (eds.), *Handbook of the Economics of Innovation, Vol. 2*. Amsterdam: Academic Press/Elsevier. pp. 873–937.

Porter, M. E., and C. van der Linde (1991). Towards a new conception of the environment-competitiveness relationship. *Journal of Economic Perspectives* 9(4): 97–118.

Poston, T., and I. Stewart (1981). Does God play dice? *Analog*, November.

Potts, J. (2000). *The New Evolutionary Microeconomics: Complexity, Competence, and Adaptive Behavior*. Cheltenham, UK: Edward Elgar.

Potts, J., J. Foster and A. Straton (2010). An entrepreneurial model of economic and environmental co-evolution. *Ecological Economics* 70(2): 375–383.

Potts, R. (1996). *Humanity's Descent: The Consequences of Ecological Instability*. New York: Morrow.

Poznik, G. D., B. M. Henn, M. C. Yee et al. (2013). Sequencing Y chromosomes resolves discrepancy in time to common ancestor of males versus females. *Science* 341(6145): 562–565.

Pred, A. R. (1966). *The Spatial Dynamics of Urban-Industrial Growth 1800–1914*. Cambridge, MA: MIT Press.

Price, G. (1970). Selection and covariance. *Nature* 227, 520–521.

Price, G. (1972). Extension of covariance selection mathematics. *Annals of Human Genetics* 35, 485–490.

Prigogine, I., and I. Stengers (1984). *Order Out of Chaos: Man's New Dialogue with Nature*. London: Heinemann.

Profet, M. (1988). The evolution of pregnancy sickness as protection to the embryo against pleistocene teratogens. *Evolutionary Theory* 8: 177–190.

Provine, R. R. (2000). *Laughter: A Scientific Investigation*. London: Viking/Penguin.

Pugh, D. S. (ed.) (1997). *Organization Theory: Selected Readings* (4th edn). London: Penguin Books.

Putman, R. J., and S. D. Wratten (1984). *Principles of Ecology*. London: Croom Helm.

Putterman, L., and D. N. Weil (2010). Post-1500 population flows and the long-run determinants of economic growth and inequality. *Quarterly Journal of Economics* 125(4): 1627–1682.

PWC (2012). Too late for two degrees? Low carbon economy index 2012, http://www.eco-business.com/research/pwc-low-carbon-economy-index-2012-too-late-for-two-degrees/

Pyne, S. (2001). *Fire: A Brief History*. Washington DC: University of Washington Press.

Queller, D. (1992). Quantitative genetics, inclusive fitness, and group selection. *The American Naturalist* 139: 540–558.

Quine, W. V. (1969). Epistemology naturalized. *In Ontological Relativity and Other Essays*. New York: Columbia University Press.

Raichle, M. E., and D. A. Gusnard (2002). Appraising the brain's energy budget. *Proceedings of the National Academy of Sciences of the USA* 99(16): 10 237–10 239.

Rammel C., and J. C. J. M. van den Bergh (2003) Evolutionary policies for sustainable development: adaptive flexibility and risk minimising. *Ecological Economics* 47(2): 121–133.

Rampino, M. R, and S. Self (1993). Climate–volcanism feedback and the Toba eruption of ~74,000 years ago. *Quaternary Research* 40: 269–280.

Rappaport, R. A. (1999). *Ritual and Religion in the Making of Humanity*. Cambridge: Cambridge University Press.

Raskin, P. (2016). *Journey to Earthland: The Great Transition to Planetary Civilization*. Boston, MA: Tellus Institute.

Raup, D. M. (1983). The geological and paleontological arguments of creationism. *In* L.R. Godfrey (ed.). *Scientists Confront Creationism*. New York: Norton, pp. 147–162.

Raup, D. M., and J. J. Sepkoski, Jr. (1984). Periodicity of extinctions in the geologic past. *Proceedings of the National Academy of Sciences of the USA* 81: 801–805.

Reader, S. M., and K. N. Laland (1999). Do animals have memes? *Journal of Memetics* 3, http://cfpm.org/jom-emit/1999/vol3/reader_sm&laland_kn.html

Rees, M. (1999). *Just Six Numbers: The Deep Forces That Shape The Universe*. New York: Basic Books.

Reyes-García, V., D. Zurro, J. Caro and M. Madella 2017. Small-scale societies and environmental transformations: coevolutionary dynamics. *Ecology and Society* 22(1), article 15. https://doi.org/10.5751/ES-09066-220115

Richardson, M., J. Hanken, L. Selwood et al. (1998). Haeckel, embryos and evolution. *Science* 280(5366): 983–984.

Richerson, P. J. (2016). Recent critiques of dual inheritance theory. http://bit.ly/2jmQVlO

Richerson, P. J., and R. Boyd (2005). *Not By Genes Alone: How Culture Transformed Human Evolution*. Chicago, IL: University of Chicago Press.

Richerson, P. J., R. Boyd and R. L. Bettinger (2001). Was agriculture impossible during the Pleistocene but mandatory during the Holocene? A climate change hypothesis. *American Antiquity* 66(3): 387–411.

Richman, B. D. (2006). How community institutions create economic advantage: Jewish diamond merchants in New York. *Law and Social Inquiry* 31(2): 383–420.

Richmond, M. (2013). What if Darwin hadn't written 'On the Origin of Species'? Darwin's methodology may equal his ideas in scientific importance. http://www.nsf.gov/news/special_reports/darwin/textonly/darwin_essay1.jsp

Ridley, M. (1995). *The Red Queen: Sex and the Evolution of Human Nature*. London: Penguin Books.

Ridley, M. (ed.) (1997). *Evolution*. Oxford: Oxford University Press.

Roberts, A., and M. Maslin (2016). Sorry David Attenborough, we didn't evolve from 'aquatic apes' – here's why. *The Conversation*, http://theconversation.com/uk

Robson, A.J. (2003). A 'bioeconomic' view of the neolithic and recent demographic transitions. Working paper, Dept. of Economics, University of Western Ontario, London, Ontario, Canada.

Roebroeks, W., and P. Villa (2011). On the earliest evidence for habitual use of fire in Europe. *Proceedings of the National Academy of Sciences of the USA* 108(13): 5209–5214.

Rogers Ackermann, R., A. Mackay and M. L. Arnold (2016). The hybrid origin of "modern" humans. *Evolutionary Biology* 43(1): 1–11.

Rogers, E. (1962). *Diffusion of Innovations*. New York: Free Press.

Romanelli, E. (1991). The evolution of new organizational forms. *Annual Review of Sociology* 17: 79–103.

Romer, P. M. (1994). The origins of endogenous growth. *The Journal of Economic Perspectives* 8(1): 3–22.

Romero, J., and P. Machado (eds.) (2007). *The Art of Artificial Evolution: A Handbook on Evolutionary Art and Music*. Berlin: Springer.

Rorty, R. (1991). *Objectivity, Relativism and Truth: Philosophical Papers, Vol. 1*. Cambridge: Cambridge University Press.

Rose, H., and S. Rose (eds.) (2000). *Alas Poor Darwin: Arguments Against Evolutionary Psychology*. New York: Harmony Books.

Rosenberg, N. (1982). *Inside the Black Box: Technology and Economics*. Cambridge: Cambridge University Press.

Rotmans, J., and D. Loorbach (2009). Complexity and transition management. *Journal of Industrial Ecology* 13(2): 184–196.

Roughgarden, J. (2004). *Evolution's Rainbow: Diversity, Gender, and Sexuality in Nature and People*. Berkeley, CA: University of California Press.

Runciman, W. G. (1998). *The Theory of Social and Cultural Section*. Cambridge: Cambridge University Press.

Ruse, M. (1979). *Sociobiology: Sense or Nonsense*. Dordrecht, The Netherlands: Reidel Publishing Co.

Ruse, M. (1999). *Mystery of Mysteries: Is Evolution a Social Construction*. Cambridge, MA: Harvard University Press.

Ruse, M. (2000). *Can a Darwinian Be a Christian? The Relationship Between Science and Religion*. New York: Cambridge University Press.

Rutherford, A. (2015). Beware the pseudo gene genies. *The Guardian*, 19 July. www.theguardian.com/science/2015/jul/19/epigenetics-dna–darwin-adam-rutherford

Sabatier, P. A. (1988). An advocacy coalition framework of policy change and the role of policy-oriented learning therein. *Policy Sciences* 21: 129–168.

Sachs, J. D., and A. M. Warner (2001). The curse of natural resources. *European Economic Review* 45: 827–838.

Safarzynska, K., and J. C. J. M. van den Bergh (2010a). Evolving power and environmental policy: explaining institutional change with group selection. *Ecological Economics* 69(4): 743–752.

Safarzynska, K., and J. C. J. M. van den Bergh (2010b). Demand–supply coevolution with multiple increasing returns: policy analysis for unlocking and system transitions. *Technological Forecasting and Social Change* 77(2): 297–317.

Safarzynska, K., and J. C. J. M. van den Bergh (2010c). Evolutionary modelling in economics: A survey of methods and building blocks. *Journal of Evolutionary Economics* 20(3): 329–373.

Safarzynska, K., and J. C. J. M. van den Bergh (2011a). Beyond replicator dynamics: Innovation-selection dynamics and optimal diversity. *Journal of Economic Behavior and Organization* 78(3): 229–245.

Safarzynska, K., and J. C. J. M. van den Bergh (2011b). Industry evolution, rationality, and electricity transitions. *Energy Policy* 3: 6440–6452.

Safarzynska, K., and J. C. J. M. van den Bergh (2013). An evolutionary model of energy transitions with interactive innovation-selection dynamics. *Journal of Evolutionary Economics* 23: 271–293.

Safarzynska, K., and J. C. J. M. van den Bergh (2016). Integrated crisis-climate policy: macro-evolutionary modelling of interactions between technology, finance and energy systems. *Technological Forecasting and Social Change* 114: 119–137.

Safarzynska, K., K. Frenken and J. C. J. M. van den Bergh (2012). Evolutionary theorizing and modelling of sustainability transitions. *Research Policy* 41: 1011–1024.

Sahi, S. K., A. P. Arora, and N. Dhameja (2013). An exploratory inquiry into the psychological biases in financial investment behavior. *Journal of Behavioral Finance* 14: 94–103.

Sahlins, M. D., and R. S. Elman (eds.) (1960). *Evolution and Culture*. Ann Arbor, MI: University of Michigan Press.

Samuelson, L. (1997). *Evolutionary Games and Equilibrium Selection*. Cambridge, MA: MIT Press.

Samuelson, P., (1993). Altruism as a problem involving group versus individual selection in economics and biology. *American Economic Review* 83, 143–148.

Sartorius, C., and S. Zundel (eds), 2005. *Time Strategies, Innovation and Environmental Policy*. Cheltenham, UK: Edward Elgar.

Saviotti, P. (1996). *Technological Evolution, Variety and the Economy*. Cheltenham, UK: Edward Elgar.

Sawyer, R. K. (2004). The mechanisms of emergence. *Philosophy of the Social Sciences* 34(2): 260–282.

Sawyer, R. K. (2005). *Social Emergence: Societies as Complex Systems*. Cambridge: Cambridge University Press.

Schadewald, R. J. (1983). The evolution of Bible-science. *In* Godfrey L.R. (ed.). *Scientists Confront Creationism*. New York: Norton, pp. 283–298.

Schelling, T. C. (1969). Models of segregation. *American Economic Review, Papers and Proceedings* 59(2): 488–493.

Schelling, T. C. (1978). *Micromotives and Macrobehaviour*. New York: Norton.

Scherder, E. (2017). *Singing in the Brain: Over de Unieke Samenwerking tussen Muziek en de Hersenen (About the Unique Cooperation between Music and the Brain)*. Athenaeum – Polak & Van Gennep, Amsterdam (in Dutch).

Schlosser, G., and G. P. Wagner (2004). *Modularity in Development and Evolution*. Chicago, IL: University of Chicago Press.

Schneider, E. D., and J. J. Kay (1994). Life as a manifestation of the second law of thermodynamics. *Mathematical and Computer Modelling* 19(6–8): 25–48.

Schot, J. W., and F. W. Geels (2007). Niches in evolutionary theories of technical change: A critical survey of the literature. *Journal of Evolutionary Economics* 17(5): 605–622.

Schowalter, T. (2011). *Insect Ecology: An Ecosystem Approach*. Amsterdam: Elsevier.

Schrödinger, E. (1945). *What is Life?* Cambridge: Cambridge University Press.

Schubert, C. (2012). Is novelty always a good thing? Towards an evolutionary welfare economics. *Journal of Evolutionary Economics* 22(3): 585–619.

Schumpeter, J. A. (1934). *The Theory of Economic Development*. Cambridge, MA: Harvard University Press.

Schumpeter, J. A. (1939). *Business Cycles: A Theoretical, Historical and Statistical Analysis of the Capitalist Process*. New York: McGraw-Hill.

Schumpeter, J. A. (1942). *Capitalism, Socialism and Democracy*. New York: Harper and Brothers Publishers.

Schumpeter, J. A. (1954). *History of Economic Analysis*. London: Allen & Unwin.

Schwoon, M. (2006). Simulating the adoption of fuel cell vehicles. *Journal of Evolutionary Economics* 16(4): 435–472.

Scott, R. A., R. Irving, L. Irwin et al. (2010). ACTN3 and ACE genotypes in elite Jamaican and US sprinters. *Medicine & Science in Sports & Exercise* 42(1): 107–112.

Scott-Phillips, T. C., T. E. Dickins and S. A. Wes (2011). Evolutionary theory and the ultimate–proximate distinction in the human behavioral sciences. *Perspectives on Psychological Science* 6(1): 38–47.

Sen, A. (1993). On the Darwinian view of progress. *Population and Development Review* 19(1): 123–137.

Service, E. R. (1963). *Primitive Social Organization: An Evolutionary Perspective*. New York: Random House.

Sethi, R., and E. Somanathan (1996). The evolution of social norms in common property resource use. *American Economic Review* 86(4): 766–788.

Shapiro, C., and H. Varian (1999). *Information Rules: A Strategic Guide to the Network Economy*. Boston: Harvard Business School Press.

Shemelev, S., and J. van den Bergh (2016). Optimal diversity of renewable energy alternatives under multiple criteria: an application to the UK. *Renewable and Sustainable Energy Reviews* 60: 679–691.

Sherman, P. W. (2008). Allergies: their role in cancer prevention. *Quarterly Review of Biology* 83(4): 339–362.

Shiller, R. J. (2003). From efficient markets theory to behavioral finance. *Journal of Economic Perspectives* 17: 83–104.

Silk, J., S. Alberts and J. Altmann (2003). Social bonds of female baboons enhance infant survival. *Science* 302: 1231–1234.

Silk, J., S. Brosnan, J. Vonk et al. (2005). Chimpanzees are indifferent to the welfare of unrelated group members. *Nature* 437: 1357–1359.

Silverberg, G. (1988). Modelling economic dynamics and technical change: mathematical approaches to self-organisation and evolution. *In* G. Dosi, C. Freeman, R. Nelson, G. Silverberg and L. Soete (eds), *Technical Change and Economic Theory*. London: Pinter Publishers.

Silverberg, G., and B. Verspagen (1994). Collective learning, innovation and growth in a boundedly rational, evolutionary world. *Journal of Evolutionary Economics* 4: 207–226.

Simmons, R., and L. Scheepers (1996). Winning by a neck: sexual selection in the evolution of the giraffe. *The American Naturalist* 148: 771–786.

Simon, H. (1990). A mechanism for social selection and successful altruism. *Science* 250: 1665–1668.

Simon, H. (1993). Altruism and economics. *American Economic Review* 83: 156–161.

Simpson, E. (1951). The interpretation of interaction in contingency tables. *Journal of the Royal Statistical Society Series B* 13: 238–241.

Simpson, G. G. (1944). *Tempo and Mode in Evolution*. New York: Columbia University Press.

Simpson, G. G. (1953). The Baldwin effect. *Evolution* 7: 110–117.

Sinn, H. W. (2008). Public policies against global warming: A supply-side approach. *International Tax and Public Finance* 15(4): 360–394.

Skea, J. (2010). Valuing diversity in energy supply. *Energy Policy* 38: 3608–3621.

Skinner, B. F. (1953). *Science and Human Behavior*. New York: Macmillan.

Slatkin, M., and M. Wade (1978). Group selection on a quantitative character. *Proceedings of the National Academy of Sciences of the USA* 75: 3531–3534.

Smead, R., R. L. Sandler, P. Forber and J. Basl (2014). A bargaining game analysis of international climate negotiations. *Nature Climate Change* 4: 442–445.

Smith, E. A. (2003). Ecological Anthropology. Lecture notes (ANTH 457), Department of Anthropology, University of Washington, Seattle.

Smith, E. A., M. Borgerhoff Mulder and K. Hill (2001). Controversies in the evolutionary social sciences: a guide for the perplexed. *Trends in Ecology and Evolution* 16(3): 128–135.

Smolin, L. (1997). *The Life of the Cosmos*. Oxford: Oxford University Press.

Smuts, B. (1995). The evolutionary origins of patriarchy. *Human Nature* 6(1): 1–32.

Sniegowski, P. D., and R. E. Lenski (1995). Mutation and adaptation: The directed mutation controversy in evolutionary perspective. *Annual Review of Ecology and Systematics* 26: 553–578.

Snowdon, C. T., E. Zimmermann and E. Altenmüller (2015). Music evolution and neuroscience. *Progress in Brain Research* 217: 17–34.

Snowdon C. T., D. Teie and M. Savage (2015). Cats prefer species-appropriate music. *Applied Animal Behaviour Science* 166: 106–111.

Sober, E. (1981). Holism, individualism, and the units of selection. *Proceedings of the Philosophy of Science Association* 2: 93–121.

Sober, E., and D. S. Wilson (1998). *Unto Others: The Evolution and Psychology of Unselfish Behavior*. Cambridge, MA: Harvard University Press.

Solow, R. M. (1957). Technical change and the aggregate production function. *Review of Economics and Statistics* 39: 312–320.

Soltis, J., R. Boyd and P. Richerson (1995). Can group-functional behaviors evolve by cultural group selection? *Current Anthropology* 36: 473–494.

Somit, A., and S. Peterson (eds) (1992). *The Dynamics of Evolution: The Punctuated Equilibrium Debate in the Natural and Social Sciences*. Ithaca, NY: Cornell University Press.

Soon, C. S., M. Brass, H.-J. Heinze and J.-D. Haynes (2008). Unconscious determinants of free decisions in the human brain. *Nature Neuroscience* 11: 543–545.

Sorrell, S. (2009). Jevons' Paradox revisited: The evidence for backfire from improved energy efficiency. *Energy Policy* 37: 1456–1469.

Speidell, L. S. (2009). Investing in the unknown and the unknowable–Behavioral finance in frontier markets. *Journal of Behavioral Finance* 10: 1–8.

Sperber, D. (1996). *Explaining Culture: A Naturalistic Approach*. Oxford: Blackwell.

Sperber, D. (2006). Why a deep understanding of cultural evolution is incompatible with shallow psychology. *In* S. Levinson and N. Enfield (eds), *Roots of Human Sociality: Culture, Cognition and Interaction*. London: Bloomsbury, pp. 431–451.

Spolaore, E., and R. Wacziarg (2009). The diffusion of development. *Quarterly Journal of Economics* 124: 469–529.

Spolaore, E., and R. Wacziarg (2013). How deep are the roots of economic development? *Journal of Economic Literature* 51(2): 325–369.

Spolaore, E., and R. Wacziarg (2014). Fertility and modernity. https://ideas.repec.org/p/tuf/tuftec/0779.html

Spolaore, E., and R. Wacziarg (2016). The diffusion of institutions. *In* D. S. Wilson and A. Kirman (eds), *Complexity and Evolution: A New Synthesis for Economics*. Cambridge, MA: MIT Press, pp. 147–166.

Spoor, F., J. J. Hublin, M. Braun and F. Zonneveld (2003). The bony labyrinth of Neanderthals. *Journal of Human Evolution* 44: 141–165.

Stake, J. E. (2004). The property `instinct'. *Philosophical Transactions of the Royal Society of London B* 359: 1763–1774.

Stavins, R. N. (2009). Yes: The transition can be gradual – and affordable. *Wall Street Journal*, 21 September 2009. http://belfercenter.ksg.harvard.edu/publication/19564/yes.html

Steele, D. (1987). Hayek's theory of cultural group selection. *Journal of Libertarian Studies* 8: 171–195.

Steele, E. J. (1981). *Somatic Selection and Adaptive Evolution: On the Inheritance of Acquired Characters* (2nd edn). Chicago, IL: University of Chicago Press.

Steele, E. J., Lindley, R. A. and Blanden, R. V. (1998). *Lamarck's Signature: How Retrogenes are Changing Darwin's Natural Selection Paradigm*. Reading, MA: Addison-Wesley-Longman.

Stern, D. I. (2011). The role of energy in economic growth. *Ecological Economics Reviews, Annals of the New York Academy of Sciences* 1219: 6–51.

Sterner, T. (2002). *Policy Instruments for Environmental and Natural Resource Management*. Washington DC: Resources for the Future Press.

Steward, J.H. (1955). *Theory of Cultural Change: The Methodology of Multilinear Evolution*. Urbana, IL: University of Illinois Press.

Stewart, O.C. (1956). Fire as the first great force employed by man. *In* W. L. Thomas (ed.), *Man's Role in the Changing Face of the Earth*. Chicago, IL: University of Chicago Press, pp. 115–133.

Stigler, G. (1958). The economies of scale. *Journal of Law and Economics* 1(Oct.): 54–71.

Stirling, A. (2007). A general framework for analysing diversity in science, technology and society. *Journal of the Royal Society Interface* 4(15): 707–719.

Stirling, A. (2010). Multicriteria diversity analysis: a novel heuristic framework for appraising energy portfolios. *Energy Policy* 38: 1622–1634.

Stoneman, P. (ed.) (1995). *Handbook of the Economics of Innovation and Technological Change*. New York: Wiley-Blackwell.

Stoneman, P. (2011). *Soft Innovation: Economics, Product Aesthetics, and the Creative Industries*. Oxford: Oxford University Press.

Strickberger, M. W. (1996). *Evolution* (2nd edn). Sudbury, MA: Jones and Bartlett Publishers.

Stuart Mill, J. (1848). *Principles of Political Economy with Some of Their Applications to Social Philosophy*. London: Longmans, Green and Co.

Stulp, G., L. Barrett, F.C. Tropf and M. Mills (2015). Does natural selection favour taller stature among the tallest people on Earth? *Proceedings of the Royal Society B, Biological Sciences* 282(1806): 20150211.

Sulloway, F. (1996). *Born to Rebel: Birth Order, Family Dynamics, and Creative Lives*. New York: Pantheon Books.

Sun, L. (2016) When democracy meets the ghost of evolution: why short presidents have vanished, *This View of Life Magazine*, https://evolution-institute.org/article/when-democracy-meets-the-ghost-of-evolution-why-short-presidents-have-vanished/

Surowecki, J. (2004). *The Wisdom of Crowds*. New York: Doubleday.

Swaab, D. (2014). *We Are Our Brains: From the Womb to Alzheimer's*. London: Penguin.

Swenson, W., Wilson, D., and Elias, R. (2000). Artificial ecosystem selection. *Proceedings of the National Academy of Sciences of the USA* 97: 9110–9114.

Taçon, P., and C. Pardoe (2002). Dogs make us human. *Nature Australia* 27(4): 52–61.

Tainter, J. A. (1988). *The Collapse of Complex Societies*. Cambridge: Cambridge University Press.

Tainter, J. A. (2011). Energy, complexity, and sustainability: a historical perspective. *Environmental Innovation and Societal Transitions* 1: 89–95.

Tarnita, C. E., T. Antala, H. Ohtsukib and M. A. Nowak (2009). Evolutionary dynamics in set structured populations. *Proceedings of the National Academy of Sciences of the USA* 106: 8601–8604.

Taylor, P., and L. Jonker (1978). Evolutionary stable strategies and game dynamics. *Mathematical Biosciences* 40: 145–156.

Tëmkin, I., and Eldredge, N. (2007). Phylogenetics and material cultural evolution. *Current Anthropology* 48(1): 146–153.

Temme, S., M. Zacharias, J. Neumann et al. (2014). A novel family of human leukocyte antigen class II receptors may have its origin in archaic human species. *Journal of Biological Chemistry* 289: 639–653.

Teubner, G. (1993). *Law as an Autopoietic System*. Oxford: Blackwell.

Teulings, C., and R. Baldwin (eds.) (2014). *Secular Stagnation: Facts, Causes, and Cures*. London: CEPR Press.

Thaler, R. H., and C. R. Sunstein (2008). *Nudge: Improving Decisions About Health, Wealth, and Happiness*. New Haven, CT: Yale University Press.

Thorne, A. G., and M. H. Wolpoff (2003). The multiregional evolution of humans. *Scientific American* 13(2): 46–53.

Tiggemann, M., and S. Hodgson (2008). The hairlessness norm extended: reasons for and predictors of women's body hair removal at different body sites. *Sex Roles* 59(11–12): 889–897.

Tinbergen, N. (1963). On aims and methods of ethology. *Zeitschrift für Tierpsychologie* 20: 410–433.

Tol, R. S. J. (2008), The social cost of carbon: trends, outliers, and catastrophes. *Economics, the Open-Access, Open-Assessment E-Journal* 2 (25): 1–24.

Toman, M. A., J. Pezzey and J. Krautkraemer (1995). Neoclassical economic growth theory and 'sustainability'. *In* D. Bromley (ed.), *Handbook of Environmental Economics*. Oxford: Blackwell, pp. 139–165.

Traulsen, A., and M. Nowak (2006). Evolution of cooperation by multilevel selection. *Proceedings of the National Academy of Sciences of the USA* 103: 10 952–10 955.

Trentmann, F. (2016). *Empire of Things: How we Became a World of Consumers, from the Fifteenth Century to the Twenty First*. London: Allen Lane.

Trigger, B. G. (1998). *Sociocultural Evolution: Calculation and Contingency*. Oxford: Blackwell.

Trivers, R. L. (1971). The evolution of reciprocal altruism. *Quarterly Review of Biology* 46: 35–57.

Troitzsch, K. G., U. Mueller, N. Gilbert and J. E. Doran (eds) (1996). *Social Science Microsimulation*. Berlin: Springer.

Turchin, P. (2015). *Ultrasociety: How 10,000 Years of War Made Humans the Greatest Cooperators on Earth*. Chaplin, CT: Beresta Books.

Tylecote, A. (1992). *The Long Wave in the World Economy: The Current Crisis in Historical Perspective*. London: Routledge.

Tylor, E. B. (1871). *Primitive Culture*. London: Murray.

Ulanowicz, R. E. (1981). A unified theory of self-organization. *In* W. J. Mitsch, R. W. Bosserman and J. M. Klopatek (eds), *Energy and Ecological Modeling*. New York: Elsevier, pp. 649–652.

Unruh, G. C. (2002). Escaping carbon lock-in. *Energy Policy* 30: 317–325.

van den Bergh, J. C. J. M. (2005). Evolutionary analysis of the relationship between economic growth, environmental quality and resource scarcity. *In* D. Simpson, M. Toman and R. U. Ayres (eds), *Scarcity and Growth in the New Millennium*. Washington DC: Resources for the Future Press, pp. 177–197.

van den Bergh, J. C. J. M. (2007). Evolutionary thinking in environmental economics. *Journal of Evolutionary Economics* 17(5), 521–549.

van den Bergh, J. C. J. M. (2008). Optimal diversity: increasing returns versus recombinant innovation. *Journal of Economic Behavior and Organization* 68(3–4): 565–580.

van den Bergh, J. C. J. M. (2009). The GDP paradox. *Journal of Economic Psychology* 30(2): 117–135.

van den Bergh, J. C. J. M. (2010). Safe climate policy is affordable: 12 reasons. *Climatic Change* 101(3): 339–385.

van den Bergh, J. C. J. M. (2011). Energy conservation more effective with rebound policy. *Environmental and Resource Economics* 48(1): 43–58.

van den Bergh, J. C. J. M. (2012). Effective climate-energy solutions, escape routes and peak oil. *Energy Policy* 46: 530–536.

van den Bergh, J. C. J. M. (2013). Environmental and climate innovation: limitations, policies and prices. *Technological Forecasting and Social Change* 80(1):11–23.

van den Bergh, J. C. J. M. (2015). Climate treaty: pricing would limit carbon rebound. *Nature* 526: 195.

van den Bergh, J. C. J. M. (2016). Rebound policy in Paris Agreement: Instrument comparison and climate-club revenue offsets. *Climate Policy*, http://www.tandfonline.com/doi/full/ 10.1080/14693062.2016.1169499.

van den Bergh, J. C. J. M. (2017). A third option for climate policy within potential limits to growth. *Nature Climate Change* 7: 107–112.

van den Bergh, J. C. J. M., and W. Botzen (2014). A lower bound to the social cost of CO_2 emissions. *Nature Climate Change* 4(April): 253–258.

van den Bergh, J. C. J. M., and J. M. Gowdy (2000). Evolutionary theories in environmental and resource economics: Approaches and applications. *Environmental and Resource Economics* 17: 37–57.

van den Bergh, J. C. J. M., and J. M. Gowdy (2003). The microfoundations of macroeconomics: an evolutionary perspective. *Cambridge Journal of Economics* 27(1): 65–84.

van den Bergh, J. C. J. M., and J. M. Holley (2002). An environmental-economic assessment of genetic modification of agricultural crops. *Futures* 34(9–10): 802–822.

van den Bergh, J. C. J. M., and M. A. Janssen (eds.) (2005). *Economics of Industrial Ecology: Use of Materials, Structural Change and Spatial Scales*. Cambridge, MA: MIT Press.

van den Bergh, J. C. J. M., and R.A. de Mooij (1999) An assessment of the growth debate. *In* van den Bergh J. C. J. M. (ed.), *Handbook of Environmental and Resource Economics*. Cheltenham, UK: Edward Elgar, pp. 643–655.

van den Bergh, J. C. J. M., and P. Rietveld (2004). Reconsidering the limits to world population: meta-analysis and meta-prediction. *BioScience* 54(3): 195–204.

van den Bergh, J. C. J. M., and S. Stagl (2003). Coevolution of economic behaviour and institutions: towards a theory of institutional change. *Journal of Evolutionary Economics* 13(3): 289–317.

van den Bergh, J. C. J. M., A. Faber, A. M. Idenburg and F. H. Oosterhuis (2006). Survival of the greenest: evolutionary economics and policies for energy innovation. *Environmental Sciences* 3(1): 57–71.

van den Bergh, J. C. J. M., A. Faber, A. M. Idenburg and F. H. Oosterhuis (2007). *Evolutionary Economics and Environmental Policy: Survival of the Greenest.* Cheltenham, UK: Edward Elgar.

van den Bergh, J. C. J. M., C. Folke, S. Polasky, M. Scheffer and W. Steffen (2015). What if solar energy becomes really cheap? A thought experiment on environmental problem shifting. *Current Opinion in Environmental Sustainability* 14: 170–179.

van den Bergh, J. C. J. M., B. Truffer and G. Kallis (2011). Environmental innovation and societal transitions: introduction and overview. *Environmental Innovation and Societal Transitions* 1(1): 1–23.

van den Heuvel S. T. A., and J. C. J. M. van den Bergh (2009). Multilevel assessment of diversity, innovation and selection in the solar photovoltaic industry. *Structural Change and Economic Dynamics* 20(1): 50–60.

van der Heide C. M., J. C. J. M. van den Bergh and E. C. van Ierland (2005). Extending Weitzman's economic ranking of biodiversity protection: combining ecological and genetic considerations. *Ecological Economics* 55: 218–223.

van der Ploeg, R. (2011). Macroeconomics of sustainability transitions: second-best climate policy, Green Paradox, and renewables subsidies. *Environmental Innnovation and Societal Transitions* 1(1): 130–134.

van der Steen, W. J. (2000). *Evolution as Natural History: A Philosophical Analysis.* Praeger, Westport, CT: Thomas Reydon.

van der Voet, E., R. Huele and R. Stevers (2000). Evolutionary predators, prey, pets and pests: evolutionary reactions on human dominance in the biosphere. Unpublished mimeo.

van der Zwaan, B., and A. Rabl (2003). Prospects for PV: a learning curve experience. *Solar Energy* 74(1): 19–31.

van Hoof, T. B., F. P. M. Bunnik, J. G. M. Waucomont, W. M. Kürschner and H. Visscher (2006). Forest re-growth on medieval farmland after the Black Death pandemic: implications for atmospheric CO_2 levels. *Palaeogeography, Palaeoclimatology, Palaeoecology* 237: 396–411.

van Hooff, J. A. R. A. M. (1972). A comparative approach to the phylogeny of laughter and smiling. In R.A. Hinde (ed.), *Nonverbal Communication.* Cambridge: Cambridge University Press, pp. 209–241.

van Hooff, J. A. R. A. M. (2002). *De mens, een primaat net zo 'eigenaardig' als de andere primaten [Man, a primate just as odd as the other primates].* The Hague: Conference of NWO/ Huygenslezing 2002.

Van Rhijn, J. (2013). *Darwin's Dating Show.* Diemen: Veen Media.

van Rueden, C. (2016). The conversation about Trump should consider the evolution of men's political psychology. *This View of Life Magazine*, https://evolution-institute.org/article/tvol-special-edition-whats-wrong-and-right-about-evolutionary-psychology/

van Schaik, C. P., and J. M. Burkart (2011). Social learning and evolution: the cultural intelligence hypothesis. *Philosophical Transactions of the Royal Society B: Biological Sciences* 366 (1567): 1008–1016.

Van Straalen, N. M. and J. Stein (2003) Evolutionary views on the biological basis of religion. In W. B. Drees (ed.), *Is Nature Ever Evil? Religion, Science and Value*. London: Routledge, pp. 321–329.

van Veelen, M. (2009). Group selection, kin selection, altruism and cooperation: when inclusive fitness is right and when it can be wrong. *Journal of Theoretical Biology* 259: 589–600.

van Veelen, M., and Hopfensitz, A. (2007). In love and war: altruism, norm formation, and two different types of group selection. *Journal of Theoretical Biology* 249: 667–680.

van Vugt, M., A. Ahuja and M. van Vugt (2011). *Naturally Selected: The Evolutionary Science of Leadership*. New York: Harper Business.

Veblen, T. (1898). Why is economics not an evolutionary science? *Quarterly Journal of Economics* 4: 393–397.

Veblen, T. (1899). *The Theory of the Leisure Class*. London: Penguin

Vermeij, G. J. (2006). *Nature: An Economic History*. Princeton, NJ: Princeton University Press.

Vermeij, G.J. (2010). *The Evolutionary World: How Adaptation Explains Everything from Seashells to Civilization*. New York: Thomas Dunn Books, St. Martin's Press.

Vernon, R. (1966). International investment and international trade in the product cycle. *Quarterly Journal of Economics* 80(May): 190–207.

Verpooten, J. (2011). Brian Boyd's evolutionary account of art: fiction or future? *Biological Theory* 6(2): 176–183.

Verspagen, B. (2009). The use of modeling tools for policy in evolutionary environments. *Technological Forecasting and Social Change* 76(4): 453–461.

von Neumann, J., and O. Morgensern (1944). *Theory of Games and Economic Behavior*. Princeton, NJ: Princeton University Press.

Wade, M. (1976). Group selection among laboratory populations of *Tribolium*. *Proceedings of the National Academy of Sciences of the USA* 73: 4604–4607.

Wagner, M. (2007). On the relationship between environmental management, environmental innovation and patenting: evidence from German manufacturing firms. *Research Policy* 36(10): 1587–1602.

Walker, B., S. Carpenter, J. Anderies et al. (2002). Resilience management in social–ecological systems: a working hypothesis for a participatory approach. *Conservation Ecology* 6(1), article 14, http://www.consecol.org/vol6/iss1/art14.

Wang, G. (2013). The genomics of selection in dogs and the parallel evolution between dogs and humans. *Nature Communications* 4: 1860.

Ward, H. (2003). The co-evolution of regimes of accumulation and patterns of rule: State autonomy and the possibility of functional responses to crisis. *New Political Economy* 8(2): 179–202.

Ward, P. (2001). *Future Evolution*. New York: Times Books.

Ward, W. H. (1967). The sailing ship effect. *Bulletin of the Institute of Physics and Physical Society* 18: 169.

Wardle, D. A., R. D Bardgett, J. N Klironomos et al. (2004). Ecological linkages between aboveground and belowground biota. *Science* 304(5677): 1629–1633.

Watkins, S. C. (1990). From local to national communities: The transformation of demographic regimes in Western Europe, 1870–1960. *Population and Development Review* 16(2): 241–272.

Watson, A. J., and J. E. Lovelock (1983). Biological homeostasis of the global environment: the parable of Daisyworld. *Tellus Series B* 35 (4): 286–289.

Watson, R. A. (2006). *Compositional Evolution: The Impact of Sex, Symbiosis, and Modularity on the Gradualist Framework of Evolution*. Cambridge, MA: MIT Press.

Watt Smith, T. (2015). *The Book of Human Emotions*. London: Profile Books.

Weber, B. H., and D. J. Depew (eds) (2003). *Evolution and Learning. The Baldwin Effect Reconsidered*. Cambridge, MA: MIT Press.

Wegner, G. (1997). Policy from an evolutionary perspective. A new approach. *Journal of Institutional and Theoretical Economics* 153: 485–509.

Weibull, J. W. (1995). *Evolutionary Game Theory*. Cambridge, MA: MIT Press.

Weischer, L., J. Morgan, and M. Patel (2012). Climate clubs: can small groups of countries make a big difference in addressing climate change? *Review of European, Comparative and International Environmental Law* 21: 177–192.

Weisman, A. (2007). *The World Without Us*. New York: St. Martin's Press.

Weißbach, D., G. Ruprechta, A. Hukea et al. (2013). Energy intensities, EROIs (energy returned on invested), and energy payback times of electricity generating power plants. *Energy* 52(1): 210–221.

Weitzman, M. L. (1992). On diversity. *Quarterly Journal of Economics* 107: 363–405.

Weitzman M. L. (1993). What to preserve? An application of diversity theory to crane conservation. *Quarterly Journal of Economics* 108(1): 157–183.

Weitzman, M. L. (1998a). Recombinant growth. *Quarterly Journal of Economics* 113(2): 331–360.

Weitzman M. L. (1998b). The Noah's ark problem. *Econometrica* 66: 1279–1298.

Werner, G. D. A., J. E. Strassman, A. B. F. Ivens, et al. (2014). Evolution of microbial markets. *Proceedings of the National Academy of Sciences of the USA* 111(4): 1237–1244.

Westbroek, P. (1991). *Life as a Geological Force: Dynamics of the Earth*. New York: W.W. Norton.

White, L. A. (1943). Energy and the evolution of culture. *American Anthropologist* 45:345–356.

White, L. A. (1959). *The Evolution of Culture: The Development of Civilization to the Fall of Rome*. New York: McGraw-Hill.

White, R. (1998). Toward the new synthesis: evolution, human nature, and the social sciences. *Choice Magazine*, Sept.: 1–20. http://faculty.msj.edu/whiter/choice.htm

Whitehouse, H., P. François, and P. Turchin (2015). The role of ritual in the evolution of social complexity: five predictions and a drum roll. *Cliodynamics* 6(2): 199–210.

Wilkins, J. S. (1998). So you want to be an anti-Darwinian: varieties of opposition to Darwinism. The TalkOrigins Archive, http://www.talkorigins.org/faqs/anti-darwin.html.

Wilkins, J. S. (2005). Is 'meme' a new 'idea'? Reflections on Aunger. *Biology and Philosophy* 20: 585–598.

Wilkinson, R. (1973). *Poverty and Progress: An Ecological Model of Economic Development*. London: Methuen and Co.

Williams, G. C. (1966). *Adaptation and Natural Selection*. Princeton, NJ: Princeton University Press.

Wills, C. (1993). *The Runaway Brain: The Evolution of Human Uniqueness*. New York: Basic Books.

Wilson, D. S. (1975). A theory of group selection. *Proceedings of the National Academy of Sciences of the USA* 72: 143–146.

Wilson, D. S. (1983). The group selection controversy: history and current status. *Annual Review of Ecology and Systematics* 14: 159–187.

Wilson, D. S. (1997). Human groups as units of selection. *Science* 276(5320): 1816–1817.

Wilson, D. S. (2000). The challenge of understanding complexity. *Behavioral and Brain Sciences* 23, 163–164.

Wilson, D. S. (2002). *Darwin's Cathedral: Evolution, Religion, and the Nature of Society.* Chicago, IL: University of Chicago Press.

Wilson, D. S. (2005). Human groups as adaptive units: toward a permanent consensus. *In* Carruthers, P., Laurence, S., and Stich, S. (eds), *The Innate Mind: Culture and Cognition.* New York: Oxford University Press, pp. 69–88.

Wilson, D. S. (2007). *Evolution for Everyone: How Darwin's Theory can Change the Way We Think about our Lives.* New York: Delacorte Press.

Wilson D. S., and J. M. Gowdy (2013). Evolution as a general framework for economics and public policy. *Journal of Economic Behavior and Organization* 90: S3–10.

Wilson, D. S., and A. Kirman (2016). *Complexity and Evolution: Towards a New Synthesis for Economics.* Cambridge, MA: MIT Press.

Wilson, D. S., and Sober, E. (1994). Re-introducing group selection to the human behavioral sciences. *Behavioral and Brain Sciences* 17: 585–654.

Wilson, D. S., and Wilson, E. O. (2007). Rethinking the theoretical foundation of sociobiology. *The Quarterly Review of Biology* 82: 327–348.

Wilson, E. O. (1975). *Sociobiology: The New Synthesis.* Cambridge, MA: Harvard University Press.

Wilson, E. O. (1978). *On Human Nature.* Cambridge, MA: Harvard University Press.

Wilson, E. O. (1984) *Biophilia: The Human Bond with Other Species.* Cambridge, MA: Harvard University Press.

Wilson, E. O. (1998). *Consilience: The Unity of Knowledge.* New York: Alfred Knopf.

Wilson, E. O. (2009). An intellectual entente. *Harvard Magazine,* 9 October 2009. https://harvardmagazine.com/breaking-news/james-watson-edward-o-wilson-intellectual-entente

Wilson, E. O. (2016). *Half-Earth: Our Planet's Fight for Life.* New York: W.W. Norton.

Wilson, E.O., and B. Hölldobler (2005). Eusociality: origin and consequences. *Proceedings of the National Academy of Sciences of the USA* 102: 13 367–13 371.

Wilthagen, T. (1992). Recht in een gesloten samenleving: het debat over reflexief recht en autopoiesis. *Recht der Werkelijkheid* 13(1): 119–138.

Winder N., B. S. McIntosh and P. Jeffrey (2005). The origin, diagnostic attributes and practical application of coevolutionary theory. *Ecological Economics* 54(4): 347–361.

Windrum, P., and C. Birchenhall, (2005). Structural change in the presence of network externalities: a coevolutionary model of technological successions. *Journal of Evolutionary Economics* 15: 123–148.

Windrum, P., T. Carli and C. Birchenhall (2009). Consumer trade-offs and the development of environmentally friendly technologies. *Technological Forecasting and Social Change* 76: 552–566.

Winter, S. G. (1964). Economic 'natural selection' and the theory of the firm. *Yale Economic Essays* 4: 225–272.

Witt, U. (1992). The endogenous public choice theorist. *Public Choice* 73: 117–129.

Witt, U. (ed.) (1993). *Evolutionary Economics.* Cheltenham, UK: Edward Elgar.

Witt, U. (1996). Innovations, externalities and the problem of economic progress. *Public Choice* 89: 113–130.

Witt, U. (1997). Self-organization and economics – what is new? *Structural Change and Economic Dynamics* 8: 489–507.

Witt, U. (2003). Economic policy making in evolutionary perspective. *Journal of Evolutionary Economics* 13: 77–94.

Witt, U. (2011). The dynamics of consumer behavior and the transition to sustainable consumption patterns. *Environmental Innovation and Societal Transitions* 1(1): 101–108.

Wohlgemuth, M. (2002). Evolutionary approaches to politics. *Kyklos* 55: 121–155.

Wolfe, N., C. Panosian Dunavan and J. Diamond (2007). Origins of major human infectious diseases. *Nature* 447: 279–283.

World Bank (2006). *Where is the Wealth of Nations? Measuring Capital for the 21st Century.* The World Bank, Washington, DC.

Wrangham, R. (2009). *Catching Fire: How Cooking Made Us Human.* New York: Basic Books.

Wright, D.K. (2017). Humans as agents in the termination of the African Humid Period. *Frontiers in Earth Science* 5: 4, https://doi.org/10.3389/feart.2017.00004

Wright, S. (1932). The roles of mutation, inbreeding, crossbreeding, and selection in evolution. *Proceedings of the Sixth International Congress of Genetics* 1: 356–366

Wrigley, E.A. (1988). *Continuity, Chance, and Change: The Character of the Industrial Revolution in England.* Cambridge: Cambridge University Press.

Wynne-Edwards, V. (1962). *Amimal Dispersion in Relation to Social Behavior.* Oliver and Boyd, Edinburgh.

Xu Y., A. Lee, W.-L. Wu, X. Liu and P. Birkholz (2013). Human vocal attractiveness as signaled by body size projection. *PLoS ONE* 8(4): e62397. doi:10.1371/journal.pone.0062397

Young, J. Z. (1965). *A Model of the Brain.* Oxford: Clarendon.

Zahavi, A., and Zahavi, A. (1997). *The Handicap Principle: A Missing Piece of Darwin's Puzzle.* Oxford: Oxford University Press.

Zeppini, P., and J. C. J. M. van den Bergh (2011). Competing recombinant technologies for environmental innovation: extending Arthur's model of lock-in. *Industry and Innovation* 18: 317–334.

Zeppini, P., and J. C. J. M. van den Bergh (2013). Optimal diversity in investments with recombinant innovation. *Structural Change and Economic Dynamics* 24: 141–156.

Zink, T., and R. Geyer (2017). Circular economy rebound. *Journal of Industrial Ecology* 21(3): 593–602.

Zipf, G. K. (1949). *Human Behavior and the Principle of Least Effort.* Cambridge, MA: Addison-Wesley Press.

Zurek, W. (2009). Quantum Darwinism. *Nature Physics* 5: 181–188.

Index

Printed in the United States
by Baker & Taylor Publisher Services